Annotated Bibliographies in Combinatorial Optimization

WILEY-INTERSCIENCE
SERIES IN DISCRETE MATHEMATICS AND OPTIMIZATION

AARTS AND KORST
Simulated Annealing and Boltzmann Machines: A Stochastic Approach to Combinatorial Optimization and Neural Computing

AARTS AND LENSTRA, EDITORS
Local Search in Combinatorial Optimization

ALON, SPENCER, AND ERDOS
The Probabilistic Method

ANDERSON AND NASH
Linear Programming in Infinite-Dimensional Spaces: Theory and Application

BARTHÉLEMY AND GUÉNOCHE
Trees and Proximity Representations

BAZARRA, JARVIS, AND SHERALI
Linear Programming and Network Flows

COFFMAN AND LUEKER
Probabilistic Analysis of Packing and Partitioning Algorithms

DASKIN
Network and Discrete Location: Models, Algorithms and Applications

DELL'AMICO, MAFFIOLI, AND MARTELLO, EDITORS
Annotated Bibliographies in Combinatorial Optimization

DINITZ AND STINSON
Contemporary Design Theory: A Collection of Surveys

Continued at the end of this volume

Annotated Bibliographies in Combinatorial Optimization

Edited by

Mauro Dell'Amico
Università di Modena, Italy

Francesco Maffioli
Politecnico di Milano, Italy

and

Silvano Martello
Università di Bologna, Italy

A Wiley-Interscience Publication

JOHN WILEY & SONS
Chichester • New York • Weinheim • Brisbane • Singapore • Toronto

Baffins Lane, Chichester,
West Sussex PO19 1UD, England

National 01243 779777
International (+44) 1243 779777

Cover photograph supplied courtesy of the Civico Museo Bibliografico Musicale

e-mail (for orders and customer service enquiries): cs-books@wiley.co.uk.
Visit our Home Page on http://www.wiley.co.uk
or http://www.wiley.com

Other Wiley Editorial Offices

John Wiley & Sons, Inc., 605 Third Avenue,
New York, NY 10158-0012, USA

WILEY-VCH Verlag GmbH, Pappelallee 3,
D-69469 Weinheim, Germany

Jacaranda Wiley Ltd, 33 Park Road, Milton,
Queensland 4064, Australia

John Wiley & Sons (Asia) Pte Ltd, 2 Clementi Loop #02-01,
Jin Xing Distripark, Singapore 129809

John Wiley & Sons (Canada) Ltd, 22 Worcester Road,
Rexdale, Ontario M9W 1L1, Canada

British Library Cataloguing in Publication Data

A catalogue record for this book is available from the British Library

ISBN 0 471 96574 X

Typeset in 10/12pt Computer Modern Roman by Sunrise Setting Ltd, Torquay
Printed and bound in Great Britain by Bookcraft (Bath) Ltd
This book is printed on acid-free paper responsibly manufactured from sustainable forestation, for which at least two trees are planted for each one used for paper production.

Contents

Preface

In 1985 Wiley published a book, edited by M. O'hEigeartaigh, J.K. Lenstra and A.H.G. Rinnooy Kan, presenting annotated bibliographies in twelve important areas of Combinatorial Optimization, which soon became a very handy tool for researchers in the field. The present book of annotated bibliographies in Combinatorial Optimization adopts the same idea, covering the huge developments that have come about in this field in the last decade (in almost 3000 references). It offers much more than a pure bibliography and provides the ideal starting point for any researcher who needs to become familiar with some application and/or theory oriented aspects of this large field in a short time. Within the academic community, both young and more experienced researchers will find it invaluable in orienting their work. The large part dedicated to applications and the many sections providing pointers to the available software will also appeal to professionals in the field. It is finally hoped to be of great use to teachers as a guide for students in their thesis work. In fact, each chapter provides, besides a list of annotated references, a brief survey of the corresponding area.

There are 24 chapters on both method and application oriented subjects, plus an initial chapter reviewing what the authors believe to be the most influential texts that have appeared during the last decade. We may roughly say that chapters 2 to 11 are more method oriented, whereas the remaining chapters are on specific topics or particular application areas. All chapters have been thoroughly refereed: our thanks go to this anonymous group of colleagues for their help in improving the quality of the final product. As each of them is a well-known expert in the specific subject he has refereed, we prefer not to reveal their names since matching referees to chapters would become an easy task.

The idea that it would have been useful to have such a tool occurred to the three editors in late 1995. After producing a still provisional list of chapters, the search for contributors began: we were very satisfied by the enthusiastic response from colleagues during this contact phase. The final set of contributors is composed of well-known experts in the area they have been asked to cover. We thank them all for their dedication and their patience in adapting their style of writing so as to obtain a uniform presentation throughout the book.

We like to acknowledge the advice of Hans Bodlaender, Giulia Galbiati, Giorgio Gallo, Dan Gusfield, Stefano Pallottino, David Shmoys, Paolo Toth, Daniele Vigo and Laurence Wolsey. We feel the need to address a special word of gratitude to Jan Karel Lenstra, who has provided continuous advice and encouragement. The interface with Wiley as a publisher has been particularly smooth thanks to the efforts of Mrs. Helen Ramsey initially and later of Mr. David Ireland. Last but not least, we acknowledge the support of Human Capital and Mobility contract no. ERB CHR XCT 930087 for the project "Models, Algorithms and Systems for Decision Making" coordinated by one of the editors: the ample occasion for research contacts that this project provided was instrumental in producing many parts of this book.

April 1997

MAURO DELL'AMICO
FRANCESCO MAFFIOLI
SILVANO MARTELLO

List of Contributors

Karen Aardal
Department of Computer Science
Utrecht University
P.O. Box 80089
3508 TB Utrecht, The Netherlands
aardal@cs.ruu.nl

Emile Aarts
Department of Mathematics and Computing Science
Eindhoven University of Technology
P.O. Box 513
5600 MB Eindhoven, The Netherlands
and
Philips Research Laboratories
Prof. Holstlaan, 4
5656 AA Eindhoven, The Netherlands
aarts@natlab.research.philips.com

Ravindra K. Ahuja
Department of Industrial & Management Engineering
Indian Insitute of Technology
208 016 Kanpur, India
ahuja@iitk.ernet.in

Anantaram Balakrishnan
Smeal College of Business Administration
The Pennsylvania State University
University Park, PA 16802, U.S.A.
anantb@psu.edu

Rainer Burkard
Institut für Mathematik B
Technische Universität Graz
Steyrergasse 30
A-8010 Graz, Austria
burkard@opt.math.tu-graz.ac.at

Alberto Caprara
Dipartimento di Elettronica, Informatica e Sistemistica
Università di Bologna
Viale Risorgimento, 2
40136 Bologna, Italy
acaprara@deis.unibo.it

Eranda Çela
Institut für Mathematik B
Technische Universität Graz
Steyrergasse 30
A-8010 Graz, Austria
cela@opt.math.tu-graz.ac.at

Sebastián Ceria
417 Uris Hall
Graduate School of Business
Columbia University
New York, NY 10027, U.S.A.
sebas@cumparsita.gsb.columbia.edu

Michele Conforti
Dipartimento di Matematica Pura ed Applicata
Università di Padova
Via Belzoni, 7
35131 Padova, Italy
conforti@math.unipd.it

Gérard Cornuéjols
Graduate School of Industrial Administration
Carnegie Mellon University
Pittsburgh, PA 15213, U.S.A.
gc0v@andrew.cmu.edu

Mauro Dell'Amico
Dipartimento di Economia Politica
Università di Modena
Via Berengario, 51
41100 Modena, Italy
dellamico@unimo.it

Harald Dyckhoff
Lehrstuhl für Industriebetriebslehre
RWTH Aachen
Templergraben, 64
D-52056 Aachen, Germany
lut@sun.lut.RWTH-Aachen.de

Matteo Fischetti
Dipartimento di Elettronica e Informatica
Università di Padova
Via Gradenigo, 6/a
35131 Padova, Italy
fisch@dei.unipd.it

András Frank
Department of Operations Research
Eötvös University
Muzeum krt. 6-8
H-1088 Budapest, Hungary
frank@cs.elte.hu

Michel X. Goemans
Department of Mathematics
Room 2-392
Massachusetts Institute of Technology
Cambridge, MA 02139, U.S.A.
goemans@math.mit.edu

Han Hoogeveen
Department of Mathematics and
Computing Science
Eindhoven University of Technology
P.O. Box 513
5600 MB Eindhoven, The Netherlands
slam@win.tue.nl

Michael Jünger
Institut für Informatik
Universität zu Köln
Pohligstraße, 1
50969 Köln, Germany
mjuenger@informatik.uni-koeln.de

Viggo Kann
Department of Numerical Analysis and
Computer Science
Royal Institute of Technology
S-100 44 Stockholm, Sweden
viggo@nada.kth.se

Ajai Kapoor
Dipartimento di Matematica Pura
ed Applicata
Università di Padova
Via Belzoni, 7
35131 Padova, Italy
ajai@sfo.com

David R. Karger
MIT Laboratory for Computer Science
545 Technology Square
Cambridge, MA 02139, U.S.A.
karger@theory.lcs.mit.edu

Jon M. Kleinberg
Department of Computer Science
Cornell University
Ithaca, NY 14853, U.S.A.
kleinber@cs.cornell.edu

Martine Labbé
Service de Mathématiques de la Gestion
Université Libre de Bruxelles
Boulevard du Triomphe, C.P. 210/01
B-1050 Bruxelles, Belgium
mlabbe@smg.ulb.ac.be

Gilbert Laporte
Centre de Recherche sur les Transports
Université de Montréal
C.P. 6128 Succursale "Centre Ville"
H3C 3J7 Montreal, Canada
gilbert@crt.umontreal.ca

Monique Laurent
CNRS, DMI
École Normale Supérieure
Paris, France
and
CWI
Kruislaan, 413
1098 SJ Amsterdam, The Netherlands
monique@cwi.nl

Hans-Peter Lenhof
Max-Planck-Institut für Informatik
Im Stadtwald, Gebäude 46.1
D-66123 Saarbrücken, Germany
len@mpi-sb.mpg.de

Jan Karel Lenstra
Department of Mathematics and
Computing Science
Eindhoven University of Technology
P.O. Box 513
5600 MB Eindhoven, The Netherlands
jkl@win.tue.nl

François Louveaux
Facultés Universitaires Notre Dame
de La Paix
8, Rempart de la Vierge
B-5000 Namur, Belgium
Francois.Louveaux@fundp.ac.be

Francesco Maffioli
Dipartimento di Elettronica
e Informazione
Politecnico di Milano
Piazza Leonardo da Vinci, 32
20133 Milano, Italy
maffioli@elet.polimi.it

Thomas L. Magnanti
Sloan School of Management and
Department of Electrical Engineering and
Computer Science
Massachusetts Institute of Technology
Cambridge, MA 02139, U.S.A.
magnanti@mit.edu

Silvano Martello
Dipartimento di Elettronica, Informatica
e Sistemistica
Università di Bologna
Viale Risorgimento, 2
40136 Bologna, Italy
smartello@deis.unibo.it

Prakash Mirchandani
Katz Graduate School of Business
University of Pittsburgh
Pittsburgh, PA 15260, U.S.A.
pmirchan@vms.cis.pitt.edu

Rolf H. Möhring
Fachbereich Mathematik, Sekr. MA 6-1
Technische Universität Berlin
Straße des 17. Juni 136
10623 Berlin, Germany
moehring@math.tu-berlin.de

Petra Mutzel
Max-Plank-Institut für Informatik
Im Stadtwald, Gebäude 46.1
D-66123 Saarbrücken, Germany
mutzel@mpi-sb.mpg.de

Paolo Nobili
IASI-CNR
Viale Manzoni, 30
00185 Roma, Italy
nobili@iasi.rm.cnr.it

Alessandro Panconesi
BRICS
Department of Computer Science
University of Aarhus
Ny Munkegade, building 540
DK–8000 Aarhus C, Denmark
ale@brics.dk

Subramanian Raghavan
U S WEST Advanced Technologies
4001 Discovery Drive
Boulder, CO 80303, U.S.A.
raghavan@alum.mit.edu

Gerhard Reinelt
Institut für Angewandte Mathematik
Universität Heidelberg
Im Neuenheimer Feld, 294
69120 Heidelberg, Germany
gerhard.reinelt@iwr.uni-heidelberg.de

Giovanni Rinaldi
IASI-CNR
Viale Manzoni, 30
00185 Roma, Italy
rinaldi@iasi.rm.cnr.it

Cornelis Roos
Department of Mathematics and
Computer Science
Delft University of Technology
Mekelweg 4, room 05.300
2628 CD Delft, The Netherlands
c.roos@twi.tudelft.nl

Antonio Sassano
Dipartimento di Informatica
e Sistemistica
Università di Roma "La Sapienza"
Via Buonarroti, 12
00185 Roma, Italy
sassano@dis.uniroma1.it

Guntram Scheithauer
Institute of Numerical Mathematics
Dresden University of Technology
Mommsenstraße, 13
D-01062 Dresden, Germany
scheit@math.tu-dresden.de

François Soumis
École Polytechnique de Montréal
Chemin de Polytechnique, 2900
H3C 3A7 Montreal, Canada
soumis@crt.umontreal.ca

Leen Stougie
Department of Operations Research
Faculty of Economics and Econometrics
University of Amsterdam
Roetersstraat, 11
1018 WB Amsterdam, The Netherlands
leen@fee.uva.nl

Tamás Terlaky
Department of Mathematics and
Computer Science
Delft University of Technology
Mekelweg, 4, room 05.260
2628 CD Delft, The Netherlands
t.terlaky@twi.tudelft.nl

Johannes Terno
Institute of Numerical Mathematics
Technical University of Dresden
Mommsenstraße, 13
D-01062 Dresden, Germany
terno@math.tu-dresden.de

Steef Van de Velde
Rotterdam School of Management
Erasmus University Rotterdam
P.O. Box 1738
3000 DR Rotterdam, The Netherlands
s.velde@fac.fbk.eur.nl

Marco Verhoeven
Philips Research Laboratories
Prof. Holstlaan, 4
5656 AA Eindhoven, The Netherlands
mverhoev@natlab.research.philips.com

Martin Vingron
Deutsches Krebsforschungszentrum
Abt. Theoretische Bioinformatik (0815)
Im Neuenheimer Feld 280
D-69120 Heidelberg, Germany
m.vingron@dkfz-heidelberg.de

Maarten H. van der Vlerk
Department of Econometrics
University of Groningen
P.O. Box 800
9700 AV Groningen, The Netherlands
m.h.van.der.vlerk@eco.rug.nl

Kristina Vušković
Department of Mathematics
University of Kentucky
Lexington, KY 40506, U.S.A.
kristina@ms.uky.edu

Dorothea Wagner
Fakultät für Mathematik und Informatik
Universität Konstanz
78434 Konstanz, Germany
dorothea.wagner@uni-konstanz.de

Robert Weismantel
Konrad-Zuse-Zentrum für Informationstechnik Berlin (ZIB)
Takustraße, 7
14195 Berlin-Dahlem, Germany
weismantel@zib.de

1 Selected Books in and around Combinatorial Optimization

Francesco Maffioli
Dipartimento di Elettronica e Informazione, Politecnico di Milano

Silvano Martello
Dipartimento di Elettronica, Informatica e Sistemistica, Università di Bologna

CONTENTS

This book of annotated bibliographies in combinatorial optimization has been conceived as a service to the scientific community. It appeared that in this spirit a first chapter reviewing the available books in the field was appropriate. Indeed this should help both young researchers and students looking for a text in one or the other part of this field, as well as colleagues looking for references in book form for their professional use or for teaching activity. Were the number of available works smaller, this chapter would have not been justified; as it appears from the following this is not the case and, even without pretending to have been able to identify all potentially interesting references, the large choice reported here is in the authors' opinion a good first selection.

Following the general policy of this volume, our bibliography starts from the mid 1980's, with two exceptions: we mention a handful of works of doubtless historical value in the first section, and we begin most of the following sections with an "old" but still relevant text.

We had to decide about where to put the frontier as far as neighbouring fields are concerned. These include quite a number of more or less advanced mathematical

Annotated Bibliographies in Combinatorial Optimization, edited by M. Dell'Amico, F. Maffioli and S. Martello

topics, as well as areas of computer science or management science, or even other parts of mathematical programming, such as linear programming and non-differentiable optimization. Our choice has been to offer only a very limited selection of books in some of these areas, this being due to the need to keep the length of the chapter within control. We also remind the reader that some more specialized references are to be found in the other chapters of this book: only some of them have been repeated here, when considered of more general interest.

Another disclaimer concerns edited books: we have excluded proceedings but included collections of well thought monographic surveys. When the covered topic is very broad, these are to be found in §6.

1 Historical

R.E. Bellman (1957). *Dynamic Programming*, Princeton University Press, Princeton, NJ.

The first complete text on the mathematical theory of multi-stage decision processes.

L.R. Ford, D.R. Fulkerson (1959). *Flows in Networks*, Princeton University Press, Princeton, NJ.

The classic in network flow as much as [Dantzig 1963] (see below) is for linear programming. Obviously out-of-date, but a very good example of clear scientific writing. Also important for the famous max-flow min-cut result.

G.B. Dantzig (1963). *Linear Programming and Extensions*, Princeton University Press, Princeton, NJ.

The most famous book on the most important optimization problem. The inventor of the simplex method treats with elegance theoretical and algorithmic aspects.

R.S. Garfinkel, G.L. Nemhauser (1973). *Integer Programming*, Wiley, New York.

A clear and comprehensive textbook, with excellent chapters on branch-and-bound and cutting planes.

2 Complexity

M.R. Garey, D.S. Johnson (1979). *Computers and Intractability: A Guide to the Theory of NP-Completeness*, Freeman, San Francisco.

The first complete treatment of computational complexity. It includes an extensive theoretical presentation and an invaluable catalogue of hundreds of NP-complete and NP-hard problems. Still an invaluable working tool and an example of clear writing style.

J. van Leeuwen (ed.) (1990). *Algorithms and Complexity*, volume A of Handbook of Theoretical Computer Science, North-Holland, Amsterdam.

The first of this two volume collection of competently written monographs on TCS contains many chapters not only dedicated to complexity issues, but also to combinatorial optimization and data structures. Highly recommended as a reference book, the only drawbacks being the fact that it is a few years old and it costs too much!

C.H. Papadimitriou (1994). *Computational Complexity*, Addison-Wesley, Reading, MA.
D.P. Bovet, P. Crescenzi (1994). *Introduction to the Theory of Complexity*, Prentice-Hall, Englewood Cliffs, NJ.
The two available up-to-date texts on computational complexity. The former is probably the best reference in book form now available, also because of a large section on logic implications. Both are well suited for teaching purposes.

3 Graphs and Networks

N. Christofides (1975). *Graph Theory: an Algorithmic Approach*, Academic Press, New York.
Partially out-of-date, but the presentation of algorithms for many graph problems is still an example of clear didactic style.

M. Gondran, M. Minoux (1984). *Graphs and Algorithms*, Wiley, New York.
Translated from the original French, this comprehensive text is now dated, mainly for what concerns exact methods. Still interesting is the part devoted to the algebra of paths, a specialty of the first author.

R.T. Rockafellar (1984). *Network Flows and Monotropic Optimization*, Wiley, New York.
A new look at the approaches based on linear programming and duality to network problems leads to a generalization called monotropic optimization by the author.

L. Lovász, M. Plummer (1986). *Matching Theory*, North-Holland, Amsterdam.
The reference text on matching theory and extensions. Very comprehensive, written with the lucidity one expects from such authors, it is a clear fundamental element of any library. Quite advanced, but also suitable with some selection for a monographic course.

U. Derigs (1988). *Programming in Networks and Graphs*, volume 300 of Lecture Notes in Economics and Mathematical Systems, Springer-Verlag, Berlin.
A careful analysis of successive shortest path algorithms for bipartite and non-bipartite matching problems.

W. Chen (1990). *Theory of Nets: Flows in Networks*, Wiley, New York.
This textbook is not as comprehensive as the one of [Ahuja, Magnanti and Orlin 1993] (see below), but more classical and easier for undergraduates.

L.R. Foulds (1992). *Graph Theory Applications*, Springer-Verlag, Berlin.
This elementary introduction to the basic notions and algorithms of graph theory includes the description of applications in different fields. Suitable for teaching elementary graph theory courses in several disciplines.

F. Glover, D. Klingman, N. Phillips (1992). *Network Models in Optimization and Their Applications in Practice*, Wiley, New York.

An effective combination of modelling techniques, algorithms and applications for pure, generalized, and integer networks.

R.K. Ahuja, T.L. Magnanti, J.B. Orlin (1993). *Network Flows*, Prentice-Hall, Englewood Cliffs, NJ.

A voluminous text (850 pages) on shortest paths, max-flow and min-cost flow problems, as well as on several additional topics. Being based on original approaches developed by the authors, the reader may have some difficulty in connecting the material to the presentation usually followed in the literature. Selected chapters are suitable for advanced courses.

V.K. Balakrishnan (1995). *Network Optimization*, Chapman & Hall, London.

A compact monograph mainly focusing on mathematical tools for the development of polynomial-time algorithms for network problems.

M.O. Ball, T.L. Magnanti, C.L. Monma, G.L. Nemhauser (eds.) (1995). *Network Models*, volume 7 of Handbooks in Operations Research and Management Science, North-Holland, Amsterdam.
M.O. Ball, T.L. Magnanti, C.L. Monma, G.L. Nemhauser (eds.) (1995). *Network Routing*, volume 8 of Handbooks in Operations Research and Management Science, North-Holland, Amsterdam.

Two collections of a total of twenty monographs on all main facets of network optimization, written by leading experts. Both algorithmic and modelling aspects are deeply surveyed.

4 Integer Programming

E.L. Lawler (1976). *Combinatorial Optimization: Networks and Matroids*, Holt, Rinehart and Winston, New York.

A milestone of combinatorial optimization. The unified presentation of augmentation algorithms for network and matroid problems is still very interesting.

C.H. Papadimitriou, K. Steiglitz (1982). *Combinatorial Optimization: Algorithms and Complexity*, Prentice-Hall, Englewood Cliffs, NJ.

This is one of the best didactical texts on combinatorial optimization, combining mathematical programming and the theory of design and analysis of algorithms. It includes an original study of the primal-dual method, which results in a unifying approach to the design of direct algorithms for many combinatorial optimization problems.

H.P. Williams (1985). *Model Building in Mathematical Programming*, Wiley, New York.

A book dedicated to the art of modelling is a good idea, since this is often neglected and not so obvious as it may appear at first glance. A good suggestion as a side book to

consider for students to consult, if not as the only text. The same author has recently issued a companion text:
H.P. Williams (1993). *Problem Solving in Mathematical Programming*, Wiley, New York.

A. Schrijver (1986). *Theory of Linear and Integer Programming*, Wiley, New York.
Probably the most complete text on linear and integer programming, with emphasis on the theory. A must for any researcher.

T. Ibaraki (1987). *Enumerative Approaches to Combinatorial Optimization -Part I*, volume 10 of Annals of Operations Research, Baltzer, Basel.
T. Ibaraki (1987). *Enumerative Approaches to Combinatorial Optimization -Part II*, volume 11 of Annals of Operations Research, Baltzer, Basel.
Everything you always wanted to know about branch-and-bound, but were afraid to ask! A comprehensive reference on enumerative approaches, including dynamic programming.

M. Grötschel, L. Lovász, A. Schrijver (1988). *Geometric Algorithms and Combinatorial Optimization*, Springer-Verlag, Berlin.
A thorough study of the geometry of integer linear programming, emphasizing theoretical aspects, such as the deep implications of the ellipsoid algorithm, basis reduction, and more. It contains the proof of the equivalence between separation and optimization in integer linear programming. A fundamental text for polyhedral combinatorics.

R.G. Parker, R.L. Rardin (1988). *Discrete Optimization*, Academic Press, New York.
Very good choice of overall organization and material, but not always successful from a didactic point of view.

G.L. Nemhauser, L.A. Wolsey (1988). *Integer and Combinatorial Optimization*, Wiley, New York.
A comprehensive book on all mathematical aspects of discrete optimization. Excellent as advanced textbook and an invaluable desk reference for researchers.

H.M. Salkin, K. Mathur (1989). *Foundations of Integer Programming*, North-Holland, Amsterdam.
More than a textbook, this volume is several textbooks together. It gives a comprehensive treatment of integer programming, perhaps no longer up-to-date, but comparatively easy to read. The last sections are devoted to important applications.

S. Walukiewicz (1991). *Integer Programming*, Kluwer Academic Publishers, Boston, MA.
A compact text on algorithmic approaches to the solution of integer programming problems, with special emphasis on relaxations and branch-and-bound.

G. Sierksma (1996). *Integer and Linear Programming: Theory and Practice*, Marcel Dekker, New York.
A textbook on classical as well as more recent solution techniques. It is accompanied

by a software package for solving small linear programming problems, also including Karmarkar's interior point method.

W. Cook, W.H. Cunningham, W.R. Pulleyblank, A. Schrijver (1997). *Combinatorial Optimization*, Wiley, New York (to appear).

There is considerable expectation for this forthcoming book.

We conclude this section with two relevant books on linear programming.

Chvátal (1983). *Linear Programming*, Freeman, New York.
M. Padberg (1994). *Linear Optimization and Extensions*, Springer-Verlag, Berlin.

The first of these is one of the best classical textbooks in linear programming, highly recommended both for teaching and as a reference. The latter is a recent textbook for an advanced course at the Ph.D. level, with emphasis on geometry of simplex algorithms, projective methods and ellipsoid algorithms.

5 Heuristics

L. Davis (ed.) (1987). *Genetic Algorithms and Simulated Annealing*, Morgan Kaufmann, Los Altos, CA.

A collection of thirteen papers, mostly on applications of genetic algorithms and simulated annealing to artificial intelligence problems.

P.J.M. Van Laarhoven, E.H.L. Aarts (1987). *Simulated Annealing: Theory and Applications*, Kluwer Academic Publishers, Boston, MA.
E.H.L. Aarts, J. Korst (1989). *Simulated Annealing and Boltzmann Machines: A Stochastic Approach to Combinatorial Optimization and Neural Computing*, Wiley, New York.

The second book contains an up-to-date account of simulated annealing from two leading experts in the field, with extension to the parallelized version and its relation with neural computing. The first book has a similar target, but no parallelism. One wonders why the authors did not decide to merge the two books into a single one.

D.E. Goldberg (1989). *Genetic Algorithms in Search, Optimization, and Machine Learning*, Addison-Wesley, Reading, MA.
M. Mitchell (1996). *An Introduction to Genetic Algorithms*, MIT Press, Cambridge, MA.

The first book is a classical introduction to genetic algorithms, illustrated through numerous useful examples. The second one provides an excellent introduction into the subject, covering traditional methods, recent variants, implementation aspects, theoretical foundations, and applications. Both may be recommended as textbooks at the graduate level.

Z. Michalewicz (1992). *Genetic Algorithms + Data Structures = Evolutionary Programs*, Springer-Verlag, Berlin.

A detailed introduction to the theory and implementation of genetic algorithms, including several applications in combinatorial optimization.

C.R. Reeves (ed.) (1993). *Modern Heuristic Techniques for Combinatorial Problems*, Halsted Press, New York.
E.H.L. Aarts, J.K. Lenstra (eds.) (1997). *Local Search in Combinatorial Optimization*, Wiley, New York.

The first of these two collections presents a good introduction to simulated annealing (by K.A. Dowsland), tabu search (by F. Glover and M. Laguna), genetic algorithms (by C.R. Reeves), neural networks (by C. Peterson and B. Söderberg) and Lagrangian relaxation (by J.E. Beasley). In the second book seven methodological surveys and six problem oriented ones provide an excellent up-to-date overview of modern heuristic techniques.

R. Motwani, P. Raghavan (1995). *Randomized Algorithms*, Cambridge University Press, Cambridge.

The leading reference on the use of randomization as a computing tool and the various, often sophisticated ways of exploiting it in the design of algorithms. Applications are to be found in various fields, one of them being combinatorial optimization.

G. Laporte, I.H. Osman (eds.) (1996). *Meta-Heuristics in Combinatorial Optimization*, volume 63 of Annals of Operations Research, Baltzer, Basel.
V.J. Rayward-Smith, I.H. Osman, C.R. Reeves, G.D. Smith (eds.) (1996). *Modern Heuristic Search Methods*, Wiley, New York.

Two recent collections on applications of modern heuristic techniques. The second book includes six interesting survey chapters.

D.S. Hochbaum (ed.) (1996). *Approximation Algorithms for NP-hard Problems*, PWS Publishing Co., Boston, MA.

An invaluable collection of well-written surveys in the fascinating field of algorithms for NP-hard problems with worst-case performance guarantees. The list of contributors could hardly be more impressive, including only leading experts in the field. A great achievement.

6 Collections

N. Christofides, A. Mingozzi, P. Toth, C. Sandi (eds.) (1979). *Combinatorial Optimization*, Wiley, New York.

One of the first modern collections of surveys in combinatorial optimization. Fifteen monographs based on the lectures given at a summer school held in Urbino in 1977.

A. Bachem, M. Grötschel, B. Korte (eds.) (1983). *Mathematical Programming: The State of the Art*, Springer-Verlag, Berlin.

A collection of twenty-one tutorial lectures given by leading experts at the Bonn Mathematical Programming Symposium.

F. Archetti, F. Maffioli (eds.) (1984). *Stochastics and Optimization*, volume 1 of Annals of Operations Research, Baltzer, Basel.

The first book of the highly successful Annals of OR series. An offspring of a workshop by the same title held in Gargnano in 1982, it is one of the first attempts to see the various aspects of stochasticity as a resource in computational optimization methods. It also tried to construct a bridge between the continous and combinatorial optimization communities.

W.R. Pulleyblank (ed.) (1984). *Progress in Combinatorial Optimization*, Academic Press, New York.

A very interesting collection of papers on advanced topics of combinatorial optimization. Style of contributions varies, from survey-like to more extensive specialized presentations.

M. O'hEigeartaigh, J.K. Lenstra, A.H.G. Rinnooy Kan (eds.) (1985). *Combinatorial Optimization: Annotated Bibliographies*, Wiley, New York.

The "father" of the present volume. Twelve annotated bibliographies on fundamental research topics, still very useful for the older references that are only partially covered here.

S. Martello, G. Laporte, M. Minoux, C. Ribeiro (eds.) (1987). *Surveys in Combinatorial Optimization*, North-Holland, Amsterdam.

A collection of eleven survey papers on a number of important theoretical and application oriented areas. The monographs are based on the lectures given at a summer school on combinatorial optimization held in Rio de Janeiro in 1985.

G.L. Nemhauser, A.H.G Rinnooy Kan, M.J. Todd (eds.) (1988). *Optimization*, volume 1 of Handbooks in Operations Research and Management Science, North-Holland, Amsterdam.

The first volume of a successful series presents ten excellent chapters on linear and nonlinear optimization.

J.K. Lenstra, A.H.G. Rinnooy Kan, A. Schrijver (eds.) (1991). *History of Mathematical Programming*, North-Holland, Amsterdam.

A unique work, prepared on the occasion of the Amsterdam Mathematical Programming Symposium. The reminescences of the founders of this discipline constitute exciting and pleasant reading.

M. Akgül, H.W. Hamacher, S. Tüfekçi (eds.) (1992). *Combinatorial Optimization: New Frontiers in Theory and Practice*, volume 82 of NATO ASI Series F, Springer-Verlag, Berlin.

Twelve tutorials from a NATO Advanced Study Institute held in Ankara in 1990. The volume also includes extended abstracts of participants' presentations.

J.R. Birge, K.G. Murty (eds.) (1994). *Mathematical Programming: State of the Art 1994*, The University of Michigan, Ann Arbor, Michigan.

A collection of sixteen interesting tutorial lectures given at the Ann Arbor Mathematical Programming Symposium.

R. Graham, M. Grötschel, L. Lovász (eds.) (1995). *Handbook of Combinatorics*, North-Holland, Amsterdam.
The best up-to-date reference on combinatorics, most chapters being directly relevant for combinatorial optimization.

7 Data Structures

A.V. Aho, J.E. Hopcroft, J.D. Ullman (1974). *The Design and Analysis of Computer Algorithms*, Addison-Wesley, Reading, MA.
A classic text on fundamental algorithms, evaluated through asymptotic performance analysis. A more complete treatment is in [Cormen, Leiserson and Rivest 1990] (see below).

R.E. Trjan (1982). *Data Structures and Network Algorithms*, SIAM, Philadelphia, PA.
A rigorous, deep and elegant analysis of the data structures needed for implementing efficient algorithms for the basic network problems (spanning trees, paths, flows, matchings).

T.H. Cormen, C.E. Leiserson, R.L. Rivest (1990). *Introduction to Algorithms*, MIT Press, Cambridge, MA.
A huge (over 1000 pages) but well organized presentation of basic algorithms (for sorting, data structures, fundamental graph problems and other topics). The detailed pseudocode descriptions are easily translatable into recursive programming languages. Excellent as a textbook.

8 Problem Oriented

8.1 Graph and network problems

E.L. Lawler, J.K. Lenstra, A.H.G. Rinnooy Kan, D.B. Shmoys (eds.) (1985). *The Traveling Salesman Problem: A Guided Tour of Combinatorial Optimization*, Wiley, New York.
The sub-title clarifies the main feature of this book. Twelve monographs, written by foremost authorities, use a famous problem for providing a unifying overview of the fundamental areas of combinatorial optimization.

M. Jünger (1985). *Polyhedral Combinatorics and the Acyclic Subdigraph Problem.* Heldermann Verlag Berlin.
G. Reinelt (1985). *The Linear Ordering Problem: Algorithms and Applications.* Heldermann Verlag Berlin.
Derived from the authors' Ph.D. theses, these two books provide an easily readable introduction to the main ideas of polyhedral combinatorics and an extensive study of the problem of finding an acyclic subdigraph of maximum weight in an edge-weighted digraph.

C.J. Colbourn (1987). *The Combinatorics of Network Reliability*, Oxford University Press, New York.

The basis for the study of network reliability.

B.L. Golden, A.A. Assad (eds.) (1988). *Vehicle Routing: Methods and Studies*, volume 16 of Studies in Management Science and Systems, North-Holland, Amsterdam.

Twenty-one surveys on algorithmic techniques, models and practical applications relative to a rich and important optimization area.

M. Stoer (1992). *Design of Survivable Networks*, volume 1531 of Lecture Notes in Mathematics, Springer-Verlag, Berlin.

An in-depth monograph on the application of polyhedral combinatorics to the design of branch-and-cut algorithms for network design problems arising in the design of modern telecommunication networks.

8.2 Location

P.B. Mirchandani, R.L. Francis (eds.) (1990). *Discrete Location Theory*, Wiley, New York.

A collection of twelve surveys on the various aspects of location problems, very much oriented toward the discrete side and covering several key theoretical issues. All contributors are well-known experts in the respective fields and several chapters are still worth reading, even if no longer up-to-date.

R.L. Francis, J.L.F. McGinnis, J.A. White (1992). *Facility Layout and Location: An Analytical Approach*, Prentice-Hall, Englewood Cliffs, NJ.

This is the second edition of one of the first books on facility location.

Z. Drezner (ed.) (1995). *Facility Location. A Survey of Applications and Methods*, Springer-Verlag, Berlin.

A collection of twenty surveys providing an in-depth overview of many aspects of facility location methods and applications.

M.S. Daskin (1995). *Network and Discrete Location*, Wiley, New York.

Although the description is not very deep, this text is useful for teaching use at the early graduate level.

8.3 Scheduling

J. Błażewicz, K.H. Ecker, G. Schmidt, J. Węglarz (1993). *Scheduling in Computer and Manufacturing Systems*, Springer-Verlag, Berlin.

An introduction to classical scheduling problems with interesting chapters on applications in flexible manufacturing systems.

M.L. Pinedo (1995). *Scheduling: Theory, Algorithms, and Systems*, Prentice-Hall, Englewood Cliffs, NJ.

The best undergraduate text in the area, covering deterministic scheduling, stochastic scheduling, and practice.

M.W. Padberg, M.P. Rijal (1996). *Location, Scheduling, Design and Integer Programming*, Kluwer Academic Publishers, Boston, MA.
A thorough study on polyhedral approaches to assignment-type problems with quadratic cost functions. It includes the listing of a FORTRAN program to solve small symmetric quadratic assignment problems.

8.4 Cutting and packing

S. Martello, P. Toth (1990). *Knapsack Problems: Algorithms and Computer Implementations*, Wiley, New York.
A survey book on knapsack and related problems (bin packing, generalized assignment, ...), with special emphasis on the development of exact algorithms. An included diskette contains the corresponding FORTRAN listings.

H. Dyckhoff, U. Finke (1992). *Cutting and Packing in Production and Distribution*, Springer-Verlag, Berlin.
The first systematic treatment of a huge field. One wonders why the four-field classification developed by the first author is not used here.

8.5 Matroids and extensions

D.J.A. Welsh (1976). *Matroid Theory*, Academic Press, New York.
The classic reference in matroid theory. Not oriented to combinatorial optimization, but aiming at a panorama of this important part of discrete mathematics, it is still worth having as a reference book.

B. Korte, L. Lovász, R. Schrader (1991). *Greedoids*, Springer-Verlag, Berlin.
The only text available on greedoids (a generalization of matroids). Well written and valuable as a reference work in this very specialized field.

J.G. Oxley (1992). *Matroid Theory*, Oxford University Press, New York.
A modern introduction to the field, blending the graph-theoretic approach of Welsh with a more geometric one. The coverage of the field is not as complete as in the older book by Welsh. Clear, rigorous, but not difficult to read: an obvious choice for teaching purposes and a good reference to keep on the shelves.

9 Software Oriented

M. Sysło, N. Deo, M. Kowalick (1983). *Discrete Optimization Algorithms*, Prentice-Hall, Englewood Cliffs, NJ.
This volume on linear and integer programming, packing and covering, network optimization, colouring and scheduling includes the Pascal listings of twenty-six programs implementing the most common algorithms in each area.

L. Schrage (1986). *Linear, Integer and Quadratic Programming with LINDO*, Scientific Press, San Francisco, CA.

An introduction to the use of a very well-known computer package for linear and integer programming.

B. Simeone, P. Toth, G. Gallo, F. Maffioli, S. Pallottino (eds.) (1988). *FORTRAN Codes for Network Optimization*, Baltzer, Basel.

Seven monographs on efficient implementations of algorithms for network problems: shortest paths, trees, max-flow, min-cost flow, assignment, matching. The codes are provided both as listings and on diskette.

D.P. Bertsekas (1991). *Linear Network Optimization: Algorithms and Codes*, MIT Press, Cambridge, MA.

A state-of-the-art book on the author's specialties: relaxation and auction techniques, applied to shortest paths, assignments, max-flow and min-cost flow problems. The corresponding codes, supplied as FORTRAN listings, are also available on diskette.

J.R. Evans, E. Minieka (1992). *Optimization Algorithms for Networks and Graphs*, Marcel Dekker, New York.

This revised version of a famous book published by Minieka in 1987 is mainly interesting for a diskette containing NETSOLVE, a package developed by J. Jarvis and D. Shier for the solution of several important network problems.

Acknowledgements

This work was supported by Ministero dell'Università e della Ricerca Scientifica e Tecnologica, and by Consiglio Nazionale delle Ricerche, Italy. We are grateful to Jan Karel Lenstra for many insightful comments on an earlier version of this chapter.

2 Hardness of Approximation

Viggo Kann
Royal Institute of Technology, Stockholm

Alessandro Panconesi
University of Aarhus

CONTENTS

The aim of computational complexity theory is to devise tools that enable us to classify computable functions according to how expensive it is to compute them. The cost is usually (but by no means always) given by the *time* needed for the computation, a notion that can be defined in a technology independent manner by means of Turing machines or other equivalent abstractions.

A turning point in the development of computational complexity was the introduction of the theory of NP-completeness. This theory has now found a stable place in introductory books and courses on algorithms and computational complexity and we assume familiarity with it. The standard reference on the subject is the book [Garey and Johnson 1979] (see §1.1). The other textbooks cited in §1.1 are also good introductions to the subject and to the other notion that we assume familiar to the reader, that of *optimization problems.*

An optimization problem can be seen as a function $f : I \rightarrow N$ between the set of input instances and the natural numbers. Ideally, we would like to determine the function $t_f(n)$ describing the running time needed to compute f on inputs of size n.

NP-completeness can be used as a first rough approximation to this classification problem. The standard way to proceed is as follows: instead of considering the problem "given x, compute $f(x)$" one introduces an integer parameter k and considers the problem "given (x, k), is $f(x) > k$?" Note that the first problem asks for computing a function, whereas the second is a problem of *language membership.* That is, for a given pair (x, k) the answer can be *yes* or *no.* In this fashion, corresponding to each function f, one can define the language L_f as the set of all pairs (x, k) that have a *yes* answer.

The theory of NP-completeness deals with languages. The well-known conjecture P $\neq$ NP states that if a language is NP-complete then it cannot be recognized in polynomial time. In particular, if one can prove for a certain function f that the corresponding language L_f is NP-complete then one has a lower bound result: f

Annotated Bibliographies in Combinatorial Optimization, edited by M. Dell'Amico, F. Maffioli and S. Martello

cannot be computed in polynomial time, unless P = NP (because as soon as you know $f(x)$ you also know if $f(x) > k$). When this happens we say that computing f is NP-hard.

This mathematical fact is important because it *may* tell us a lot about the world. Literally hundreds, perhaps thousands, of optimization problems arising naturally in the most diverse contexts—economics, molecular biology, algebra, logic, computer science, all sorts of engineering, etc—are now known to be NP-hard. So, if the conjecture P $\neq$ NP is correct we must compute in a world that is inherently computationally intractable.

This is the starting point of the theory of approximation algorithms. Optimization problems arise in practical contexts and we need to do something about them. A natural suggestion is the following. An optimization problem is the task of maximizing (or minimizing) a certain objective function over a space of feasible solutions that usually cannot be searched efficiently because of its huge size. Given that computing the optimum would take too long (if P $\neq$ NP) then one could settle for a fast algorithm that is *guaranteed to always deliver* good solutions, say, solutions that, for maximization problems, are always at least 69% of the best possible.

This leads to the following definition: a maximization problem F is *approximable within* $1 + \rho$ if there exists a polynomial time algorithm A such that, for all inputs I,

$$\frac{opt_F(I)}{A(I)} \leq 1 + \rho,$$

where $A(I)$ denotes the output of the algorithm and $opt_F(I)$ corresponds to $f(I)$, i.e. denotes the optimum value on input I. For minimization problems one considers the inverse ratio. The $1 + \rho$ notation is suggestive of the fact that the closer we are to 1, the better the approximation algorithm. This definition poses a new computational complexity question: *given an optimization problem F, what is the smallest ρ for which it is approximable within $1 + \rho$ in polynomial time?*

We remark that in this definition of approximation it is required that the approximation be constructive, i.e. that a feasible solution be delivered, instead of just a numerical value. Under this assumption one always has $A(I) \leq opt_F(I)$ for maximization problems and $A(I) \geq opt_F(I)$ for minimization problems.

It was quickly observed that NP optimization problems (i.e. functions f whose language L_f is in NP) display different kinds of behaviour:

- Some problems are not approximable within any $1 + \rho$, unless P = NP. This is for instance the case with the travelling salesperson problem and with integer programming.

- Some problems are approximable within *some* constant $1+\rho$. The class consisting of these problems is called APX. For instance, MAX CUT is approximable within 2 (one can always find a cut whose size is at least half of the edges).

- Some problems are approximable within $1 + \rho$ *for all* ρ's greater than 0. In other words, for each $\rho > 0$ there is a $(1 + \rho)$-approximate polynomial time algorithm A_ρ. The infinite sequence $\{A_\rho : \rho > 0\}$ is called a *polynomial time approximation scheme* (ptas). The class of problems admitting such a scheme is called PTAS. Examples of problems in PTAS are MAX INDEPENDENT SET for planar graphs and finding the minimum length schedule on parallel machines. Notice that each A_ρ must run in time polynomial in the size of the input, but could depend exponentially (or worse) in $1/\rho$.

- The last inconvenience is obviated if we can find a *fully polynomial approximation scheme*, that is, a *ptas* such that the running time of each A_ρ is polynomial in both $|I|$ and $1/\rho$. The resulting class is denoted as FPTAS. An example of a problem in FPTAS is MAX KNAPSACK.

The theory of approximation algorithms tries to place problems in these classes by proving upper bounds—i.e. the existence of an approximation algorithm or scheme—and lower bounds—by showing that a problem is not approximable within a certain accuracy. Sometimes it is possible to prove that a problem is approximable or non-approximable within $1+\rho(n)$, where $\rho(n)$ is a function of the input size. In these cases, upper bounds are of particular interest when $\rho(n)$ is a slowly growing function such as $\log n$, whereas lower bounds are particularly depressing when $\rho(n)$ grows as fast as $\sqrt{n}$ or worse.

The field of approximation algorithms has developed tremendously in the 20 years that have passed since its birth. In this survey we first give a list of books and lecture notes dealing with the field in general, and then focus on hardness results, i.e. proving that approximating within some accuracy is impossible. So far, no absolute hardness results for approximation have been proven or are in sight. Rather, hardness results are established assuming some widely held complexity theoretic conjecture; typically, but not exclusively, that P $\neq$ NP. Even then, proving hardness results for approximation has proved next to impossible for a long time, but the situation has changed dramatically in the past 3 or 4 years with the advent of the PCP revolution, still in progress. This is still very much an open field with a lot of frenetic activity and plenty of challenging open problems.

When available, we shall always give the journal reference. The reader should be advised that this might substantially alter the chronology of the various contributions, since many results are published in journals only several years after their first conference appearance. When applicable, we shall also add a pointer to a web page or a short reference to the two premier algorithm conferences in the computer science community, namely the ACM Symposium on Theory of Computing (STOC) and the IEEE conference on the Foundations of Computer Science (FOCS). These extra references might be useful to the reader since they are often easily available.

1 Books and Surveys

1.1 Books

We list some books, starting with some suitable for the newcomer to the area of approximation, and proceeding with increasingly specialized ones.

T. Cormen, C. Leiserson, R. Rivest (1989). *Introduction to Algorithms*, The MIT Press, Cambridge, Mass.

E. Horowitz, S. Sahni (1978). *Fundamentals of Computer Algorithms*, Pitman, London.

C.H. Papadimitriou, K. Steiglitz (1982). *Combinatorial Optimization, Algorithms and Complexity*, Prentice-Hall, Englewood Cliffs, New Jersey.

These are three good undergraduate level books in algorithms which contain some clearly written introductory chapters to the field of approximation algorithms.

C.H. Papadimitriou (1994). *Computational Complexity*, Addison-Wesley, Amsterdam.

D.P. Bovet, P. Crescenzi (1994). *Introduction to the Theory of Complexity*, Prentice-Hall, Englewood Cliffs, New Jersey.

These books deal with complexity theory and therefore approach approximation from this angle. Roughly speaking, while the books cited earlier focus on algorithms, these ones focus on lower bounds. The treatment in the book by Papadimitriou is more extensive and covers up-to-date topics such as MAX SNP in some detail.

M.R. Garey, D.S. Johnson (1979). *Computers and Intractability: a guide to the theory of NP-completeness*, W. H. Freeman and Company, San Francisco.

This book is the encyclopedia of NP-completeness, but it also contains a chapter on approximation with definitions of how to measure approximability and examples of how to show upper and lower bounds on the approximability of some problems. The notation introduced here is still used.

D.S. Hochbaum (ed.) (1997). *Approximation Algorithms for NP-Hard Problems*, PWS Publishing Company, Boston.

This is a great new book covering more advanced and up-to-date topics, which makes it perhaps suitable for a graduate course. It consists of 13 chapters, each a self-contained survey on some aspects of approximation algorithms. Most chapters cover algorithms for different types of problems together with general techniques for approximating those problems. Noteworthy is the chapter about hardness of approximation by Arora and Lund. It surveys recent results while at the same time serving as a tutorial for anyone who wants to learn how to construct lower bounds on approximability. This book should be of interest to students and researchers alike.

Finally, once you are ready to engage in research this resource might prove valuable:

P. Crescenzi, V. Kann (1995). *A compendium of NP optimization problems*, Technical Report SI/RR-95/02, Dipartimento di Scienze dell'Informazione, Università di Roma "La Sapienza". The list is updated continuously. The latest version is available as `http://www.nada.kth.se/theory/problemlist.html`.

This is a Garey-and-Johnson style list of about two hundred NP-hard optimization problems and several variations thereof. For each problem the best currently known upper and lower bounds on approximability are given (to the best of the authors' knowledge), together with references.

1.2 Surveys and lecture notes

There are several surveys that give a good overview of the field.

D.B. Shmoys (1995). Computing near-optimal solutions to combinatorial optimization problems. W. Cook, L. Lovász, P. Seymour (eds.). *Special year on combinatorial optimization*, DIMACS Series in Discrete Math. and Theoretical Comput. Sci., 20, American Math. Soc.

This is a comprehensive, clear and insightful survey, and perhaps the broadest of the surveys we mention. Although its viewpoint is more algorithmic than structural, a good deal about lower bounds can be learned from it, including recent developments following the PCP theorem.

R. Motwani (1992). *Lecture notes on approximation algorithms – volume I*, Technical Report STAN-CS-92-1435, Department of Comput. Sci., Stanford University. Available as `http://theory.stanford.edu/people/motwani/postscripts/approximations.ps.Z`.

These lecture notes are a clear introduction to approximation algorithms and cover a large part of the basic general results, but also contain some clearly explained advanced material such as asymptotic approximation schemes for MIN BIN PACKING. Recent hardness topics such as MAX SNP or PCP are not covered.

M.X. Goemans (1994). *Lecture notes*, Technical report, Department of Math., MIT. Available as `ftp://theory.lcs.mit.edu/pub/classes/18.415/notes-approx.ps`.

The lecture notes by Goemans concentrate on recent algorithmic results, and complement Motwani's lecture notes quite well.

There are also surveys dealing with structural issues instead of methods for finding upper and lower bounds:

D. Bruschi, D. Joseph, P. Young (1989). A structural overview of NP optimization problems. *Proc. Optimal Algorithms*, Springer-Verlag, Berlin, 205–231, Lecture Notes Comput. Sci. 401. It also appeared 1991 in *Algorithms Rev.* 2, 1–26.

This survey is quite detailed and covers most known structural results up to 1989, but as much has happened since, it is now rather outdated.

V. Kann (1992). *On the Approximability of NP-complete Optimization Problems*, Technical Report TRITA-NA-9206, Department of Numer. Anal. and Computing Sci., Royal Institute of Technology, Stockholm.

This thesis defines, discusses and compares different measures of approximation, types of approximation preserving reductions, and approximation complexity classes, including the syntactically defined classes. It also contains some hardness results.

G. Ausiello, P. Crescenzi, M. Protasi (1995). Approximate solution of NP optimization problems. *Theoret. Comput. Sci.* 150, 1–55.

This is a survey dealing with completeness in approximation classes and related topics. It also discusses the difference between computing an approximate value and constructing a solution with that value.

Finally, we cite some surveys describing the interplay between interactive proofs and approximation:

D.S. Johnson (1992). The NP-completeness column: an ongoing guide; The tale of the second prover. *J. Algorithms* 13, 502–524.

L. Babai (1994). Transparent proofs and limits to approximation. *Proc. First European Congress of Mathematicians.* Birkhäuser, Basel. Expanded version available as `ftp://ftp.cs.uchicago.edu/pub/publications/tech-reports/TR-94-07.ps`.
J. Håstad (1994). Recent results in hardness of approximation. *Proc. 4th Scand. Workshop Algorithm Theory.* Springer-Verlag, Berlin, 231–239. Lecture Notes Comput. Sci. 824.
M. Bellare (March 1996). Complexity theory column 12: Proof checking and approximation: towards tight results. *SIGACT News*, 27, The Association for Computing Machinery, 2–13. Expanded version available as
`http://www-cse.ucsd.edu/users/mihir/papers/pcp-survey.ps.gz`.

The first three are good introductions to the PCP theorem and the role of interactive proofs in proving lower bounds for approximation. The fourth is a good introduction to further improvements of the PCP theorem and their consequences for approximation.

2 Early Lower Bounds

Although the notion of approximation algorithms for combinatorial optimization problems had appeared before, to our knowledge the organized study of algorithms with guaranteed approximation began in the seventies with the following paper:

D.S. Johnson (1974). Approximation algorithms for combinatorial problems. *J. Comput. Syst. Sci.* 9, 256–278.

Johnson here presents and analyzes some important (and still useful) approximation algorithms for common optimization problems, such as MAX SATISFIABILITY, MIN DOMINATING SET, MAX CLIQUE and MIN GRAPH COLORING. The paper emphasizes the notion of *performance guarantee* as opposed to a heuristic that is expected to work fine in practice but without any proven guarantee.

The field developed quite rapidly and many clever approximation algorithms and schemes were developed. A survey on early approximation algorithms is the following:

M.R. Garey, D.S. Johnson (1976). *Approximation algorithms for combinatorial problems: an annotated bibliography*, Academic Press, New York, 41–52.

Of course this annotated bibliography is outdated now, but in any case it shows that already in 1976 quite a large number of approximation results were known.

While many approximation algorithms were discovered, lower bounds seemed to be extremely difficult to come by. The situation changed radically in the last few years with the advent of the PCP theorem, to be discussed in §4. In the rest of this section we list some of the most interesting early lower bounds which do not rely on it.

S.K. Sahni, T.F. Gonzalez (1976). P-complete approximation problems. *J. ACM* 23, 555–565.

The quest for lower bounds starts with this paper, where hardness results under the hypothesis P $\neq$ NP are given for several problems, among them the travelling salesperson problem.

M.R. Garey, D.S. Johnson (1976). The complexity of near-optimal graph coloring. *J. ACM* 23, 43–49.

Among the first lower bounds is the result that graph coloring cannot be approximated within a factor better than 2, unless P = NP. This result was unthreatened for almost 20 years, until dramatically harsher lower bounds were found using the new PCP theorem, see §4.

Much research on lower bounds has focused on the clique problem, not only in the case of approximation. For instance, a famous result of Razborov states that monotone (without NOT-gates) boolean circuits that compute the clique function must have super-polynomial size. (See, for instance, chapter 14 in the Handbook of Theoretical Computer Science, volume A, Elsevier, Amsterdam.) But the clique problem has also been central to the study of approximation algorithms. [Garey and Johnson 1979] (see §1.1) contains a nice proof of a puzzling result known as the self-improvability of MAX CLIQUE: this problem is either non-approximable within any constant or it is approximable within all constants. Although we now probably know the ultimate answer concerning the non-approximability of MAX CLIQUE (see §4), there are some very interesting lower bound results whose proof does not make use of the PCP theorem.

P. Berman, G. Schnitger (1992). On the complexity of approximating the independent set problem. *Inform. and Comput.* 96, 77–94.

By making use of randomized reductions, this paper establishes an interesting connection between MAX 3-SAT and MAX CLIQUE: even an $O(n^\delta)$-approximation algorithm for MAX CLIQUE (for some small enough $1 > \delta > 0$) would yield a *randomized ptas* for MAX 3-SAT.

N. Alon, U. Feige, A. Wigderson, D. Zuckerman (1995). Derandomized graph products. *Computational Complexity* 5, 60–75.

Here, the construction of [Berman and Schnitger 1992] is derandomized by making use of *expander graphs*, to which we return in §4. Thanks to the derandomization, the result can be restated as: there exists $\delta > 0$ such that, if MAX CLIQUE is approximable within $O(n^\delta)$ then MAX 3-SAT is in PTAS.

An important class of problems for which lower bounds could be shown using pre-PCP methods is that of scheduling problems. See the chapter "Nasty Gaps" in the survey of [Shmoys 1995] (see §1.2). Among papers containing early lower bound results the following is worth mentioning:

D.S. Hochbaum, D.B. Shmoys (1986). A unified approach to approximation algorithms for bottleneck problems. *J. ACM* 33, 533–550.

Very nice matching lower and upper bounds are shown for several natural problems in APX including MIN K-CLUSTERING, MIN K-CENTER, MIN K-SUPPLIER, and MIN K-SWITCHING NETWORK. See also the references cited there for a few more hardness results.

In spite of these early efforts, the field of approximation algorithms was characterized by a lack of a unifying framework and was made of many scattered, albeit clever, results. The situation was to change with the introduction of the notion of MAX SNP-hardness, which successfully captured a large class of problems. The PCP theorem later showed that none of these problems are in PTAS, unless P = NP. These two important developments are the topic of the next sections.

3 Structural Results

A successful way to characterize a class is to give a set of intuitive, "easy-to-use", necessary and/or sufficient conditions for membership in the class. In this respect, the notion of NP-completeness can be considered quite successful, as in many cases one can "smell" that a problem is NP-complete and quickly prove it. As far as approximation is concerned, the quest for intuitively compelling characterizations and easy-to-use tools of problem classification started in the late seventies.

M.R. Garey, D.S. Johnson (1978). Strong NP-completeness results: motivation, examples, and implications. *J. ACM* 25, 499–508.

This paper gives a very successful characterization of the class FPTAS by formalizing the idea that some problems, e.g. knapsack, are NP-complete only if exponentially large weights are allowed in the input and become polynomially solvable whenever the weights are polynomial in the input size. Such problems are called solvable in *pseudo-polynomial time.* The complementary class is comprised of problems whose NP-completeness does not depend on the presence of large weights, e.g. clique or the travelling salesperson problem. Problems of this kind are called *strongly NP-complete.* Garey and Johnson prove that, if the optimum value $opt_F(I)$ is polynomially bounded in both the number of objects in the input and the largest weight, and the corresponding decision problem is strongly NP-complete, then F cannot be in FPTAS (unless P = NP). These assumptions on the input are satisfied by all known natural examples. This implies that if a problem with bounded optimum value is in FPTAS then it must be solvable in pseudo-polynomial time.

A. Paz, S. Moran (1981). Non deterministic polynomial optimization problems and their approximations. *Theoret. Comput. Sci.* 15, 251–277.

Under assumptions verified by many natural problems, this paper shows the converse: if a problem is solvable in pseudo-polynomial time then it is in FPTAS. An attempt to characterize PTAS is also done, and perhaps the first approximation preserving reduction is defined.

G. Ausiello, A. D'Atri, M. Protasi (1980). Structure preserving reductions among convex optimization problems. *J. Comput. Syst. Sci.* 21, 136–153.

A more restrictive type of reduction preserving also some structure of the solution is introduced.

In spite of these early efforts the question of how to prove that a problem is not in PTAS or APX was as open as before. A tempting way to proceed is to introduce approximation complexity classes and to find problems that are complete in these classes (with respect to suitable reductions), mimicking the complexity theory for decision problems.

P. Orponen, H. Mannila (1987). *On approximation preserving reductions: complete problems and robust measures*, Technical Report C-1987-28, Department of Comput. Sci., University of Helsinki.

In this much underappreciated paper some approximation preserving reductions are introduced, and the travelling salesperson problem is shown to be complete in the class of NP minimization problems.

P. Crescenzi, A. Panconesi (1991). Completeness in approximation classes. *Inform. and Comput.* 93, 241–262.

This paper extends the approach of Orponen and Mannila to the classes APX and PTAS and shows that several problems, none of them very natural though, are complete in these classes. Moreover, it shows the existence of intermediate (non-complete) problems in these classes.

V. Kann (1994). Polynomially bounded minimization problems that are hard to approximate. *Nordic J. Comput.* 1, 317–331. Presented at ICALP '93.

Here, natural complete problems for MIN NPO PB, the class of minimization problems whose objective function is bounded by a polynomial in the size of the input, are found. The cited [Berman and Schnitger 1992] (see §2) contains similar results for MAX NPO PB.

A turning point in this line of research was reached thanks to this seminal paper:

C.H. Papadimitriou, M. Yannakakis (1991). Optimization, approximation, and complexity classes. *J. Comput. Syst. Sci.* 43, 425–440. See also STOC '88: 229–234.

In an attempt to avoid the difficulty of capturing the notion of *approximately correct computation*, Papadimitriou and Yannakakis use a logical (machine independent) characterization of NP given by Fagin. They define a class of problems called MAX SNP and show that MAX SNP $\subseteq$ APX. Moreover, they show that many natural problems such as MAX 3-SAT, and MAX CUT are MAX SNP-complete under an approximation preserving reduction called the L-reduction. The class MAX SNP was introduced in an attempt to group together APX problems that seemed (and were conjectured) not to be in PTAS. The definition of MAX SNP-hardness is crafted in such a way that if one could show that just one MAX SNP-complete problem does not belong to PTAS it would follow automatically that no MAX SNP-hard problem does. This is precisely what the PCP theorem does (see §4); it shows that one such problem, MAX 3-SAT, does not belong to PTAS. Therefore, no MAX SNP-hard problem is in PTAS (unless P = NP).

That MAX SNP captures the "right" level of completeness is attested to the fact that dozens of problems have been shown to be MAX SNP-complete or MAX SNP-hard (see the problem list [Crescenzi and Kann 1995] in §1.1). This notion is so successful that nowadays standard MAX SNP-completeness results are considered routine and are therefore rejected by journals!

The definition of MAX SNP via logic raised the possibility of an unexpected connection between logic and approximation. This problem has been investigated rather thoroughly:

A. Panconesi, D. Ranjan (1993). Quantifiers and approximation. *Theoret. Comput. Sci.* 107, 145–163. See also STOC '90: 446–456.

This paper defines other logical classes and shows completeness of many natural problems for them, e.g. MAX CLIQUE and MAX SET PACKING. It also shows by simple model theoretic arguments that, under some natural assumptions on the encoding, lots of natural optimization problems, e.g. MAXIMUM MATCHING, are not in MAX SNP (*without* any complexity theoretic assumption). This shows that MAX SNP does not contain P. This disappointing fact is however to be expected. If one could find a logically defined class containing P and prove by model theoretic arguments that it is a proper subclass of NP optimization problems, one would have separated P from NP.

P.G. Kolaitis, M.N. Thakur (1994). Logical definability of NP optimization problems. *Inform. and Comput.* 115, 321–353.
P.G. Kolaitis, M.N. Thakur (1995). Approximation properties of NP minimization classes. *J. Comput. Syst. Sci.* 50, 391–411.
T. Behrendt, K.J. Compton, E. Grädel (1993). Optimization problems: expressibility, approximation properties and expected asymptotic growth of optimal solutions. *Proc. 6th Workshop Comput. Sci. Logic.* Springer-Verlag, Berlin, 43–60. Lecture Notes Comput. Sci. 702.

The first two papers systematically investigate the relationship between logic and approximation for, respectively, maximization and minimization problems. Unfortunately the relationship between syntax and approximation seem to be rather weak in the latter case. The third paper extends the logical approach by introducing more sophisticated logical notions such as fix point logic.

S. Khanna, R. Motwani (1996). Towards a syntactic characterization of PTAS. *Proc. 28th Annual ACM Symp. Theory of Comput.* See also STOC '96: 329–337.

Here Khanna and Motwani find a way to capture a large chunk of PTAS via a logical characterization. The idea is to specify a problem syntactically in such a way that the input instances exhibit a planar structure so that powerful graph separator theorems can be invoked to derive *ptas*'s in a uniform manner (i.e. for all problems in the class).

N. Creignou (1995). A dichotomy theorem for maximum generalized satisfiability problems. *J. Comput. Syst. Sci.* 51, 511–522.
S. Khanna, M. Sudan, D.P. Williamson (1997). A complete classification of the approximability of maximization problems derived from boolean constraint satisfaction. *Proc. 29th Annual ACM Symp. Theory of Comput.* To appear.
S. Khanna, M. Sudan, L. Trevisan (1997). Constraint satisfaction: the approximability of minimization problems. *Proc. 12th Annual IEEE Conf. Comput. Complexity.* To appear.

These recent papers show that the logical approach is quite successful for the restricted but rich class of constraint satisfaction problems (CSP's). Interestingly, they show a restatement of an earlier result by Schaefer in the context approximation (see Schaefer (1978), The complexity of satisfiability problems, STOC '78, 216–226). Namely, these classes are neatly divided into a finite number of levels, i.e. a CSP problem can only have a certain (finite) number of behaviours w.r.t. approximation.

Returning to MAX SNP we can see that rather than MAX SNP itself the class of interest is the *closure* of MAX SNP w.r.t. approximation preserving reductions, i.e. the set of problems that are reducible to some MAX SNP problem. Several papers investigate the closure of MAX SNP w.r.t. various reducibilities:

S. Khanna, R. Motwani, M. Sudan, U. Vazirani (1994). On syntactic versus computational views of approximability. *Proc. 35th Annual IEEE Symp. Found. Comput. Sci.*, 819–830. Full version available as ECCC Report TR95-023, `ftp://ftp.eccc.uni-trier.de/pub/eccc/reports/1995/TR95-023/index.html`.

Surprisingly, using a slightly less restrictive notion of approximation preserving reduction than the L-reduction, called E-reduction, this closure turns out to be the whole of APX PB—the class of approximable problems whose objective function is polynomially bounded. This means, for example, that MAX 3-SAT is APX PB-complete. The proof relies on the PCP theorem (see §4).

P. Crescenzi, L. Trevisan (1994). On approximation scheme preserving reducibility and its applications. *Proc. 14th Annual Conf. Found. Software Tech. and Theoret. Comput. Sci.* Springer-Verlag, Berlin, 330–341. Lecture Notes Comput. Sci. 880.

Using a less restrictive version of the E-reduction this paper finally captures the whole of APX as the closure of MAX SNP. Thus MAX SNP captures the core of hardest problems in APX, and therefore we can now prove APX-completeness, as opposed to just MAX SNP-completeness, for many natural problems.

In a different line of research, people have used tools from structural complexity theory, which can be roughly described as an offspring of Recursion Theory, to investigate approximation properties:

P. Crescenzi, V. Kann, R. Silvestri, L. Trevisan (1995). Structure in approximation classes. *Proc. 1st Annual Int. Conf. Comput. and Combin.* Springer-Verlag, Berlin, 539–548. Lecture Notes Comput. Sci. 959.

Among other things, this paper shows that if an optimization problem has an *asymptotic ptas*, then it cannot be APX-complete under the complexity theoretic assumption that the Polynomial Hierachy (PH) does not collapse, a fact which is widely conjectured to be true. (An asymptotic *ptas* for, say, a minimization problem F is an infinite collection $\{A_\epsilon\}$ of polynomial time algorithms such that there exists a constant k such that, for all inputs I and $\epsilon > 0$, $A_\epsilon(I) \leq \epsilon opt_F(I) + k$.) Two well-known problems which have asymptotic *ptas*'s are MIN BIN PACKING and MIN EDGE COLORING. This result shows the existence of natural intermediate problems within approximation classes (unless the PH collapses). The result is established by using connections between the approximability properties and the *query complexity* (the number of queries to an NP oracle).

Relating the query complexity to approximation has given rise to several other interesting papers. For instance, it is possible to show that, under suitable complexity theoretic assumptions, computing an approximate feasible solution is harder than just computing an approximate numerical value. The paper:

R. Chang (October 1994). The structural complexity column: a machine model for NP-approximation problems and the revenge of the boolean hierarchy. *Bulletin of the EATCS*, 54, 166–182.

... contains an excellent discussion of these results with the relevant bibliography.

4 The PCP Revolution

The recent dramatic developments in the theory of approximation are based on the so-called PCP theorem, which establishes an unexpected connection between interactive proofs and approximation algorithms. The PCP theorem is a deep result in many ways. From a technical point of view, the result makes use of new algebraic techniques for the testing of polynomial identities and combines them with a host of tools brought from diverse areas within and outside of theoretical computer science, such as interactive proof systems, self testing and correcting of polynomials, mathematical logic and coding theory. This result also intensifies the philosophical debate concerning the role of randomness in computing. Moreover, as we shall see, the result has far reaching consequences in the area of approximation algorithms.

S. Goldwasser, S. Micali, C. Rackoff (1989). The knowledge complexity of interactive proof-systems. *SIAM J. Comput.* 18, 186–208. See also STOC '85: 291–304.
L. Babai, S. Moran (1988). Arthur-Merlin games: a randomized proof system, and a hierarchy of complexity classes. *J. Comput. Syst. Sci.* 36, 254–276. See also STOC '85: 421–429.

Unexpectedly, the paradigm that gave birth to the PCP theorem is the seemingly unrelated field of interactive proof systems which was introduced at about the same time in the first of these papers in the context of cryptography, and in the second as a game theoretic extension of NP. Two insightful and readable accounts describing the role of interactive proofs in the development of the PCP and related theorems are [Johnson 1992] and [Babai 1994] (see §1.2).

In order to understand the implications for approximation, a brief description of a probabilistic checker of proof is in order. One way to understand the class NP is that it deals with statements (theorems), made in some formal system, whose proof might be computationally hard to come by, but which has fast verification procedures of correctness. The typical example is Satisfiability: deciding whether a given formula is satisfiable might take exponential time in the length of the formula, but to verify that a given truth assignment satisfies it can be quickly checked in polynomial time. In this context, a satisfying assignment constitutes the proof that the formula is satisfiable.

Suppose now that in order to decide whether a given bit string is indeed a correct proof of a statement "$x \in L$", for some language $L \in$ NP, we do not want to look at the whole proof. Rather, we would like to perform a few random checks and then conclude with reasonable confidence that the proof is correct. This bold plan can be formalized as follows. An $(r(n), q(n))$-verifier for a language L is a Turing machine V, whose task is to determine the correctness of *alleged proofs* of statements of the form "$x \in L$". The machine is given the input x and an alleged proof of the statement. The alleged proof is contained in a large *database* D, which is simply a bit string. The verifier must work in time polynomial in $n = |x|$, but can take advantage of two extra computational resources: (a) $r(n)$-many *random bits* and (b) $q(n)$-many accesses to the database D, i.e. to the bits of the alleged proof. Intuitively, the random bits are used to generate random addresses of the $2^{r(n)}$-many database locations. This allows the verifier to sample the alleged proof at random places. The verifier must try to check whether the purported proof contained in the database is correct, in which case it accepts x. If it is incorrect the verifier rejects (regardless of whether $x \in L$). More precisely, we say that the verifier accepts a language L if the following two conditions are met:

- $x \in L \Rightarrow \exists D \ \Pr[V(D, x) \text{ accepts}] = 1$, i.e. when the proof is correct the verifier accepts for all random strings.
- $x \notin L \Rightarrow \forall D \ \Pr[V(D, x) \text{ accepts}] \leq 1/2$, i.e. the probability that the verifier accepts a fallacious proof is small.

Note that by repeating the verification procedure k times, where k is a constant, the error probability becomes as small as 2^{-k}.

The class $\mathrm{PCP}(r(n), q(n))$ is the class of languages accepted by some $(O(r(n)), O(q(n)))$-verifier. The PCP theorem states that

$$\mathrm{PCP}(\log n, 1) = \mathrm{NP}.$$

This result establishes the amazing fact that for all languages in NP there is a way of writing proofs (databases) such that a verifier need only probe a *constant number* of locations in order to conclude with error probability 1/2 whether the proof is correct or wrong. We now know that probing 9 bits suffices (see [Håstad 1997] cited at the end of §4)! Moreover, by running the protocol a few times the verifier can boost this probability to $1 - o(1)$.

What does the PCP theorem have to do with optimization? Given a verifier V for a language $L \in \mathrm{NP}$, the following optimization problem can be defined: on input x, the feasible solutions are all possible, exponentially many, databases $D \in \{0,1\}^{2^{r(n)}}$; the value of the objective function on a given feasible solution D is the number of random strings $r \in \{0,1\}^{r(n)}$ for which the verifier accepts. The optimum of this problem is

$$opt_V(x) = \max_D |\{r : V(D, x) \text{ accepts if the random string is } r\}|.$$

This is an NP optimization problem, because, crucially, $r(n) = O(\log n)$. This implies that the feasible solutions are polynomial size bit strings, and that the objective function is polynomially computable. Indeed, R, the total number of random strings, is $2^{r(n)}$—a polynomial—and checking each one of them for acceptance requires simulating the verifier, which can be done in polynomial time. Note now that this problem exhibits a *gap*: if the formula is satisfiable then $opt_V(x) = R$. Otherwise, $opt_V(x) \leq R/2$. It follows that no polynomial time algorithm can approximate the problem within a factor smaller than 2, unless P = NP. Indeed, if $x \notin L$, such an algorithm must deliver a value $\leq R/2$ whereas, if $x \in L$, the algorithm must deliver a value $> R/2$. Hence, one can use the approximation algorithm to decide whether $x \in L$, where L can be any NP-complete language.

A crucial fact about the PCP theorem is that it makes it possible to create instances of natural optimization problems (as opposed to artificial examples like the one above) that exhibit such gaps, the construction of which eluded researchers for about two decades.

An interesting fact about the PCP theorem, that might be relevant for sociologists of science and of the information age, is that it is a collective achievement. The number of people that authored or co-authored papers containing substantial or crucial steps toward the complete solution is more than a dozen. Scientific progress has always relied on constant interaction among people, but the rate of exchange and personal contacts now made possible by the Pentagon-supported information and aviation revolution confronts us with a seemingly new sociological phenomenon in the realm of science. (See for instance, Babai (1990), E-mail and the unexpected power of interaction, *Proc. 5th IEEE Symp. on Structure in Complexity Theory*, 30–44.) In what follows, we do not attempt to sort out the different contributions. Rather, we list those papers that seem to us more directly relevant for their consequences to approximation.

A. Condon (1993). The complexity of the max-word problem and the power of one-way interactive proof systems. *Computational Complexity* 3, 292–305.

This is the first paper to show a connection between interactive proofs and approximation, but the turning point was a paper that appeared a few months later.

U. Feige, S. Goldwasser, L. Lovász, S. Safra, M. Szegedy (1991). Approximating clique is almost NP-complete. *Proc. 32nd Annual IEEE Symp. Found. Comput. Sci.*, 2–12.

The connection between interactive proofs and approximation is here clearly shown by exhibiting a crisp reduction between any $(r(n), q(n))$-verifier and the clique problem. The reduction establishes a bijection between databases for which the verifier accepts on k different random strings and cliques of size k. The reduction is so clean that it can be explained in a few lines. See for example the surveys of [Johnson 1992], [Håstad 1994], and [Shmoys 1995] cited in §1.2.

L. Babai, L. Fortnow, C. Lund (1991). Non-deterministic exponential time has two-prover interactive protocols. *Computational Complexity* 1, 3–40. See also FOCS '90: 16–25.

This paper contains the beautiful result which motivated the Feige et al. paper. Namely, $\mathrm{PCP}(p(n), p(n)) = \textsc{Nexptime}$, where $p(n)$ is a polynomial. The Feige et al. paper "scales down" this construction from $\textsc{Nexptime}$ to NP to get $\mathrm{PCP}(c(n), c(n)) \supseteq$ NP, with $c(n) = \log n \log\log n$. Together with the clique reduction, this implies that clique is not approximable within any constant unless $\mathrm{NP} \subset \textsc{Dtime}[n^{\log\log n}]$, the class of languages that can be recognized in time $O(n^{\log\log n})$ by a deterministic Turing machine, a fact that appears unlikely and is conjectured to be false.

These papers do not explicitly frame their result in terms of PCP. This class was formally introduced in a subsequent paper:

S. Arora, S. Safra (1992). Probabilistic checking of proofs; a new characterization of NP. *Proc. 33rd Annual IEEE Symp. Found. Comput. Sci.*, 2–13.

This breakthrough paper introduces the class PCP and shows that $\mathrm{PCP}(\log n, \log n)$ = NP, thereby solving a 20 year old open problem: approximating clique within any constant is NP-hard. Like many papers in the area, it is a technical *tour-de-force.* Its main technical contribution is probably a sophisticated recursive procedure to check polynomial identities very quickly and with high confidence.

Another important contribution of the Arora and Safra paper is the use of *expander graphs* to boost the hardness result for clique. Using later improvements, this technique eventually yields the following: there exists an ϵ such that approximating clique within n^ϵ is NP-hard.

Expanders seem to play an important role in the derivation of hardness results and have already proven to be extremely useful in other areas of computer science such as parallel architectures (a celebrated result of Ajtai–Komlós–Szemerédi showing the existence of small depth sorting networks is based on the existence of such graphs, see for instance, chapter 15 in the Handbook of Theoretical Computer Science, volume A, Elsevier, Amsterdam). A nice introduction to expanders is contained in the book by Alon and Spencer (1991), *The Probabilistic Method*, Wiley & Sons, New York. As far as hardness results are concerned, expanders were used in the seminal paper of Papadimitriou and Yannakakis (see §3) to show MAX SNP–hardness results of several problems, including MAX 2-SAT, MAX CUT and MAX 3-SAT-B, the variant of MAX SAT where each variable appears at most B times. Two more interesting results using expanders are:

E. Amaldi, V. Kann (1995). The complexity and approximability of finding maximum feasible subsystems of linear relations. *Theoret. Comput. Sci.* 147, 181–210.
P. Crescenzi, R. Silvestri, L. Trevisan (1996). To weight or not to weight: Where is the question? *Proc. 4th Israel Symp. Theory Comput. and Syst.* IEEE, 68–77.

In the first of the two, expanders are used to prove an n^ϵ hardness bound for approximating the feasibility problem for systems of linear equations (given a set of n linear equations, what is the largest subset of feasible equations?). The second paper shows that, for a reasonably large class of optimization problems, the approximation threshold for the weighted and unweighted case is the same. Roughly speaking, the class in question consists of those problems for which the feasibility of a solution does not depend on its weight and includes, among others, MAX CUT and MAX SAT.

After the PCP$(\log n, \log n)$ = NP result by Arora and Safra, it was observed that a gap preserving reduction exists between any $(r(n), q(n))$-verifier and 3-SAT formulas. Given a verifier V and an input x, the reduction produces a boolean formula $\varphi_{V,x}$ of size $O(p(n)Q(n))$, where $|x| = n$, $p(n)$ is a polynomial and $Q(n) = O(2^{2^{q(n)}})$, such that: (a) if $x \in L$ then all clauses of $\varphi_{V,x}$ are satisfiable; and (b) if $x \in L$ then at most a $1 - 1/Q(n)$ fraction of the clauses of $\varphi_{V,x}$ is satisfiable (the reduction is not hard and can be found in the cited [Shmoys 1995] and [Johnson 1992], see §1.2).

This suggests a bizarre but exciting possibility: if one could bring the number of queries $q(n)$ down to a constant, a (roughly) $(1 + 1/Q(n))$-hardness result for MAX 3-SAT would follow because $Q(n)$ would be huge, but constant. Sure enough, after a few months the result was announced in this paper:

S. Arora, C. Lund, R. Motwani, M. Sudan, M. Szegedy (1992). Proof verification and hardness of approximation problems. *Proc. 33rd Annual IEEE Symp. Found. Comput. Sci.*, 14–23.

This paper puts the last nail in the coffin of the PCP theorem, which states that:

$$\text{PCP}(\log n, 1) = \text{NP}$$

The consequences of this characterization of NP are momentous. Recall that after Papadimitriou and Yannakakis introduced the class MAX SNP dozens of problems, including MAX 3-SAT, were shown MAX SNP-complete w.r.t. L-reductions. The PCP theorem shows that MAX 3-SAT is not in PTAS (unless P = NP) which implies that no MAX SNP-hard problem is in PTAS. In particular, now we have a very effective and easy-to-use tool to show dozens of hardness results: exhibit an approximation preserving reduction from a problem known to be MAX SNP-hard to the problem at hand. This task often turns out to be just slightly more difficult than an average NP-completeness proof.

The complete proof of the PCP theorem can be found in this important PhD thesis of Sanjeev Arora:

S. Arora (1994). *Probabilistic Checking of Proofs and Hardness of Approximation Problems*, Technical Report CS-TR-476-94, Department of Comput. Sci., Princeton University. Revised version available as ECCC Book 1995, `http://www.eccc.uni-trier.de/eccc-local/ECCC-Books/sanjeev_book_readme.html`.

This thesis contains a clear and detailed exposition of the proof of the PCP theorem, written by one of its foremost contributors. Because of its thoroughness, this is the best reference for those interested in studying this long and difficult proof.

An important ingredient in the proof of the PCP theorem is the use of techniques for checking polynomial identities developed in the context of the theory of self-testing and correcting. The original motivation of this theory was to give a satisfactory answer, at least partially, to one of the most fundamental problems of computer science: how can we effectively and feasibly check that a given program computes what it is supposed to compute?

A very nice discussion of the relationship between self-testing and correcting and the PCP theorem can be found in this wonderful PhD thesis:

M. Sudan (1995). *Efficient checking of polynomials and proofs and the hardness of approximation problems*, Springer-Verlag, Berlin. Lecture Notes Comput. Sci. 1001.

This thesis contains a description and the proof of the PCP theorem from a different angle. It is, however, not as complete as that contained in the thesis of Arora. This thesis is also a good starting point for those who want to venture into the theory of self-testing and correcting.

S. Hougardy, H.J. Prömel, A. Steger (1994). Probabilistically checkable proofs and their consequences for approximation algorithms. *Disc. Math.* 136, 175–223.

This is a survey containing a proof of the PCP theorem and a discussion of some of its applications to approximation. It is important to stress that the proof contained in the paper is not the authors' (so far there is a single known proof of the PCP theorem). This survey was intended to publicize these important results within the discrete mathematics community by presenting in an organized and compact fashion what at the time was only available in scattered and fragmented sources. Since the authors had to rely on incomplete materials such as drafts and conference abstracts, their work can be considered as an independent verification of this difficult proof. In this sense it represents a valuable service to the community.

After the Arora et al. paper, problems that had been open for years continued to fall one after the other.

C. Lund, M. Yannakakis (1994). On the hardness of approximating minimization problems. *J. ACM* 41, 960–981. See also STOC '93: 286–293.

In this beautiful paper two long standing open problems are solved. The first result is a gap preserving reduction between clique and chromatic number which shows that there exists a $\delta > 0$ such that $\chi(G)$ is not approximable within $O(n^\delta)$ unless P = NP—quite a dramatic improvement from the previous best lower bound of 2 (see §2).

The second result is a proof that MIN SET COVERING cannot be approximated better than $(1/4)\log n$, unless NP $\subseteq$ DTIME$[2^{\text{polylog}(n)}]$. This result is important in many respects. First, quite magically, it matches the upper bound of $O(\log n)$ given by the greedy algorithm (see for instance Chvatál (1979), A greedy heuristic for the set covering problem, *Math. Oper. Res.* 4, 233–235). Second, rather than the PCP theorem, it uses a different characterization of NP due to Feige and Lovász in terms of so-called multiprover interactive proof systems, another construct developed in the context of cryptography.

S. Khanna, N. Linial, S. Safra (1993). On the hardness of approximating the chromatic number. *Proc. 2nd Israel Symp. Theory Comput. and Syst.* IEEE, 250–260.

Here, a very elegant and much simpler proof of the chromatic number lower bound of [Lund and Yannakakis 1994] can be found.

M. Bellare (1993). Interactive proofs and approximation: reductions from two provers in one round. *Proc. 2nd Israel Symp. Theory Comput. and Syst.* IEEE, 266–274.

Although this paper does not contain any big improvements, it was perhaps the first to notice a fact fully exploited in [Lund and Yannakakis 1994], namely that some interactive proof characterizations of NP can be better suited than others for the problem at hand.

The first hardness results using the new methods had an existential flavour. For instance, the result for MAX CLIQUE stated that: there exists *some* $\delta > 0$ such that $\omega(G)$ is not approximable within n^δ. The value of δ was small and hard to compute. The next step, quite naturally, was to look into the PCP technology and try to make some parts of this complex machinery more efficient.

M. Bellare, S. Goldwasser, C. Lund, A. Russell (1993). Efficient probabilistically checkable proofs and applications to approximation. *Proc. 25th Annual ACM Symp. Theory of Comput.*, 294–304.

This paper initiates what is to this day the current trend. By looking at PCP parameters like average query complexity, query size, answer size, etc. Bellare et al. were able to prove hardness results for quite reasonable constants. For instance, for MAX CLIQUE they get $n^{1/30}$.

A stream of highly technical papers introduced and put to fruition more sophisticated parameters such as free bits, average free bits and the like, obtaining better and better constants.

M. Bellare, O. Goldreich, M. Sudan (1995). Free bits, PCPs and non-approximability – towards tight results. *Proc. 36th Annual IEEE Symp. Found. Comput. Sci.*, 422–431. Full version available as ECCC Report TR95-024, `ftp://ftp.eccc.uni-trier.de/pub/eccc/reports/1995/TR95-024/index.html`.

This treatise systematizes and improves upon the most recent developments and gives stronger hardness bounds than ever before for several problems. It also contains a detailed and enjoyable history of the various contributions.

After this stream of results we are now in a position unimaginable a few years back: the best known lower and upper bounds for some fundamental combinatorial problems are not too far apart. For instance, for the MAX 3-SAT problem the best algorithm delivers a solution whose value is about 80.0 % of the optimum whereas no algorithm can deliver more than 87.5 % of the optimum (unless P = NP). For MAX CUT the bounds are, respectively, 87.8 % and 94.2 %.

One feels hesitant in stating these results because the pace at which improvements are taking place can make these statements obsolete. For instance, Feige has recently announced a $(1-\epsilon)\ln n$ lower bound for MIN SET COVERING, for any $\epsilon > 0$, under the improved (weaker) complexity theoretic assumption NP $\not\subseteq$ DTIME$[n^{O(\log\log n)}]$. For MAX CLIQUE, the lower bound increased rapidly, going from n^{ϵ} to $n^{1/30}, n^{1/6}, n^{1/5}, n^{1/4}, n^{1/3}$ until Håstad announced recently a $n^{1/2-o(1)}$ lower bound under the hypothesis NP $\not\subseteq$ coRP (i.e. NP-complete problems cannot be solved by randomized polynomial time algorithms). By the time this paper was published—it was presented at STOC '96 together with Feige's set covering result—it was already obsolete!

J. Håstad (1996). Clique is hard to approximate within $n^{1-\epsilon}$. *Proc. 37th Annual IEEE Symp. Found. Comput. Sci.*, 627–636.
U. Feige, J. Kilian (1996). Zero knowledge and the chromatic number. *Proc. 11th Annual IEEE Conf. Comput. Complexity*, 278–287.

Here Håstad announced the stunning tight hardness result of $n^{1-o(1)}$ for the MAX CLIQUE problem under the assumption NP $\not\subseteq$ coRP. In turn, this was immediately used by Feige and Kilian to give a tight $n^{1-o(1)}$ lower bound for chromatic number.

As for 1997, we can look forward to this result:

J. Håstad (1997). Some optimal inapproximability results. *Proc. 29th Annual ACM Symp. Theory of Comput.* To appear.

Here matching lower bounds for MAX Ek-SAT (the variant of MAX SAT where each clause has exactly k literals) are presented together with improved lower bounds for, among others, MAX CUT and MIN VERTEX COVER.

And more is yet to come...

Acknowledgements

We would like to thank the following persons for useful comments on the text: Gunnar Andersson, Pierluigi Crescenzi, Devdatt Dubhashi, Thomas Emden-Weinert, Lars Engebretsen, Johan Håstad, Stefan Hougardy, Sanjeev Khanna, Hans Jurgen Prömel, David Shmoys, Aravind Srinivasan, Madhu Sudan, Luca Trevisan.

3 Polyhedral Combinatorics

Karen Aardal
Utrecht University, Utrecht

Robert Weismantel
Konrad-Zuse-Zentrum, Berlin

CONTENTS

Polyhedral combinatorics is the study of the integer programming polyhedron

$$P = \text{conv}(X),$$

where X is given as a subset of the integer lattice $\mathbb{Z}^n$ and conv denotes the convex hull operator. Typically, we have $X = \{x \in \mathbb{Z}^n : Ax \leq b\}$.

H. Weyl (1935). Elementare Theorie der konvexen Polyeder. *Comentarii Mathematici Helvetici* 7, 290–306.

By Weyl's Theorem there exists a matrix $D \in \mathbb{R}^{m \times n}$ and a vector of right hand sides $d \in \mathbb{R}^m$ such that

$$P = \{x \in \mathbb{R}^n : Dx \leq d\}.$$

The system $Dx \leq d$ is said to *describe* P, and each hyperplane $\{x \in \mathbb{R}^n : D_{i.}^T x = d_i\}$ is called a *cutting plane.* One of the central questions in polyhedral combinatorics is to find the cutting planes that describe P. The issues surrounding this question are discussed in this chapter.

We start with a section on books and collections of survey articles that treat polyhedral combinatorics in detail. In §2 on integer programming by linear programming we discuss general schemes by which all cutting planes are generated. We treat the separation problem, the concepts of total unimodularity and total dual integrality, and give a reference to the computational complexity of deriving an explicit description of conv(X). For NP-hard problems, such as the knapsack, covering, packing, partitioning, and variable bound flow problems, one cannot expect to derive such a description. Then it is of interest to describe the associated polyhedra partially. Articles on this issue are presented in §3.

Our policy in selecting references has been as follows. We have chosen books that give a modern account of polyhedral combinatorics. The purpose of §2 is to review

Annotated Bibliographies in Combinatorial Optimization, edited by M. Dell'Amico, F. Maffioli and S. Martello

the most important theoretical results. When selecting problems for §3 we chose basic combinatorial structures that form substructures of a large collection of combinatorial optimization problems. Some prominent problems of this type are treated in separate chapters of this book, such as the traveling salesman problem, and the maximum cut problem, and are therefore not included here.

1 Books

Here we present a selection of books that are often used as references, and that contain an in-depth treatment of polyhedral combinatorics.

A. Schrijver (1986). *Theory of Linear and Integer Programming*, Wiley, Chichester.
This is a broad book directed to researchers. It contains much more than polyhedral combinatorics, and is particularly useful as it puts polyhedral combinatorics in the general context of linear and integer programming.

M. Grötschel, L. Lovász, A. Schrijver (1988). *Geometric Algorithms and Combinatorial Optimization*, Springer-Verlag, Berlin.
The authors derive algorithmic versions of results from geometry and number theory, and link them to combinatorial optimization. One of the outstanding results in polyhedral combinatorics, namely that the separation problem and the optimization problem for a family of polyhedra are polynomially equivalent, is discussed extensively.

G.L. Nemhauser, L.A. Wolsey (1988). *Integer and Combinatorial Optimization*, Wiley, New York.
All aspects of polyhedral combinatorics are treated. Next to the general theory, it also gives examples of problem-specific results, both with respect to families of strong valid inequalities, and separation.

M. Padberg (1995). *Linear Optimization and Extensions*, Springer, Berlin.
The book has a comprehensive chapter on the theory of polyhedra. It treats all central issues in polyhedral combinatorics, and the links between optimization and separation.

W. Cook, P.D. Seymour (eds.) (1990). *Polyhedral Combinatorics*. DIMACS Series in Discrete Mathematics and Theoretical Computer Science 1, AMS, Providence, ACM, New York.
R.L. Graham, M. Grötschel, L. Lovàsz (eds.) (1995). *Handbook of Combinatorics*, Vol. II, North-Holland, Amsterdam.
These books contain selections of survey papers related to polyhedral combinatorics.

E.L. Lawler (1976). *Combinatorial Optimization: Networks and Matroids*, Holt, Rinehart and Winston, New York.
L. Lovász, M.D. Plummer (1986). *Matching Theory*, Akadémiai Kiadó, Budapest.
K. Trümper (1992). *Matroid Decomposition*, Academic Press, San Diego.
R.K. Ahuja, T.L. Magnanti, J.B. Orlin (1993). *Network Flows*, Prentice-Hall,

New Jersey.
G.M. Ziegler (1995). *Lectures on Polytopes*, Springer-Verlag, New York.

Even though the central theme of the above books is not polyhedral combinatorics, we still want to mention them as they give considerable insight in the study of polyhedra.

2 Integer Programming by Linear Programming

If a linear description of conv(X) is given explicitly, then one can solve the problem $\min\{c^Tx : x \in X\} = \min\{c^Tx : x \in \text{conv}(X)\}$ by linear programming techniques, which is computationally "easy". There is one special case where, for every integral vector $b \in \mathbb{R}^n$, the integrality of the polyhedron $\{x \in \mathbb{R}^n : \; Ax \leq b\}$ is guaranteed. This situation arises when the matrix A is *totally unimodular*, i.e., each subdeterminant of A is either -1, 0 or 1. Within the last 40 years a deep theory on totally unimodular matrices has emerged that we cannot discuss here. The interested reader is referred to the books of Schrijver and Trümper listed in §1 for references and surveys on this subject.

If the constraint matrix A is not totally unimodular, then the integrality of the linear programming relaxation, i.e., the relaxation obtained by dropping the integrality requirements on the variables, is quite rare and depends on the right-hand side vector b. A linear description of the convex hull of all the feasible integer points of the problem can however always be constructed. This topic is discussed next.

R.E. Gomory (1958). Outline of an algorithm for integer solutions to linear programs. *Bull. American Math. Soc.* 64, 275–278.

Consider the set $X = \{x \in \mathbb{Z}^n_+ : \; Ax \leq b\}$. If the constraint matrix A, and the vector of right-hand sides b are integral, and if the set X is bounded, then there exists an implementation of Gomory's cutting plane algorithm, such that for every objective function the procedure terminates after a finite number of iterations with an integral optimum solution.

V. Chvátal (1973). Edmonds polytopes and a hierarchy of combinatorial problems. *Discr. Math.* 4, 305–337.

If the coefficients of A and b are real numbers, and if the feasible set is bounded, then Chvátal's rounding scheme will produce conv(X) after a finite number of iterations.

A. Schrijver (1980). On cutting planes. M. Deza, I.G. Rosenberg (eds.). *Combinatorics 79 Part II*, Ann. Discr. Math. 9, North-Holland, Amsterdam, 291–296.

An elegant way to formulate, and even generalize Chvátal's result was presented by Schrijver, who considered the case where the set of feasible solutions is not necessarily bounded, and where the entries of A and b are rational numbers. In each step of the algorithm he derives a system of linear inequalities $A'x \leq b'$ that is *totally dual integral* (TDI), and where all entries of A' are integral, and rounds down the elements of the vector b'. After a final number of iterations conv(X) is obtained. In the special case that X is bounded, Chvátal's result for not necessarily rational polytopes is obtained.

In comparison to Gomory's procedure the step of adding up linear combinations of current inequalities and rounding down the left hand sides becomes redundant if one resorts to a TDI representation of the current polyhedron.

J. Edmonds, R. Giles (1977). A min-max relation for submodular functions on graphs. P.L. Hammer, E.L. Johnson, B.H. Korte, G.L. Nemhauser (eds.). *Studies in Integer Programming*, Ann. Discr. Math. 1, North-Holland, Amsterdam, 185–204.
R. Giles, W.R. Pulleyblank (1979). Total dual integrality and integer polyhedra. *Linear Algebra and Appl.* 25, 191–196.
A. Schrijver (1981). On total dual integrality. *Linear Algebra and Appl.* 38, 27–32.

A rational system $A'x \leq b'$ of linear inequalities is called TDI if for each integral vector c such that $\min\{y^T b' : \ y \in \mathbb{R}^n_+, \ y^T A' = c\}$ is finite, the minimum is attained by an integral vector. The notion of TDI-ness was introduced by Edmonds and Giles, and it was proved by Giles and Pulleyblank that every rational polyhedron can be described by a TDI system. Schrijver showed that such a TDI system is unique, if the rational polyhedron is full-dimensional. This and many other beautiful results, such as a relation between TDI-ness and Hilbert bases that allows one to derive an integer analogue of Carathéodory's Theorem, can be found in the book of Schrijver, see §1, and in the following articles.

R. Chandrasekaran (1981). Polynomial algorithms for totally dual integral systems and extensions. P. Hansen (ed.). *Studies on Graphs and Discrete Programming*, Ann. Discr. Math. 11, 39–51.
W. Cook (1983). Operations that preserve total dual integrality. *Oper. Res. Lett.* 2, 31–35.
J. Edmonds, R. Giles (1984). Total dual integrality of linear inequality systems. W.R. Pulleyblank (ed.). *Progress in Combinatorial Optimization*, Academic Press, New York, 117–129.
W. Cook (1986). On box totally dual integral polyhedra. *Math. Program.* 34, 48–61.
W. Cook, J. Fonlupt, A. Schrijver (1986). An integer analogue of Carathéodory's theorem. *J. Combin. Theory B* 40, 63–70.

R. Gomory (1969). Some polyhedra related to combinatorial problems. *Linear Algebra and Appl.* 2, 451–558.

Related to the question of describing conv(X) by a system of linear inequalities is the study, initiated by Gomory, of the so-called *corner polyhedra* that builds a bridge between linear programming and the group problem in integer programming.

E. Balas (1975). Disjunctive programming: Cutting planes from logical conditions. O.L. Mangasarian et al. (eds.) *Nonlinear Programming 2*, Academic Press, 279–312.
C.E. Blair (1976). Two rules for deducing valid inequalities for 0-1 problems. *SIAM J. Appl. Math.* 31, 614–617.
R.G. Jeroslow (1977). Cutting plane theory: Disjunctive methods. P.L. Hammer, E.L. Johnson, B.H. Korte, G.L. Nemhauser (eds.). *Studies in Integer Programming*, Ann. Discr. Math. 1, North-Holland, Amsterdam, 293–330.
E. Balas (1979). Disjunctive programming. P.L. Hammer, E.L. Johnson, B.H. Korte (eds.). *Discrete Optimization II*, Ann. Discr. Math. 5, North-Holland,

Amsterdam, 3–51.

If we restrict the variables to take values zero or one only, then there is an alternative procedure, based on *disjunctive programming*, for generating the convex hull of all 0/1-vectors satisfying $Ax - b \geq 0$ with $A \in \mathbb{R}^{m \times n}, b \in \mathbb{R}^m$. The basic theory was developed by Balas.

H. Sherali, W. Adams (1990). A hierarchy of relaxations between the continuous and convex hull representations for zero-one programming problems. *SIAM J. Discr. Math.* 3, 411–430.
L. Lovász, A. Schrijver (1991). Cones of matrices and set-functions and 0-1 optimization. *SIAM J. Optim.* 1, 166–190.
E. Balas, S. Ceria, G. Cornuéjols (1993). A lift-and-project cutting plane algorithm for mixed 0-1 programs. *Math. Program.* 58, 295–324.

These references deal with generating the convex hull of a 0-1 integer program by disjunctive programming techniques. The idea of Lovász and Schrijver can be described as follows. Every constraint $a^T x - \beta \geq 0$ in the current system of inequalities is multiplied by $(1 - x_j) \geq 0$ and $x_j \geq 0$ for $j = 1, ..., n$, where n is the number of variables in the original formulation. This gives rise to a linear formulation in n^2 variables and $2mn$ constraints under the substitution $x_i x_j := y_{ij}$, $y_{ij} := y_{ji}$ and $x_i^2 := x_i$. The resulting polyhedron is projected down onto the original space of the x-variables. Lovász and Schrijver showed that this process needs to be repeated at most n times before the convex hull of feasible solutions is obtained. Balas, Ceria, and Cornuéjols showed that it is sufficient to multiply each constraint by one single variable x_j and its complement at a time. In this way, the inequality system in the lifted space consists of at most $2n$ variables and $2m$ constraints.

M. Grötschel, L. Lovász, A. Schrijver (1981). The ellipsoid method and its consequences in combinatorial optimization. *Combinatorica* 1, 169–197 [Corrigendum: 4, 291–295].

Here, one of the fundamental results in polyhedral combinatorics, namely the equivalence between the *optimization problem* $\min\{c^T x : x \in X\}$ and the *separation problem* for the polyhedron $\text{conv}(X)$ in terms of computational complexity, is described. The latter problem is to find a hyperplane separating a given point x^* from the polyhedron $\text{conv}(X)$, or to assert that no such hyperplane exists.

Often one is interested in a solution of the separation problem for a specific family $\mathcal{F}$ of inequalities. This is the problem of finding an inequality in $\mathcal{F}$ that violates x^*, or asserting that no separating hyperplane in this family exists. For certain families of inequalities this problem is solvable in polynomial time, although the optimization problem for which they are valid is NP-hard, and the number of inequalities in the family is exponential in the encoding length of the optimization problem. This is one of the explanations behind the computational success of polyhedral techniques, see Chapter 4 for further details.

J. Edmonds (1965). Maximum matchings and a polyhedron with 0,1-vertices. *Journal of Research of the National Bureau of Standards (B)* 69, 9–14.
M.W. Padberg, M.R. Rao (1982). Odd minimum cut-sets and b-matchings. *Math. Oper. Res.* 7, 67–80.

An important example of a polynomially separable family of inequalities that is valid for an NP-hard problem, is the separation problem for the family of 2-matching constraints that is valid for the traveling salesman polytope. This family was originally invented by Edmonds for the 2-matching polytope, and gives, together with the original constraints (the degree constraints and the bounds on the variables), a complete description of the convex hull of 2-matchings. The result that the separation problem for the family of 2-matching constraints can be solved in polynomial time was derived by Padberg and Rao. Their algorithm can also be used to solve the following more general problem. Let $A \in \{0,1\}^{m\times n}$ be a matrix with at most two 1's per column and $b \in \mathbb{Z}^m$. For every point $y \in \mathbb{R}^n$, the $0-1/2$-cut separation problem $\min\{\lfloor\lambda^T b\rfloor - \lambda^T Ay : \lambda \in \{0,\frac{1}{2}\}^m,\ \lambda^T A \text{ integer}\}$ can be solved in polynomial time. In other words, the minimum here is taken over all Chvátal-Gomory cuts with dual multipliers in $\{0,\frac{1}{2}\}$. The result implies that if there exists a Chvátal-Gomory cut with multipliers in $\{0,\frac{1}{2}\}$ that separates y from the polyhedron $\text{conv}(\{x \in \mathbb{Z}^n : Ax \leq b\})$, then the most violated inequality in this family can be found in polynomial time. A result in the same vein was obtained recently by Caprara and Fischetti.

A. Caprara, M. Fischetti (1996). $\{0,\frac{1}{2}\}$-Chvátal-Gomory cuts. *Math. Program.* 74, 221–235.

The authors show that the $0-1/2$-cut separation problem is polynomially solvable if the constraint matrix modulo 2 is related to the edge-path incidence matrix of a tree. This result is, in particular, applicable to matrices with at most two odd entries for each row, or at most two odd entries for each column.

There is an important technique that can be used to increase the dimension of a face induced by an inequality that is valid for the polyhedron $P_S = \text{conv}(\{x \in \mathbb{Z}^n_+ : Ax \leq b\} \cap \{x_j = \xi_j,\ j \in S\})$ where S is a proper subset of $\{1,\ldots,n\}$. For ease of explanation we assume that P_Q is full-dimensional for all $Q \subseteq S$, and that $\xi_j = 0$ for all $j \in S$. For $a \geq 0$, $\beta > 0$, let $a^T x \leq \beta$ be an inequality that is valid for P_S such that the dimension of the face $\{x \in P_S : a^T x = \beta\}$ is equal to t. At each iteration of the so-called *sequential lifting* algorithm we choose a variable x_k, $k \in S$, set $S := S \setminus \{k\}$, and compute a coefficient $\gamma \geq 0$ such that $a^T x + \gamma x_k \leq \beta$ is valid for P_S. Let γ_0 be the maximum value of γ such that the inequality is valid for P_S. For any choice of $\gamma \leq \gamma_0$ the resulting inequality is valid, and if we choose $\gamma = \gamma_0$, then the face induced by $\{x \in P_S : a^T x + \gamma x_k = \beta\}$ has dimension $t+1$.

M.W. Padberg (1975). A note on zero-one programming. *Oper. Res.* 23, 833–937.
L.A. Wolsey (1976). Facets and strong valid inequalities for integer programs. *Oper. Res.* 24, 367–372.
E. Zemel (1978). Lifting the facets of zero-one polytopes. *Math. Program.* 15, 268–277.

Sequential lifting was first applied by Padberg to the vertex packing problem, see §3.2. Sequential lifting was stated in a general form by Wolsey. Zemel developed a more general technique called *simultaneous lifting*. Here any subset of the variables in S can be lifted simultaneously, yielding inequalities that in general cannot be obtained by lifting the variables sequentially.

In this chapter we have indicated that it is possible, in principle, to describe the polyhedron conv(X) by means of linear inequalities. The descriptions are, however, implicit and can only be constructed in an iterative fashion. For quite a few polyhedra, such as the matching polyhedron, an explicit description is at hand. A natural question to ask is under which conditions we can expect to derive an explicit description of the convex hull of feasible solutions.

R.M. Karp, C.H. Papadimitriou (1980). On linear characterizations of combinatorial optimization problems. *Proc. 21st Annual IEEE Symp. Found. Comput. Sci.*, IEEE, New York, 1–9.

The answer was given by Karp and Papadimitriou who proved that if the optimization problem under consideration is NP-hard, then one cannot find an explicit description of the convex hull of feasible solutions, unless NP=co-NP. More precisely, if a certain optimization problem is NP-hard, e.g., the traveling salesman problem, and if the problem to decide whether a valid inequality defines a facet for the associated class of polyhedra is in NP, then this would imply that NP=co-NP. If NP=co-NP, then there exists a compact certificate for the no-answer for all problems in NP, which is unlikely. Despite this negative answer, polyhedral techniques can be effective for NP-hard integer programming problems in the sense that we can find good partial descriptions of the convex hull of feasible solutions. This is the topic of the next section.

3 Selected Combinatorial Problems

Here we study polyhedra associated with a few basic NP-hard combinatorial optimization problems that form relaxations of several more complex problems. If a family of inequalities is valid for a relaxation of a problem Π, then the family is also valid for Π. It is therefore important to analyze the polyhedral structure of such basic problems in order to understand the polyhedral structure of the more complex ones.

3.1 The knapsack problem

The *knapsack problem* is the basic version of a data dependent problem, and is defined as follows. For a capacity $a_0 \in \mathbb{Z}_+$ and a set $N = \{1, ..., n\}$ of items, where each item $j \in N$ has a weight a_j and a profit c_j, the knapsack problem is the problem of finding a subset of items, with total weight less than or equal to the capacity a_0, that maximizes the total profit. The mathematical programming formulation is as follows.

$$\max\{\sum_{j \in N} c_j x_j : \sum_{j \in N} a_j x_j \leq a_0, \; x_j \in \{0, 1\} \text{ for all } j \in N\}.$$

Since a slight change in the weights of the items might drastically change the inequalities that describe the polyhedron, it seems important to understand the principles by which valid inequalities are constructed.

Most of the polyhedral studies presented so far involve the basic object of minimal covers, see for instance the following article.

L.A. Wolsey (1975). Faces of linear inequalities in 0-1 variables. *Math. Program.* 8, 165–178.

A subset $S \subseteq N$ is a *cover* (or dependent set) if its weight exceeds the capacity. With the cover S one can associate the *cover inequality* $\sum_{j \in S} x_j \leq |S| - 1$ that is valid for the knapsack polyhedron. If the cover S is minimal with respect to inclusion, the associated inequality is called a *minimal cover inequality.*

M. Laurent, A. Sassano (1992). A characterization of knapsacks with the max-flow-min-cut property. *Oper. Res. Lett.* 11, 105–110.

An interesting question is to characterize weight vectors $a = (a_1, \ldots, a_n) \in \mathbb{N}^n$ for which the minimal cover inequalities describe the knapsack polyhedron. Laurent and Sassano showed that n minimal cover inequalities suffice to describe the knapsack polytope when $a = (a_1, \ldots, a_n)$ is a weakly superdecreasing sequence, i.e., $\sum_{j \geq q} a_j \leq a_{q-1}$, for all $q = 2, \ldots, n$.

M.W. Padberg (1980). (1,k)-configurations and facets for packing problems. *Math. Program.* 18, 94–99.
Y. Pochet, R. Weismantel (1994). The sequential knapsack polytope. Preprint SC 94-30, Konrad-Zuse-Zentrum, Berlin.
R. Weismantel (1994). On the 0/1 knapsack polytope. *Math. Program.* (to appear).
E. Balas (1975). Facets of the knapsack polytope. *Math. Program.* 8, 146–164.
E. Zemel (1989). Easily computable facets of the knapsack polytope. *Math. Oper. Res.* 14, 760–764.

A slightly more general object than minimal covers are $(1,k)$-configurations that were introduced by Padberg. A $(1,k)$-*configuration* consists of an independent set S, i.e., a set S such that $\sum_{j \in S} a_j \leq a_0$, plus one additional item z, such that every subset of S of cardinality k, together with z, forms a minimal cover. A $(1,k)$-configuration gives rise to the inequality $\sum_{j \in S} x_j + (|S| - k + 1)x_z \leq |S|$. Padberg showed that if the set N of items defines a $(1,k)$-configuration, then the convex hull of the associated knapsack polyhedron is given by the lower and upper bound constraints and the set of all $(1,l)$-configuration inequalities defined by $T \subseteq S$, where $T \cup \{z\}$ is a $(1,l)$-configuration for some $l \leq k$. This result was generalized by Pochet and Weismantel to knapsack problems where the weights of the items have the divisibility property, i.e., for every pair of weights, the bigger one is an integer multiple of the smaller one. Inequalities derived from both covers and $(1,k)$-configurations are special cases of extended weight inequalities that have been introduced by Weismantel who describes the knapsack polyhedron when $a_j = 1$, or $a_j \geq \lfloor a_0/2 \rfloor + 1$, for all $j \in N$. Weismantel also showed that, independent of the lifting sequence, the lifting coefficient of a variable x_j in the extended weight inequality is either equal to the value α_j that this variable would obtain if it was the first one in the sequence, or it equals $\alpha_j - 1$. The correct value of the lifting coefficient for a given sequence can be computed in polynomial time. For cover inequalities these results can be found in the articles by Balas and Zemel.

E.J. Johnson (1980). Subadditive lifting methods for partitioning and knapsack problems. *J. Algorithms* 1, 75–96.

For knapsack type problems there are other techniques for lifting lower-dimensional faces of the associated polyhedra. One such techniques, developed by Johnson, is based on an analysis of subadditive functions.

H.P. Crowder, E.L. Johnson, M.W. Padberg (1983). Solving large-scale zero-one linear programming problems. *Oper. Res.* 31, 803–834.
T.J. Van Roy, L.A. Wolsey (1987) Solving mixed 0-1 programs by automatic reformulation. *Oper. Res.* 35, 45–57.

The above articles describe successful implementations of automatically generating violated inequalities for the knapsack polytope. They are mentioned here since they represent major breakthroughs in the use of polyhedral techniques. For more references on computational aspects and results, see Chapter 4.

3.2 Packing, covering, and partitioning problems

For a matrix $A \in \mathbb{R}_+^{m \times n}$ and a vector $b \in \mathbb{R}_+^m$ a *packing problem* is of the form

$$\max\{c^T x : \; Ax \leq b, \; x \in \mathbb{Z}_+^n\},$$

where c is a m-dimensional vector of real coefficients. By replacing the $\leq$-sign by the $\geq$-sign and the max-operator by the min-operator we obtain a *covering problem.* A *partitioning problem* is of the form $\min\{c^T x : \; Ax = b, x \in \mathbb{Z}_+^n\}$.

D.R. Fulkerson (1971). Blocking and anti-blocking pairs of polyhedra. *Math. Program.* 1, 168–194.
L.R. Ford, Jr., D.R. Fulkerson (1956). Maximal flow through a network. *Canadian J. Math.* 8, 399–404.
A. Schrijver (1983). Min-max results in combinatorial optimization. A Bachem, M. Grötschel, B. Korte (eds.). *Mathematical Programming: The State of the Art*, Springer, Berlin, 439–500.

From the theory of linear programming it follows that the dual of the linear programming relaxation of the packing problem is a covering problem. This allows us to derive min-max results for the linear programming relaxations of covering and packing models. In case that b is the m-dimensional vector of all ones, $\mathbf{1}$, we can also use the theory of *blocking and anti-blocking polyhedra*, developed by Fulkerson, to derive min-max results for the linear programming relaxations of the packing and covering models. Let A be a matrix with rows $a^1, \ldots, a^m$, and D a matrix with rows $d^1, \ldots, d^t$. If $\{x \in \mathbb{R}_+^n : Ax \geq \mathbf{1}\} = \text{conv}\{d^1, \ldots, d^t\} + \mathbb{R}_+^n := \{x + y : \; x \in \text{conv}\{d^1, \ldots, d^t\}, y \in R_n^+\}$, then the blocking polyhedron is equal to $\{x \in \mathbb{R}_+^n : \; Dx \geq \mathbf{1}\} = \text{conv}\{a^1, \ldots, a^m\} + \mathbb{R}_+^n$. One application of the theory of blocking and anti-blocking polyhedra is an elegant proof of the max-flow-min-cut theorem of Ford and Fulkerson. For further details on blocking and anti-blocking polyhedra we refer to the survey of Schrijver in the Handbook of Combinatorics, see §1. Packing and covering models have been extensively surveyed by Schrijver.

A. Schrijver (1984). Total dual integrality from directed graphs, crossing families, and sub- and supermodular functions. W.R. Pulleyblank (ed.). *Progress in Combinatorial Optimization*, Academic Press, New York, 315–361.

Often min-max results for discrete packing and covering models can be derived for totally dual integral systems. Examples are integrality results for crossing families defined on the set of vertices of a digraph and the blocking collection of covers of the crossing family. A comprehensive survey on results in this spirit is provided by Schrijver.

C. Berge (1970). Sur certains hypergraphes généralisant les graphes bipartites. P. Erdös, A. Rényi, V.T. Sós (eds.). *Combinatorial Theory and Its Applications I*, Colloquia Mathematica Societatis János Bolyai, Vol. 4, North-Holland, Amsterdam, 119–133.

C. Berge (1972). Balanced matrices. *Math. Program.* 2, 19–31.

M. Conforti, G. Cornuéjols, A. Kapoor, M. R. Rao, K. Vušković (1994). Balanced matrices. J.R. Birge, K.G. Murty (eds.). *Mathematical Programming: State of the Art 1994*, University of Michigan, 1–33.

M. Conforti, G. Cornuéjols, A. Kapoor, K. Vušković (1994). Recognizing balanced 0 ± 1 matrices, *Proc. 5-th Annual ACM-SIAM Symp. Discr. Algorithms*, ACM, New York, SIAM, Philadelphia, 103–111.

Special, but particularly important cases of packing and covering models arise when the vector b of right hand sides is equal to $\mathbf{1}$, when the elements of the matrix A are either 0 or 1, and when the variables are binary. Then one speaks of *set packing, set covering*, and *set partitioning* problems, respectively. For the associated polyhedra, explicit descriptions are sometimes known. implying model. In particular, if the 0-1 matrix A is *balanced*, i.e., if A does not contain a square submatrix of odd order with two ones per row or columns, then the set packing as well as the the set covering polyhedra are integral, see Berge. For more details on balanced matrices we refer to Chapter 11, and to the articles by Conforti et al.

M. Padberg (1973). On the facial structure of set packing polyhedra. *Math. Program.* 5, 199–215.

F. Barahona, A.R. Mahjoub. On the cut polytope. *Math. Program.* 36, 157–173.

Every set packing problem can be interpreted as the problem of finding a maximum *stable set* in the graph whose nodes correspond to the columns of the matrix and whose edges represent the pairs of columns with intersecting support. The stable set polyhedron has been studied extensively in the literature during the last 20 years starting with the work of Padberg, who introduced clique inequalities of the form $\sum_{j \in C} x_j \leq 1$, where C is a clique, i.e., a node set of a complete subgraph of the given graph. Odd circuit constraints $\sum_{j \in H} x_j \leq (|H| - 1)/2$, with H being a subset of the set of nodes of odd cardinality whose induced subgraph is a cycle without chords, as well as the constraints based on the complements of odd circuits, can be found in this reference, too. Odd circuit constraints can be separated in polynomial time by adapting the odd cycle separation algorithm for the max-cut problem that was introduced by Barahona and Mahjoub. We also refer to Grötschel et al. (1988) listed in §1.

V. Chvátal (1975). On certain polytopes associated with graphs. *Journal of Combinatorial Theory B* 18, 305–337.

The graphs for which the set of all odd circuit constraints, the nonnegativity constraints, and the constraints $x_i + x_j \leq 1$ for all edges $\{i, j\}$ in the graph, suffice to describe the set packing polyhedron, are called *t-perfect*. This notion was introduced by Chvátal. Similarly, one can ask for the integrality of the polyhedron that is defined by all clique constraints together with the nonnegativity constraints and the constraints $x_i + x_j \leq 1$ for all edges $\{i, j\}$ in the graph. Such graphs are called *perfect*, and are discussed in Chapter 11.

A. Sassano (1989). On the facial structure of the set-covering polytope. *Math. Program.* 44, 181–202.
E. Balas, S.M. Ng (1989). On the set covering polytope I: all the facets with coefficients in $\{0, 1, 2\}$. *Math. Program.* 43, 1–20.

For a thorough treatment of the mathematics underlying the constraint generation for the set covering polyhedron we refer the readers to Sassano, and Balas and Ng.

R. Euler, M. Jünger, G. Reinelt (1987). Generalizations of odd cycles and anticycles and their relation to independence system polyhedra. *Math. Oper. Res.* 12, 451–462.
M. Laurent (1989). A generalization of antiwebs to independence systems and their canonical facets. *Math. Program.* 45, 97–108.
E. Balas (1979). Disjunctive programming. P.L. Hammer, E.L. Johnson, B.H. Korte (eds.). *Discrete Optimization II*, Ann. Discr. Math. 5, North-Holland, Amsterdam, 3–51.
R. Müller, A.S. Schulz (1996). Transitive packing. W.H. Cunningham, S.T. McCormick, M.Queyranne (eds.). *Integer Programming and Combinatorial Optimization*, 5th International IPCO Conference, Vancouver, British Columbia, Canada, June 1996, Proceedings, Springer Lecture Notes in Computer Science 1084, Springer, Berlin, 430–444.

Both the set packing and the set covering polyhedron are equivalent to special cases of independence system polyhedra that have been studied by Euler, Jünger, and G. Reinelt. Polyhedra associated with independence systems are included in the family of transitive packing polyhedra introduced by Balas. Müller and Schulz describe a common framework for valid inequalities induced by graphic structures such as cliques, odd cycles, odd anticycles, webs, antiwebs etc. It also generalizes polyhedral results for certain graph partitioning problems. A collection of papers on this subject is given below.

E. Balas, M. Padberg (1976). Set partitioning: A survey. *SIAM Review* 18, 710–760.
M. Grötschel, Y. Wakabayashi (1990). Facets for the clique partitioning polytope. *Math. Program.* 47, 367–388.
S. Chopra, M.R. Rao (1993). The partition problem. *Math. Program.* 59, 87–115.
C.E. Ferreira, A. Martin, C. de Souza, R. Weismantel, L. Wolsey (1996). Formulations and valid inequalities for the node capacitated graph partitioning problem. *Math. Program.* 74, 247–266.

3.3 Network design and flows

L.R. Ford, Jr., D.R. Fulkerson (1956). Maximal flow through a network. *Canadian J. Math.* 8, 399–404.
B. Korte, L. Lovász, H.J. Prömel, A. Schrijver (eds.) (1990). *Paths, Flows, and VLSI-*

Layout, Springer, Berlin.
M.V. Lomonosov (1985). Combinatorial approaches to multiflow problems. *Discr. Appl. Math.* 11, 1 – 94.

When considering any kind of flow problem in integer programming, the paper by Ford and Fulkerson that contains the well-known max-flow-min-cut theorem is the fundamental reference. Many extensions of this result have been formulated in the setting of multicommodity flows including duality results for the problem of packing paths and cuts in graphs under capacity restrictions. In order to cover this and related topics in detail, it would require a book on its own. We refer here to the excellent survey articles contained in the book by Korte, Lovász, Prömel, and Schrijver. Another comprehensive survey can be found in the article by Lomonosov.

Each flow in a graph can be decomposed into paths. Since a subgraph that is induced by a path is node- and edge-connected, a flow can be viewed as a special graph structure that requires connectivity. Besides paths, there are other connectivity structures in graphs that have become important. Such structures are discussed next.

M.X. Goemans (1994). The Steiner tree polytope and related polyhedra. *Math. Program.* 63, 157–182.
S. Chopra, M.R. Rao (1994). The Steiner tree problem I: formulations, compositions and extensions of facets. *Math. Program.* 64, 209–229.
S. Chopra, M.R. Rao (1994). The Steiner tree problem II: properties and classes of facets. *Math. Program.* 64, 231–246.

Consider an undirected graph $G = (V, E)$ and a subset T of V. A *Steiner tree* in G is a subgraph that spans T and possibly vertices in $V \setminus T$. Polyhedral results regarding various versions of the Steiner tree problem can be found in the articles given above. In the paper by Goemans a characterization of the convex hull of all incidence vectors of Steiner trees (in the space of the number of nodes plus the number of edges) is given when the underlying graph is series parallel. Chopra and Rao use a directed formulation for the problem of finding a minimum weighted Steiner tree in a graph with weights on the edges. They show that the linear relaxation of the directed formulation is stronger than the linear relaxation of the undirected one. This result is obtained by projecting the polyhedron associated with the directed formulation onto the subspace defined by the variables associated with the undirected formulation.

M. Grötschel, A. Martin, R. Weismantel (1996). Packing Steiner trees: polyhedral investigations. *Math. Program.* 72, 101–123.

Similarly as for flows and multicommodity flows, it is interesting to study the problem of packing Steiner trees under capacity restrictions. We do not want to go into details here, but refer the readers to the article by Grötschel, Martin, and Weismantel.

M. Stoer (1992). Design of survivability networks. *Lecture Notes Math.* 1531, Springer, Heidelberg.
M. Grötschel, C.L. Monma, M. Stoer (1995). Polyhedral and computational investigations for designing communication networks with high survivability requirements. *Oper. Res.* 43, 1012–1024.

In communication network design it is essential to design networks that are reliable in the sense that a failure at any component of the network does not disconnect important clients. This requirement is taken into account by designing networks that are k-node connected or k-edge connected, where the number k has to be specified by the customers. Two recent references on this topic are given above.

Until now we have briefly sketched results associated with purely integer connectivity type requirements. There is a lot of ongoing research on mixed integer problems that have a flow structure.

M.W. Padberg, T.J. Van Roy, L.A. Wolsey (1985). Valid inequalities for fixed charge problems. *Oper. Res.* 33, 842–861.
K. Aardal, Y. Pochet, L.A. Wolsey (1995). Capacitated facility location: valid inequalities and facets. *Math. Oper. Res.* 20, 562–582 [Erratum: 21, 253–256].

One basic form of a mixed integer flow structure is the so-called *single-node flow polytope*,

$$X_F = \{(x, y) \in \mathbb{R}^n_+ \times \{0,1\}^n : \sum_{j=1}^{n} x_j = b,\ x_j \leq u_j y_j, \text{ for all } j = 1, \ldots, n\},$$

where we have a single node with a fixed outflow b, and a set $N = \{1, \ldots, n\}$ of arcs with variable upper bounds entering the node. The associated single-node flow polytope is a relaxation of several polyhedra associated with fixed-charge planning and distribution problems, such as lot-sizing and location problems, see further Chapter 15. Let J be a subset of N such that $\sum_{j \in J} u_j = b + \lambda$, $\lambda > 0$. The set J is called a *flow cover.* Let $(m)^+$ denote $\max\{0, m\}$. The *flow cover inequalities* $\sum_{j \in J} x_j \leq b - \sum_{j \in J}(u_j - \lambda)^+(1 - y_j)$ were developed by Padberg, Van Roy, and Wolsey.

One way of extending the flow cover inequalities is to include variables x_j for $j \in L \subseteq N \setminus J$ in the inequality. This yields an inequality of the form $\sum_{j \in J \cup L} x_j \leq b - \sum_{j \in J}(u_j - \lambda)^+(1 - y_j) + \sum_{j \in L}(\bar{u}_j - \lambda)y_j$, where $\bar{u}_l = \max\{\max_{j \in J}\{u_j\}, u_l\}$ for all $l \in L$. Padberg et al. showed that if $u_j = u$ for all $j \in N$, then $\mathrm{conv}(X_F)$ is described by all the constraints in the mixed integer programming formulation and the family of extended flow cover inequalities. It was observed by Aardal, Pochet, and Wolsey that the separation problem for the family of extended flow cover inequalities can be solved in polynomial time when $u_j = u$, for all $j \in N$.

T.J. Van Roy, L.A. Wolsey (1986). Valid inequalities for mixed 0-1 programs. *Discr. Appl. Math.* 14, 199–213.
L.A. Wolsey (1989). Submodularity and valid inequalities in capacitated fixed charge networks. *Oper. Res. Lett.* 8, 119–124 [Erratum: 8, 295].

A slightly more complicated model arises when arcs having variable upper and lower bounds on the flow can both enter and leave a node. For this model various versions of generalized flow cover inequalities were developed by Van Roy and Wolsey. The authors also discuss the separation problems based on these inequalities. Later, Wolsey generalized several of the families of inequalities mentioned above by introducing the family of submodular inequalities.

T.J. Van Roy, L.A. Wolsey (1986). Valid inequalities and separation for uncapacitated fixed-charge networks. *Oper. Res. Lett.* 4, 105–112.
Y. Pochet, L.A. Wolsey (1995). Algorithms and reformulations for lot sizing problems. W. Cook, L. Lovász, P. Seymour (eds.). *Combinatorial Optimization*, DIMACS Series in Discrete Mathematics and Theoretical Computer Science 20, AMS, Providence, ACM, New York, 245–293.

For uncapacitated directed fixed-charge networks a general class of inequalities was developed by Van Roy and Wolsey. The inequalities are based on the idea of using the 0-1 variables when bounding the continuous flow that can pass along a subset of the arcs that form a directed cut in the network. Such inequalities have been particularly useful when solving uncapacitated lot-sizing problems, as described by Pochet and Wolsey.

Acknowledgement

Large parts of this chapter was written while the first author was visiting the Konrad-Zuse-Zentrum für Informationstechnik, Berlin (ZIB). The financial support provided by ZIB is greatly acknowledged. We are most grateful for the comments provided by Oktay Günlük, Laurence Wolsey, and an anonymous referee.

4 Branch-and-Cut Algorithms

Alberto Caprara
DEIS, University of Bologna

Matteo Fischetti
DMI, University of Udine

CONTENTS

Branch-and-cut is rapidly becoming one of the most popular techniques for the solution of (mixed) integer linear programs.

An *Integer Linear Program* (ILP) is a problem of the form $\min\{c^T x : Ax \leq b, x \geq 0, x \text{ integer}\}$. A *Mixed* ILP (MILP) is a generalization of an ILP in which some variables may not be restricted to be integer. A 0-1 MILP (ILP) is a MILP (ILP) where the integer variables are restricted to be 0 or 1. The *Linear Programming* (LP) relaxation of a given MILP is the problem obtained by disregarding the integrality requirements.

A common approach to a MILP consists of solving its LP relaxation in the hope of finding an optimal solution x^* which happens to be integer. If this is not the case, there are two main ways to proceed. In the *cutting-plane* approach, one enters the *separation phase*, where a linear inequality (*cut*) $\alpha^T x \leq \alpha_0$ is identified which separates x^* from the feasible solutions of the MILP. The cut is appended to the current LP relaxation, and the procedure is iterated. In the *branch-and-bound* approach, instead, the MILP is replaced by two subproblems obtained, e.g., by imposing an additional restriction of the type $x_j \leq \lfloor x_j^* \rfloor$ and $x_j \geq \lceil x_j^* \rceil$, respectively, where x_j is an integer-constrained variable with fractional value in x^*. The procedure is then recursively applied to each of the two subproblems.

In the late 50's, Gomory pioneered the cutting-plane approach, proposing a very elegant and simple way to derive cuts for an ILP by using information associated with an optimal LP basis. Practical experience with Gomory's algorithm shows however

Annotated Bibliographies in Combinatorial Optimization, edited by M. Dell'Amico, F. Maffioli and S. Martello

that the quality of the cuts generated becomes rather poor after a few iterations, which causes the so called *tailing-off* phenomenon—a long sequence of iterations without significant improvements towards integrality. On the other hand, the branch-and-bound scheme can generate a huge number of subproblems in case the LP relaxation does not approximate tightly the convex hull of the integer feasible points.

Combination of cutting-plane and branch-and-bound techniques has been attempted since the early 70's. In those days, however, the constraint generator was used as a simple preprocessor to obtain a tighter LP relaxation of the initial MILP formulation. In the mid 80's, Padberg and Rinaldi (see §2) introduced a new methodology for an effective integration of the two techniques, which they named *branch-and-cut*. This is an overall solution scheme whose main ingredients include: the generation at every node of the branching tree of (facet-defining) cuts globally valid along the tree; efficient cut management by means of a constraint pool structure; column/row insertion and deletion from the current LP; variable fixing and setting; and the treatment of inconsistent LP's.

Branch-and-cut has a number of advantages over pure cutting-plane and branch-and-bound schemes. With respect to the branch-and-bound approach, the addition of new cuts improves the LP relaxation at every branching node. With respect to the pure cutting-plane technique, one can resort to branching as soon as tailing-off is detected. As the overall convergence is ensured by branching, the cut separation can be of heuristic type, and/or can restrict to subfamilies of problem-specific cuts which capture some structures of the problem in hand. Moreover, the run-time variable pricing and cut generation/storing mechanisms allow one to deal effectively with tight LP relaxations having in principle a huge number of variables and constraints.

The effectiveness of the branch-and-cut techniques depends heavily on the quality of the cuts generated. One theoretical quality certificate is the property of being facet defining for the convex hull of the feasible points (or at least of some relaxation). This gives motivation for studying the facet structure of the convex hull of the feasible solutions to general and/or structured MILP's, an important issue in *polyhedral theory* (see Chapter 3 in this book).

Although rather new, branch-and-cut is already used in quite a wide spectrum of applications. The method involves several techniques of both theoretical and practical flavor. As a result, it was not completely clear how to select the references to be included in this chapter, and how to deal with possible overlapping with other chapters in this book. We decided to include only papers presenting computational results and to leave background theoretical papers to other chapters. Furthermore, we extended the scope of this chapter so as to also cover cutting-plane schemes which can be embedded within a modern branch-and-cut framework. In order to avoid a substantial overlapping with Chapter 8, we decided not to deal with papers on the *branch-and-price* scheme, which combines branch-and-bound and column generation.

1 Surveys and Books

M. Padberg, M. Grötschel (1985). Polyhedral computations. E. Lawler, J.K. Lenstra, A.H.G. Rinnooy Kan, D.B. Shmoys (eds.). *Traveling Salesman Problem: A Guided Tour of Combinatorial Optimization*, J. Wiley and Sons, New York, 307–359.

G.L. Nemhauser, L.A. Wolsey (1988). *Integer and Combinatorial Optimization*, J. Wiley and Sons, New York.

These references describe the state of the art on the use of polyhedral techniques for the solution of hard combinatorial problems as it was in the mid 80's, before the introduction of branch-and-cut.

M. Jünger, G. Reinelt, S. Thienel (1995). Practical problem solving with cutting plane algorithms in combinatorial optimization. W. Cook, L. Lovász, P. Seymour (eds.). *Combinatorial Optimization. DIMACS Series in Discrete Mathematics and Theoretical Computer Science*, AMS, 111–152.
K. Aardal, S. van Hoesel (1995). *Polyhedral techniques in combinatorial optimization*, CentER Discussion Paper 9557, Tilburg University, Tilburg.
A. Lucena, J.E. Beasley (1996). Branch and cut algorithms. J.E. Beasley (ed.). *Advances in Linear and Integer Programming*, Oxford University Press, 187–221.

These papers give a detailed description of theoretical and practical issues in the design of branch-and-cut algorithms for solving combinatorial optimization problems. Applications to several problems are surveyed.

2 Methodological Papers

G.B. Dantzig, D.R. Fulkerson, S.M. Johnson (1954). Solution of a large-scale traveling-salesman problem. *Oper. Res.* 2, 393–410.
G.B. Dantzig, D.R. Fulkerson, S.M. Johnson (1959). On a linear-programming, combinatorial approach to the traveling-salesman problem. *Oper. Res.* 7, 58–66.

For the first time, polyhedral information on the solution structure is used to attack a hard combinatorial optimization problem. The optimal solution to a (large scale, at that time) 49-city traveling salesman problem is obtained by solving a sequence of LP relaxations. Problem-specific cuts are identified visually, and added "by hand" to the current LP relaxation.

P. Miliotis (1976). Integer programming approaches to the traveling salesman problem. *Math. Program.* 10, 367–378.

Problem-specific valid inequalities for the traveling salesman problem are used within a branch-and-bound framework. The method allows for the addition of cuts in any node of the branching tree, but only when the solution of the associated LP relaxation is integer. In other words, the inequalities are added only to make sure that the branch-and-bound uses a correct ILP formulation of the problem.

H.P. Crowder, E.L. Johnson, M.W. Padberg (1983). Solving large-scale zero-one linear programming problems. *Oper. Res.* 31, 803–834.

The paper reports on the solution of 10 real-world large-scale 0-1 integer programs. Each constraint of the problem is viewed as a knapsack constraint, for which strong cuts are derived. Preprocessing and clever branch-and-bound techniques are also described. The seminal idea behind this paper is that effective structured cuts can be exploited even for 0-1 integer programs with no apparent special structure.

M. Grötschel, M. Jünger, G. Reinelt (1984). A cutting plane algorithm for the linear ordering problem. *Oper. Res.* 32, 1195–1220.

The linear ordering problem is solved by means of a cutting-plane/branch-and-bound algorithm based on facet-defining inequalities and heuristics. The paper introduces one of the main ideas of the branch-and-cut approach—the cut generation at every node of the branching tree.

M.W. Padberg, G. Rinaldi (1987). Optimization of a 532-city symmetric traveling salesman problem by branch and cut. *Oper. Res. Lett.* 6, 1–7.
M.W. Padberg, G. Rinaldi (1991). A branch-and-cut algorithm for the resolution of large-scale symmetric traveling salesman problems. *SIAM Rev.* 33, 60–100.

The birth of branch-and-cut, as intended today. These seminal papers give a detailed description of the overall branch-and-cut framework and of its main components, which include the generation at every branching node of polyhedral cuts globally valid along the tree, the variable pricing and fixing, the treatment of inconsistent LP's, and the use of a constraint pool to handle the cuts generated.

K.L. Hoffman, M.W. Padberg (1985). LP-based combinatorial problem solving. *Ann. Oper. Res.* 4, 145–194.
K.L. Hoffman, M.W. Padberg (1991). Improving LP-representation of zero-one linear programs for branch-and-cut. *ORSA J. Comput.* 3, 121–134.

Two computational studies done before and after the introduction of branch-and-cut methodology which well illustrate the transition from the 2-phase approach (cutting-plane followed by branch-and-bound) to branch-and-cut. The same set of 0-1 programs is solved, by the same authors and with the same theoretical tools, by using the two approaches.

M. Grötschel, O. Holland (1991). Solution of large-scale symmetric traveling salesman problems. *Math. Program.* 51, 141–202.
J.M. Clochard, D. Naddef (1993). Using path inequalities in a branch and cut code for the symmetric traveling salesman problem. G. Rinaldi, L.A. Wolsey (eds.). *Proc. of the Third IPCO Conf.*, Erice. CIACO, Louvain-la-Neuve, 291–311.
M. Jünger, G. Reinelt, S. Thienel (1994). Provably good solutions for the traveling salesman problem. *Z. Oper. Res.* 40, 183–217.
D. Applegate, R.E. Bixby, V. Chvátal, W. Cook (1995). *Finding cuts in the TSP (a preliminary report)*, Technical Report 95-05, DIMACS, Rutgers University, New Brunswick.
M. Fischetti, J.J. Salazar, P. Toth (1995). *Solving the orienteering problem through branch-and-cut*, Research Report OR-95-19, DEIS, University of Bologna.
T. Christof, G. Reinelt (1996). Combinatorial optimization and small polytopes. *TOP* 4, 1–64.

These papers describe techniques for a fruitful interaction between enumeration and cutting-plane generation. The first reference introduces Lagrangian techniques to select the first set of active LP variables. The second and fourth papers propose sophisticated branching strategies. The third one discusses how to exploit LP information to drive a heuristic procedure. The fifth reference addresses the use of conditional cuts (i.e., cuts which may cut off the optimal solution) and their integration in the enumerative scheme. The last paper addresses the use of a complete description of small problem instances as a separation tool.

E. Balas, S. Ceria, G. Cornuéjols (1993). A lift-and-project cutting plane algorithm for mixed 0-1 programs. *Math. Program.* 58, 295–324.
E. Balas, S. Ceria, G. Cornuéjols (1996). Mixed 0-1 programming by lift-and-project in a branch-and-cut framework. *Management Sci.* 42, 1229–1246.

Disjunctive cuts are used to solve mixed 0-1 programs in a finitely-convergent procedure. As in the Gomory's approach, no structural information on the problem is required. Separation amounts to the solution of a linear program; this allows the identification of the deepest violated cut in the family—a property not shared by Gomory's procedure. Computational results analyze the performance of these cuts in a branch-and-cut framework.

E. Balas, S. Ceria, G. Cornuéjols, G. Pataki (1994). *Polyhedral methods for the maximum clique problem*, Management Science Research Report 602, Graduate School of Industrial Administration, Carnegie-Mellon University, Pittsburgh.
C. Helmberg (1994). *An interior point method for semidefinite programming and max-cut bounds*, Doctoral thesis, Institut für Mathematik, Technische Universität Graz.
C. Helmberg, S. Poljak, F. Rendl, H. Wolkowicz (1995). Combining semidefinite and polyhedral relaxations for integer programs. E. Balas, J. Clausen (eds.). *Proc. of the Fourth IPCO Conf.*, Lecture Notes Comput. Sci. 920, Copenhagen. Springer, Berlin, 124–134.
C. Helmberg, F. Rendl (1995). *Solving quadratic (0,1) problems by semidefinite programming and cutting planes*, Preprint, Institut für Mathematik, Technische Universität Graz.
C. Helmberg, F. Rendl, R. Weismantel (1996). Quadratic knapsack relaxations using cutting planes and semidefinite programming. W.H. Cunningham, S.T. McCormick, M. Queyranne (eds.). *Proc. of the Fifth IPCO Conf.*, Lecture Notes Comput. Sci. 1084, Vancouver. Springer, Berlin, 175–189.

These papers describe the integration of semidefinite programming with polyhedral techniques within the branch-and-cut framework. Some promising results are reported.

E. Balas, S. Ceria, G. Cornuéjols, N. Natraj (1996). Gomory cuts revisited. *Oper. Res. Lett.* 19, 1–9.

More than 30 years after their discovery, Gomory cuts are used within a modern branch-and-cut framework based on robust LP solvers. Unlike the original proposal, cuts are generated in rounds rather than one at a time. Moreover, Gomory cuts are shown to be globally valid along the branching tree under appropriate conditions.

3 Applications

3.1 General (mixed) integer programs

M. Guignard, K. Spielberg (1981). Logical reduction methods in zero-one programming. *Oper. Res.* 29, 49–74.
[Hoffman and Padberg 1991] (see §2)
M.W.P. Savelsbergh (1994). Preprocessing and probing techniques for mixed integer programming. *ORSA J. Comput.* 6, 445–454.
L. Escudero, S. Martello, P. Toth (1995). A framework for tightening 0-1 programs based on extensions of pure 0-1 KP and SS problems. E. Balas, J. Clausen (eds.). *Proc. of the Fourth IPCO Conf.*, Lecture Notes Comput. Sci. 920, Copenhagen. Springer, Berlin, 110–123.

These papers address the possibility of improving the LP representation of a given MILP by fixing the value of some variables and/or strengthening the inequality coefficients. This is typically obtained by a sort of what-if analysis in which one explores the consequences of tentatively setting a variable to a given value.

T.J. Van Roy, L.A. Wolsey (1987). Solving mixed integer programming problems using automatic reformulation. *Oper. Res.* 35, 45–57.
M.X. Goemans (1989). Valid inequalities and separation for mixed 0-1 constraints with variable upper bounds. *Oper. Res. Lett.* 8, 315–322.

0-1 MILP's are solved using strong valid inequalities, which are automatically generated so as to obtain a tighter reformulation of the problem. Cuts are derived from pure 0-1 and fixed-charge network substructures associated with single constraints and variable upper bounds.

J.E. Mitchell, M.J. Todd (1992). Solving combinatorial optimization problems using Karmarkar's algorithm. *Math. Program.* 56, 245–284.
J.E. Mitchell, B. Borchers (1993). *Solving real-world linear ordering problems using a primal-dual interior point cutting plane method*, Report 207, Rensselaer Polytechnic Institute.

The authors describe the use of Karmarkar's algorithm for linear programming within a cutting-plane framework. Computational results are reported for applications to the matching problem (first paper) and to the linear ordering problem (second paper).

E.A. Boyd (1993). Solving integer programs by Fenchel cutting planes and preprocessing. G. Rinaldi, L.A. Wolsey (eds.). *Proc. of the Third IPCO Conf.*, Erice. CIACO, Louvain-la-Neuve, 209–220.
E.A. Boyd (1994). Fenchel cutting planes for integer programs. *Oper. Res.* 42, 53–64.

The feasible set of an ILP is enlarged so as to obtain a relaxation which can be solved for every linear objective function by an effective (but not necessarily polynomial) algorithm. Deep cuts for the relaxation, and hence for the original problem, are then obtained by solving a Lagrangian dual problem.

Z. Gu, G.L. Nemhauser, M.W.P. Savelsbergh (1994). *Inequalities for 0-1 programs:*

computation, Report COC-94-09, Georgia Institute of Technology, Atlanta.
Z. Gu, G.L. Nemhauser, M.W.P. Savelsbergh (1995). Sequence independent lifting of cover inequalities. E. Balas, J. Clausen (eds.). *Proc. of the Fourth IPCO Conf.*, Lecture Notes Comput. Sci. 920, Copenhagen. Springer, Berlin, 452–461.

The seminal work of Crowder, Johnson and Padberg shows that knapsack-specific constraints can be very useful in tightening the formulation of a generic 0-1 ILP. These papers address a prominent class of knapsack constraints—the cover inequalities. Several lifting procedures for strengthening the basic cover inequality are analyzed.

S. Ceria, G. Cornuéjols, M. Dawande (1995). Combining and strengthening Gomory cuts. E. Balas, J. Clausen (eds.). *Proc. of the Fourth IPCO Conf.*, Lecture Notes Comput. Sci. 920, Copenhagen. Springer, Berlin, 438–451.
S. Ceria, C. Cordier, H. Marchand, L.A. Wolsey (1995). *Cutting planes for integer programs with general integer variables*, Technical report, CORE, Université Catholique de Louvain.
[Balas, Ceria, Cornuéjols and Natraj 1996] (see §2)

These papers use Gomory cuts within a branch-and-cut scheme. Computational results show that a general-purpose code based on Gomory cuts can be competitive with special-purpose codes, on some classes of problems.

3.2 Routing

H.P. Crowder, M.W. Padberg (1980). Solving large-scale symmetric traveling salesman problems to optimality. *Management Sci.* 26, 459–509.
M. Grötschel (1980). On the symmetric traveling salesman problem: solution of a 120 city problem. *Math. Program. Study* 12, 61–77.
M.W. Padberg, S. Hong (1980). On the symmetric traveling salesman problem. *Math. Program. Study* 12, 78–107.
[Grötschel and Holland 1991] (see §2)

These papers, among others, address the solution of the traveling salesman problem by using problem-specific cuts. All of them allow cut generation only at the root node. Solution of instances with up to 1000 nodes is reported in the last paper.

B. Fleischmann (1985). A cutting plane procedure for the travelling salesman problem on road networks. *European J. Oper. Res.* 21, 307–317.
M.W. Padberg, G. Rinaldi (1989). A branch and cut approach to a traveling salesman problem with side constraints. *Management Sci.* 35, 1393–1412.
M. Fischetti, J.J. Salazar, P. Toth (1994). *A branch-and-cut algorithm for the symmetric generalized traveling salesman problem*, Research Report OR-94-2, DEIS, University of Bologna. (To appear in *Oper. Res.*).
[Fischetti, Salazar and Toth 1995] (see §2)
M. Gendreau, G. Laporte, F. Semet (1995). *The Covering Tour Problem*, Publication CRT-95-8, Centre de Recherche sur le Transports, Université de Montréal.

Some non-spanning versions of the traveling salesman problem are considered, in which the tour is not required to cover all the vertices. Specific cuts are described along with the corresponding separation procedures. Instances with several hundred nodes are solved within reasonable computing time.

J.R. Araque (1989). *Solution of a 48-city vehicle routing problem by branch-and-cut*, Working paper, Department of Applied Mathematics and Statistics, State University of New York, Stony Brook.
G. Cornuéjols, F. Harche (1993). Polyhedral study of the capacitated vehicle routing problem. *Math. Program.* 60, 21–52.
J.R. Araque, G. Kudva, T.L. Morin, J.F. Pekny (1994). A branch-and-cut algorithm for vehicle routing problems. *Ann. Oper. Res.* 50, 37–59.
P. Augerat, J.M. Belenguer, E. Benavent, A. Corberán, D. Naddef, G. Rinaldi (1995). *Computational results with a branch-and-cut code for the capacitated vehicle routing problem*, Rapport de recherche, ARTEMIS–IMAG, Grenoble.
N.R. Achuthan, L. Caccetta, S.P. Hill (1996). A new subtour elimination constraint for the vehicle routing problem. *European J. Oper. Res.* 91, 573–586.

The basic vehicle routing problem is attacked by means of branch-and-cut algorithms based on polyhedral cuts. Difficult instances with up to 135 nodes are solved to optimality by the Augerat et al. approach.

M.W. Padberg, G. Rinaldi (1990). Facet identification for the traveling salesman polytope. *Math. Program.* 47, 219–257.
[Clochard and Naddef 1993] (see §2)
[Jünger, Reinelt and Thienel 1994] (see §2)
[Applegate, Bixby, Chvátal and Cook 1995] (see §2)

A detailed description of effective separation procedures for the traveling salesman problem, to be used within a branch-and-cut framework. The algorithm of Applegate et al. solved to proven optimality the so far largest traveling salesman problem instance, involving more than 7397 nodes.

M. Grötschel, Z. Win (1992). A cutting plane algorithm for the windy postman problem. *Math. Program.* 55, 339–358.
A. Corberán, J.M. Sanchis (1994). A polyhedral approach to the rural postman problem. *European J. Oper. Res.* 79, 95–114.
Y. Nobert, J.-C. Picard (1996). An optimal algorithm for the mixed Chinese postman problem. *Networks* 27, 95–108.

Branch-and-cut algorithms for two different NP-hard variants of the Chinese postman problem are presented. Optimal solutions for instances with up to 26 nodes and 489 edges (first paper), 50 nodes and 184 edges (second paper), and 225 nodes, 341 arcs and 633 edges (third paper) have been obtained.

G. Laporte, F. Louveaux, H. Mercure (1992). The vehicle routing problem with stochastic travel times. *Transportation Sci.* 26, 161–170.
G. Laporte, F. Louveaux, H. Mercure (1994). A priori optimization of the probabilistic traveling salesman problem. *Oper. Res.* 42, 543–549.

Two branch-and-cut algorithms for the vehicle routing problem with stochastic service and travel times (first paper) and for a traveling salesman problem in which each vertex has a given probability of being present (second paper). The largest instances whose solution is reported in the two papers involve 20 and 50 nodes, respectively.

N. Ascheuer, L.F. Escudero, M. Grötschel, M. Stoer (1993). A cutting plane approach to the sequential ordering problem (with applications to job scheduling in manufacturing). *SIAM J. Optim.* 3, 25–42.
N. Ascheuer (1995). *Hamiltonian path problems in the on-line optimization of flexible manufacturing systems*, Doctoral thesis, Konrad-Zuse-Zentrum für Informationstechnik, Berlin.

The asymmetric traveling salesman problem with precedence and time-window constraints is approached by branch-and-cut. For the precedence-constrained case, instances with up to 176 nodes have been solved within short computing time. The case with time windows is instead much harder for this approach.

M. Fischetti, P. Toth (1994). *A polyhedral approach to the asymmetric traveling salesman problem*, Research Report OR-94-3, DEIS, University of Bologna. (To appear in *Management Sci.*).

The asymmetric version of the traveling salesman problem is considered. New separation procedures and a modified variable pricing scheme are described and analyzed computationally. Solution of hard instances with up to 171 nodes is reported.

M. Jünger, P. Störmer (1995). *Solving large-scale traveling salesman problems with parallel branch-and-cut*, Report 95.191, Universität zu Köln.

A sophisticated parallelization of a branch-and-cut code for the traveling salesman problem is described.

3.3 Scheduling

D. Applegate, W. Cook (1991). A computational study of the job-shop scheduling problem. *ORSA J. Comput.* 3, 149–156.

Lower bounds for the job shop scheduling problem are obtained by solving an LP relaxation amended by additional constraints. For 8 out of the 10 problems considered, the cutting-plane bounds were superior to those obtainable by standard methods, but required an excessively large computational effort.

M. Queyranne, Y. Wang (1991). *A cutting plane procedure for precedence-constrained single machine scheduling*, Technical Report, Faculty of Commerce, University of British Columbia.

A single machine scheduling problem with precedence constraints is addressed, in which the objective is to minimize the total weighted completion time. A cutting-plane algorithm based on a compact linear formulation with one variable per job is developed. Solution of instances with up to 160 jobs require a very short time.

J.B. Lasserre, M. Queyranne (1992). Generic scheduling polyhedra and a new mixed-integer formulation for single-machine scheduling. E. Balas, G. Cornuéjols, R. Kannan (eds.). *Proc. of the Second IPCO Conf.*, Pittsburgh. Carnegie-Mellon University Press, 136–149.
G.L. Nemhauser, M.W.P. Savelsbergh (1992). A cutting plane algorithm for the single machine scheduling problem with release times. M. Akgul, H. Hamacher, S. Tufecki

(eds.). *Combinatorial Optimization: New Frontiers in Theory and Practice*, NATO ASI Series F: Computer and System Sciences 82. Springer, 63–84.

The basic formulation here involves continuous start-time variables and 0-1 sequence-determining variables. Polyhedral cuts and separation procedures are presented. A computational analysis for problems with 10-20 jobs is given.

J.P. de Sousa, L.A. Wolsey (1993). A time-indexed formulation of non-preemptive scheduling. *Math. Program.* 54, 353–367.
J.M. van den Akker, C.A.J. Hurkens, M.W.P. Savelsbergh (1995). *A time-indexed formulation for single machine scheduling problems: branch-and-cut*, Report COC-95-02, Georgia Institute of Technology, Atlanta.
Y. Crama, F.C.R. Spieksma (1996). Scheduling jobs of equal length: complexity, facets and computational results. *Math. Program.* 72, 207–227.

In order to obtain tighter LP relaxations, the authors model the single machine scheduling problem in a high dimensional space. Decision variables are of the type x_{it}, where $x_{it} = 1$ if the i-th job starts at time t, $= 0$ otherwise. Computational experience on problems with 20-30 jobs is reported.

S. Chaudhuri, R.A. Walker, J.E. Mitchell (1994). Analyzing and exploiting the structure of the constraints in the ILP approach to the scheduling problem. *IEEE Trans. VLSI Sys.* 2, 456–471.

The paper addresses ILP formulations of a family of scheduling problems, discussing in particular how to model assignment, timing, and resource constraints. The formulations proposed are enhanced by means of valid inequalities, derived from the problem structure. Computational results for different objective functions are given.

3.4 Graph partitioning

M. Grötschel, M. Jünger, G. Reinelt (1987). Calculating exact ground states of spin glasses: a polyhedral approach. J. L. van Hemmen, I. Morgenstern (eds.). *Proc. of the Heilderbrg Colloquium on "Glassy Dynamics"*, Lecture Notes Physics 275, Heidelberg. Springer, Berlin, 325–353.
F. Barahona, M. Grötschel, M. Jünger, G. Reinelt (1988). An application of combinatorial optimization to statistical physics and circuits layout design. *Oper. Res.* 36, 493–513.
F. Barahona, M. Jünger, G. Reinelt (1989). Experiments in quadratic 0-1 programming. *Math. Program.* 44, 127–137.
F. Barahona (1994). Ground state magnetization of Ising spin glasses. *Physical Review B* 49, 12864–12867.
C. De Simone, G. Rinaldi (1994). A cutting plane algorithm for the max-cut problem. *Optimization Methods and Software* 3, 195–214.
C. De Simone, M. Diehl, M. Jünger, P. Mutzel, G. Rinaldi (1995). Exact ground states of Ising spin glasses: new experimental results with a branch-and-cut algorithm. *J. Statistical Physics* 80, 487–496.
C. De Simone, M. Diehl, M. Jünger, P. Mutzel, G. Rinaldi (1996). *Exact ground states of $2D \pm J$ Ising spin glasses*, Report 96.217, Universität zu Köln. (To appear in J.

Statistical Physics).

The max cut problem is a basic graph partitioning problem. It finds important applications in physics as a model for ground states of spin glasses with exterior magnetic field. The listed papers, among others, describe successful polyhedral approaches to this problem. State of the art codes can solve instances in dense graphs with up to 25-50 nodes. For sparse graphs, instead, the problem is much easier, in that instances with several thousand nodes can be solved in moderate computation time.

M. Grötschel, Y. Wakabayashi (1989). A cutting plane algorithm for a clustering problem. *Math. Program.* 45, 59–96.

The clustering problem consists of finding a partition of the nodes of a complete graph into cliques, with the objective of minimizing the sum of the weights of the edges inside the cliques. A cutting-plane algorithm based on exact and heuristic separation procedures for facet-defining cuts is described. Exact solution of instances with up to 158 nodes is reported. In all cases, branching is not required.

G.L. Nemhauser, S. Park (1991). A polyhedral approach to edge coloring. *Oper. Res. Lett.* 10, 315–322.

The edge coloring problem is viewed as the problem of covering the edges of a graph by matchings. This problem is solved by using column generation and cutting planes. The computational section addresses randomly-generated 3-regular graphs ranging from 20 to 60 nodes.

S. Chopra, M.R. Rao (1991). On the multiway cut polyhedron. *Networks* 21, 51–89.
S. Chopra, J.H. Owen (1995). *A note on formulations for the A-partition problem on hypergraphs*, Report TR-95-67, Department of Industrial Engineering and Management Sciences, Northwestern University.
S. Chopra, J.H. Owen (1996). Extended formulations for the A-cut problem. *Math. Program.* 73, 7–30.

The A-cut problem amounts to finding a minimum-weight subset C of edges in a graph such that no two nodes of a given subset A are in the same component of the graph obtained by removing the edges in C. The first paper considers a formulation with edge variables only, while extended formulations with edge and node variables are introduced and analyzed in the last paper, which reports exact solution of instances with up to 320 nodes and $|A| = 11$. The second paper discusses an analogous problem defined on hypergraphs. Different formulation are discussed and compared both theoretically and computationally.

M. Grötschel, A. Martin, R. Weismantel (1993). Routing in grid graphs by cutting planes. G. Rinaldi, L.A. Wolsey (eds.). *Proc. of the Third IPCO Conf.*, Erice. CIACO, 447–461.
M. Grötschel, A. Martin, R. Weismantel (1996). Packing Steiner trees: a cutting plane algorithm and computational results. *Math. Program.* 72, 125–145.

A problem arising in VLSI design is modeled as the problem of packing a given number of Steiner trees in a graph with edge capacities, with the objective of minimizing the sum of the weights of the trees. The papers describe sophisticated separation procedures for important families of cuts.

E.L. Johnson, A. Mehrotra, G.L. Nemhauser (1993). Min-cut clustering. *Math. Program. (B)* 62, 133–151.

Min cut clustering amounts to finding a partition of the node set of a graph into k subsets, so as to minimize the overall cost of the edges joining nodes in different sets of the partition. The authors propose a column generation approach, in which columns are generated by solving an NP-hard optimization problem by branch-and-cut.

L. Brunetta, M. Conforti, G. Rinaldi (1994). *A branch-and-cut algorithm for the equicut problem*, Preprint 6/94, Dipartimento di Matematica Pura e Applicata, University of Padova. (To appear in *Math. Program. B*).

The equicut problem is a variant of the max cut problem in which the two shores of the cut must contain the same number of nodes. A branch-and-cut algorithm based on several classes of valid inequalities is proposed and analyzed computationally on a library containing 200 instances. The authors report exact solutions for dense graphs with up to 50 nodes and for sparse graphs with about 2500 edges.

C.E. Ferreira, A. Martin, C. de Sousa, R. Weismantel, L.A. Wolsey (1994). *The node capacitated graph partitioning problem: a computational study*, Preprint SC 94-17, Konrad-Zuse-Zentrum für Informationstechnik, Berlin.

The problem addressed in this paper is a variant of the min cut clustering problem, in which the sum of the node weights in each subset cannot exceed a given bound. The problem is modeled by means of edge and node variables, and solved through branch-and-cut.

H. Lee, G.L. Nemhauser, Y. Wang (1996). Maximizing a submodular function by integer programming: polyhedral results for the quadratic case. *European J. Oper. Res.* 94, 154–166.

The paper considers the quadratic cost partition problem, a generalization of the max-cut problem, which is formulated as the problem of maximizing a submodular function. Several classes of valid inequalities are derived and used within a branch-and-cut algorithm.

3.5 Set packing, partitioning and covering

G.L. Nemhauser, G. Sigismondi (1992). A strong cutting plane/branch-and-bound algorithm for node packing. *J. Oper. Res. Soc.* 43, 443–457.

The node packing problem is solved by means of a cutting-plane algorithm based on exact separation and lifting of odd-hole inequalities, and heuristic separation of clique inequalities. Solution of low-density instances with up to 120 nodes, and medium-density instances with up to 60 nodes, is reported.

P. Nobili, A. Sassano (1992). A separation routine for the set covering polytope. E. Balas, G. Cornuéjols, R. Kannan (eds.). *Proc. of the Second IPCO Conf.*, Pittsburgh. Carnegie-Mellon University Press, 201–219.

The paper describes a sophisticated procedure for the separation of rank inequalities for the set covering polytope. A branch-and-cut algorithm based on this procedure is capable of solving set covering instances with up to 400 rows and columns (after an initial reduction).

K.L. Hoffman, M.W. Padberg (1993). Solving airline crew scheduling problems by branch and cut. *Management Sci.* 39, 657–682.

A detailed description of the implementation of a branch-and-cut algorithm for the solution of set partitioning problems arising in airline crew scheduling. Solution of real-world problems with hundreds of rows and hundreds of thousands of columns is reported.

L. Qi, E. Balas, G. Gwan (1994). A new facet class and a polyhedral method for the three-index assignment problem. D.-Z. Du, J. Sun (eds.). *Advances in Optimization and Approximation*, Kluwer Academic Publishers, Dordrecht.

The paper describes a polyhedral approach to the three-index assignment problem, based on linear-time separation procedures. The largest instance solved has order 26.

3.6 Network design

M. Grötschel, C.L. Monma, M. Stoer (1992). Computational results with a cutting plane algorithm for designing communication networks with low-connectivity constraints. *Oper. Res.* 40, 309–330.

M. Grötschel, C.L. Monma, M. Stoer (1995). Polyhedral and computational investigations for designing communication networks with high survivability requirements. *Oper. Res.* 43, 1012–1024.

M. Grötschel, C.L. Monma, M. Stoer (1995). Design of survivable networks. M.O. Ball, T.L. Magnanti, C.L. Monma, G.L. Nemhauser (eds.). *Network Routing*, Handbooks in Operations Research and Management Science, 8, North-Holland, Amsterdam.

The problem addressed was originated from fiber optic communication networks design. Given an edge-weighted undirected graph, the problem calls for finding a minimum-weight subset of edges which induces a graph satisfying some edge and node connectivity requirements. In particular, the first and second paper address the case of low and high connectivity requirements, respectively. Branch-and-cut codes are used to solve to optimality real-world instances.

D. Bienstock, O. Günlük (1995). Computational experience with a difficult mixed-integer multicommodity flow problem. *Math. Program.* 68, 213–237.

The problem considered here arises in the design of lightwave networks. Given a demand matrix and an integer p, the problem amounts to designing a network such that each node has in- and out-degree equal to p, and the flow between each pair of nodes (excluding a source and a sink) is equal to the corresponding entry in the demand matrix. The objective is to minimize the maximum total flow on single edges. A cutting plane algorithm using several classes of valid inequalities is proposed, which allows computation of tight lower bounds on the optimal solution value.

A. Balakrishnan, T.L. Magnanti, R.T. Wong (1995). A decomposition algorithm for local access telecommunications network expansion planning. *Oper. Res.* 43, 58–76.

T.L. Magnanti, P. Mirchandani, R. Vachani (1995). Modeling and solving the two-facility capacitated network loading problem. *Oper. Res.* 43, 142–157.
F. Barahona (1995). *Network design using cut inequalities*, Working paper, IBM T.J. Watson Research Center. (To appear in *SIAM J. Optim.*).
D. Bienstock, S. Chopra, O. Günlük, C.-Y. Tsai (1995). *Minimum capacity installation for multicommodity network flows*, Working paper, Department of IEOR, Columbia University, New York. (To appear in *Math. Program.*).
D. Bienstock, O. Günlük (1995). *Capacitated network design – polyhedral structure and computation*, Working paper, Department of IEOR, Columbia University, New York. (To appear in *INFORMS J. Comput.*).
G. Dahl, M. Stoer (1995). *A cutting plane algorithm for multicommodity survivable network design problems*, Preprint 3, Department of Informatics, University of Oslo.

These papers present cutting plane algorithms for the following problem, arising in telecommunications: given point-to-point traffic demands in a network along with specified survivability requirements and a discrete cost/capacity function for each link, find minimum-cost capacity expansions satisfying the given demands. Each paper addresses a special case of this general problem, and proposes classes of valid cuts. Computational results on real-world instances are given.

L. Brunetta, M. Conforti, M. Fischetti (1995). *A polyhedral approach to a multi-commodity flow problem*, Preprint 1/95, Dipartimento di Matematica Pura e Applicata, University of Padova.

The polyhedral structure of a multi-commodity flow problem is studied, and new classes of inequalities as well as facet composition procedures are introduced. Computational results and a comparison between two different formulations are given.

3.7 Location

K. Aardal (1994). *Capacitated facility location: separation algorithms and computational experience*, CentER Discussion Paper 9480, Tilburg University, Tilburg. (To appear in *Math. Program.*).

The paper describes a polyhedral approach to the capacitated facility location problem, based on exact and heuristic separation of various classes of inequalities. Computational results on instances involving up to 100 clients and 75 facilities are given.

G. Laporte, F. Louveaux, L.V. Hamme (1994). Exact solution to a location problem with stochastic demands. *Transportation Sci.* 28, 95–103.

A capacitated facility location problem with stochastic customer demands is solved by branch-and-cut. Exact solution of instances with up to 40 clients and 10 facilities is reported.

A. Caprara, M. Fischetti, D. Maio (1995). Exact and approximate algorithms for the index selection problem in physical database design. *IEEE Trans. Knowledge and Data Eng.* 7, 955–967.
A. Caprara, J.J. Salazar (1995). *A branch-and-cut algorithm for the index selection problem*, Research Report OR-95-9, DEIS, University of Bologna.

The index selection problem is a variant of the uncapacitated facility location problem arising in the physical design of databases. The papers describe branch-and-cut algorithms based on knapsack cover inequalities (first paper) and generalized odd hole inequalities (second paper).

3.8 Miscellaneous

I. Barany, T.J. Van Roy, L.A. Wolsey (1984). Strong formulations for multi-item capacitated lot sizing. *Management Sci.* 30, 1255–1261.
G.D. Eppen, R. Kipp Martin (1987). Solving multi-item capacitated lot-sizing problems using variable redefinition. *Oper. Res.* 35, 832–848.
Y. Pochet (1988). Valid inequalities and separation for capacitated economic lot sizing. *Oper. Res. Lett.* 7, 109–115.
Y. Pochet, L.A. Wolsey (1988). Lot-size models with backlogging: strong reformulations and cutting planes. *Math. Program.* 40, 317–335.
J.M.Y. Leung, T.L. Magnanti, R. Vachani (1989). Facets and algorithms for capacitated lot sizing. *Math. Program. (B)* 45, 331–359.
T.L. Magnanti, R. Vachani (1990). A strong cutting plane algorithm for production scheduling with changeover costs. *Oper. Res.* 38, 456–473.
M. Constantino (1996). A cutting plane approach to capacitated lot sizing with start-up costs. *Math. Program.* 75, 353–376.

The papers address different versions of the lot-sizing problem, all of which are approached polyhedrally. The definition of an initial strong MILP formulation and the use of facet-defining inequalities in a cutting plane approach yields optimal (or near-optimal) solutions in reasonable computing time.

M. Grötschel, O. Holland (1985). Solving matching problems with linear programming. *Math. Program.* 33, 243–259.

The minimum-weight perfect matching problem is solved by means of a cutting-plane algorithm based on the sparsification of the original (complete) graph and on the extensive use of heuristic separation procedures. The algorithm is shown to be competitive with the fastest combinatorial matching algorithms.

M. Grötschel, M. Jünger, G. Reinelt (1987). A cutting plane algorithm for minimum perfect 2-matchings. *Computing* 39, 327–344.

The paper describes a cutting plane algorithm for the minimum-weight 2-matching problem, a well-known polynomially-solvable relaxation of the traveling salesman problem.

S. Chopra, E.R. Gorres, M.R. Rao (1992). Solving the Steiner tree problem on a graph using branch and cut. *ORSA J. Comput.* 3, 149–156.

A branch-and-cut algorithm for the Steiner tree problem is described, with computational results on a wide set of instances. The largest instances solved involve 500 nodes for the complete graph case, and 2500 nodes and 250,000 edges for the sparse graph case.

C.E. Ferreira, A. Martin, R. Weismantel (1993). *Solving multiple knapsack problems by cutting planes*, Preprint SC 93-4, Konrad-Zuse-Zentrum für Informationstechnik, Berlin. (To appear in *SIAM J. Optim.*).

The paper contains a polyhedral study of the multiple knapsack problem, along with a discussion on separation procedures and on the implementation of a cutting-plane procedure. Computational experience with instances arising from main-frame processor design and electronic circuit layout is illustrated.

G. Dijkhuizen, U. Faigle (1993). A cutting-plane approach to the edge-weighted maximal clique problem. *European J. Oper. Res.* 69, 121–130.
K. Park, K. Lee, S. Park (1996). An extended formulation approach to the edge-weighted maximal clique problem. *European J. Oper. Res.* 95, 671–682.

These papers address the problem of finding a clique with at most b nodes in an edge-weighted graph, such that the sum of the weights of the edges in the clique is maximized. Different ILP formulations are analyzed and tested within a cutting-plane approach, capable of finding optimal solutions for instances with up to 30 nodes.

M. Jünger, P. Mutzel (1993). *Maximum planar subgraph and nice embeddings: practical layout tools*, Report 93.145, Universität zu Köln. (To appear in *Algorithmica*).
M. Jünger, P. Mutzel (1993). Solving the maximum weight planar subgraph. G. Rinaldi, L.A. Wolsey (eds.). *Proc. of the Third IPCO Conf.*, Erice. CIACO, Louvain-la-Neuve, 479–492.
M. Jünger, P. Mutzel (1995). The polyhedral approach to the maximum planar subgraph: new chances for related problems. R. Tamassia, I.G. Tollis (eds.). *Proc. of the DIMACS International Workshop on Graph Drawing '94*, Lecture Notes Comput. Sci. 894, Princeton. Springer, Berlin, 119–130.
P. Mutzel (1995). A polyhedral approach to planar augmentation and related problems. P. Spirakis (ed.). *Proc. of the Third Annual European Symposium on Algorithms*, Lecture Notes Comput. Sci. 979, Corfú. Springer, Berlin, 494–507.

The problem of finding a maximum-weight planar subgraph of a given edge weighted graph is addressed in the first two papers. Facet defining inequalities are described and used within a branch-and-cut algorithm, capable of solving instances with up to 100 nodes. The last two papers discuss a polyhedral approach for solving planar subgraph problems with additional constraints and planar augmentation problems, respectively.

K. Aardal, A. Hippolito, C.P.M. van Hoesel, B. Jansen (1995). *A branch-and-cut algorithm for the frequency allocation problem*, METEOR Research Memoranda RM/96/011, Limburg University, Maastricht.
M. Fischetti, C. Lepschy, G. Minerva, G. Romanin Jacur, E. Toto (1996). *Frequency assignment in mobile radio systems using branch-and-cut techniques*, Working paper 96/33, DEI, University of Padova.

Frequency assignment problems occur when a network of radio links has to be established. The objective is to assign an operating frequency to each link, so as to satisfy interference restrictions while minimizing the number of frequencies used. The first paper describes a branch-and-cut algorithm based on effective preprocessing techniques and on the use of suitably-defined clique inequalities. The second paper addresses the case of cumulative noise due to several interferences, and gives an ILP model whose solution is achieved through branch-and-cut. Both papers report the exact solution of real-world instances.

M. Jünger, P. Mutzel (1995). Exact and heuristic algorithms for 2-layer straightline crossing minimization. F. Brandenburg (ed.). *Proc. of the International Workshop on Graph Drawing '95*, Lecture Notes Comput. Sci., Princeton. Springer, Berlin. (To appear).

The problem considered arises in automatic graph drawing, and requires the alignment of the two shores of a bipartite graph on two parallel straight lines such that the number of crossings between the edges—drawn as straight lines connecting their endnodes—is minimized. A branch-and-cut algorithm is described for the case in which the alignment of one shore is fixed. The algorithm is capable of solving within short computing time instances on dense and sparse graphs with up to 20 and 100 nodes per shore, respectively.

G. Gallo, C. Gentile, D. Pretolani, G. Rago (1995). *Max Horn SAT and the minimum cut problem in directed hypergraphs*, Technical report TR-15/95, Dipartimento di Informatica, Università di Pisa.

The paper discusses different ILP formulations of the max Horn satisfiability problem. Among these, a set covering formulation with an exponential number of constraints is solved by branch-and-cut, using a heuristic procedure to solve the corresponding (NP-complete) separation problem.

M. Fischetti, J.J. Salazar (1995). *Models and algorithms for the cell suppression problem in statistical disclosure control*, Working paper 95-13, DEIOC, Universidad de La Laguna.

Given are a statistical 2-dimensional table which has to be published along with the marginal values of each row and column, and a set of entries in the table corresponding to private information. The problem amounts to finding a minimum-weight set of entries to be deleted from the table before publication, so as to ensure that the private information cannot be inferred from the published entries and the marginal values. Tables with up to 500 rows and columns are solved by a branch-and-cut algorithm based on several classes of cuts.

J. Cheriyan, W.H. Cunningham, L. Tunçel, Y. Wang (1996). A linear programming and rounding approach to MAX 2-SAT. D.S. Johnson, M.A. Trick (eds.). *Second DIMACS Challenge: Cliques, Coloring and Satisfiability. DIMACS Series in Discrete Mathematics and Theoretical Computer Science*, AMS. (To appear).

The paper considers an ILP formulation of the max 2-satisfiability problem, which is strengthened by means of cycle and wheel inequalities. Efficient separation routines are described. The optimal solution of the LP relaxation is rounded so as to obtain near-optimal feasible solutions.

A. Billionnet, F. Calmels (1996). Linear programming for the 0-1 quadratic knapsack problem. *European J. Oper. Res.* 92, 310–325.

[Helmberg, Rendl and Weismantel 1996] (see §2)

The 0-1 quadratic knapsack problem is formulated as an integer linear program and solved by a cutting-plane procedure. Solution of instances with up to 61 binary variables (in the original nonlinear formulation) is reported.

D. Bienstock (1996). Computational study of a family of mixed-integer quadratic programming problems. *Math. Program.* 74, 121–140.

The paper describes a branch-and-cut algorithm for the solution of quadratic programming problems in which an upper bound on the number of nonzero variables is imposed. Computational experience with the use of Gomory, intersection, and disjunctive cuts is reported.

M. Fischetti, D. Vigo (1996). A branch-and-cut algorithm for the resource-constrained minimum-weight arborescence problem. *Networks* 29, 55–67.

An NP-hard variant of the shortest spanning arborescence problem is considered, where for each vertex the total weight of the selected out-going arcs cannot exceed a given vertex capacity. The problem is solved by branch-and-cut, using knapsack cover and subtour elimination constraints. A specialized pricing procedure is also proposed.

M. Funke, G. Reinelt (1996). A polyhedral approach to the feedback vertex set problem. W.H. Cunningham, S.T. McCormick, M. Queyranne (eds.). *Proc. of the Fifth IPCO Conf.*, Lecture Notes Comput. Sci. 1084, Vancouver. Springer, Berlin, 445–459.

The feedback vertex set problem calls for a minimum weight acyclic subgraph of a directed graph. The authors consider valid (facet-defining) inequalities associated with paths of length three, triangles, and cycles, and use them in a branch-and-cut algorithm, reporting optimal solutions for instances with up to 35 nodes.

T. Christof, M. Jünger, J. Kececioglu, P. Mutzel, G. Reinelt (1997). A branch-and-cut approach to physical mapping with end-probes. S. Istrail, P. Pevzner, M. Waterman (eds.). *Proc. of the First Ann. Conf. on Computational Molecular Biology*, Santa Fe. ACM Press, New York, 84–92.
K. Reinert, H.-P. Lenhof, P. Mutzel, K. Mehlhorn, J. Kececioglu (1997). A branch-and-cut algorithm for multiple sequence alignment. S. Istrail, P. Pevzner, M. Waterman (eds.). *Proc. of the First Ann. Conf. on Computational Molecular Biology*, Santa Fe. ACM Press, New York, 241–250.

These papers use branch-and-cut to attack NP-hard optimization problems in computational molecular biology. The first paper deals with a variant of the linear ordering problem arising in the construction of physical maps of chromosomes. The second one considers the alignment of a set of strings, a problem of interest in DNA sequence assembly.

3.9 General purpose software

Most papers on branch-and-cut describe software implementing the proposed algorithms. Moreover, some commercial LP solvers give branch-and-cut as an additional feature. The following packages have the advantage of providing the user with an "open" branch-and-cut framework.

G.L. Nemhauser, M.W.P. Savelsbergh, G. Sigismondi (1994). MINTO, a mixed integer optimizer. *Oper. Res. Lett.* 15, 47–58.

MINTO is a general-purpose branch-and-cut code for MILP's that automatically provides constraint classification, preprocessing, primal heuristics and constraint generation. A basic feature is that the user can enrich the algorithm by adding his own special-purpose application routines.

S. Thienel (1995). *ABACUS – A Branch-And-Cut System*, Doctoral thesis, Universität zu Köln.

Along the same line as MINTO, ABACUS provides a framework for the implementation of branch-and-cut algorithms, allowing the user to concentrate on problem-specific parts such as cutting plane, column generation and primal heuristics. ABACUS is implemented in the C++ programming language, whose object-oriented features are used extensively.

Acknowledgments

Work supported by M.U.R.S.T., Italy. We thank an anonymous referee for his/her valuable suggestions, which led to an improvement of the paper organization.

5 Matroids and Submodular Functions[1]

András Frank
Eötvös University, Budapest

CONTENTS

In 1935 H. Whitney defined the notion of matroid in:
H. Whitney (1935). On the abstract properties of linear dependence. *American J. Math.* 57, 509–533.

His main motivation was to capture the fundamental properties of linear independence by setting up an axiomatization for abstract independence. From this approach the following definition is quite natural. A *matroid* M is a pair $(S, \mathcal{I})$ consisting of a finite ground-set S and a non-empty family $\mathcal{I}$ of subsets of S satisfying the following axioms.

(I1) Each subset of every member of $\mathcal{I}$ belongs to $\mathcal{I}$,
(I2) for every subset X of S, all maximal subsets of X belonging to $\mathcal{I}$ have the same cardinality $r(X)$, called the rank of X.

(A member I of $\mathcal{F}$ is called maximal in X if $I \subseteq X$ and there is no $J \in \mathcal{I}$ with $I \subset J \subseteq X$.) The members of $\mathcal{F}$ are called *independent* sets (while all other subsets of S are *dependent.*)

Since the rank-functions of two distinct matroids are distinct, by investigating the properties of r one may obtain information about the structure of the matroid. A rank-

[1]Work supported by the Hungarian National Foundation for Scientific Research Grant, OTKA T17580.
Annotated Bibliographies in Combinatorial Optimization, edited by M. Dell'Amico, F. Maffioli and S. Martello

function r is clearly non-negative, integer-valued and monotonously increasing in the sense that $r(X) \geq r(Y)$ whenever $X \supset Y$. It is also subcardinal, that is, $r(X) \leq |X|$ for every $X \subseteq S$. Moreover, it is not difficult to prove that r is *submodular,* that is,

$$r(X) + r(Y) \geq r(X \cup Y) + r(X \cap Y) \text{ for every } X, Y \subseteq S. \tag{1}$$

Conversely, Whitney proved that any set-fuction admitting these properties is the rank-function of a matroid.

Later, researchers realized that among the properties of a matroid rank-function the submodular inequality is the most important one and they started to investigate submodular set-functions which are not necessarily non-negative or monotonous or sub-cardinal. Though submodular functions were introduced earlier, supermodular functions are not less important in applications. (A set-function is called *supermodular* if the reverse inequality holds in (1), in other words, the negative of a submodular function is supermodular.)

J. Edmonds (1970). Submodular functions, matroids, and certain polyhedra. R. Guy, H. Hanani, N. Sauer, J. Schonheim (eds.). *Combinatorial Structures and their Applications*, Gordon and Breach, New York, 69–87.

This is a seminal paper including a systematic study of submodular functions. As a main contribution, Edmonds recognized that certain polyhedra, called polymatroids, can be associated with submodular functions, and then linear programming ideas may be used. He also showed that there is a one-to-one correspondence between polymatroids and polymatroid functions (a set-function having all the properties of a matroid rank-function except subcardinality).

Relying on Edmonds' ideas, the notion of polymatroids was later extended to basis polyhedra, submodular polyhedra and g-polymatroids. Another early paper on the use of submodular function is the following.

L. Lovász (1970). A generalization of König's theorem. *Acta. Math.* 21, 443–446.

This is the first place where a general framework concerning graphs and submodular functions is introduced. Later several other abstract models have been set up to bring graphs and sub- or supermodular functions together.

Most notable is the notion of submodular flows due to Edmonds and Giles (see §§3.1). Let $G = (V, E)$ be a directed graph, b a submodular function on the subsets of V, and $f : V \to \Re$, $g : V \to \Re$ two functions with $f \leq g$. A vector $x : E \to R$ is called a *submodular flow* if $f \leq x \leq g$ and $\sum(x(e) : e$ enters $Z)$ - $\sum(x(e) : e$ leaves $Z) \leq b(Z)$ for every $Z \subseteq V$. The fundamental result of Edmonds and Giles is that this system is totally dual integral (TDI).

Another successful way of abstraction has been to bring partially ordered sets and sub- or supermodular functions together. For example, lattice polyhedra, a notion due to A. Hoffman and his co-workers, belong to this category [Hoffman 1982] (see §§1.2). Another rich area where partially ordered sets and submodular functions are combined is the theory of greedoids [Korte, Lovász and Schrader 1991] (see §§1.1). A greedoid may be defined by a pair $(S, \mathcal{F})$ where $\mathcal{F}$ is a family of subsets of S, called feasible sets, satisfying the second independence axiom of matroids.

Before proceeding to a more systematic overview of the material let me mention a predecessor of this work, an annotated bibliography on submodular functions and related topics, written by E. Lawler (who passed away untimely in 1994).

E. Lawler (1985). Submodular functions and polymatroidal optimization. M. O'hEgearteigh, J.K. Lenstra, A.H.G. Rinnooy Kan (eds.). *Combinatorial Optimization: Annotated Bibliography*, John Wiley and Sons, New York, 32–39.
In this paper Lawler finishes his overview with the year 1984.

Let us take up the thread with this year. That is, we will concentrate on works which have appeared since 1984 but some important earlier papers will also be mentioned. Readers interested in recent developments of the topic are in a good position since the last dozen of years produced quite a few excellent books and survey papers.

1 Books and Surveys

1.1 Books

A. Recski (1989). *Matroid Theory and its Applications in Electric Network Theory and Static*, Springer Verlag, Berlin, and Akadémiai Kiadó, Budapest.
This is a basic textbook that can be useful for beginners and teachers of the topic, as well. Researchers will enjoy the book being a rich source of engineering applications of matroids and submodular functions. Rigidity problems and automatized recognition of line-drawings are among the applied areas discussed in Recski's book.

A more detailed discussion of these two topics can be found in the following two books, respectively.
J. Graver, B. Servatius, H. Servatius (1993). *Combinatorial Rigidity*, Graduate Studies in Mathematics 2, Amer. Math. Soc.
K. Sugihara (1986). *Machine Interpretation of Line-drawings*, MIT Press, Cambridge, Mass.

T. Ibaraki, N. Katoh (1988). *Resource Allocation Problems – Algorithmic Approaches*, The MIT Press Series in the Foundation of Computing.
A third, interesting application-oriented book that includes submodular functions. Section 8 of this book summarizes the basics of submodular functions while Section 9 is an exciting exhibition of resource allocations problems and their solutions under submodular constraints.

K. Murota (1987). *System analysis by graphs and matroids. Structural solvability and controllability*, Algorithms and Combinatorics 3, Springer-Verlag, Berlin.
One more recommendable application-guided monograph concerning matroids, graphs and linear algebra along with their applications in engineering.

More theoretically oriented readers are recommended the following works:
K. Trümper (1992). *Matroid Decomposition*, Academic Press, San Diego.
This book covers a great number of the deep results concerning matroid minors,

decompositions, and representations, including classical theorems of Tutte and of Seymour, as well as the fundamental research achievements due to the author of the book.

J.G. Oxley (1992). *Matroid Theory,* Oxford Science Publications, Oxford.
This is another highly recommendable thorough treatment concentrating on minors, decompositions and representations.

L. Lovász, A. Recski (eds.) (1985). *Matroid Theory*, North-Holland, Amsterdam.
N. White (ed.) (1986). *Theory of Matroids*, Cambridge University Press, Cambridge
N. White (ed.) (1992). *Matroid Applications,* Cambridge University Press, Cambridge
These books are collections of papers on matroids. Some of them will also be mentioned in the sequel.

M. Grötschel, L. Lovász, A. Schrijver (1988). *Geometric Algorithms and Combinatorial Optimization*, Springer Verlag, Berlin.
This monograph describes in detail general-purpose algorithms, such as the ellipsoid methods and the basis reduction algorithm, and shows several applications of these methods concerning matroids and submodular functions. Perhaps the most important of these applications is a polynomial time algorithm to minimize a submodular function. Algorithms concerning submodular flows are also exhibited.

Among the many possible generalizations of matroids we refer here to two: greedoids and oriented matroids. The books concerning these topics are:
B. Korte, L. Lovász, R. Schrader (1991). *Greedoids*, Springer-Verlag, Berlin.
A. Björner, M. Las Vergnas, B. Sturmfels, N. White, G. Ziegler (1993). *Oriented Matroids,* Cambridge University Press, Cambridge.
A. Bachem and W. Kern (1992). *Linear Programming Duality: An Introduction to Oriented Matroids,* Springer, Berlin.

Multimatroids and Δ-matroids form yet another generalization of matroids. (A. Bouchet, who developed a great part of this flourishing theory, is writing a book on the topic.) Last but not at all least comes the following indispensible monograph on submodular functions:
S. Fujishige (1991). Submodular functions and optimization. *Ann. Discr. Math.* 47.
Several models concerning submodular functions and graphs along with their relationship are analyzed. The reader may also read about the algorithms which are far-reaching extensions of classical max-flow algorithms. Some non-linear optimization problems concerning submodular functions are also discussed.

G.L. Nemhauser, L.A. Wolsey (1988). *Integer and Cominatorial Optimization*, Wiley, New York.
This book includes a chapter on the optimization aspects of matroids and submodular functions.

R. Graham, M. Grötschel, L. Lovász (eds.) (1995). *Handbook of Combinatorics*, Elsevier Science, Amsterdam.
This monumental handbook covers the whole body of combinatorics and includes

several survey papers concerning our topics (see the beginning of next subsection).

1.2 Survey papers

D.J.A. Welsh (1995). Matroid theory: fundamental concepts. R. Graham, M. Grötschel, L. Lovász (eds.). *Handbook of Combinatorics*, Elsevier Science, Amsterdam, 481–526.

This paper provides a concise introduction starting from the basic concepts and ending with an outline of the most recent developments.

P.D. Seymour (1995). Matroid minors. R. Graham, M. Grötschel, L. Lovász (eds.). *Handbook of Combinatorics*, Elsevier Science, Amsterdam, 527-550.

This paper will be unavoidable for those studying or investigating matroid minors and decompositions. Likewise, the following article will serve a similar role in the area of matroid optimization.

R.E. Bixby, W.H. Cunningham (1995). Matroid optimization and algorithms. R. Graham, M. Grötschel, L. Lovász (eds.). *Handbook of Combinatorics*, Elsevier Science, Amsterdam, 551–609.

U. Faigle (1987). Matroids in combinatorial optimization. N. White (ed.). *Combinatorial Geometries*, Encyclopedia of Mathematics and its Applications, 29, Cambridge University Press, Cambridge, 161–210.

One more survey paper from the same area.

A. Bouchet (1995). Covering and Δ-covering. E. Balas, J. Clausen (eds.). *Integer Programming and Combinatorial Optimization*, Lecture Notes in Computer Science, 920 Springer, Berlin 228–244.

A good introduction to Δ-matroids.

A. Frank, É. Tardos (1988). Generalized polymatroids and submodular flows. *Math. Program. (B)* 42, 489–563.

As far as submodular functions are concerned, one may obtain some insight of submodular flows, polymatroids and their relationship from this paper.

A. Schrijver (1984). Total dual integrality from directed graphs, crossing families and sub- and supermodular functions. W. R. Pulleyblank (ed.). *Progress in Combinatorial Optimization,* Academic Press, New York, 315–361.

An excellent cross-section for a reader interested in the relationship between the several models and frameworks concerning submodular functions and graphs.

The following two survey papers concentrate on the applications of submodular functions in graph theory.

A. Frank (1993). Submodular functions in graph theory. *Discr. Math.* 111, 231–243.

A. Frank (1993). Applications of submodular functions. K. Walker (ed.). *Surveys in Combinatorics,* London Math. Soc. Lecture Note Series 187, Cambridge University Press, Cambridge, 85–136.

A. Frank (1990). Packing paths, circuits, and cuts – a survey. B. Korte, L. Lovász, H-J. Prömel, A. Schrijver (eds.). *Paths, Flows and VLSI-Layouts*, Springer Verlag, Berlin, 47–100.
A. Frank (1994). Connectivity augmentation problems in network design. J.R. Birge, K.G. Murty (eds.). *Mathematical Programming: State of the Art*, The University of Michigan, 34–63.
Submodular functions turned out to be a basic proof technique in edge-disjoint paths problems as well as in connectivity augmentation problems. These areas are surveyed, respectively, in these papers.

I promised to mention only few works appeared before 1984, but there are two fundamental works which certainly cannot be left out from any overview.
A. Hoffman (1982). Ordered sets and linear programming. I. Rival (ed.). *Ordered Sets,* D. Reidel Publishing Comp., 619–654.
This is a very exciting survey on *lattice polyhedra* and other models introduced and analyzed by Hoffman and his co-workers. In my view this area deserves more attention than it has got in the last decade. One beautiful application of lattice polyhedra is the strikingly simple derivation of fundamental theorems of Greene and of Greene and Kleitman on optimal families of chains and antichains of a partially ordered set.

L. Lovász (1983). Submodular functions and convexity. A. Bachem, M. Grötschel, B. Korte (eds.). *Mathematical Programming: the State of the Art,* Springer, Berlin, 235–257.
This is another basic survey on the close parallel of convex functions and their discrete counter-parts, submodular functions. For brand-new developments in this direction, see §3.

M. Iri (1983). Applications of matroid theory. A. Bachem, M. Grötschel, B. Korte (eds.). *Mathematical Programming: the State of the Art,* Springer, Berlin, 160–201.
Another important overview on the applicability of matroids.

After mentioning these older papers, let me finish the list of surveys with a very new one.
R.E. Burkard, B. Klinz, R. Rudolf (1996). Perspectives of Monge properties in optimization. *Discr. Appl. Math.* 70, 95–162.
A superb survey (with 130 items in its reference list),which is highly recommendable for those interested in greedy algorithms, a method strongly related to submodular functions.

2 Submodular Functions

2.1 Minimization

One of the most exciting problems of the area is minimizing a submodular function. The only existing polynomial time algorithm for this purpose uses the ellipsoid method [Grötschel, Lovász and Schrijver 1988] (see §1.1). However there are important special cases when purely combinatorial algorithms are available.

W.H. Cunningham (1985). On submodular function minimization. *Combinatorica* 5, 185–192.

This solves the problem for "small" submodular functions. The method relies on the theory of polymatroids and may be considered as a clever refinement of Edmonds' matroid-partition algorithm.

M. Queyranne (1996). A combinatorial algorithm to minimize symmetric submodular functions. *Math. Program. (B)*, to appear.

This recent result settles completely the minimization problem for symmetric submodular functions: the algorithm is strongly polynomial. The surprising thing here is that no polymatroid theory is needed at all. The method is an extension of the revolutionary algorithm of [Nagamochi and Ibaraki 1995] (see §5) to compute the edge-connectivity of an undirected graph. Queyranne's algorithm may also be specialized to compute the minimum cut of a hypergraph.

M.X. Goemans, V.S. Ramakrishnan (1995). Minimizing submodular functions over families of sets. *Combinatorica* 15, 499–541.

In some cases one is interested in the minimum of a submodular function over the members of a certain family of sets. Already the book of [Grötschel, Lovász and Schrijver 1988] (see §1.1) includes non-trivial results in this direction and an elegant extension is found in this paper.

2.2 Intersection theorem

Perhaps the most important central result of the whole theory is the matroid intersection theorem of Edmonds. Its non-weighted special case asserts that *two matroids have a k-element independent set in common if and only if there is no bi-partition $\{X_1, X_2\}$ of the ground-set for which $r_1(X_1) + r_2(X_2) < k$ where r_1 and r_2 are the rank-functions of the matroids.* This theorem has a great number of applications and serves as a root to many extensions. One of them, the polymatroid intersection theorem, was already found by Edmonds in his paper of 1970 mentioned above.

A. Frank (1984). Finding feasible vectors of Edmonds-Giles polyhedra. *J. Combin. Theory B* 36, 221–239.

This paper consider an interesting version of the polymatroid intersection theorem, the discrete separation theorem. It asserts that, *given integer-valued super- and submodular functions p and b, there is an integer-valued modular function m for which $p \leq m \leq b$.* (A set function satisfying (1) with equality is called **modular**.) Related results are investigated in:
J. Kindler (1988). Sandwich theorems for set functions. *J. Math. Analysis Appl.* 133, 529–544.

F.D.J. Dunstan, A.W. Ingleton, D.J.A. Welsh (1972). Supermatroids. D.J.A Welsh, D.R. Woodall (eds.). *Combinatorics*, The Institute of Mathematics and its Applications, London, 72–122.

A difficult extension of Edmonds' matroid intersection theorem concerns distributive supermatroids which have been introduced in the following paper. A distributive supermatroid is a family $\mathcal{F}$ of ideals of a partially ordered set satisfying the following axioms. (1) $\emptyset \in \mathcal{F}$, (2) if $X \subset Y \in \mathcal{F}$ and X is an ideal, then $X \in \mathcal{F}$, (3) if $X \subset Y \in \mathcal{F}$ and $|X| < |Y|$, then there is an element $x \in X - Y$ such that $Y \cup \{x\} \in \mathcal{F}$.

É. Tardos (1990). An intersection theorem for supermatroids. *J. Combin Theory B* 50, 150–159.

In this paper an intersection theorem is described concerning two distributive supermatroids defined on the same partially ordered set.

A. Schrijver (1985). Supermodular colorings. L. Lovász, A. Recski (eds.). *Matroid Theory*, North-Holland, Amsterdam.

This paper present another type of intersection theorem concerning common supermodular colourings of two supermodular functions.

É. Tardos (1985). Generalized matroids and supermodular colorings. L. Lovász, A. Recski (eds.). *Matroid Theory*, North-Holland, Amsterdam.

Tardos discovered a connection between supermodular colourings and generalized polymatroid intersection and provided this way an alternative proof of Schrijver's theorem.

2.3 Greedy algorithms

Any introductory course on matroid theory certainly discusses the greedy algorithm. For a given weight-function w on the ground-set of the matroid, the greedy algorithm consists of choosing step by step a not-yet-chosen element of largest positive weight in such a way that the set of chosen elements form an independent set of the matroid. The greedy algorithm theorem asserts that the final independent set is of maximum weight. It is useful to realize that the independence axioms of matroids may be interpreted as requiring that for every $0-1$ weight-function w the greedy algorithm computes correctly a maximum weight independent set. The basic algorithm has been extended in several directions and there is a vast literature of work on greedy algorithms related to matroids and extensions. For a survey, see the paper of [Burkard, Klinz, Rudolf 1996] (see §1.2). An early interesting extension of the matroid greedy algorithm is the following.

D. Kornblum (1978). *Greedy algorithms for some optimization problems on a lattice polyhedron*, Ph. D. Thesis, Graduate Center of the City University of New York.

This thesis was written under the guidance of A. Hoffman who has several papers on greedy algorithms. See, for example:
A. Hoffman (1985). On greedy algorithms that succeed. I. Anderson (ed.). *Surveys in Combinatorics*, London Math. Soc. Lecture Notes Series, 103, Cambridge University Press, Cambridge, 97–112.

In these papers the original matroid greedy algorithm is generalized.

There are other greedy-type algorithms, primarily those concerning the so-called

Monge-property, which have not been known until recently to have any connection to submodularity. The following enlightening paper found a bridge between the two large classes of greedy algorithms.

M. Queyranne, F. Spieksma, F. Tardella (1993). A general class of greedily solvable linear programs. G. Rinaldi, L. Wolsey (eds.). *Proc. of the 3rd IPCO Conf.*, Erice, 385–399.

This work contains a common generalization of Edmonds' polymatroid greedy algorithm and the greedy algorithm of Hoffman for transportation problems when the cost function satisfies the Monge property. The algorithm has been further generalized in the following paper.

U. Faigle, W. Kern (1996). Submodular linear programs on forests, *Math. Program.* 72, 195–206.

Actually this is a two-phase greedy algorithm in the sense that first the primal linear program is solved in a greedy way and then the dual is solved greedily. A similar type of two-phase greedy approach is described for another model in the next paper.

A. Frank (1996). *Increasing the rooted connectivity of a digraph by one*, submitted to *Mathematical Programming Ser. B.*

This algorithm may be considered as a common generalization of Fulkerson's minimum cost arborescence algorithm and Kornblum's algorithm for (supermodular) lattice polyhedra.

3 Frameworks for Sub- and Supermodular Functions

Sub- and supermodular functions are often considered in connection with other structures like directed or undirected graphs, partially ordered sets. Several models have been developed to incorporate the various phenomena appearing in special cases.

3.1 Submodular flows

J. Edmonds, R. Giles (1977). A min-max relation for submodular functions on graphs. *Ann. Discr. Math.* 1, 185–204.

The model of submodular flows is perhaps the most convenient and flexible among the several equivalent models (such as, independent flows, polymatroidal flows, kernel systems, etc).

Successful efforts have been made to carry over the known techniques of the classical network flows to submodular flows. There follows a selection of papers of this type: their titles already indicates the flow technique they have extended to submodular flows.

W. Cunningham, A. Frank (1985). A primal-dual algorithm for submodular flows. *Math. Oper. Res.* 10, 251–261.

U. Zimmermann (1985). Augmenting circuit methods for submodular flow problems. L. Lovász, A. Recski (eds.). *Matroid Theory*, North-Holland, Amsterdam.
S. Fujishige (1987). An out-of-kilter method for submodular flows. *Discr. Appl. Math.* 17, 3–16.
W. Cui, S. Fujishige (1988). A primal algorithm for the submodular flow problem with minimum-mean cycle selection. *J. Oper. Res. Soc. Japan* 31, 431–441.
U. Zimmermann (1992). Negative circuits for flows and submodular flows. *Discr. Appl. Math.* 36, 179–189.
S.T. McCormick, T.R. Ervolina (1993). Canceling most helpful total submodular cuts for submodular flow. G. Rinaldi, L. Wolsey (eds.). *Proc. of the 3rd IPCO Conf.*, Erice. CIACO, 343–353.
H.N. Gabow (1993). A framework for cost-scaling algorithms for submodular flow problems. *Proc. 34th Annual IEEE Symp. Found. Comput. Sci.*, 449–458.
S. Fujishige, H. Röck, U. Zimmermann (1989). A strongly polynomial algorithm for minimum cost submodular flow problems. *Math. Oper. Res.* 14, 60–69.

H.N. Gabow (1995). Centroids, representations, and submodular flows. *J. Algorithms* 18, 586–628.
This paper presents a general technique to speed up several of the above algorithms.

3.2 Delta-matroids and submodular functions in two variables

A. Bouchet (1987). Greedy algorithm and symmetric matroids. *Math. Program.* 38 147–159,
R. Chandrasekaran, S.N. Kabadi (1988). Pseudomatroids. *Discr. Math.* 71, 206–217.
Various possible applications led researchers to investigate structures that may be associated with sub- or supermodular functions in two variables. A Δ-matroid is a pair $D = (S, \mathcal{F})$ where S is a finite ground-set and $\mathcal{F}$ is a family of subsets of S, called feasible sets, satisfying the following symmetric exchange axiom: for $F_1, F_2 \in \mathcal{F}$ and for $x \in F_1 \Delta F_2$, there is an element $y \in F_1 \Delta F_2$ with $F_1 \Delta \{x, y\} \in \mathcal{F}$. (Here Δ denotes the symmetric difference.) This was introduced independently, under different names, in these two papers.

By now the name Δ-matroid seems to become the generally accepted one. It can be shown that a Δ-matroid is a matroid (given by its bases) if and only if all the feasible sets have the same cardinality. Dress and Havel introduced a slightly weaker notion, called *metroid.*

A. Bouchet, A. Dress, T. Havel (1992). Δ-matroids and metroids. *Adv. Math.* 91, 136–142.
It was pointed out that metroids are exactly those Δ-matroids for which the empty set is feasible.

One important feature of Δ-matroids is that the greedy algorithm computes correctly a maximum weight feasible set.
M. Nakamura (1988). A characterization of those polytopes in which the greedy algorithm works (abstract). *Proc. 13th Int. Symp. on Math. Program.*, Tokyo.

S.N. Kabadi, R. Chandrasekaran (1990). On totally dual integral systems. *Discr. Appl. Math.* 26, 87–104.

L. Qi (1988). Directed submodularity, ditroids, and directed submodular flows. *Math. Program.* 42, 579–599.

The convex hull P of characteristic vectors of feasible sets has been described independently in these three papers. Namely, $P = \{x : x(A) - x(B) \leq f(A, B),\ A, B \subseteq S, A \cap B = \emptyset\}$ where $f(A, B) := \max(|F \cap A| - |F \cap B| : F \in \mathcal{F})$. Function f can be shown to be bisubmodular, that is, $f(A, B) + f(A', B') \geq f(A \cap A', B \cap B') + f((A \cup A') - (B \cup B'), (B \cup B') - (A \cup A'))$. Actually, the linear system $x(A) - x(B) \leq f(A, B),\ A, B \subseteq S, A \cap B = \emptyset\}$ is totally dual integral whenever f is a bisubmodular function.

A. Bouchet, W. Cunningham (1995). Delta-matroids, jump systems, and bisubmodular polyhedra. *SIAM J. Discr. Math.* 8, 17–32.

This paper presents more structural results on bisubmodular functions.

A. Frank, T. Jordán (1995). Minimal edge-coverings of pairs of sets. *J. Combin. Theory B*, 65, 73–110.

Another type of two-variable supermodular set-function is considered.

A non-negative integer-valued function p defined on the pairs of disjoint subsets of S is said to be *crossing bi-supermodular* if $p(X, Y) + p(X', Y') \leq p(X \cap X', Y \cup Y') + p(X \cup X', Y \cap Y')$ holds whenever $p(X, Y), p(X', Y') > 0,\ X \cap X', Y \cap Y' \neq \emptyset$. The problem is to find an integer-valued function $x \geq 0$ defined on the ordered pairs (u, v) $(u, v \in S)$ so that $\sum(x(u, v) : u \in X, v \in Y) \geq p(X, Y)$ holds for every $X, Y \subseteq S$ and so that $(\sum x(u, v) : u, v \in S)$ is as small as possible. The main result is a general min-max theorem that implies a characterization on the minimum number of new edges to be added to a given directed graph to make it k-node-connected or k-edge-connected. Other special cases are an extension of a deep theorem of E. Győri on intervals, W. Mader's theorem on splitting off edges in directed graphs, and J. Edmonds' theorem on matroid partitions.

3.3 Valuated matroids

Valuated matroids were introduced by Dress and Wenzel. The idea is that the greedy algorithm works not only for linear cost-functions but some more general ones, as well, provided the cost function satisfies an exchange-type property.

A. Dress, W. Wenzel (1992). Valuated matroids. *Adv. Math.* 93, 214–250.

By a valuation w on a matroid Dress and Wenzel mean a function $w : \mathcal{B} \rightarrow \Re$ on the family of bases $\mathcal{B}$ so that for any two bases B, B' and element $u \in B - B'$, there is an element $v \in B' - B$ for which $w(B) + w(B') \leq w(B - u + v) + w(B' - v + v)$. A matroid with a valuation is called a *valuated matroid.* Dress and Wenzel proved that a suitable version of the greedy algorithm works for valuated matroids.

K. Murota (1996). Valuated matroid intersection I: optimality criteria. *SIAM J. Discr. Math.* 9, 545–561.

K. Murota (1996). Valuated matroid intersection II: algorithms. *SIAM J. Discr. Math.* 9, 561-576.

Murota showed that not only the greedy algorithm but the matroid intersection algorithm (and theory) as well extends nicely to valuated matroids. Moreover, Murota made an important discovery. He realized that valuated matroids may be viewed as a tool combining ideas from matroid theory and non-linear programming.

K. Murota (1996). Convexity and Steinitz's exchange property. *Proc. 5th Int. Conf. on Integer Program. and Comb. Optim.*, Vancouver.

3.4 Matroid parity

Perhaps the most difficult optimization problem concerning matroids is the matroid parity problem.

L. Lovász, M. Plummer (1986). *Matching Theory,* North-Holland, Amsterdam.

This book summarizes the early fundamental results on the subject.

H. Gabow, M. Stallmann (1986). An augmenting path algorithm from the linear matroid parity problem. *Combinatorica* 6, 123–150.

J. Orlin, J. VandeVate (1996). Solving the linear matroid parity problem as a sequence of problems. *Math. Program.* to appear.

One of the main concerns is the algorithmic side of the matroid parity problem. Lovász' original polynomial-time algorithm for the (linear) matroid parity is extremely complicated. Here are two simplified versions.

Gammoids, a special class of matroids, are defined on a node-set of a directed graph $G = (V, E)$, as follows. Let S be a specified subset of k nodes of a G and let $\mathcal{B}$ be the family of k-element subsets T of V for which there are k node-disjoint paths from S to T. Then it can be proved that $\mathcal{B}$ is the basis set of a matroid and the matroids and submatroids arising this way are called *gammoids.* (The smallest matroid that is not a gammoid is the circuit matroid of K_4.)

Po Tong, E.L. Lawler, V.V. Vazirani (1984). Solving the weighted parity problem for gammoids by reduction to graphic matching. W.R. Pulleyblank (ed.). *Progress in Combinatorial Optimization*, Academic Press, New York, 363–374.

This paper does exactly what its title indicates.

There is a very exciting probabilistic approach to solve algorithmically the matroid parity problem. The underlying idea is simple and perhaps best undersood in the special case of deciding whether a bipartite graph G has a perfect matching. Assign a matrix M to G with columns corresponding to one of the two colour-classes of G and with rows corresponding to the other colour-class. Define entry a_{ij} zero if v_i and v_j are not incident and x_{ij} if they are incident. Here the non-zero entries are considered independent indeterminates. It is not difficult to see that G has a perfect matching if and only if the determinant of M is not the zero polynomial. It may not be easy to answer this question algorithmically since the naive way to compute the determinant may lead to far too many expansion terms. There is however another natural approach: substitute random integers for each indeterminate and compute the determinant of the arising integer matrix. If this number is not zero, then the polynomial is not identically zero and hence the graph has a perfect matching. If this number is zero, then we cannot be sure that the polynomial is identically zero because we just may have hit a root of this polynomial. But intuitively the chance of such an event is clearly small and it can actually be proved that the probability of hitting a root is smaller than a fixed number $\varepsilon < 1$. Therefore if we repeat the same procedure several times and every time the resulting determinant is zero, then we can be sure, with arbitrarily high probability, that the graph has no perfect matching.

L. Lovász, (1979). On determinants, matchings and random algorithms. L. Budach (ed.). *Fundamentals of Comput. Th.*, Akademia Verlag, Berlin.

Of course, there are very efficient deterministic algorithm for bipartite matching but the nice thing is that the probabilistic algorithm above can be extended to matroid parity problems as was shown in this paper.

P.M. Camerini, G. Galbiati, F. Maffioli (1992). Random pseudo-polinomial algorithms for exact matroid problems. *J. Algorithms* 13, 258–273.

This paper extends the idea to related topics. Among others, it describes a polynomial time random algorithm to determine whether two matroids with a red and blue coloured common ground-set have a common basis with exactly k red elements. No polynomial-time deterministic algorithm is known even for the special case of this problem when one wants to determine if a red-blue edge-coloured bipartite graph has a perfect matching with exactly k red edges.

J.H. VandeVate (1992). Fractional matroid matchings. *J. Comb. Theory B* 5,5 133-145.

J.H. VandeVate (1992). Structural properties of matroid matchings. *Discr. Appl. Math.* 39, 69-85,

Some interesting structural results concerning the matroid parity problem can be found in these works.

M.L. Furst, J.L. Gross, L.A. McGeogh (1988). Finding a minimum-genus graph embedding. *J. ACM* 35, 523–534.

Matroid parity finds a nice application in graph embedding problems.

L. Nebesky (1983). A Note on Upper Embeddable Graphs. *Czechslovak Math. Journal,* 33, 37–40.

This remarkable paper is concerned with the co-graphic matroid parity problem and finds a very neat characterization for graphs having a spanning tree T for which each component arising from G by deleting the edges in T has an odd number of edges.

4 Minor, Decompositions, and Representations

Let A be a matrix with entries from an arbitrary field F. The set S of columns of A forms a subset of a vector space over F and linear independence defines a matroid on

S. A fundamental question is to decide whether a given matroid can be represented in such a form. Very often the answer relies on the notion of minors. For example, a classic result of Tutte asserts that a matroid M is binary (i.e., M can be represented over GF(2)) if and only if M does not include the uniform matroid $U_{4,2}$ as a minor. As I mentioned earlier, excellent recent books and survey papers are available concerning this area therefore here I list only a few papers to provide a flavour of the type of theorems. For example, Tutte characterized matroids which are representable over every field (the so-called regular matroids) by proving that regular matroids are those not containing $U_{4,2}$, the Fano matroid, and the dual Fano matroid.

A.M.H. Gerards (1989). A short proof of Tutte's characterization of totally unimodular matrices. *Linear Algebra and its Appl.* 114, 217-222.

The original proof is very difficult but this paper turns Tutte's result accessible for everyone familiar with the basics of matroid theory.

J. Kahn, P.D. Seymour (1988). On forbidden minors of GF(3), *Proc. American Math. Soc.* 102, 437-440.

This includes a simple proof of a characterization for ternary matroids, a result proved first by R. Reid (unpublished). Structural descriptions are often related to connectivity properties.

R.E. Bixby (1974). l-marices and a characterization of binary matroids. *Discr. Math.* 8, 139–145,

This work shows that a non-binary matroid not only includes a $U_{4,2}$-minor, but every element belongs to such a minor provided that the matroid is connected (i.e. every two elements are contained in a circuit).

P.D. Seymour (1985). On minors of 3-connected matroids. *European J. Combinatorics* 6, 375–382.

It is shown that if M is 3-connected and not binary, then even every pair of elements belong to a $U_{4,2}$-minor. As far as decomposition of matroids are concerned we mention two far reaching results. Tutte's above-mentioned characterization of regular matroids provides a certificate (namely the occurrence of certain minors) to show that a certain matroid is not regular.

P.D. Seymour (1980). Decomposition of regular matroids. *J. Comb. Theory B* 28, 305–359.

This paper provides a certificate to show that a matroid is regular. The certificate is a way to build up the matroid from graphic and cographic matroids and from a specific regular matroid on ten elements, where the building operations are the so called 1-, 2- and 3-sums.

F.T. Cheng, K. Truemper (1986). A decomposition of the matroids with the max-flow min-cut property. *Discr. Appl. Math.* 15, 329–364.

Another important decomposition theorem desribes how to construct matroids with the max-flow min-cut property.

P.D. Seymour (1977). The matroids with the max-flow min-cut property. *J. Comb. Theory B* 23, 189–222.
These matroids are characterized in terms of forbidden minors.

5 Applications to Networks, Scheduling, Game Theory, and Allocation Problems

A. Frank (1992). On a theorem of Mader. *Ann. Discr. Math.* 101, 49–57.
One of the main application area of submodular and related functions concerns connectivity of graphs. For example, skew-supermodular functions are used in this paper to provide a relatively simple proof of a deep theorem of Mader on splitting of edges without decreasing the local edge-connectivity. (A set-function p is called skew-supermodular if the following inequality holds for every pair of sets X, Y: $p(X) + p(Y) \leq \max(p(X \cap Y)p(X \cup Y),\ p(X - Y) + p(Y - X))$.

D.P. Williamson, M.X. Goemans, M. Mihail, V.V. Vazirani (1995). A primal dual algorithm for generalized Steiner network problems. *Combinatorica* 15, 435-454.
Here another class of submodular-type functions is used in a fundamental new technique of the primal-dual approximation methods.

A. Frank, E. Tardos (1989). An application of submodular flows. *Linear Algebra and its Appl.* 114/115, 329–348.
In this paper the minimum cost rooted connectivity augmentation problem is reduced to submodular flows.

H.N. Gabow (1991). A matroid approach to finding edge-connectivity and packing arborescences. *Proc. 23rd ACM Symp. Theory of Comput.*, 112–122.
H.N. Gabow, K.S. Manu (1995). Packing algorithms for arborescences (and spanning trees) in capacitated graphs. E. Balas, J. Clausen (eds.). *Integer Programming and Combinatorial Optimization*, Lecture Notes in Computer Science, 920 Springer, Berlin 388–402.
H.N. Gabow (1991). Applications of a poset representation to edge connectivity and graph rigidity. *Proc. 32nd Annual IEEE Symp. Found. Comput. Sci.*, 812–820.
H.N. Gabow (1994). Efficient splitting off algorithms for graphs. *Proc. 26th ACM Symp. Theory of Comput.*, Montreal, 696–706.
These important works of H. Gabow concern algorithmic aspects of connectivity-related problems.

L.A. Wolsey (1989). Submodularity and valid inequalities in capacitated fixed charged networks. *Oper. Res. Lett.* 8, 119–124.
Another role of submodular functions is utilized here. The observation that the flow values of certain fixed charged networks are submodular is used to derive valid inequalities.

H. Nagamochi, T. Ibaraki (1995). A faster edge splitting algorithm in multigraphs and its application to the edge-connectivity augmentation problem. E. Balas, J. Clausen

(eds.). *Integer Programming and Combinatorial Optimization*, Lecture Notes in Computer Science, 920 Springer, Berlin, 403-413.
H. Nagamochi, K. Nishima, T. Ibaraki (1995). Computing all small cuts in undirected networks. D-Z.Du, X-S. Zhang (eds.). *Algorithms and Computation*, Lecture Notes in Computer Science 834, 190-198.

In connection with Queyranne's algorithm for minimizing symmetric submodular functions, I have already referred to a fundamental paper of Nagamochi and Ibaraki. These works describe interesting extensions. The submodular technique again plays an important role.

Submodular functions proved to be useful in the area of scheduling. See for example the following two papers.
F. Blanchini, M. Queyranne, F. Rinaldi, W. Ukovich (1996). A feedback strategy for periodic network flows. *Networks* 27, 25–34.
M. Queyranne, A.S. Schulz (1995). Scheduling unit jobs with compatible release dates on parallel machines with nonstationary speed. E. Balas, J. Clausen (eds.). *Integer Programming and Combinatorial Optimization*, Lecture Notes in Computer Science, 920 Springer, Berlin, 307-320.

Another large and interesting topic of applications of submodular functions is the area of resource allocations. Here I mention some of the basic results.
N. Katoh, T. Ibaraki, H. Mine (1985). An algorithm for the equipollent resource allocation problem. *Math. Oper. Res.* 10, 44–53.
A. Federgruen, H. Gronevelt(1986). Optimal flows in networks with multiple sources and sinks, with applications to oil and gas lease investment programs. *Oper. Res.* 34, 218–225.
A. Federgruen, H. Gronevelt (1986). The greedy procedure for resource allocation problems: Necessary and sufficient conditions for optimality. *Oper. Res.* 34, 909–918.
S. Fujishige, N. Katoh, T. Ichimori (1988). The fair resource allocation problem. *Math. Oper. Res.* 13, 164–173.

Finally, let me draw attention to the applicability of supermodular functions in game theory. This connection was already discovered in
L.S. Shapley (1971). Cores of convex games. *Int. J. Game Theory* 1, 11–26.

T. Ichiishi (1981). Super-modularity: Applications to convex games and to the greedy algorithm for LP. *J. Economic Theory* 25, 283–286.

This paper analyzes the relationship of greedy algorithms and convex games.

H. Kanekom, M. Fushimi (1986). A polymatroid associated with convex games. *Discr. Appl. Math.* 14, 33–45.

This work makes use of the theory of principal partition of polymatroids.

M. Iri (1979). A review of recent work in Japan on principal partitions of matroids and their applications. *Annals of the New York Academy of Sciences* 319, 306–319.

Principal partitions form a basic tool for describing the fine structure of submodular functions.

6 Perfect, Ideal and Balanced Matrices

Michele Conforti
Università di Padova

Gérard Cornuéjols
Carnegie Mellon University

Ajai Kapoor
Università di Padova

Kristina Vušković
University of Kentucky

Contents

Polyhedral Combinatorics is an important research area in Integer Programming. Loosely speaking, polyhedral combinatorics addresses the following problem:

Given a finite family $\mathcal{F}$ of integer-valued vectors, describe the convex hull of the vectors in $\mathcal{F}$ as the intersection of finitely many halfspaces.

A fundamental theorem in polyhedral theory shows that such a description exists and can (in theory) be computed. The success of Linear Programming provides the main motivation for this approach. For, suppose one wants to select a vector in $\mathcal{F}$ that optimizes a given linear objective function, then one has to optimize this objective function, subject to the linear constraints provided by such a description, and this optimization problem is a linear program.

Most of the time, the family $\mathcal{F}$ is the set of integral vectors that satisfy a linear system of inequalities. That is,

$$\mathcal{F} = \{x \text{ integral} : Bx \leq b\}$$

In this case, we can address the following problem:

When is it true that the extreme vectors in the convex hull of $\mathcal{F}$ are exactly the extreme vectors in the polyhedron $\{x : Bx \leq b\}$?

If this happens, $\{x : Bx \leq b\}$ is an *integral polyhedron*. Here we survey the results and the bibliography concerning the integrality of *set packing polytopes* and

Annotated Bibliographies in Combinatorial Optimization, edited by M. Dell'Amico, F. Maffioli and S. Martello

set covering polyhedra. We also cover recent results concerning *generalized* set packing and covering polytopes.

C. Berge (1961). Färbung von Graphen deren sämtliche bzw. deren ungerade Kreise starr sind (Zusammenfassung). *Wissenschaftliche Zeitschrift*, Martin Luther Universität Halle-Wittenberg, Mathematisch-Naturwissenschaftliche Reihe, 114–115.
C. Berge(1970). Sur Certains Hypergraphes Généralisant les Graphes Bipartis. P. Erdös, A. Rényi and V. Sós (eds.). *Combinatorial Theory and its Applications I*, Colloq. Math. Soc. János Bolyai, North Holland, Amsterdam, 119–133.
C. Berge (1972). Balanced Matrices. *Math. Program.* 2, 19–31.
C. Berge (1989). *Hypergraphs*, North Holland, Amsterdam.

Let A be a $0,1$ matrix. Such a matrix is *perfect* if the set packing polytope $\{x \geq 0 : \; Ax \leq 1\}$ is integral. It is *ideal* if the set covering polyhedron $\{x \geq 0 : \; Ax \geq 1\}$ is integral. It is *balanced* if no square submatrix of odd order contains exactly two 1's per row and per column. The concepts of perfection and balancedness are due to Berge.

A. Lehman (1979). On the Width-Length Inequality. Mimeographic notes. Published: *Math. Program.* 17 (1979), 403–417.

The concept of idealness is introduced by Lehman under the name of width-length property. [Berge 1972] (see above) proves that a matrix A being balanced is equivalent to A and all its submatrices being perfect which is equivalent to A and all its submatrices being ideal.

Therefore, perfect, ideal and balanced matrices give rise to integer programs that can be solved as linear programs, for all objective functions.

J. Edmonds, R. Giles (1977). A Min-Max Relation for Submodular Functions on Graphs. *Ann. Discr. Math.* 1, 185–204.
A. Hoffman (1974). A generalization of max flow-min cut. *Math. Program.* 6, 352–359.

An interesting situation occurs when the dual linear program has an integral optimal solution for all integral objective functions for which it has an optimal solution. Let B be a matrix with integral entries. A linear system $Bx \leq b, \; x \geq 0$ is *totally dual integral* (TDI) if the linear program max $\{cx : \; Bx \leq b, \; x \geq 0\}$ has an integral optimal dual solution y for all $c \in Z^n$ for which it has an optimal solution. The first paper proves that if the linear system $Bx \leq b, \; x \geq 0$ is TDI, then the polyhedron $\{x \geq 0 : \; Bx \leq b\}$ only has integral extreme points. This generalizes an earlier result presented in the second paper.

1 Perfect Matrices

L. Lovász (1972). Normal hypergraphs and the perfect graph conjecture. *Discr. Math.* 2, 253-267.

It is shown that, for a $0,1$ matrix A, the following statements are equivalent:

(i) the linear system $Ax \leq 1, \; x \geq 0$ is TDI,

(ii) the matrix A is perfect,

(iii) $\max\{cx : \; Ax \leq 1, \; x \geq 0\}$ has an integral optimal solution x for all $c \in R^n$,

(iv) $\max\{cx : Ax \leq 1, x \geq 0\}$ has an integral optimal solution x for all $c \in \{0,1\}^n$.

Clearly (i) implies (ii) implies (iii) implies (iv), where the first implication is the Edmonds-Giles property and the other two are immediate. What is surprising is that (iv) implies (i) and, in fact, that (ii) implies (i). This result follows from, and in fact is equivalent to, [Lovász 1972]'s famous perfect graph theorem.

V. Chvátal (1975). On Certain Polytopes Associated with Graphs. *J. Combin. Theory B* 13, 138–154.

In a graph, a *clique* is a set of pairwise adjacent nodes. The *chromatic number* is the smallest number of colours needed to colour the nodes so that adjacent nodes have distinct colours. Since all nodes of a clique must have a distinct colour, the chromatic number is always at least as large as the size of a largest clique. A graph G is *perfect* if, for every node induced subgraph of G, the chromatic number equals the size of a largest clique. The perfect graph theorem states that a graph G is perfect if and only if its complement $\bar{G}$ is perfect. (The *complement* of graph G is the graph $\bar{G}$ having same node set and complement edge set.) The relation between perfect graphs and perfect matrices is obtained through the concept of clique-matrix of a graph. A *clique-matrix* of a graph G is a $0,1$ matrix whose columns are indexed by the nodes of G and whose rows are the incidence vectors of the maximal cliques of G. Chv́atal proves that a $0,1$ matrix is perfect if and only if its nonredundant rows form the clique-matrix of a perfect graph.

A major open question is to characterize the graphs which are not perfect but all their proper node induced subgraphs are. These graphs are called *minimally imperfect*. A *hole* is a chordless cycle of length greater than three and it is *odd* if it contains an odd number of edges. Odd holes are minimally imperfect since their chromatic number is three and the size of the largest clique is two, but all proper induced subgraphs are bipartite and therefore perfect. If G is minimally imperfect, then so is its complement, by Lovász's perfect graph theorem. The *strong perfect graph conjecture* of [Berge 1961] (see above) states that the odd holes and their complements are the only minimally imperfect graphs.

M. W. Padberg (1974). Perfect zero-one matrices. *Math. Program.* 6, 180–196.

M. W. Padberg (1976). Almost integral polyhedra related to some combinatorial optimization problems. *Linear Algebra and its Applications* 15, 69–88.

The above conjecture can be formulated in terms of $0,1$ matrices as follows: A $0,1$ matrix A is *minimally imperfect* if A is not perfect but all its proper column submatrices are. Or, equivalently, if the polyhedron $\{x \geq 0 : Ax \leq 1\}$ has a fractional extreme point but all polyhedra obtained from it by setting a variable x_j equal to 0 only have integral extreme points. In this context, we assume that A has no redundant rows. A $0,1$ matrix that is not a clique-matrix but all its column submatrices are *minimally not a clique-matrix*. Then Berge's conjecture says that A is minimally imperfect if and only if either A is minimally not a clique-matrix or A is the clique-matrix of an odd hole or its complement.

An $m \times n$ $0,1$ matrix A has the property $\pi_{\beta,n}$ if A contains an $n \times n$ nonsingular submatrix B whose row and column sums are all equal to β and each row of A not in B has row sum strictly less than β. Padberg shows that if an $m \times n$ $0,1$ matrix A is minimally imperfect, then A has the property $\pi_{\beta,n}$ for some $\beta \geq 2$.

A 0,1 matrix which is minimally not a clique-matrix must be a square matrix of the form $E - I$, where E is a matrix of all 1's and I an identity. This corresponds to the property $\pi_{\beta,n}$ with $\beta = n - 1$ and $n \geq 3$. An alternative characterization of clique-matrices is due to Gilmore, see above [Berge 1989]. It implies that we can test in polytime whether A is a clique-matrix.

C. Berge, V. Chvátal (eds.) (1984). *Topics on perfect graphs*, volume 21 of Annals of Discrete Mathematics, North Holland, Amsterdam.

Important papers on perfect matrices and graphs are collected.

2 Ideal Matrices

A. Lehman (1990). On the width-length inequality and degenerate projective planes. Unpublished manuscript (1981), published: W. Cook and P.D. Seymour (eds.). *Polyhedral Combinatorics*, DIMACS Series in Discrete Mathematics and Theoretical Computer Science 1, American Mathematical Society, Providence, R.I. 101–105.

For a 0,1 matrix A, the following statements are equivalent:

(i) the matrix A is ideal,

(ii) $\min\{cx : Ax \geq 1,\ x \geq 0\}$ has an integral optimal solution x for all $c \in R^n$,

(iii) $\min\{cx : Ax \geq 1,\ x \geq 0\}$ has an integral optimal solution x for all $c \in \{0, 1, +\infty\}^n$.

Statements (i) and (ii) are equivalent by the definition of idealness, and the fact that (ii) implies (iii) is immediate. The difficult part of Lehman's theorem is that (iii) implies (ii). Here, contrary to the situation for perfection, idealness of A does not imply TDIness of the linear system $Ax \geq 1,\ x \geq 0$. This can be seen using $A = Q_6$ as defined below.

$$Q_6 = \begin{pmatrix} 1 & 1 & 0 & 1 & 0 & 0 \\ 1 & 0 & 1 & 0 & 1 & 0 \\ 0 & 1 & 1 & 0 & 0 & 1 \\ 0 & 0 & 0 & 1 & 1 & 1 \end{pmatrix}$$

Choosing $c = (1, 1, 1, 1, 1, 1)$, the unique optimal dual solution is $y = (\frac{1}{2}, \frac{1}{2}, \frac{1}{2}, \frac{1}{2})$. Yet, it is easy to check that the polyhedron $\{x \geq 0 : Q_6 x \geq 1\}$ only has integral extreme points.

Therefore, we are led to define another class of matrices: a 0,1 matrix A has the *max flow min cut property* (MFMC property) if the linear system $Ax \geq 1,\ x \geq 0$ is TDI.

M. Conforti, G. Cornuéjols (1993). *Clutters that Pack and the Max Flow Min Cut Property: A Conjecture*, Preprint, Graduate School of Industrial Administration, Carnegie Mellon University, Pittsburgh.

It is conjectured that for a 0,1 matrix A, the following statements are equivalent:

(i) the matrix A has the MFMC property,

(ii) $\min\{cx : Ax \geq 1,\ x \geq 0\}$ has an integral optimal dual solution y for all $c \in Z^n_+$,

(iii) $\min\{cx : Ax \geq 1, x \geq 0\}$ has an integral optimal dual solution y for all $c \in \{0, 1, +\infty\}^n$.

Statements (i) and (ii) are equivalent by the definition of the MFMC property, and the fact that (ii) implies (iii) is immediate. The difficult part of the conjecture is to show that (iii) implies (ii). Several minmax theorems and colouring theorems in graph theory can be seen as instances of problems of this type or problems where min $\{cx : Ax \geq 1, x \geq 0\}$ has an integral optimal dual solution y when $c = (1, 1, \ldots, 1)$.

A *minor* of A is obtained by repeatedly performing either one of the following two operations:

- delete column j
- delete column j and all rows i such that $a_{ij} = 1$.

These operations correspond to setting x_j to 0 or to 1 in the polyhedron $\{x \geq 0 : Ax \geq 1\}$.

A 0,1 matrix A is *minimally nonideal* if A is not ideal but all its proper minors are. Equivalently, if the polyhedron $\{x \geq 0 : Ax \geq 1\}$ has a fractional extreme point but all polyhedra obtained from it by setting a variable x_j to 0 or to 1 only have integral extreme points. Note that A is nonideal if and only if A contains a minimally nonideal minor.

For $n \geq 3$, the following $n \times n$ matrices, denoted J_n, are minimally nonideal:

$$J_n = \begin{pmatrix} 0 & 1 & 1 & 1 & 1 & 1 \\ 1 & 1 & 0 & 0 & 0 & 0 \\ 1 & 0 & 1 & 0 & 0 & 0 \\ 1 & 0 & 0 & 1 & 0 & 0 \\ 1 & 0 & 0 & 0 & 1 & 0 \\ 1 & 0 & 0 & 0 & 0 & 1 \end{pmatrix}$$

For $(x_1 = \frac{n-2}{n-1}, x_j = \frac{1}{n-1}, j = 2, \ldots, n)$ uniquely minimizes an objective function of all 1's and it is easy to see that all proper minors of J_n are ideal.

An $m \times n$ 0,1 matrix A has the property $\phi_{\alpha,n}$ if A contains an $n \times n$ nonsingular submatrix B whose row and column sums are all equal to α and each row of A not in B has row sum strictly greater than α. It is shown in [Lehman 1990] (see above) that if an $m \times n$ 0,1 matrix A is minimally nonideal, then either after permutation of rows and columns $A = J_n$ or A has the property $\phi_{\alpha,n}$ for some $\alpha \geq 2$.

The above result has the same spirit as Padberg's theorem for perfection and the following papers contain different proofs of Lehman's theorem.

M. Padberg (1993). Lehman's forbidden minor characterization of ideal 0,1 matrices. *Discr. Math.* 111, 409–420.

P.D. Seymour (1990). On Lehman's width-length characterization. W. Cook and P.D. Seymour (eds.). *Polyhedral Combinatorics*, DIMACS Series in Discrete Mathematics and Theoretical Computer Science 1, AMS, Providence, R.I. 75–78.

In [Lehman 1965] (see the introduction), three infinite classes of minimally nonideal matrices are given. But, as for minimally imperfect graphs, it is an open problem to list all the minimally nonideal matrices. In fact, the situation appears more complicated than for minimally imperfect graphs since, in addition to Lehman's three infinite classes, there is a host of small examples that are known.

G. Cornuéjols, B.A. Novick (1994). Ideal 0,1 Matrices. *J. Combin. Theory B* 60, 145–157.

It is proved that there are exactly 10 square minimally nonideal circulant matrices with k consecutive 1's, $k \geq 3$. A square minimally nonideal circulant matrix where the 1's are not consecutive is the Fano matrix:

$$F_7 = \begin{pmatrix} 1 & 1 & 0 & 1 & 0 & 0 & 0 \\ 0 & 1 & 1 & 0 & 1 & 0 & 0 \\ 0 & 0 & 1 & 1 & 0 & 1 & 0 \\ 0 & 0 & 0 & 1 & 1 & 0 & 1 \\ 1 & 0 & 0 & 0 & 1 & 1 & 0 \\ 0 & 1 & 0 & 0 & 0 & 1 & 1 \\ 1 & 0 & 1 & 0 & 0 & 0 & 1 \end{pmatrix}$$

This example was already known to [Lehman 1965]. Two additional small examples which are not circulant matrices have 8 and 10 columns, respectively.

C. Lütolf, F. Margot (1995). *A Catalog of Minimally Nonideal Matrices*, Preprint, Graduate School of Industrial Administration, Carnegie Mellon University, Pittsburgh.

All minimally nonideal matrices with up to 12 columns are enumerated. Lütolf and Margot also find quite a number of examples with 14 and with 17 columns.

P.D. Seymour (1977). The matroids with the max-flow min-cut property. *J. Combin. Theory B* 28, 189–222.

A class of $0,1$ matrices that has received much attention are those with the *binary property*. This happens when any minimal $0,1$ vector that satisfies $Ax \geq 1$ has an odd intersection with every minimal row of A. A classical example occurs when A is the $0,1$ incidence matrix of st-paths versus edges in a graph. In this case, the minimal $0,1$ vectors that satisfy $Ax \geq 1$ are the st-cuts of the graph and, since they intersect all st-paths in an odd number of edges, A has the binary property. The Ford–Fulkerson theorem shows that A also has the MFMC property in this case. Now consider the matrix Q_6 introduced earlier. It is easy to check that it has the binary property, but we saw that it does not have the MFMC property. Seymour shows that, in fact, this is the only obstruction, i.e. that given a $0,1$ matrix A with the binary property, then A has the MFMC property if and only if it contains no Q_6 minor.

P.D. Seymour (1981). Matroids and multicommodity flows. *European J. of Combinatorics* 2, 257–290.

A conjecture about the obstructions to idealness for the $0,1$ matrices with the binary property.

3 Perfect and Ideal $0, \pm 1$ Matrices

The concepts of perfect and of ideal 0,1 matrices can be extended to $0, \pm 1$ matrices. Given a $0, \pm 1$ matrix A, denote by $n(A)$ the column vector whose i^{th} component is the number of 1's in the ith row of matrix A. The $0, \pm 1$ matrix A is *perfect* if its generalized set packing polytope

$$P(A) = \{x : \; Ax \leq 1 - n(A), \; 0 \leq x \leq 1\}$$

only has integral extreme points. Similarly, the $0, \pm 1$ matrix A is *ideal* if its generalized set covering polytope

$$Q(A) = \{x : \ Ax \geq 1 - n(A), \ 0 \leq x \leq 1\}$$

only has integral extreme points. The *switching* on a subset J of columns of A is the matrix A_J obtained from A by changing the signs of all nonzero entries in the columns of J. Obviously the transformation $\{y_j = 1 - x_j \ j \in J, \ y_j = x_j, j \notin J\}$ maps $P(A)$ into $P(A_J)$ and $Q(A)$ into $Q(A_J)$. So A is perfect or ideal if and only if all switchings of A are.

M. Conforti, G. Cornuéjols (1995). A class of logic problems solvable by linear programming. *J. ACM* 42 n. 5, 1107–1113.

It is well known that several problems in propositional logic, such as SAT, MAXSAT and logical inference, can be written as integer programs of the form

$$\min\{cx : \ Ax \geq 1 - n(A), \ x \in \{0,1\}^n\}.$$

These problems are NP-hard in general but they can be solved in polytime by linear programming or by ad hoc methods when the corresponding $0, \pm 1$ matrix A is ideal.

E. Boros, O. Čepek (1995). *Perfect* $0, \pm 1$ *matrices*, Preprint, RUTCOR, Rutgers University. To appear in *Proceedings of the Marseille conference on Graph Theory* (C. Berge, J.-L. Fouquet, and P. Rosenstiehl eds.).
M. Conforti, G. Cornuéjols, C. de Francesco (1997). *Perfect* $0, \pm 1$ *Matrices Linear Algebra and its Applications* 253, 299–309.
B. Guenin (1994). *Perfect and Ideal* $0, \pm 1$ *Matrices*, Preprint, Graduate School of Industrial Administration, Carnegie Mellon University, Pittsburgh. To appear in *Math. Oper. Res.*
J. Hooker (1996). Resolution and the Integrality of Satisfiability Polytopes. *Math. Program. 74*, 1-10.
P. Nobili, A. Sassano (1994). $(0, \pm 1)$ *Ideal Matrices*, Preprint, University of Rome. To appear in *Math. Program.*

The remainder of this section presents the results contained in these five papers. Hooker was the first to relate idealness of a $0, \pm 1$ matrix to that of a family of 0,1 matrices. A similar result for perfection was obtained in Conforti, Cornuéjols and De Francesco. These results were strengthened by Guenin and by Boros and Čepek for perfection, by Nobili and Sassano for idealness.

The key tool for these results is the following:

Given a $0, \pm 1$ matrix A, let P and R be $0, 1$ matrices of the same dimension as A, such that $P_{ij} = 1$ if and only if $A_{ij} = 1$, and $R_{ij} = 1$ if and only if $A_{ij} = -1$. The matrix:

$$D_A = \left[\begin{array}{c|c} P & R \\ \hline I & I \end{array}\right],$$

is the $0, 1$ *extension* of A. Note that the transformation $x^+ = x$ and $x^- = 1 - x$ maps every vector x in $P(A)$ into a vector in $\{(x^+, x^-) \geq 0 : \ Px^+ + Rx^- \leq 1, \ x^+ + x^- = 1\}$ and every vector x in $Q(A)$ into a vector in $\{(x^+, x^-) \geq 0 : \ Px^+ + Rx^- \geq 1, \ x^+ + x^- = 1\}$. So $P(A)$ and $Q(A)$ are respectively the faces of $P(D_A)$ and $Q(D_A)$, obtained by setting the inequalites $x^+ + x^- \leq 1$ and $x^+ + x^- \geq 1$ at equality.

We say that a polytope Q contained in the unit hypercube $[0,1]^n$ is *irreducible* if, for each j, both polytopes $Q \cap \{x : x_j = 0\}$ and $Q \cap \{x : x_j = 1\}$ are nonempty. This property can be checked by linear programming. If A is an $m \times n$ $0, \pm 1$ matrix, [Boros, Čepek 1995] gives a combinatorial algorithm to test whether $P(A)$ is irreducible in $O(mn^2)$ steps.

The *completion* of a $0, \pm 1$ matrix A is the matrix A^* obtained by adding to A all row vectors a that induce a generalized set packing inequality $ax \leq 1 - n(a)$ which is valid for $P(A)$ and not dominated by any other inequality in A^*. The size of A^* may be exponential in the size of A. A *monotone completion* of A is a $0,1$ matrix obtained from A^* by multiplying through some columns by -1 and replacing all negative entries of the resulting matrix by 0. In [Conforti, Cornuejols, De Francesco 1993] it is proved that if A is a $0, \pm 1$ matrix such that $P(A)$ is irreducible, then A is perfect if and only if all the monotone completions of A are perfect $0,1$ matrices.

Given a $0, \pm 1$ matrix A, let a^1 and a^2 be two rows of A, such that there is one index k such that $a^1_k a^2_k = -1$ and, for all $j \neq k$, $a^1_j a^2_j = 0$. A *disjoint implication* of A is the $0, \pm 1$ vector $a^1 + a^2$. The matrix A^+ obtained by recursively adding all disjoint implications and removing all dominated rows is called the *disjoint completion* of A. In [Guenin 1994] it is proved that if A is a $0, \pm 1$ matrix, where $P(A)$ is not contained in any of the hyperplanes $\{x : x_j = 0\}$ or $\{x : x_j = 1\}$, then A is perfect if and only if the $0,1$ matrix D_{A^+} is perfect. Guenin also shows that a $0, \pm 1$ matrix A is perfect if and only if $\max\{cx : x \in P(A)\}$ admits an integral optimal solution for every $c \in \{0, \pm 1\}^n$. Moreover, if A is perfect, the linear system $Ax \leq 1 - n(A)$, $0 \leq x \leq 1$ is TDI. This is the natural extension of the theorem in [Lovász 1972] for perfect $0,1$ matrices.

A $0, \pm 1$ matrix A is *Horn* if no row of A has more that one negative entry. In [Boros, Čepek 1995] it is shown that, for every $0, \pm 1$ matrix A, one of the following holds:

- $P(A)$ is either empty or contained in a hyperplane $\{x : x_j = \frac{1}{2}\}$ for some j.
- A can be made Horn by multiplying by -1 a certain subset of its columns.

Moreover [Boros, Čepek 1995] give an $O(mn)$ algorithm that checks which alternative is true, and in the second case, the algorithm provides the subset of columns to be multiplied.

Consider now the case in which A is Horn and let V represent the column set of A. We say that a subset S of V is *covered* if the polytope

$$Z(A,S) = \left\{ (x, x_{n+1}) \;\middle|\; \begin{array}{rl} Ax \leq (1 - n(A))x_{n+1} & \\ x_j = 1 & j \in S, \\ 0 \leq x_j \leq |S| - 1 & j \notin S \end{array} \right\}$$

is empty.

Define now an undirected graph $R(A) = (V, E)$ where a pair ij is an edge in E whenever the set $\{i, j\}$ is covered. Finally, let $\tilde{A}$ be the matrix

$$\left[\begin{array}{c|c} A & 0 \\ \hline P & R \\ \hline I & I \end{array} \right]$$

In [Boros, Čepek 1995] it is shown that if A is an irreducible Horn matrix, then A is perfect if and only if $R(\tilde{A})$ is a perfect graph and every maximal clique of $R(\tilde{A})$ is covered.

The above result relates the perfection of a $0, \pm 1$ matrix to the perfection of a "small" graph. The last condition parallels the condition of A being a clique-matrix of [Chvátal 1975] (see Section 1).

We now turn to ideal $0, \pm 1$ matrices. A *prime implication* of $Q(A)$ is a generalized set covering inequality $ax \geq 1 - n(a)$ which is satisfied by all the $0, 1$ vectors in $Q(A)$ but is not dominated by any other such generalized set covering inequality. A *row monotonization* of A is any $0, 1$ matrix obtained from a row submatrix of A by multiplying some of its columns by -1. A row monotonization of A is *maximal* if it is not a proper submatrix of any row monotonization of A.

In [Hooker 1996] it is proved that if A is a $0, \pm 1$ matrix such that $Q(A)$ contains all of its prime implications, then A is ideal if and only if all the maximal row monotonizations of A are ideal $0, 1$ matrices.

In [Guenin 1994], the idealness of a $0, \pm 1$ matrix A is linked to the idealness of a single $0, 1$ matrix as follows: Let A be a $0, \pm 1$ matrix such that $Q(A)$ contains all of its prime implications. Then A is ideal if and only if the $0, 1$ matrix D_A is ideal.

Furthermore A is ideal if and only if $\min\{cx : x \in Q(A)\}$ has an integer optimum for every vector $c \in \{0, \pm 1, \pm\infty\}^n$.

In [Nobili, Sassano 1994], a condition for a $0, \pm 1$ matrix A to be ideal, without assuming that $Q(A)$ contains all of its prime implications is given as follows: Let A be a $0, \pm 1$ matrix. Then A is ideal if and only if D_{A^+} is an ideal $0, 1$ matrix, where A^+ is the disjoint completion of A.

Let J be a subset of columns of a $0, \pm 1$ matrix A. The *deletion* of J consists of removing all columns in J, all rows with at least one 1 in a column of J and rows that become dominated. The *contraction* of J consists of removing all columns in J, all rows with at least one -1 in a column of J and rows that become dominated. The *semi-deletion* of J consists of removing all rows with a 1 in at least one column of J and then all zero columns. The *semi-contraction* of J consists of removing all rows with at least one -1 in a column of J and then all zero columns.

Nobili and Sassano define a *weak minor* of a $0, \pm 1$ matrix A to be any submatrix that can be obtained from A by a sequence of deletions, contractions, semi-deletions and semi-contractions. They define A to be *minimally nonideal* if A is not ideal but every weak minor of A is ideal. The usefulness of this concept comes from the fact that a $0, \pm 1$ matrix A is minimally nonideal if and only D_A is a minimally nonideal $0, 1$ matrix.

For $n \geq 3$, the following $n \times n$ $0, \pm 1$ matrix, denoted $\tilde{J}_n$, is minimally nonideal:

$$\tilde{J}_n = \begin{pmatrix} -1 & 1 & 1 & 1 & 1 & 1 \\ 1 & 1 & 0 & 0 & 0 & 0 \\ 1 & 0 & 1 & 0 & 0 & 0 \\ 1 & 0 & 0 & 1 & 0 & 0 \\ 1 & 0 & 0 & 0 & 1 & 0 \\ 1 & 0 & 0 & 0 & 0 & 1 \end{pmatrix}$$

The following characterization of minimally nonideal $0, \pm 1$ matrices is given in [Nobili, Sassano 1994]:

Let A be a $0, \pm 1$ matrix with n columns. Then A is minimally nonideal if and only if A is a switching of $\tilde{J}_n$, after permutation of rows and columns, or A is a switching of a minimally nonideal $0,1$ matrix or A contains an $n \times n$ submatrix B with two nonzeroes per row and per column and $det(B) = \pm 2$ and all rows in A but not in B have at least three nonzeroes.

4 Perfect $0, \pm 1$ Matrices, Bidirected Graphs and Related Conjectures

E. L. Johnson, M. W. Padberg (1985). Degree-two inequalities, clique facets and biperfect graphs. *Ann. Discr. Math.* 16, 169–187.
M. W. Padberg (1973). On the facial structure of set packing polyhedra. *Math. Program.* 5, 199–251.

The *intersection graph* G_A of a $0,1$ matrix A is the graph whose node set is the column set of A and nodes j, k are adjacent if A contains a row i with $a_{ij} = a_{ik} = 1$. The $0,1$ vectors of $P(A)$ are exactly the incidence vectors of the stable sets of G_A. So they are the $0,1$ vectors satisfying the system:

$$x_j + x_k \leq 1, \ \forall jk \in E(G_A)$$

Let K be a maximal clique of G_A. The inequality $\sum_{i \in K} x_i \leq 1$ is clearly satisfied by the above $0,1$ vectors. In fact this is the only such inequality that is facet-inducing and has a right-hand side of 1. So, if A' is the clique matrix of G_A, then the inequalities defining $P(A')$ give a strengthening, with respect to the edge inequalities, for the convex hull of such $0,1$ vectors, see [Padberg 1973].

Johnson and Padberg show that the situation is similar when A is a $0, \pm 1$ matrix:

An inequality $ax \leq 1 - n(a)$ of a generalized set packing problem can be written also as $\sum_{i \in P} x_i + \sum_{j \in N}(1 - x_j) \leq 1$, where P is the set of indices i where $a_i = 1$ and N is the set of indices j where $a_j = -1$. Therefore the set of $0,1$ solutions of the above inequality is exactly the set of $0,1$ solutions of the following system of boolean inequalities with two variables:

$$x_i + x_j \leq 1, \ \forall i, j \in P$$

$$x_i + x_j \geq 1, \ \forall i, j \in N$$

$$x_i \leq x_j, \ \forall i \in P, j \in N$$

Therefore, more generally, if A is a $0, \pm 1$ matrix, the set of $0,1$ vectors in $P(A)$ is also the set of $0,1$ solutions of a system of 2-variable boolean inequalities (2SBI).

We can model these inequalities with a *bigraph* $B = (V; P, N, M)$ where the set of nodes V represents the variables, edges in P represent the inequalities of the first type and have two $+$ signs at their ends, edges in N represent the inequalities of the second type and have two $-$ signs at their ends, and edges in M represent the inequalities of the third type and have a $+$ in correspondence to the endnode i and a $-$ sign for endnode j. So $S \subseteq V$ is such that its representative vector is a solution of a 2SBI if and only if S is an independent set of $B = (V; P, \emptyset, \emptyset)$, a node cover of $B = (V; \emptyset, N, \emptyset)$ and satisfies the precedences of $B = (V; \emptyset, \emptyset, M)$. However, there are some differences with the 0, 1 case: for instance, the inequalities $x_i + x_j \leq 1$ and $x_k \leq x_j$ clearly imply the inequality $x_i + x_k \leq 1$, which is another boolean inequality on two variables, which corresponds to an additional edge of the bigraph. Therefore we can define the *transitive closure* of a 2SBI as the set of boolean constraints satisfied by the 0, 1 solutions of the original system. A 2SBI is *closed* if it coincides with its transitive closure.

Also, unlike the 0, 1 case, a 2SBI may fix the value of a variable to 0 or 1 or to be equal to another variable. A 2SBI is *reduced* if no variable is fixed to 0 or to 1 and no pair of variables must be identically equal or sum to 1. When a 2SBI is not reduced, some variables may be eliminated. It is obvious that if a 2SBI has only inequalities of the first type (the 0, 1 case described above), then this system is always closed and reduced.

Johnson and Padberg show how to compute the closure of 2SBI and to test if an 2SBI is reducible. (This is also known in the context of boolean optimization.)

A *biclique* K in a bigraph is a subset of mutually adjacent nodes, such that no two edges of K meet the same endnode with distinct signs. So the nodes of K are partitioned in K^+ and K^-, according to the signs of the edges of K. The inequality:

$$\sum_{i \in K^+} x_i + \sum_{j \in K^-} (1 - x_j) \leq 1$$

is clearly valid for the 0, 1 vectors in a 2SBI.

Biclique K is *strong* if there exists no node u not in K, that is joined to every node v in K^+ with an edge with sign $+$ for endnode v and to every node w in K^- with an edge with sign $-$ for endnode w. Johnson and Padberg show that an inequality:

$$\sum_{i \in K^+} x_i + \sum_{j \in K^-} (1 - x_j) \leq 1$$

associated to a biclique K is facet-inducing for the convex hull of the 0, 1 solutions of 2SBI if and only if K is a strong biclique.

Johnson and Padberg define a bigraph with set of strong bicliques $\mathcal{K}$ to be perfect if both the linear program

$$\max\{bx : \sum_{i \in K^+} x_i + \sum_{j \in K^-} (1 - x_j) \leq 1 \; K \in \mathcal{K}, \; 0 \leq x \leq 1\}$$

and its dual admit optimal solutions that are integral whenever b is a $0, \pm 1$-valued vector.

Bidirecting the edges of an undirected graph G means assigning $+$ or $-$ signs to the endnodes of each edge. G is *biperfect* if every bigraph that is closed and reduced and is obtained by bidirecting the edges of G, is perfect.

Johnson and Padberg conjecture that a graph G is biperfect if and only if G is perfect (Conjecture 1) and that if there exists a bidirection of G that gives a closed, reduced bigraph that is perfect, then G is perfect (Conjecture 2).

Since bidirecting the edges with $+$, $+$ at their endnodes gives a closed and reduced graph, then every biperfect graph is perfect. Therefore Conjecture 2 implies Conjecture 1.

E.C. Sewell (1996). Binary integer programs with two variables per inequality. *Math. Program.* 75, 467–476.

Both conjectures are proven here: An important tool is the following result: Let

$$x_i + x_j \leq 1, \ \forall i, j \in P$$

$$x_i + x_j \geq 1, \ \forall i, j \in N$$

$$x_i \leq x_j, \ \forall i \in P, j \in N$$

be a 2SBI such that the associated bigraph $B = (V; P, N, M)$ is closed and reduced. Then every maximal vector satisfying:

$$x_i + x_j \leq 1, \ \forall i, j \in P$$

$$0 \leq x \leq 1$$

satisfies also the other two sets of inequalities.

Y. Ikebe, A. Tamura (1996). *Perfect bidirected graphs*, Report CSIM–2 Dept of Computer Science and Information Mathematics, University of Electro-Communications, Tokyo.

W.J. Li (1995) *Degree two inequalities and biperfect graphs*, PhD thesis. State University of New York, Stony Brook.

Together, the two references give another, independent proof of Johnson and Padberg's conjectures. The polyhedral approach of [Guenin 1994] is also relevant here.

5 Balanced Matrices

Berge's motivation for introducing balancedness was to extend to hypergraphs the notion of bipartite graph. A 0, 1 matrix A can be viewed as the *node-edge matrix* of a hypergraph H: the nodes of the hypergraph H correspond to the rows of A and the edges correspond to the columns with edge j containing node i if and only if $a_{ij} = 1$. A hypergraph is *balanced* if its node-edge matrix is balanced. Therefore, a hypergraph is balanced if it has no odd cycle in which every edge contains exactly two nodes of the cycle. Berge defined a hypergraph to be *bicolourable* if its nodes can be partitioned into two classes, say red and blue, in such a way that every edge of cardinality two or greater contains at least one blue and at least one red node. For a hypergraph H with node set V and edge set $\{E_1, \ldots, E_n\}$, the *restriction* of H to $U \subseteq V$ is the hypergraph with node set U and edge set $\{E_1 \cap U, \ldots, E_n \cap U\}$. [Berge 1970] (see above) shows that a hypergraph is balanced if and only if all its restrictions are bicolorable.

When specializing this theorem to graphs we get the well known and easy to prove result that a graph has no odd cycle if and only if it is bipartite. Several properties of bipartite graphs extend to balanced hypergraphs. For example, Berge and Las Vergnas generalize König's theorem. A *matching* is a set of pairwise nonintersecting edges and a *transversal* is a node set that intersects all the edges.

C. Berge, M. Las Vergnas (1970). Sur un Théorème du Type König pour les Hypergraphes. *Annals of the New York Academy of Sciences 175*, 32–40.

In a balanced hypergraph, the maximum cardinality of a matching equals the minimum cardinality of a transversal.

M. Conforti, G. Cornuéjols, A. Kapoor, K. Vušković (1996). Perfect Matchings in Balanced Hypergraphs. *Combinatorica* 16, 325–329.

The celebrated theorem of Hall about the existence of a perfect matching in a bipartite graph also extends to balanced hypergraphs as follows: (A matching is *perfect* if every node belongs to an edge of the matching.) A balanced hypergraph has no perfect matching if and only if there exist disjoint node sets R and B with $|R| > |B|$ and such that every edge contains at least as many nodes in B as in R.

Interestingly, the notion of balanced $0,1$ matrix (motivated as a generalization of bipartite graphs) can itself be extended to $0,\pm1$ matrices, while still preserving many important properties.

Specifically, a $0,\pm1$ matrix is *balanced* if, in every square submatrix with two nonzero entries per row and per column, the sum of the entries is a multiple of four. The class of balanced $0,\pm1$ matrices also properly includes totally unimodular $0,\pm1$ matrices (a matrix is *totally unimodular* if every square submatrix has determinant equal to $0,\pm1$).

M. Conforti, G. Cornuéjols (1995). Balanced $0,\pm1$ Matrices, Bicoloring and Total Dual Integrality. *Math. Program.* 71, 249–258.

A $0,\pm1$ matrix A is *bicolourable* if its rows can be partitioned into blue rows and red rows in such a way that every column with two or more nonzero entries, either contains two entries of opposite signs in rows of the same colour, or contains two entries of the same sign in rows of different colours. For a $0,1$ matrix, this definition coincides with Berge's notion of bicolourable hypergraph. The above paper shows that a $0,\pm1$ matrix is balanced if and only if all its row submatrices are bicolourable.

Berge's theorem stated in the introduction also extends as follows: For a $0,\pm1$ matrix A, the following statements are equivalent:

(i) the matrix A is balanced,

(ii) every submatrix of A is perfect,

(iii) every submatrix of A is ideal.

6 Recognition

Given a $0,1$ matrix A, are there polytime algorithms to recognize whether A is perfect, ideal or balanced? These problems are related:

M. Conforti, M.R. Rao (1992). Properties of balanced and perfect matrices. *Math. Program.* 55, 35–47.

The existence of a polytime algorithm that checks if a $0,1$ matrix is perfect implies the existence of a polytime algorithm that checks if a $0,1$ matrix is balanced. However, no such algorithm is known for perfection or idealness. For balancedness, the situation is quite different:

M. Conforti, G. Cornuéjols, M.R. Rao (1991). *Decomposition of Balanced* $0,1$ *Matrices, Parts I-VII*, Preprints, Graduate School of Industrial Administration, Carnegie Mellon

University, Pittsburgh.

Balancedness of a $0,1$ matrix can be recognized in polytime. The situation is the same for $0,\pm1$ matrices: no algorithm is known for checking that a $0,\pm1$ matrix is perfect or ideal.

M. Conforti, G. Cornuéjols, A. Kapoor, K. Vušković (1993). *Balanced* $0,\pm1$ *Matrices, Parts I and II*, Preprints, Graduate School of Industrial Administration, Carnegie Mellon University, Pittsburgh.

Balancedness of a $0,\pm1$ matrix can be recognized in polytime. The algorithm is complicated and its computational complexity, although polynomial, is rather high. Nevertheless the basic idea is very simple. The algorithm is based on a theorem stating that, if a matrix is balanced, then either it belongs to a "basic class" or else it can be decomposed into smaller matrices using a well-defined "decomposition operation". Based on this theorem, the recognition algorithm recursively decomposes the matrix until no further decomposition exists. Then each of the final blocks is checked for balancedness. For this approach to work, one has to be able to recognize "basic" balanced matrices in polytime and the decomposition operation must have three properties: the fact that a matrix can be decomposed must be detected in polytime, the two blocks of the decomposition should be balanced if and only if the original matrix is balanced and third, the total number of blocks generated in the algorithm must be polynomial. Although no polytime algorithm is known for checking perfection of a graph G, the following approach might be envisioned, assuming Berge's strong perfect graph conjecture holds true: verify that G contains no odd hole (and do the same for $\bar{G}$). This approach is the object of ongoing research. Indeed, checking that every hole of a graph has length 0 $(mod\ 2)$ has much of the same flavour as the problem of checking that every hole of a bipartite graph has length 0 $(mod\ 4)$. And the latter question is equivalent to checking that a $0,1$ matrix is balanced. This is so because a $0,1$ matrix A can be viewed as the node-node matrix of a bipartite graph $B(A)$ having an edge connecting nodes i and j if and only if $a_{ij} = 1$. From the definition, it follows that A is balanced if and only if every hole of the bipartite graph $B(A)$ has length 0 $(mod\ 4)$. Since balancedness can be checked in polytime, a similar approach might be tried for finding an odd hole in a graph. As an added bonus, one might even hope to prove Berge's conjecture!

Acknowledgment: We thank Ondrei Čepek, Paolo Nobili and Antonio Sassano for explaining us the most recent results on perfect and ideal $0,\pm1$ matrices.

7 Advances in Linear Optimization[1]

Cornelis Roos
Delft University of Technology, Delft

Tamás Terlaky
Delft University of Technology, Delft

CONTENTS

Linear optimization as a research topic has existed for fifty years. The first 35 years were dominated by research on simplex-type pivot methods [Lenstra et al. 1991; Padberg 1995; Terlaky and Zhang 1993](see §1). By designing an ellipsoid method, Khachiyan proved in 1979 that the linear optimization problem is solvable in polynomial time [Khachiyan 1979; Padberg 1995](see §1).

Since Karmarkar presented his polynomial algorithm for linear optimization in 1984 [Karmarkar 1984](see §3.6.1), the research in linear optimization concentrated on *Interior Point Methods* (IPMs). By now over 3000 research papers were written on the subject, several powerful scientific and commercial implementations have been designed. This review concentrates on these turbulent developments on IPMs and just gives a brief glimpse on general and pivoting results in §1.

The current success of IMPs is remarkable as one notices that this type of method has already been analyzed and used in the fifties and sixties (see §2.1). The decreasing interest in IPMs in the seventies was due to the negative theoretical and computational results for these methods when applied to nonlinear optimization problems. The significant new results compared to what was already known in the sixties are:

[1]We prefer to use the name *Linear Optimization* instead of the traditional name *Linear Programming*. Nowadays the word 'programming' usually refers to the activity of writing computer programs and, as a consequence its use instead of the more natural word 'optimization' gives rise to confusion.

Annotated Bibliographies in Combinatorial Optimization, edited by M. Dell'Amico, F. Maffioli and S. Martello

1. Introduction and thorough analysis of the central path, duality theory (see §3.2) and sensitivity analysis (see §3.4) based on IPM concepts;
2. Strongly polynomial optimal basis identification procedure (see §3.3);
3. Derivation of polynomial complexity bounds (see §3.6, §3.7 and §3.8) by inventing proper proximities to measure distance to an exact minimizer, and the analysis of Newton's method for minimizing barrier functions (see §3.7);
4. Development of primal-dual methods (see §3.5.1, §3.6.1, §3.7.1 and §3.8);
5. Results on asymptotic convergence behavior (see §3.8);
6. Design of infeasible interior point methods (see §3.9);
7. Preprocessing techniques, efficient sparse matrix technology to solve Newton systems (see §4);
8. High performance implementations (see §5).

These developments are discussed in detail in the cited books (see §2.2), edited volumes (see §2.3) and surveys (see §3.1).

In the early years after Karmarkar's paper, IPMs were criticized for not producing an optimal basis solution and hence not allowing sensitivity analysis. Both criticisms turned out to be invalid. Megiddo [1991](see §3.3) showed that an optimal basis can be obtained from interior optimal solutions in strongly polynomial time, while this is not true for the inverse problem. Concerning sensitivity analysis some people realized that sensitivity analysis based on an optimal basis, as discussed in most text-books and implemented in packages is incorrect [Rubin, Wagner 1990](see §3.4). It turned out that it not only can be done based on interior optimal solutions [Adler, Monteiro 1992](see §3.4) but that this approach has several advantages.

To give a brief survey of IPMs let us consider the linear optimization problem in the standard form

$$(P) \quad \min_x \{ \, c^T x \; : \; Ax = b, \; x \geq 0 \, \},$$

where $c, x \in \mathbb{R}^n$, $b \in \mathbb{R}^m$ and A is an $m \times n$ matrix of rank m, and its dual problem

$$(D) \quad \max_{y,s} \{ \, b^T y \; : \; A^T y + s = c, \; s \geq 0 \, \},$$

where $y \in \mathbb{R}^m$ and $s \in \mathbb{R}^n$. The optimality conditions for (P) and (D) are given by the nonlinear system

$$\begin{aligned} Ax &= b, \quad x \geq 0, \\ A^T y + s &= c, \quad s \geq 0, \\ Xs &= 0, \end{aligned} \tag{1}$$

where X is the diagonal matrix with the elements of x on the diagonal. The fundamental idea in interior point methods is to replace the (polyhedral) nonnegativity conditions by either an ellipsoid or by some smooth potential/barrier function.

Given a strictly feasible point $\tilde{x} > 0$ (i.e. $\tilde{x}$ is feasible and $\tilde{x} > 0$), affine scaling methods replace the inequality constraints $x \geq 0$ by

$$(x - \tilde{x})^T \tilde{X}^2 (x - \tilde{x}) \leq \rho$$

where $\rho < 1$. Then one solves the resulting easy linear optimization problem over an ellipsoid and makes a step either to the boundary of the above ellipsoid or by a fixed fraction of the step to the boundary of the feasible set. Affine scaling methods were first proposed by [Dikin 1967](see §2.1) and are discussed in detail in §3.5.

Karmakar (see §3.6) considered the problem in the form

$$\min_x \{ c^T x \ : \ Ax = 0, \ e^T x = n, \ x \geq 0 \},$$

where e is the all one vector and he assumed that the optimal objective value is zero. By using a projective transformation the current interior point $\tilde{x}$ is projected into the point e, then the transformed objective is optimized over an inscribed sphere and the new point is transformed back to the original space by the inverse transformation. Karmarkar measured the progress of the algorithm by the potential function

$$\log \frac{(c^T x)^n}{\Pi_i x_i}.$$

At each step a constant reduction of this potential function is realized.

Most interior point methods explicitly or implicitly incorporate the inequality constraints in the objective function by a logarithmic term with a parameter that is gradually decreased to zero [Frisch 1955](see §2.1). The logarithmic barrier functions for (P), (D) and for the primal–dual problem together are given by

$$\begin{aligned} f_P(x;\mu) &= c^T x - \mu \sum_{i=1}^n \ln x_i, \\ f_D(y,s;\mu) &= -b^T y - \mu \sum_{i=1}^n \ln s_i, \\ f_{PD}(x,y,s;\mu) &= c^T x - b^T y - \mu \sum_{i=1}^n \ln(x_i s_i), \end{aligned}$$

with $\mu \geq 0$. To have the barrier functions well defined we need the so-called *interior point assumption*: There exists (x^0, y^0, s^0) such that

$$Ax^0 = b, \ x^0 > 0, \ A^T y^0 + s^0 = c, \ s^0 > 0. \tag{2}$$

For fixed μ the logarithmic barrier functions can be minimized over the respective feasible sets. The minimizers $x(\mu)$ of $f_P(x;\mu)$, for each $\mu > 0$, exist and form the so-called central path of the primal problem. Similarly the minimizers $y(\mu), s(\mu)$ of $f_D(y,s;\mu)$ form the dual central path and, finally the same vectors $x(\mu), y(\mu), s(\mu)$ minimize f_{PD}. The primal logarithmic barrier method, for a given value of μ, compute an approximation to $x(\mu)$, then update μ and repeat. The minimization is done by Newton steps (with or without line-search). Similar procedure holds for the dual and primal-dual methods respectively. An important difference in various path-following methods is the 'closeness to the path' for successive iterates, which is intimately related to the updates of the barrier parameter μ between successive iterations. This way one speaks about small- (full Newton-) step methods and about large-update methods. All polynomial path-following (see §3.7) and predictor-corrector methods (see §3.8) follow a (possibly weighted) central path. Further, with respect to the asymptotical behavior of the algorithms it was shown that, by applying the predictor–corrector method, the duality gap converges quadratically to zero (see §3.8).

The following results are known about the central paths. It is easy to check that for all the three barrier functions over the respective spaces, the optimality conditions are given by

$$\begin{aligned} Ax &= b, \quad x \geq 0 \\ A^T y + s &= c, \quad s \geq 0 \\ Xs &= \mu e. \end{aligned} \tag{3}$$

Comparing the systems (1) and (3) one sees that interior point methods relax the (nonlinear) complementarity condition. The system (3) has a unique solution for each $\mu > 0$, which is $(x(\mu), y(\mu), s(\mu))$. The set $\{x(\mu) : \mu \geq 0\}$ defines a smooth analytic path in the primal feasible region $\mathcal{P}$. The set $\{(y(\mu), s(\mu)) : \mu \geq 0\}$ defines a smooth analytic path in the dual feasible region $\mathcal{D}$. For fixed μ, the duality gap on the paths is $c^T x(\mu) - b^T y(\mu) = x(\mu)^T s(\mu) = n\mu$. The paths converge to optimal solutions of (P) and (D), i.e.,

$$\lim_{\mu \to 0} x(\mu) = x^*, \ \lim_{\mu \to 0} y(\mu) = y^*, \ \lim_{\mu \to 0} s(\mu) = s^*,$$

where (x^*, y^*, s^*) satisfies system (1). The solution (x^*, y^*, s^*) is strictly complementary, i.e., $x^* + s^* > 0$.

Nowadays all efficient implementations (see §4 and §5) employ a primal-dual method, i.e. at each iteration both a primal and a dual interior solutions are available and updated. Supposing $(\tilde{x}, \tilde{y}, \tilde{s})$ to be given vectors with $\tilde{x} > 0$ and $\tilde{s} > 0$, one would like to compute a displacement $(\Delta x, \Delta y, \Delta s)$ such that

$$\begin{aligned} A(\tilde{x} + \Delta x) &= b, \\ A^T(\tilde{y} + \Delta y) + \tilde{s} + \Delta s &= c, \\ (\tilde{X} + \Delta X)(\tilde{s} + \Delta s) &= \mu e, \end{aligned}$$

for some $\mu > 0$. The last equation contains the nonlinear term $\Delta X \Delta s$; hence using Newton's linearization, the search-direction is obtained by solving the linear system

$$\begin{aligned} A\Delta x &= b - A\tilde{x}, \\ A^T \Delta y + \Delta s &= c - A^T \tilde{y} - \tilde{s}, \\ \tilde{X}\Delta s + \tilde{S}\Delta x &= \mu e - \tilde{X}\tilde{s}. \end{aligned} \tag{4}$$

Initially, the methods used feasible points i.e. the first two equations have right-hand-side zero. Later, infeasible iterates were used in implementations with great success (see §4 and §3.9 for supporting theory). A different, more elegant way to handle infeasibility is to incorporate both the primal and dual problem in a new self-dual problem, for which a starting point is readily available. Moreover, a certificate is given for the infeasibility of the original problems if it applies [Ye et al. 1994](see §3.9).

In all interior point methods the major amount of work per iteration is in solving the system (4). First substituting in the second equation the value of Δs obtained from the third equation we get

$$A\Delta x = b - A\tilde{x}, \quad A^T\Delta y - \tilde{X}^{-1}\tilde{S}\Delta x = (c - A^T\tilde{y} - \tilde{s}) - \tilde{X}^{-1}(\mu e - \tilde{X}\tilde{s}). \tag{5}$$

This symmetric indefinite system, the so-called *augmented system*, can efficiently be solved by using Bunch–Parlett factorization. Alternatively, multiplying the second equation in (5) by $A\tilde{X}\,\tilde{S}^{-1}$ and using the first we have

$$A\tilde{X}\,\tilde{S}^{-1}A^T\Delta y = A\tilde{X}\,\tilde{S}^{-1}((c - A^T\tilde{y} - \tilde{s}) - \tilde{X}^{-1}(\mu e - \tilde{X}\tilde{s})) + (b - A\tilde{x}).$$

Efficient sparse Cholesky factorization techniques have been developed to solve the latter system, which is called the *normal equation.*

The theoretical developments, the efficient sparse-matrix technology and the use of extensive preprocessing to eliminate redundant constraints and variables made IPM implementations so successful. None of these three were available in the seventies when IPMs were put aside.

Finally, we want to indicate that IPMs are generalized to smooth convex programming problems. We do not discuss these results here although some items in the bibliography contain such sections. The most general results are given by

Y.E. Nesterov and A.S. Nemirovskii (1993). *Interior Point Polynomial Methods in Convex Programming : Theory and Algorithms*, SIAM Publications, Philadelphia, USA.

The so-called *self-concordance condition* of Nesterov and Nemirovskii (see also [den Hertog 1994] in §2.2) characterizes the most general problem class where polynomial results are derived. These include linear and convex quadratic programming, quadratically constrained quadratic programming, geometric programming, ℓ_p approximation, entropy and ℓ_p-optimization. One of the generalities of the work by Nesterov and Nemirovskii is that their results are not restricted to functions that are defined on the space of real numbers. Another important application is to the space of symmetric semidefinite matrices. Many problems in mathematical programming can be modeled as so-called semidefinite programs, for which only interior point methods are known to be efficient. The general statement is as follows:

$$(SP) \quad \max_y \{\, b^T y \;:\; F(y) \preceq F_0 \,\}$$

where $y, b \in \mathbb{R}^m$, $F(y) = \sum_{i=1}^m y_i F_i$ and $F_i = F_i^T \in \mathbb{R}^{n\times n}$ for $i = 0, \ldots, m$; here $F(y) \preceq F_0$ means $F_0 - F(y)$ is positive semidefinite. For a survey Dord

L. Vandenberghe and S. Boyd (1996). Semidefinite programming. *SIAM Review* 38, 49–95 and

F. Alizadeh (1991). *Combinatorial Optimization with Interior Point Methods and Semi-Definite Matrices, Ph.D. Thesis*, University of Minnesota, Minneapolis, Minnesota.

1 Linear Optimization in General

In this section we list some recent books that can be used as a general introduction to linear optimization. Also some important papers are mentioned that do not deal with IPMs.

L.G. Khachiyan (1979). A polynomial algorithm in linear programming. *Doklady Akademiia Nauk SSSR* 244, 1093–1096. Translated into English in *Soviet Mathematics Doklady* 20, 191–194.

The first polynomial algorithm, the *ellipsoid method* for solving linear optimization problems. Unfortunately, theoretical efficiency of the ellipsoid method did not come together with practical efficiency.

J.K. Lenstra and A.H.G. Rinnooy Kan and A. Schrijver (eds.) (1991). *History of mathematical programming. A collection of personal reminiscences*, CWI, North–Holland, Amsterdam.

The pioneers of mathematical programming tell us how the fundamental results of the field were established. Among others Dantzig, Gomory, Edmonds and Kuhn give their view on the origins of the field.

S.C. Fang and S. Puthenpura (1993). *Linear Optimization and Extensions*, Prentice Hall, Englewood Cliffs, NJ.

The first book that intended to give a concise introduction to all results and methods for linear optimization, including IPMs. Unfortunately it neglects complexity results and contains too many typos.

M. Padberg (1995). *Linear Optimization and Extensions*, Springer, Berlin.

This new book provides an excellent introduction to the classical theory of linear optimization and provides probably the best survey of the related polyhedral theory and the ellipsoid method. It contains a chapter on IPMs, but this part concentrates on Karmarkar's projective algorithm. Finally some combinatorial applications follow.

R. Saigal (1995). *Linear Programming: A Modern Integrated Analysis*, Kluwer Academic Publishers, Dordrecht.

After a short survey of the necessary mathematical background the duality theory of linear optimization is discussed. Discussing simplex methods a very limited attention is paid to degeneracy problems. The most important part discusses interior point methods. This covers a detailed analysis of affine scaling algorithms but also presents a projective and a path-following method. Brief instructions to implement both simplex and IPMs are given as well.

J. E. Beasley (ed.) (1996). *Advances in Linear and Integer Programming*, Oxford University Press, Oxford.

This volume gives a survey of the major advances in linear and integer optimization. Each chapter is written by recognized experts. Both the simplex method and the IPM are considered. Implementation issues, as well as applicability to solve combinatorial optimization problems are discussed.

T. Terlaky and S. Zhang (1993). Pivot rules in linear programing: A survey. *Ann. Oper. Res.* 46, 203–233.

None of the above recent books give a fair survey of recent results on pivot methods. This paper fills this gap. All recent minimal index and recursive type pivot rules are discussed in a uniform way. Pivot rules that can be related to the shadow vertex algorithm and pivot rules that were motivated by some interior point ideas are surveyed.

R.E. Bixby (1994). Progress in linear programming. *ORSA J. Comput.* 6(1), 15–22.
Describes how simplex based solvers were accelerated by several orders in the last decade.

2 Interior Point Methods

2.1 Classics of interior point methods

Here the first papers on different interior point methods from the early years of linear optimization are collected. Some of them remained unnoticed up till the recent interest in IPMs.

R. Frisch (1955). The logarithmic potential method for solving linear programming problems. *Memorandum*, University Institute of Economics, Oslo.
This is the first paper that proposes the use of logarithmic barrier functions in linear optimization.

C.W. Caroll (1961). The created response surface technique for optimizing nonlinear restrained systems. *Oper. Res.*, 9, 2, 169–184.
P. Huard (1967). Resolution of mathematical programming with nonlinear constraints by the method of centers. In J. Abadie (ed.) *Nonlinear Programming*, North-Holland, Amsterdam, 207–219.
Both papers introduce the method of centers. They propose to calculate the analytic center of the feasible set cutted by the objective plane shifted to a proper position and furnished with a proper weight.

A.V. Fiacco and G.P. McCormick (1968, 1990). *Nonlinear Programming : Sequential Unconstrained Minimization Techniques*, John Wiley & Sons, 1968, New York. Reprint: Volume 4 of *SIAM Classics in Applied Mathematics*, SIAM Publications, Philadelphia, 1990.
The classical thorough treatment of barrier methods. It is still an every day handbook for IPM people. A Sequential Unconstrained Minimization Technique (SUMT) was developed based on the results summarized in the book.

F.A. Lootsma (1974). *Boundary Properties of Penalty Functions for Constrained Minimization. Ph.D. Thesis*, Eindhoven, University of Technology.
A detailed analysis of the possible numerical, ill-conditioning problems in penalty-barrier methods.

I.I. Dikin (1967). Iterative solution of problems of linear and quadratic programming. *Doklady Akademii Nauk SSSR* 174, 747–748. Translated in: *Soviet Mathematics Doklady*, 8, 674–675.
I.I. Dikin (1974). On the convergence of an iterative process. *Upravlyaemye Sistemi* 12, 54–60, (in Russian).
These papers introduce the affine scaling algorithm, its generalization to optimization problems with nonlinear convex objective and convergence analysis under dual nondegeneracy assumption.

2.2 Books on interior point methods

The rest of this chapter is devoted to recent developments on IPMs. The interested reader can reach the most recent papers and lots of useful information in the IPM WWW–site on the following internet address:

`http://www.mcs.anl.gov:80/home/otc/InteriorPoint/`

D. den Hertog (1994). *Interior Point Approach to Linear, Quadratic and Convex Programming, Algorithms and Complexity*, Kluwer Academic Publishers, Dordrecht.

This book gives a thorough analysis of primal and dual logarithmic barrier path following methods and methods of centers for linear and smooth convex optimization.

A. Arbel (1993). *Exploring Interior Point Linear Programming: Algorithms and Software*, Foundations of Computing Series, MIT Press, Cambridge, MA.

An easy to read introduction to the basic concepts of IPMs and implementations. It comes together with a floppy disc that contains implementations of the primal, dual affine scaling and a primal-dual path following method with different sparsity handling heuristics. This tool is very useful for illustrations.

S.J. Wright (1996). *Primal-Dual Interior Point Methods*, SIAM Publications, Philadelphia.

It is common knowledge that primal-dual interior point methods have attractive theoretical properties and perform best in practice. This book is synthesizing the most important developments of the last ten years. It is an excellent handbook for all researchers and graduate students.

C. Roos, T. Terlaky and J.-Ph. Vial (1997). *Theory and Algorithms for Linear Optimization: An Interior Point Approach*, John Wiley & Sons, New York.

The approach to linear optimization in this book is new in many aspects. In particular the interior point methods based development of duality theory is surprisingly elegant. The algorithmic part contains a complete discussion of many algorithmic variants, including predictor-corrector methods, partial updating, higher order methods sensitivity and parametric analysis and implementation issues. This textbook can be used both for undergraduate and graduate courses as well.

2.3 Edited volumes

N. Megiddo (ed.) (1989). *Progress in Mathematical Programming : Interior Point and Related Methods*, Springer Verlag, New York.

This volume contains several fundamental papers belonging to the most frequently cited ones in the IPM literature.

C. Roos and J.-Ph. Vial (eds.) (1991). *Math. Program. (B)* 52. No.3.
J.-L. Goffin and J.-Ph. Vial (eds.) (1995). *Math. Program. (B)* 69. No.1.
K.M. Anstreicher and R.M. Freund (eds.) (1996). Interior Point Methods in Mathematical Programming. *Ann. Oper. Res.* 62.

These three special issues are rich collections of state of the art research papers concerning the theory and applications of IPMs.

T. Terlaky (ed.) (1996). *Interior Point Methods of Mathematical Programming*, Kluwer Academic Publishers, Dordrecht.

Most aspects of IPMs are covered by this volume. Some of the best experts survey their specific area. The linear programming part contains surveys about the theory of linear optimization by IPMs, the affine scaling, path following, potential reduction algorithms, infeasible IPMs and a quite extensive survey of implementation issues.

3 Papers

We classify the papers on IPMs according to the type of algorithm developed and analyzed. Before going into such discussions some survey papers are presented.

3.1 Reviews

C.C. Gonzaga (1991). Search directions for interior linear programming methods. *Algorithmica* 6, 153–181.
D. den Hertog and C. Roos (1991). A survey of search directions in interior point methods for linear programming. *Math. Program.*, 52, 481–509.

These two papers were crucial to the synthesis of the excessive literature of IPMs. It is shown that most interior point methods use a search direction that is an appropriate combination of the affine scaling and the centering direction.

C.C. Gonzaga (1991). Large step path-following methods for linear programming, Part I : Barrier function method. *SIAM J. Optim.* 1, 268–279.
C.C. Gonzaga (1991). Large step path-following methods for linear programming, Part II : Potential reduction method. *SIAM J. Optim.* 1, 280–292.

These papers provide the first unified survey of a large part of the interior point literature, more specifically short and long step path-following methods are considered.

O. Güler, D. den Hertog, C. Roos, T. Terlaky and T. Tsuchiya (1993). Degeneracy in interior point methods for linear programming. *Ann. Oper. Res.* 46, 107–138.

Interior point methods suffer much less from degeneracy problems than pivot methods. The complexity results are not effected by degeneracy, but it effects the convergence of the affine scaling algorithm, the efficiency of optimal basis identification techniques and the numerical stability of calculations..

3.2 Theoretical results

G. Sonnevend (1986). An "analytic center" for polyhedrons and new classes of global algorithms for linear (smooth, convex) programming. *System Modeling and Optimization: Proceedings of the 12th IFIP–Conference* held in Budapest, Hungary, September 1985, (eds.) A. Prékopa, J. Szelezsán and B. Strazicky, 866–876. Lecture Notes in Control and Information Sciences, 84, Springer Verlag, Berlin.
D.A. Bayer and J.C. Lagarias (1989). The nonlinear geometry of linear programming, Part I : Affine and projective scaling trajectories. *Trans. of the American Math. Soc.* 314, 2, 499–526.
N. Megiddo (1989). Pathways to the optimal set in linear programming. N. Megiddo

(ed.) *Progress in Mathematical Programming: Interior Point and Related Methods*, Springer Verlag, New York, 131–158.

The central path plays a crucial role in almost all IPMs. These three papers were crucial to highlighting the importance of analytic centers and the central path. Sonnevend was the first to define the analytic center of a polytope and suggested an algorithmic scheme based on this fundamental concept. The central path is the set of analytic centers of appropriate level sets. Bayer and Lagarias and Megiddo analyzed the properties of the central path in detail.

O. Güler, C. Roos, T. Terlaky and J.-Ph. Vial (1995). A survey of the implications of the behavior of the central path for the duality theory of linear programming. *Management Sci.* 41, 1922–1934.
B. Jansen, C. Roos and T. Terlaky (1993). The theory of linear programming: skew symmetric self–dual problems and the central path. *Optimization* 29, 225–233.

These two papers explore what the appealing properties of the central path imply for the duality theory of linear optimization. Based on this fundamental concept a self-contained duality theory using only elementary calculus is developed. The second paper gives also an elegant solution how to handle the initialization problem (see also § 3.9).

3.3 Crossover

N. Megiddo (1991). On finding primal- and dual-optimal bases, *ORSA J. Comput.* 3, 63–65.

Simplex methods provide an optimal basis and an optimal basic solution. IPMs provide optimal solutions close to the analytic center of the optimal faces. Megiddo showed that identifying an optimal basis from an interior point optimal solution is strongly polynomial (observe that the reverse problem, identifying an interior, i.e. strictly complementary, solution from an optimal basic solution is equivalent to solving the original problem).

E.D. Andersen and Y. Ye (1996). Combining interior-point and pivoting algorithms for linear programming. IT Report, Dept. of Management Sci., The University of Iowa, Iowa. *Management Sci.*, 42(21), 1719–1731.

Using Megiddo's procedure, a practical optimal basis identification algorithm is developed and tested.

Computational aspects of basis identification procedures are discussed in several papers in §4.

3.4 Sensitivity analysis

Up to recently sensitivity analysis was done based on a given optimal basis.

D.S. Rubin and H.M. Wagner (1910). Shadow prices: tips and traps for managers and instructors. *Interfaces* 20, 150–157.

This paper called the attention to this practically important subject. It makes explicit that dual variables given by a single optimal basis are not always shadow prices, thereby indicating the weakness of the classical approach to sensitivity analysis.

J.E. Ward and R.E. Wendell (1990). Approaches to sensitivity analysis in linear programming. *Ann. Oper. Res.* 27, 3–38.

The first but unpractical treatment of sensitivity analysis based on optimal bases which avoids all traps indicated in the previous paper.

I. Adler and R.D.C. Monteiro (1992). A geometric view of parametric linear programming. *Algorithmica* 8, 161–176.
R.D.C. Monteiro and S. Mehrotra (1996). A general parametric analysis approach and its implication to sensitivity analysis in interior point methods. *Math. Program.* 72, 65–82.
B. Jansen, J.J. de Jong, C. Roos and T. Terlaky (1993). Sensitivity analysis in linear programming: just be careful! *Report* AMER.93.022, Royal/Shell Laboratories Amsterdam, Amsterdam.
B. Jansen, C. Roos and T. Terlaky (1992). An interior point approach to postoptimal and parametric analysis in linear programming. *Report* 92–21, Dept. of Math. and Comp. Sci., T.U. Delft, Delft. To appear in *EJOR*.

The correct theory of sensitivity analysis for linear optimization problems is developed in these papers. First the importance of a central (strictly complementary) solution and the optimal partition is highlighted. Later the analysis was extended so that any description of the optimal sets suffices.

H.J. Greenberg (1994). The use of the optimal partition in a linear programming solution for postoptimal analysis. *Oper. Res. Lett.* 15, 179–186.

The practical implication of having an optimal interior solution and having the optimal partition is discussed and illustrated. Practical problems are considered when it is important to know the optimal partition.

3.5 Affine scaling algorithms

Affine scaling algorithms are probably the simplest IPMs. Their complexity is not established yet, it is conjectured that they are not polynomial.

3.5.1 Primal or dual methods

E.A. Barnes (1986). A variation on Karmarkar's algorithm for solving linear programming problems. *Math. Program.* 36, 174–182.
R.J. Vanderbei and M.S. Meketon and B.A. Freedman (1986). A modification of Karmarkar's linear programming algorithm. *Algorithmica* 1, 4, 395–407.
I.I. Dikin (1988). Letter to the editor. *Math. Program.* 41, 393–394.

In §2.1 we have seen that affine scaling methods were discovered in the sixties. This method was rediscovered, while Dikin pointed out that it is in fact his old method.

R.J. Vanderbei and J.C. Lagarias (1990). I.I. Dikin's convergence result for the affine–scaling algorithm. *Contemporary Math.* 114, J.C. Lagarias and M.J. Todd (eds.) American Mathematical Society, Providence, Rhode Island, 109–119.

Dikin's results were made available for the English speaking IPM community. Surprisingly enough, Dikin's convergence proofs used weaker nondegeneracy assumptions than the first cited two papers.

T. Tsuchiya (1991). Global convergence of the affine scaling methods for linear programming problems. *Math. Program.* 52, 377–404.
I.I. Dikin (1991). The convergence of dual variables. *Report* December, Siberian Energy Institute, Irkutsk, Russia.
T. Tsuchiya and M. Muramatsu (1995). Global convergence of the long–step affine scaling algorithm for degenerate linear programming problems. *SIAM J. Optim.* 5, 3, 525–551.

The three most important papers to establish the convergence of the affine scaling algorithm without any nondegeneracy assumption. The final result proves that with step length at most two third to the boundary the primal iterates converge to a point in the relative interior of the primal optimal set and the dual estimates converge to the analytic center of the dual optimal set. This bound is sharp.

W.F. Mascarenhas (1993). The affine scaling algorithm fails for $\lambda = 0.999$. *Report* October, Universidade Estadual de Campinas, Campinas Sao. Paolo.

A surprising counter-example that with very large step the affine scaling algorithm converges to a nonoptimal solution.

T. Terlaky and T. Tsuchiya (1996). A note on Mascarenhas' counter example about global convergence of the affine scaling algorithm. *Report* 96–45, Dept. of Math. and Comp. Sci., T.U. Delft, Delft.

Mascrenhas' counter-example is sharpened, the analysis is simplified. The step-length in the counter-example is reduced to about 0.917, smaller than the step-length used in all efficient implementations.

3.5.2 Primal–dual methods

R.D.C. Monteiro, I. Adler and M.G.C. Resende (1990). A polynomial–time primal–dual affine scaling algorithm for linear and convex quadratic programming and its power series extension. *Math. Oper. Res.* 15, 191–214.

The primal-dual affine scaling algorithm is proved to be polynomial without any nondegeneracy assumption. The iteration complexity is $O(\sqrt{n}L)$ worse than the best known complexity result.

B. Jansen, C. Roos and T. Terlaky (1996). A new and polynomial-time Dikin-type primal-dual interior point method for linear programming. *Math. Oper. Res.* 21. 341–353.

Applying Dikin's original idea a better variant of the primal-dual affine scaling algorithm is developed. The complexity is improved by a factor of $O(L)$.

3.6 Potential reduction algorithms

Potential reduction algorithms have a distinguished role in the area of interior point methods for mathematical programming because Karmarkar's original algorithm belonged to this class.

3.6.1 Primal or dual methods

N. Karmarkar (1984). A new polynomial-time algorithm for linear programming. *Combinatorica* 4, 373–395.

Karmarkar's epoch making paper that initiated the research into interior point methods. It was based on four key ingredients: a non-standard problem formulation, projective transformations, interior iterates and a potential function with which the progress of the algorithm was measured.

M. Iri and H. Imai (1986). A multiplicative barrier function method for linear programming. *Algorithmica* 1, 455–482.

A multiplicative version of Karmarkar's potential function. Polynomial complexity results for this method were established just years later.

M.J. Todd and B.P. Burrell (1986). An extension of Karmarkar's algorithm for linear programming using dual variables. *Algorithmica* 1, 409–424.
G. de Ghellinck and J.-Ph. Vial (1986). A polynomial Newton method for linear programming. *Algorithmica,* 1, 425–453.
M.J. Todd and Y. Ye (1990). A centered projective algorithm for linear programming. *Math. Oper. Res.* 15, 508–529.
K.M. Anstreicher (1986). A monotonic projective algorithm for fractional linear programming. *Algorithmica* 1, 483–498.
R.M. Freund (1991). A potential-function reduction algorithm for solving a linear program directly from an infeasible 'warm start'. *Math. Program.* 52, 441–466.

It was shown soon that the non-standard formulation could be avoided. Algorithms were developed that avoided the use of the projective transformations, but retained the use of a potential function to measure progress. Thus one may conclude that the really essential elements in Karmarkar's approach are interior iterates and the potential function. Some potential reduction algorithms allow to start the algorithm from infeasible points. Further modifications gave potential reduction algorithms having the state-of-the-art theoretical complexity of $O(\sqrt{n}L)$ iterations.

M. Kojima, S. Mizuno and A. Yoshise (1991). An $O(\sqrt{n}L)$ iteration potential reduction algorithm for linear complementarity problems. *Math. Program.* 50, 331–342.

Best theoretical complexity with a primal-dual method.

M.J. Todd (1996). Potential-reduction methods in mathematical programming, *Report* School of IE/OR, Cornell University, Ithaca, NY. *Math. Program.*, 76, 3–45.
K.M. Anstreicher (1996). Potential reduction algorithms. T. Terlaky, (ed.) *Interior Point Methods of Mathematical Programming,* Kluwer Academic Publishers, Dordrecht, 189–252.

Two excellent recent surveys of potential reduction algorithms.

R.D.C. Monteiro (1992). On the continuous trajectories for a potential reduction algorithm for linear programming. *Math. Oper. Res.* 17, 225–253.
M. Muramatsu and T. Tsuchiya (1993). A convergence analysis of the projective scaling algorithm based on a long-step homogeneous affine scaling algorithm. The Institute of Statistical Math. Tokyo, Japan. To appear in *Math. Program.*

The analyis of the trajectories of potential functions help to understand the algorithms. Such fundamental analysis is done in these papers.

3.6.2 Volumetric potential reduction algorithms

P. Vaydia (1990). An algorithm for linear programming which requires $O(((m+n)n^2 + (m+n)^{1.5}n)L)$ arithmetic operations. *Math. Program.* 114, 175–201.

Volumetric barrier methods were suggested by Nesterov and Nemirovskii (see the introduction). The surprising result is that the iteration complexity of such methods does not depend on n, the number of inequality constraints in the standard dual form but only on m the number of variables in the dual problem. Vaydia succeeded in proving this surprising result, although the constants involved are extremely large.

K.M. Anstreicher (1996). Large step volumetric potential reduction algorithms for linear programming. *Ann. Oper. Res.* 62, 521–538.

Vaydia's results are improved considerably. The analysis is simplified and the constants are reduced to reasonable size.

3.7 Path-following algorithms

The large majority of interior point methods follow the central path as a guideline. All of these methods are essentially primal-dual because if a primal (dual) iterate is sufficiently close to the respective central path, a dual (primal) interior feasible solution is always available.

3.7.1 Primal or dual methods

J. Renegar (1988). A polynomial-time algorithm, based on Newton's method, for linear programming. *Math. Program.* 40, 59–93.
C.C. Gonzaga (1989). An algorithm for solving linear programming problems in $O(n^3L)$ operations. N. Megiddo, (ed.) *Progress in Mathematical Programming : Interior Point and Related Methods*, Springer Verlag, New York, 1–28.

The first papers on short-step path-following algorithms developed independently. Both results achieve the best known $O(\sqrt{n}L)$ iteration complexity and $O(n^3L)$ computational complexity result to date. A partial updating method is used to reduce the computational cost per iteration.

C. Roos and J.-Ph. Vial (1992). A polynomial method of approximate centers for linear programming. *Math. Program.* 54, 295–305.

A simple and easy to understand analysis of path-following methods.

C. Roos and J.-Ph. Vial (1989). Long steps with the logarithmic penalty barrier function in linear programming. J. Gabszevwicz, J. F. Richard and L. Wolsey, (eds.) *Economic Decision-Making : Games, Economics and Optimization*, Elsevier Science Publisher B.V., Amsterdam, 433–441.
C. C. Gonzaga (1991). Large step path-following methods for linear programming, Part I : Barrier function method. *SIAM J. Optim.* 1, 268–279.

Long-step path following algorithms based on large updates of the barrier parameter and subsequent Newton steps. In spite the worst case complexity of these algorithms is $O(\sqrt{n})$ worse then their short-step counterparts, they are more practical. Here the practical performance is much better then the pessimistic worst case analysis indicates.

3.7.2 Primal–dual methods

M. Kojima, S. Mizuno, and A. Yoshise (1989). A primal–dual interior point algorithm for linear programming. N. Megiddo, (ed.) *Progress in Mathematical Programming, Interior–Point and Related Methods*, Springer-Verlag, New York, 29–47.
M. Kojima, S. Mizuno, and A. Yoshise (1989). A polynomial–time algorithm for a class of linear complementary problems. *Math. Program.* 44, 1–26.

It is shown that primal-dual problems are solvable simultaneously by interior point methods. It turned out that this approach has several advantages above pure primal or pure dual methods.

R. D. C. Monteiro and I. Adler (1989). Interior path following primal–dual algorithms. Part I: linear programming. *Math. Program.* 44, 27–41.

The primal-dual interior point approach is explored.

B.Jansen, C. Roos, T. Terlaky and J.-Ph. Vial (1996). Primal-dual target following algorithms for linear programming, *Ann. Oper. Res.*, 62, 197–231.
B.Jansen, C. Roos, T. Terlaky and J.-Ph. Vial (1996). A long-step primal-dual target following algorithm for linear programming. *Report* 94–46, Dept. of Math. and Comp. Sci., T.U. Delft, Delft. *Math. Meth. Oper. Res.*, 44, 11–30.

Most path-following algorithms are put in a unified frame by using the one-to-one map of primal and dual interior iterates and positive vectors in $\mathbb{R}^n_+$. Interior point methods define target sequences, the subsequent targets are approximated by (possibly damped) Newton steps.

3.8 Predictor corrector algorithms

S. Mehrotra (1992). On the implementation of a (primal-dual) interior point method. *SIAM J. Optim.* 2, 575–601.
S. Mizuno, M.J. Todd and Y. Ye (1993). On adaptive step primal-dual interior-point algorithms for linear programming. *Math. Oper. Res.* 18, 945–963.

The importance of predictor corrector methods is underlined by the fact that as the time being all state-of-the-art implementations are based on predictor corrector methods. The implementation of predictor-corrector algorithms was first suggested by Mehrotra. Several details needed for a practically efficient implementation are discussed as well.

S. Mehrotra (1991). Quadratic convergence in a primal-dual method. *Math. Oper. Res.* 18, 741–751.
Y. Ye, O. Güler, R.A. Tapia and Y. Zhang (1993). A quadratically convergent $O(\sqrt{n}L)$ iteration algorithm for linear programming. *Math. Program.* 59, 151–162.
C.C. Gonzaga and R.A. Tapia (1997). On the quadratic convergence of the simplified Mizuno-Todd-Ye algorithm for linear programming. *Report* 91–41, Dept. of Math. Sci., Rice University, Houston, revised November 1992. *SIAM J. Optim.*, 7, 47–65.

Appropriately adjusted predictor corrector interior point methods converge asymptotically quadratically to optimal solutions. This result indicates that the convergence of IPMs might accelerate as optimum is approached. The last paper showed that in certain interior point methods the iterates themselves converge to the limit-points of the respective central paths.

C.C. Gonzaga (1997). The largest step path following algorithm for monotone linear complementarity problems. *Report* 94–07, Dept. of Math. and Comp. Sci., T.U. Delft. *Math. Program.*, 76, 309–333.

It is worked out what are the largest possible steps in a predictor corrector algorithm so that the iterates remain exactly on the boundary of a predetermined neighborhood of the central path. The result is a quadratically convergent method and, with some extra safeguard, the iterates converge too.

3.9 Infeasible interior point methods

Infeasible interior point methods solve a given linear optimization problem from an arbitrary positive point by generating a sequence of feasible or infeasible positive iterates. Several variants do this directly without using any artificial problem. Another more appealing approach is to embed the linear optimization problem in a somewhat larger self-dual problem with known centered initial interior point and zero optimal value. First we consider papers based on the direct approach.

I. J. Lustig (1990/91). Feasibility issues in a primal-dual interior-point method for linear programming. *Math. Program.* 49, 145–162.
K. Tanabe (1990). Centered Newton method for linear programming: Interior and 'exterior' point method'. K. Tone (ed.), *New Methods for Linear Programming 3*, The Institute of Statistical Math., Tokyo, Japan, 98–100 (in Japanese).

The idea of initiating an IPM with a possibly infeasible point was introduced first in these papers.

I.J. Lustig, R.E. Marsten and D.F. Shanno (1991). Computational experience with a primal-dual interior point method for linear programming. *Lin. Algebra and Its Appl.* 152, 191–222.

The first encouraging computational results are reported with infeasible interior point methods.

M. Kojima, N. Megiddo and S. Mizuno (1993). A primal-dual infeasible-interior-point algorithm for linear programming. *Math. Program.* 61, 261–280.

Global convergence of an infeasible-interior-point algorithm with different step sizes for primal and dual variables is proved here.

Y. Zhang (1994). On the convergence of a class of infeasible interior-point methods for the horizontal linear complementarity problem. *SIAM J. Optim.* 4, 208–227.

Assuming that the problem is feasible, $O(n^2L)$-iteration complexity bound is proved for an infeasible-interior-point algorithm.

S. Mizuno (1994). Polynomiality of infeasible-interior-point algorithms for linear programming. *Math. Program.* 67, 109–119.

Mizuno removed the feasibility assumption that was assumed by Zhang and further showed that the complexity can be reduced to $O(nL)$.

F.A. Potra (1996). An infeasible interior-point predictor–corrector algorithm for linear programming. *SIAM J. Optim.* 6, 19–32.
J. Stoer (1994). The complexity of an infeasible interior-point path-following method for the solution of linear programs. *Optim. Methods and Software*, 3, 1–12.

While the above infeasible methods are based on the feasible central path – which might not exists – these papers are based on a path of infeasible centers in order to compute search directions. These algorithms possess an $O(nL)$-iteration complexity and superlinear convergence.

S. Mizuno, M.J. Todd and Y. Ye (1995). A surface of analytic centers and infeasible-interior-point algorithms for linear programming. *Math. Oper. Res.* 20, 135–162.

The synthesis of the above approaches by analyzing a surface of central paths where the surface is parameterized by the barrier parameter and by a parameter representing infeasibility.

Y. Ye, M.J. Todd and S. Mizuno (1994). An $O(\sqrt{n}L)$-iteration homogeneous and self-dual linear programming algorithm. *Math. Oper. Res.* 19, 53–67.

This paper gives the most elegant solution of the initialization problem. An artificial self-dual problem with known perfectly centered initial interior point and known zero optimal value is constructed. This problem can be solved by any feasible IPM that yields automatically an infeasible IPM for the original problem. Another version of the embedding problem and applications to derive the theory of linear optimization is given by [Jansen, Roos and Terlaky 1993](see §3.2).

4 Implementation Issues

Interior point methods require far less iterations in practice than predicted by theoretical worst-case analysis. Mostly only 10–40 iterations are needed and the growth in the number of iterations is very slow (conjectured to be logarithmically) in the number of variables. Whether this practical observation can be justified by the theory is still an open question.

4.1 Early papers on dual methods

R.E. Marsten and D.F. Shanno (1985). On implementing Karmarkar's algorithm. *Report*, Dept. of MIS, University of Arizona, Tucson, AZ.

R.E. Marsten, M.J. Saltzman, D.F. Shanno, J.F. Ballintijn and G.S. Pierce (1989). Implementation of a dual affine interior point algorithm for linear programming. *ORSA J. Computing* 1, 287–297.

Encouraging numerical results with the dual affine scaling algorithm.

4.2 Preprocessing

Presolving, i.e. preprocessing the problem in order to eliminate redundant constraints and variables, improve sparsity of the coefficient matrix and find a compact equivalent form is a useful technique and almost completely independent of the algorithm to be used. However the importance of preprocessing was only recognized as IPM based solvers were developed.

E.D. Andersen, K.D. Andersen (1995). Presolving in Linear Programming, *Math. Program.* 71, 221–245.

J. Gondzio (1994). Presolve analysis of linear programs prior to applying the interior point method. *Report* 3, Logilab, HEC Geneva, Section of Management Studies, University of Geneva, Revised December 1994. To appear in *ORSA J. Computing.*

These two papers give a thorough survey of the techniques, heuristics used in state-of-the-art implementations.

4.3 Implementing primal–dual methods

As the time being all state-of-the-art implementations use some variants of primal-dual predictor–corrector methods.

S. Mehrotra (1992). On the implementation of a primal-dual interior point method. *SIAM J. Optim.* 2, 575–601.
I.J. Lustig, R.E. Marsten and D.F. Shanno (1992). On implementing Mehrotra's predictor–corrector interior point method for linear programming. *SIAM J. Optim.*, 2, 435–449.

These papers has changed the course in implementations. Since their publication all implementations are based on primal-dual predictor–corrector methods.

R.E. Bixby, J.W. Gregory, I.J. Lustig, R.E. Marsten and D.F. Shanno (1992). Very large-scale linear programming : A case study in combining interior point and simplex methods. *Oper. Res.* 40, 885–897.

An extremely large problem with about 12,000,000 variables was solved.

I.J. Lustig and R.E. Marsten and D.F. Shanno (1992). The interaction of algorithms and architectures for interior point methods. P.M. Pardalos (ed.), *Advances in Optimization and Parallel Computing*, North–Holland, Amsterdam, 190–205.

The interaction of different computer architectures and sparsity handling heuristics were studied. For several widely used computers the most appropriate method is proposed.

Cs. Mészáros (1996). Fast Cholesky factorization for interior point methods of linear programming. *Computers & Math. with App.* 31, No. 4/5, 49-54.
R. Fourer and S. Mehrotra (1991). Performance of an augmented system approach for solving least-squares problems in an interior-point method for linear programming. *COAL Newsletter* 19, 26–31.
R.J. Vanderbei and T.J. Carpenter (1993). Symmetric indefinite systems for interior point methods. *Math. Program.* 58, 1–32.
I. Maros and Cs. Mészáros (1995). The role of the augmented system in interior point methods. *Report* TR/06/95, Brunel University, Dept. of Math. and Stat., London.

In the first implementations the search directions were computed by solving a symmetric positive definite system, the so-called normal equation. A successful implementation of this method is discussed in the first paper. For several problems this approach results in unacceptable fill-in. To preserve sparsity another approach based on the solution of a symmetric indefinite system, the so-called augmented system, was proposed and extensively tested as described in the latter three papers.

J. Gondzio (1996). Multiple centrality corrections in a primal-dual method for linear programming. *Report* 20, Logilab, HEC Geneva, Section of Management Studies, University of Geneva. Revised May 1995. *Computational Opt. Appl*, 6, 137–156. The importance of efficient centering is demonstrated. The method proposed and reported to be highly efficient is based on the target-following principle.

4.4 Recent surveys

I.J. Lustig, R.E. Marsten and D.F. Shanno (1994). Interior point methods for linear programming: computational state of the art. *ORSA J. Comput.* 6, 1–14.

A brief but important survey collecting the basic principles what are needed for a successful implementation of IPMs.

J. Gondzio and T. Terlaky (1996). A computational view of interior point methods for linear programming. J.E. Beasley (ed.) *Advances in Linear and Integer Programming,* Oxford University Press, Oxford.

E.D. Andersen, J. Gondzio, Cs. Mészáros and X. Xu (1996). Implementation of Interior Point Methods for Large Scale Linear Programming. T. Terlaky (ed.), *Interior Point Methods of Mathematical Programming,* Kluwer Academic Publishers, Dordrecht, 189–252.

These two papers give thorough description of most methods, heuristics, parameter selections used in state-of-the-art implementations. Topics like correct sensitivity analysis, warm start, that are not solved yet satisfactorily in current implementations are discussed as well.

5 Available Software

After ten years of intensive research IPMs are well understood both in theory and practice. IPMs have proved to be significantly more efficient than the best available simplex implementations for many linear optimization problems. As a result current codes to solve linear optimization problems are sophisticated tools, frequently containing both a simplex and an IPM solver. They are capable to solve linear problems on a PC in some minutes that were hardly solvable on a super computer ten years ago.

CPLEX(CPLEX/ BARRIER), *CPLEX user's guide,* CPLEX Optimization, Inc., Incline Village, NV, 1995. For more information contact `http://www.cplex.com`.

CPLEX is leading the market at this moment. It is the most complete, most robust package. It contains primal and dual simplex solver, efficient interior point implementation with cross-over, good mixed-integer algorithms, network and quadratic programming solver. It is supported by most modeling languages and available on most platforms.

OSL, *Optimization Subroutine Library, Guide and References,* IBM Corporation, Kingston, USA, 1991. More information at `http://www.research.ibm.com/osl/`.

IBM's large optimization library also contains both simplex and interior solvers, it is capable of solving linear, linear mixed integer and quadratic programs.

LOQO, R.J. Vanderbei, LOQO User's Manual, *Technical Report,* SOL 92–05, 1992, Dept. of Civil Engineering and Operations Research, Princeton University, Princeton, NJ 08544, USA. Available from `http://www.sor.princeton.edu/~rvdb/`.

LOQO is a robust implementation of a primal-dual infeasible IPM for convex quadratic programming. It is a commercial package, but it is free if it is used for academic purposes.

DASH XPRESS–MP. For information contact `http://www.dash .com`.
An excellent package including simplex and IPM solvers. It is almost as complete as CPLEX. For further information

HOPDM. Available from `http://ecolu-info.unige.ch/~logilab/software/`.
Gondzio's HOPDM is an implementation of a higher order primal-dual method. It is public domain in the form of FORTRAN source files for academic purposes.

BPMPD. Available from `ftp://ftp.sztaki.hu/pub/oplab/SOFTWARE/BPMPD`
Mészáros' BPMPD is an implementation of a higher order primal-dual method including both the normal and augmented system approach. The code is available in the form of FORTRAN source files only for academic purposes.

LIPSOL. Available from `http://pc5.math.umbc.edu/~yzhang/`.
Zhang's LIPSOL is written in MATLAB and FORTRAN. It is an implementation of the predictor–corrector primal-dual method. Its unique advantage is the ease of use and modify resulting from the use of the MATLAB programming language.

Who is interested in more information about codes for linear optimization, either commercial or research ones, may consult the WWW site of LP FAQ (LP Frequently Asked Questions) at

- `http://www.skypoint.com/subscribers/ashbury/linear-programming-faq`
- `ftp://rtfm.mit.edu/pub/usenet/sci.answers/linear-programming-faq`

8 Decomposition and Column Generation

François Soumis
GERAD, École Polytechnique de Montréal

CONTENTS

Decomposition methods were introduced with the first research on large-scale optimization problems. They are the fruit of researchers' ingenuity and the following observation.

Large-scale problems almost always present constraint matrices with a low density of non-zero elements. In fact, it would be practically impossible and very costly to gather the millions and even billions of coefficients of the constraint matrix if it was not sparse. Moreover, the matrices of large problems often present submatrices corresponding to particular problems that are easier to solve: block diagonal form, network flow, etc. However, these submatrices are connected by coupling constraints or variables which prevent the use of already existent efficient algorithms to solve these particular problems.

The basic idea behind decomposition methods is to treat the coupling constraints or fix certain variables at a superior level; this with the goal of obtaining a residual problem which is much easier to solve. This subproblem, treated at an inferior level, will often be separable in several independent problems. Each subproblem can be solved by specialized algorithms which exploit their particular structures.

Decomposition methods treat problems in an iterative fashion by working alternatively at the superior and inferior levels. These methods exchange information between the two levels in order to obtain an optimal solution to the global problem

Annotated Bibliographies in Combinatorial Optimization, edited by M. Dell'Amico, F. Maffioli and S. Martello

even if the solution methods are applied to one level at a time.

Since the 1960s, many decomposition methods have been developed. They distinguish themselves by the nature of the elements that are treated at the superior level and by the solution method of the superior problem.

Among the decomposition methods that treat the coupling constraints at the superior level, there are those which solve this superior level problem in its primal form. *Dantzig-Wolfe* decomposition uses the simplex algorithm at the superior level and the *analytic centers* method uses an interior point algorithm. *Lagrangean relaxation* methods solve the superior level problem in its dual form. The algorithms used to solve the Lagrangean relaxation include the subgradient, dual-ascent, augmented Lagrangean and bundle methods.

Benders decomposition and *resource-directive* decomposition are among the methods used to handle coupling variables. *Cross* decomposition treats coupling constraints as well as coupling variables.

Finally, *column generation* corresponds to the solution process used in Dantzig-Wolfe decomposition. However, this approach can also be used directly by formulating a master problem and subproblems rather than obtaining them by decomposing a global formulation of the problem.

In this study, papers presenting fundamental developments were favored. Some papers containing methodological contributions on the treatment of applications are also presented. It is advisable to recall that the treatment of large-scale applications is the reason behind decomposition methods and that fundamental and applied research are strongly connected. The section presenting each method combines the work on linear, nonlinear and integer cases.

1 Books and Surveys

1.1 Books

L.S. Lasdon (1970). *Optimization Theory for Large Systems*, Macmillan Series in Operations Research, London.

Even if it is not recent, this book, which is one of the first to present an overview of this field of research, contains a lot of know-how on the efficient use of decomposition methods as well as the theory behind them.

D.A. Wismer (1971). *Optimization Methods for Large Scale Systems*, McGraw Hill, New York.

This book assembles chapters that were prepared by different authors: Static multi-level systems and on-line multilevel systems by J.D. Schoeffler; Large-scale linear and nonlinear programming by A.M. Geoffrion; Generalized linear programming by G.B. Dantzig and R.M. Van Slyke; Dynamic decomposition techniques by J.D. Pearson; Aggregation by M. Aoki; Distributed multilevel systems by D.A. Wismer; Trajectory decomposition by E.J. Bauman.

D.M. Himmelblau (1973). *Decomposition of Large Scale Problems*, North Holland, Amsterdam.

This collection of 38 papers presents methodological developments, applications and surveys in certain areas.

M. Minoux (1985). *Mathematical Programming*, John Wiley, Chichester.
A more recent well structured synthesis of decomposition methods is presented.

1.2 Surveys and synthesis papers

A.M. Geoffrion (1970). Elements of large scale mathematical programming I & II. *Management Sci.* 16, 652–691.
These two joint papers identify many types of large-scale problems with structures allowing decomposition. The author introduces the concept of problem manipulations; the main ones being *Projection*, *Inner Linearization* and *Outer Linearization*. He also discusses the concepts of *Piecewise*, *Restriction* and *Relaxation* solution strategies. The author characterizes most decomposition methods as a combination of these concepts.

J.L. Kennington (1978). A survey of linear cost multicommodity network flows. *Oper. Res.* 26, 209–236.
This survey presents the result of the application of many decomposition methods on this problem.

A.A. Assad (1978). Multicommodity network flows—A survey. *Networks* 8, 37–91.
This survey treats not only the linear case but also the nonlinear case which is not covered in the previous paper.

R.E. Burkard, H.W. Hamacher, J. Tind (1985). On general decomposition schemes in mathematical programming. *Math. Program. Study* 24, 238–252.
A general decomposition technique in ordered structures is described. This technique can be applied to algebraic mathematical programs and reduces to Benders and Dantzig-Wolfe decomposition in the classical linear case. This approach enables the application of decomposition to problems involving bottleneck, multi-criteria and fuzzy functions with real or integer variables. The decomposition procedure is based on general duality theory. Moreover, an extension of the so-called cross-decomposition procedure is discussed.

O.E. Flippo, A.H.G. Rinnooy Kan (1993). Decomposition in general mathematical programming. *Math. Program.* 60, 361–382.
In this paper, a unifying framework is presented for the generalization of the decomposition methods originally developed by [Benders 1962] (see §3.2) and [Dantzig and Wolfe 1960] (see §2.1.). These generalizations, called Variable decomposition and Constraint decomposition respectively, are based on the general duality theory developed by Tind and Wolsey [*Math. Program.* 21 (1981) 241–261]. The framework presented is of a general nature since there are no restrictive conditions imposed on problem structure; moreover, inaccuracies and duality gaps that are encountered during computations are accounted for. The two decomposition methods are proven not to cycle if certain (fairly general) conditions are met. Furthermore, finite convergence can be ensured under the traditional finiteness conditions and asymptotic convergence can

be guaranteed once certain continuity conditions are met. Other decomposition techniques, such as Lagrangean decomposition and Cross decomposition, turn out to be captured by the general framework presented here as well.

J. Desrosiers, Y. Dumas, M.M. Solomon, F. Soumis (1995). Time constrained routing and scheduling. *Handbooks in Operations Research and Management Science 8*, Volume on Network Routing, Elsevier, 35–139.

This survey presents the theory and art of column generation in the context of routing and scheduling.

2 Problems with Coupling Constraints

Coupling constraints are treated at the superior level in a coordination problem in order to obtain separable and/or simpler subproblems. These decomposition methods are often called price-directive approaches as they ensure the coordination of the subproblems by transferring prices to them.

2.1 Primal approach for the coordination problem

2.1.1 Dantzig-Wolfe

L.R. Ford Jr., D.R. Fulkerson (1958). A suggested computation for multicommodity flows. *Management Sci.* 5, 97–101.

This paper presents, for the maximal multicommodity network flow problem, the idea of what would soon become known as Dantzig-Wolfe decomposition.

G.B. Dantzig, P. Wolfe (1960). Decomposition principle for linear programs. *Oper. Res.* 8, 101–111.

The basic idea of Dantzig-Wolfe decomposition in the context of linear programming is presented in this paper. The authors give credit to Ford and Fulkerson (1958) for their previous work.

J. Abadie, A.C. Williams (1968). Dual and parametric methods in decomposition. R.L. Graves, P. Wolfe (eds.). *Recent Advances in Mathematical Programming*, McGraw-Hill, New York.

This paper presents a dual version of Dantzig-Wolfe decomposition. This algorithm uses dual feasible solutions of the master problem and modifies them until they become primal feasible solutions. The objective of the subproblems, which generate new columns, corresponds to the pricing rule of the dual simplex algorithm.

T.L. Magnanti, J.F. Shapiro, M.H. Wagner (1976). Generalized linear programming solves the dual. *Management Sci.* 22, 1195–1203.

Dantzig-Wolfe decomposition for convex, nonconvex and integer programming problems is studied in this paper. An arbitrary mathematical programming problem can be analyzed as a sequence of linear programming approximations. The authors show that any limit point of the sequence of optimal linear programming dual prices is also optimal for the dual of the original primal problem.

K. Kim, J.L. Nazareth (1991). The decomposition principle and algorithms for linear programming. *Linear Algebra and Appl.* 152, 119–133.

A variant of Dantzig-Wolfe decomposition using interior points of subproblems as columns in the master problem is studied in this paper. These points can be obtained with an interior point algorithm. This variant alleviates certain computational difficulties such as long-tail convergence and numerical instabilities often present when subproblems include variables which are restricted to large intervals.

K.L. Jones, I.J. Lustig, J.M. Farvolden, W.B. Powell (1993). Multicommodity network flows: The impact of formulation on decomposition. *Math. Program.* 62, 95–117.

This paper investigates the impact of problem formulation on Dantzig-Wolfe decomposition for the multicommodity network flow problem. These problems are formulated in three ways: origin-destination specific, destination specific and product specific. The path-based origin-destination specific formulation is equivalent to the tree-based destination specific formulation by a simple transformation. Solving the path-based problem formulation by decomposition results in substantially fewer master problem iterations and lower CPU times than by using decomposition on the equivalent tree-based formulation.

F. Vanderbeck (1996). *On Integer Programming Decomposition and Ways to Enforce Integrality in the Master*, Research report, Judge Institute of Management Studies, Cambridge University.

This paper discusses the formulation of the integrality requirement in the master problem. Some branching strategies compatible with Dantzig-Wolfe decomposition are proposed. Computational results for the cutting stock problem are presented.

2.1.2 Analytic center

J.-L. Goffin, J.-P. Vial (1990). Cutting planes and column generation techniques with the projective algorithm. *J. Optim. Theory Appl.* 65, 409–429.

The problem studied is that of solving linear programs defined recursively by column generation techniques or cutting plane techniques using, respectively, the primal projective method or the dual projective method.

J.-L. Goffin, A. Haurie, J.-P. Vial (1992). Decomposition and nondifferentiable optimization with the projective algorithm. *Management Sci.* 38, 284–302.
J.-L. Goffin, A. Haurie, J.-P. Vial, D.L. Zhu (1993). Using central prices in the decomposition of linear programs. *EJOR* 74, 393–409.

These two papers study a decomposition technique for linear programs which proposes a new treatment of the master problem in the classical Dantzig-Wolfe decomposition algorithm. This new approach does not use prices corresponding to an optimal combination of the available columns but rather central prices (i.e. analytic centres) which combine them all. This approach seems to be a promising alternative to the classical Dantzig-Wolfe algorithm which suffers from very slow convergence in some practical problems.

J.E. Mitchell (1994). An interior point column generation method for linear programming using shifted barriers. *SIAM J. Optim.* 4, 423–440.

A column generation method for linear programming problems is described. This method is based on Freund's shifted barrier method for linear programming. The use of shifts makes it possible to give new primal variables the value zero, with the resulting point being both feasible and interior in the shifted or relaxed formulation. The algorithm takes $O(nL)$ iterations. It is shown that the direction obtained when using the shifted barrier column generation method is related to one proposed by Mitchell and Todd. Preliminary computational results are included, which show that the method may be useful when small numbers of variables are added.

2.2 Dual Approach for the Coordination Problem

2.2.1 Subgradient

M. Held, R.M. Karp (1970). The traveling-salesman problem and minimum spanning trees. *Oper. Res.* 18, 1138–1162.

This paper introduces Lagrangean relaxation where the multipliers are adjusted through subgradient optimization.

M. Held, P. Wolfe, H.P. Crowder (1974). Validation of subgradient optimization. *Math. Program.* 6, 62–88.

The subgradient method is experimentally studied on several classes of specially structured large-scale linear programming problems.

A.M. Geoffrion (1974). Lagrangean relaxation for integer programming. *Math. Program. Study* 2, 82–114.

This paper presents a systematic development of Lagrangean relaxation for integer programming. The author establishes important results on the quality of the lower bound according to subproblem properties.

J.-L. Goffin (1977). On the convergence rates of subgradient optimization methods. *Math. Program.* 13, 329–347.

Rates of convergence of subgradient optimization are studied. If the step size is chosen to be a geometric progression with ratio ρ, the convergence, if it occurs, is geometric with rate ρ. For convergence to occur, it is necessary that the initial step size be large enough, and that the ratio ρ be greater than a sustainable rate.

J.F. Shapiro (1979). A survey of Lagrangean techniques for discrete optimization. P.L. Hammer, E.L. Johnson, B.H. Korte (eds.). *Discrete Optimization II*, Ann. Discr. Math. 5, 113–138.

This survey presents a good synthesis of the theoretical results on Lagrangean techniques.

M.L. Fisher (1981). The Lagrangian relaxation method for solving integer programming problems. *Management Sci.* 27, 1–18.

This paper presents a review of Lagrangean relaxation. A number of successful applications are documented, including the quality of the lower bounds.

2.2.2 Dual-ascent (steepest-ascent)

R.C. Grinold (1972). Steepest ascent for large scale linear programs. *SIAM Rev.* 14, 447–464.

Many structured large-scale linear programming problems can be transformed into an equivalent problem of maximizing a piecewise linear, concave function subject to linear constraints. The concave problem can be solved in a finite number of steps using a steepest-ascent algorithm. This principle is applied to block diagonal systems yielding refinements of existing algorithms. An application to the multistage problem produces an entirely new algorithm.

M.S. Bazaraa, J.J. Goode, R.L. Rardin (1978). A finite steepest-ascent algorithm for maximizing piecewise-linear concave functions. *J. Optim. Theory Appl.* 25(3), 437–442.

This note describes a finitely convergent steepest-ascent scheme for maximizing piecewise-linear concave functions. Given any point, the algorithm moves along the direction of steepest ascent, that is, along the shortest subgradient, until a new ridge is reached. The overall process is then repeated by moving along the new steepest-ascent direction.

2.2.3 Augmented Lagrangean and bundle method

S.M. Robinson (1986). Bundle-based decomposition: Description and preliminary results. *Lecture Notes in Control and Information Sciences* 84, Springer-Verlag, Berlin, 571–576.

This paper transform a block-angular linear programming problem in a non-smooth concave optimization problem by dualizing the linking constraints. This concave problem is solved by the "bundle method" presented by Lemaréchal, Stradiat and Bihain, in *Nonlinear Programming* 4, edited by Mangasarian, Meyer and Robinson, Academic Press, 1981.

D. Medhi (1990). Parallel bundle-based decomposition for large-scale structured mathematical programming problems. *Ann. Oper. Res.* 22, 101–127.
D. Medhi (1994). Bundle-based decomposition for large-scale convex optimization: Error estimation and application to block-angular linear programs. *Math. Program.* 66, 79–101.

These two papers present supplementary results on theoretical questions and implementation know-how.

A. Ruszczynski (1986). A regularized decomposition method for minimizing a sum of polyhedral functions. *Math. Program.* 35, 309–333.
A. Ruszczynski (1988). Regularized decomposition and augmented Lagrangian decomposition for angular linear programming problem. Aspiration based decision support systems, p. 80–91. *Lecture Notes in Econom. and Math. Systems* 331, Springer, Berlin, 1989.

The first paper proposes a method using quadratic penalties on variations of dual variables while the second paper applies this method to block-angular problems. The second paper also presents an augmented Lagrangean relaxation of the master problem constraints. This relaxation has major advantages compared to the relaxation of linking constraints. It yields linear separable subproblems instead of introducing nonlinear linking costs.

C. Lemaréchal (1992). Lagrangian decomposition and nonsmooth optimization: Bundle algorithm, prox iteration, augmented Lagrangian. *Proc. Erice 91, Nonsmooth Optimization Methods and Applications*, F. Giannessi (ed.), Gordon & Breach, 201–216.

The main aim of this paper is clarification. They gather some results appearing in the literature, which are often overlooked or even forgotten; namely: *(i)* solving an optimization problem via Lagrangean decomposition is primarily minimizing a (nonsmooth) convex function; *(ii)* for this, the methods of Dantzig-Wolfe (column generation) and of Kelley-Cheney-Goldstein (cutting planes) are just the same; *(iii)* they are particular instances of bundle methods, a family in which is also found the subgradient method; *(iv)* these bundle methods can be also viewed as specific implementations of the dual prox-iteration or, in the convex case, of the augmented Lagrangean technique.

2.3 Combined methods

R.E. Marsten, W.W. Hogan, J.W. Blankenship (1975). The boxstep method for large-scale optimization. *Oper. Res.* 23, 389–405.

R.E. Marsten (1975). The use of boxstep method in discrete optimization. *Math. Program. Study* 3, 127–144.

These two papers propose the use of the boxstep method to reduce the oscillation of dual variables in subgradient or Dantzig-Wolfe optimization. The idea consists in solving a sequence of problems in which the dual variables are constrainted by a box around the previous solution. This method can be viewed as a precursor of analytic center and bundle methods using L^∞ norm defining constraints instead of L^2 norm defining penalties to stabilize dual variables.

3 Problems with Coupling Variables

Coupling variables are fixed temporarily at the superior level in the coordination problem in order to obtain separable and/or simpler subproblems. These decomposition methods are called resource-directive approaches as they ensure the coordination of the subproblems by fixing the sharing of certain resources.

3.1 Resource allocation

J.T. Robacker (1956). *Notes on Linear Programming: Part XXXVII Concerning Multicommodity Networks*, RM-1799, The Rand Corporation, Santa Monica, CA.

This paper introduces the idea of decomposition by resource allocation in the context of linear multicommodity network flow problems.

A.M. Geoffrion (1970). Primal resource-directive approaches for optimizing nonlinear decomposable systems. *Oper. Res.* 18, 375–403.

A systematic and structured description of decomposition by resource allocation for linear and nonlinear problems is presented in this paper.

G.J. Silverman (1972). Primal decomposition of mathematical programs by resource allocation: Part I and part II. *Oper. Res.* 20, 58–93.

A method for primal decomposition by resource allocation of large convex separable problems is studied. The main advantage of primal decomposition over Lagrangean relaxation or dual decomposition methods is that primal feasibility is maintained during the course of the iterations. The direction finding procedure for the resource allocation problem uses a piecewise linear approximation of the primal resource allocation function.

L.A. Wolsey (1981). A resource decomposition algorithm for general mathematical programs. *Math. Program. Study* 14, 244–257.

Using results on duality theory for nonconvex and integer programs, this paper shows how in theory a resource decomposition algorithm, generalizing Benders' algorithm, can be developed for general mathematical programs. As an example of this algorithm, the author obtains an algorithm for bilinear programs. In addition, he examines a question of importance for post-optimality analysis, that of finding an optimal dual solution to the original problem from the solution of the decomposed problem.

3.2 Benders

J.F. Benders (1962). Partitioning procedures for solving mixed variables programming problems. *Numer. Math.* 4, 238–252.

The author introduces the idea of the decomposition method that now bears his name.

A.M. Geoffrion (1972). Generalized Benders decomposition. *J. Optim. Theory Appl.* 10, 237–260.

In this paper, Benders' approach is generalized to a class of problems in which the parametrized subproblem need no longer be a linear program. Nonlinear convex duality theory is employed to derive the cuts corresponding to those in Benders' case.

R.M. Van Slyke, R. Wets (1969). *L*-shaped linear programs with applications to optimal control and stochastic programming. *SIAM J. Appl. Math.* 17, 638–663.

This paper presents adaptations of the Benders decomposition for these important types of problems.

A.M. Geoffrion, G.W. Graves (1974). Multicommodity distribution systems design by Benders decomposition. *Management Sci.* 20(5), 822–844.

This paper considers the optimal localization of intermediate distribution facilities between plants and customers. A solution technique based on Benders decomposition was developed, implemented and successfully applied to a real large-scale problem.

The authors give some ideas on the art of using Benders decomposition in practice.

T.L. Magnanti, T.R. Wong (1981). Accelerated Benders decomposition: algorithmic enhancement and model section criteria. *Oper. Res.* 29, 464–484.

This paper proposes methodology for improving the performance of Benders decomposition when applied to mixed integer programs. It introduces a new technique for accelerating the convergence of the algorithm and theory for distinguishing "good" model formulations of a problem that has distinct but equivalent mixed integer programming representations. The acceleration technique is based upon selecting judiciously from the alternate optima of the Benders subproblem to generate strong or pareto-optimal cuts.

4 Problems with Coupling Constraints and Variables

T.J. Van Roy (1983). Cross decomposition for mixed integer programming. *Math. Program.* 25, 46–63.

A new approach combining Lagrangean relaxation and Benders decomposition which exploits simultaneously the primal and the dual structure of the problem is presented in this paper. The author shows that the more constraints that can be included in the Lagrangean relaxation, the fewer Benders cuts one may expect to need.

T.J. Van Roy (1986). A cross decomposition algorithm for capacitated facility location. *Oper. Res.* 34, 145–163.

The author presents an implementation of the *Cross Decomposition* method to solve the capacitated facility location problem. The method unifies Benders decomposition and Lagrangean relaxation into a single framework that involves successive solutions to a transportation problem and a simple plant location problem.

K. Holmberg (1990). On the convergence of cross decomposition. *Math. Program.* 47, 269–296.

Cross decomposition is a recent method for mixed integer programming problems, exploiting simultaneously both the primal and the dual structure of the problem, thus combining the advantages of Dantzig-Wolfe decomposition and Benders decomposition. Finite convergence of the algorithm equipped with some simple convergence tests has been proved. Stronger convergence tests have been proposed, but not shown to yield finite convergence.

In this paper cross decomposition is generalized and applied to linear programming problems, mixed integer programming problems and nonlinear programming problems (with and without linear parts). Using the stronger convergence tests finite exact convergence is shown in the first cases. Unbounded cases are discussed and also included in the convergence tests. The behaviour of the algorithm when parts of the constraint matrix are zero is also discussed. The cross decomposition procedure is generalized (by using generalized Benders decomposition) in order to enable the solution of nonlinear programming problems.

K. Holmberg (1992). Generalized cross decomposition applied to nonlinear integer programming problems: Duality gaps and convexification in parts. *Optimization* 23, 341–356.

This paper studies the lower bounds on the optimal objective function value of nonlinear pure integer programming problems obtainable by convexification in parts, achieved by using generalized Benders or cross decomposition, and compare them to the best lower bounds obtainable by the convexification introduced by the Lagrangean dual, i.e. by Lagrangean relaxation together with subgradient optimization or (nonlinear) Dantzig-Wolfe decomposition. This paper shows how to obtain a number of different bounds and specify the known relations between them. In one case generalized cross decomposition can automatically yield the best of the Benders decomposition bound and the Lagrangean dual bound, without any a priori knowledge of which is the best. In another case the cross decomposition bound dominates the Lagrangean dual bound.

5 Column Generation

P.C. Gilmore, R.E. Gomory (1960). A linear programming approach to the cutting stock problem. *Oper. Res.* 9, 849–859.
P.C. Gilmore, R.E. Gomory (1963). A linear progrmming approach to the cutting stock problem: Part II. *Oper. Res.* 11, 863–888.

The first part of this paper introduces the column generation technique in the context of the cutting stock problem, whereas the second part of the paper presents the algorithm implementation and results.

P.C. Gilmore, R.E. Gomory (1965). Multistage cutting-stock problems in two or more dimensions. *Oper. Res.* 13, 94–120.

This paper presents a special case of the two-dimensional cutting-stock problem. It is a first step in dealing with complex combinatorial subproblems.

J. Desrosiers, F. Soumis, M. Desrochers (1984). Routing with time windows by column generation. *Networks* 14, 545–565.

This paper, which received the EURO prize in 1983, contributed in increasing the interest in column generation to treat integer programming problems. It shows that the integrality gap can be greatly reduced by solving an integer subproblem with dynamic programming. A branch-and-bound algorithm compatible with column generation was presented for this problem.

M. Desrochers, F. Soumis (1989). A column generation approach to the urban transit crew scheduling problem. *Transportation Sci.* 23, 1–13.

The column generation approach decomposes the problem into two parts. A set covering problem chooses a schedule from already known feasible workdays. A shortest path problem with resource constraints is used to propose new feasible workdays to improve the current solution of the set covering problem. The approach has been successfully tested on real-life problems. This paper opened the door to the treatment of problems having columns which are defined by numerous complex constraints.

C.C. Ribeiro, M. Minoux, M.C. Penna (1989). An optimal column-generation-with-ranking algorithm for very large scale set partitioning problems in traffic assignment. *European J. Oper. Res.* 41, 232–239.

This paper shows how branch-and-bound and column generation techniques can be combined very efficiently to solve to optimality some very large scale set partitioning problems with special structure, such as the matrix decomposition problem arising in the context of satellite communication systems optimization. The main contribution of this approach is the use of column generation during the tree search procedure, combined with a ranking procedure which ensures that the exact optimal integer solution is obtained. Extensive computational experiments are reported for the matrix decomposition problem encountered in the search for optimal schedules in satellite switching systems, showing the effectiveness of this approach.

The following papers apply column generation to many types of complex combinatorial problems. The main result is that column generation solves problems much larger than problems solved using previously known methods.

B. Jaumard, P. Hansen, M. Poggi de Aragão (1991). Column generation for probabilistic logic. *ORSA J. Comput.* 3, 135–148.

S. Lavoie, M. Minoux, E. Odier (1988). A new approach of crew pairing problems by column generation and application to air transportation. *European J. Oper. Res.* 35, 45–58.

C. Ribeiro, F. Soumis (1994). A column generation approach to the multiple depot vehicle scheduling problem. *Oper. Res.* 42, 41–52.

C. Barnhart, E.L. Johnson, G.L. Nemhauser, M.W.P. Savelsberg, P.H. Vance (1995). Branch-and-price: Column generation for solving huge integer programs. *Oper. Res.* (to appear).

C. Barnhart, C.A. Hane, E.L. Johnson, G. Sigismondi (1995). A column generation and partitioning approach for multi-commodity flow problems. *Telecommunication Systems* 3, 239–258.

M. Parker, J. Ryan (1994). A column generation algorithm for bandwith packing. *Telecommunication Systems* (to appear).

M. Gamache, F. Soumis, G. Marquis, J. Desrosiers (1995). A column generation approach for large scale aircrew rostering problem. *Oper. Res.* (to appear).

M. Desrochers, J. Desrosiers, M. Solomon (1992). A new optimization algorithm for the vehicle routing problem with time windows. *Oper. Res.* 40, 342–354.

9 Stochastic Integer Programming

Leen Stougie
University of Amsterdam

Maarten H. van der Vlerk
University of Groningen

CONTENTS

Stochastic programming problems are obtained as a result of modeling uncertainty about problem data by specification of *probability distributions* over these data. If the underlying deterministic problem, i.e., the problem induced by a single realization of the random data, is an integer or combinatorial optimization problem, we arrive at the field of stochastic integer programming.

The difficulties encountered in solving stochastic integer programming problems stem from both the stochastic and the combinatorial nature of the problems. It appears that the former, however, has a predominant effect on the computational complexity, at least for a major subclass of stochastic programming models [Dyer and Stougie 1996] (see §2.2).

This sets research on stochastic integer programming closer to that in traditional stochastic *linear* programming than to that in combinatorial optimization, rendering it a place in the margin of the scope of this book. Indeed, most of the theoretical research on these problems has developed along the paradigmatic lines of stochastic linear programming research. On the other hand, most practice oriented research in the field has concentrated on the use of combinatorial optimization techniques.

Historically a distinction is made between several subclasses of stochastic programming models. The first two classes are so-called 'here-and-now' decision

Annotated Bibliographies in Combinatorial Optimization, edited by M. Dell'Amico, F. Maffioli and S. Martello

models. An optimal decision is to be made now, while only later realizations of the random parameters become known. Consequently, it may turn out that the decision made is infeasible. First, there is the class of models that allow infeasibilities to occur with no more than a prefixed probability. This is the class of models usually referred to as *probabilistic constrained* or *chance constrained* problems.

In the second class of models infeasibilities arising by particular realizations of the random problem parameters have to be compensated (see below for an alternative interpretation of these models). A set of feasible recourse actions is defined, each yielding a certain cost. Thus, the problem is seen to consist of two stages. The objective is to find a first-stage decision, such that the sum of its direct costs and the expected costs of the optimal recourse action in the second stage is minimized. These problems are called *stochastic two-stage decision problems* or *stochastic recourse problems. Multi-stage* problems are a straightforward extension, reflecting situations in which subsets of realizations become available at different moments in time.

In the latter setting the various decision stages can also be thought of as reflecting decision levels, in which typically higher levels model more detailed decisions, whereas lower levels model more aggregate decisions. For instance, consider the problem of allocating budgets to various activities, which may be used in a number of projects at a later time. At the time the budgeting decision is made there is no precise information about the amounts of the activities required by each of the projects, or the possible profits of each of the projects are uncertain. This is an example of a so-called *stochastic hierarchical planning problem.* More than two levels may be involved. A survey of such problems is found in [Bitran and Tirupati 1993] (see §1).

Actually, within this class we will consider also problems that have only one stage; they can be regarded as two-stage problems with a trivial first stage. Also models with both chance constraints and recourse structure occur in the literature.

Finally, there is a class of problems in which optimal solutions are studied as random elements. An optimal solution is seen as a random strategy based on realizations of the random problem data. In this setting the problems are often referred to as 'wait-and-see' problems. The aim is to find (characteristics of) the distribution of the optimal solutions or values given the probability distributions of the random problem parameters. Problems in this class are known as *distribution problems.* In combinatorial optimization there exists extensive literature on such problems under the name *probabilistic analysis.* Here the aim is to derive asymptotic characterizations and other concentration results of optimal solution values. Mostly, the results are used for (asymptotic) probabilistic performance analysis of heuristics for solving combinatorial optimization problems. Almost all work on stochastic knapsack problems and stochastic bin-packing problems belong to this category. Historically, probabilistic analysis has never been regarded as part of stochastic programming. Be this correct or not, references to this type of research are not included here, but appear in several other contributions in this volume. However, results from probabilistic analysis are useful in solving problems belonging to the other two stochastic programming models (see [Lenstra, Rinnooy Kan and Stougie 1984] §2.2), and in analyzing performance of heuristic recourse rules (see [Bertsimas, Jaillet and Odoni 1990] and [Jaillet 1993] §4.2). In these contexts we refer to a few results belonging to this field.

In this bibliography we have deliberately omitted the literature on stochastic scheduling. Although these problems fit in the framework of stochastic integer programming, research on this topic has developed independently from stochastic integer programming and is more related to queuing theory.

Although this volume outlines literature that has appeared since 1985, we have included also references to less recent literature. The reason is that stochastic integer programming has not been covered in the previous volume of annotated bibliographies. This was mainly due to the fact that the amount of literature on this subject originating before 1985 is very limited. It is only over the last ten years that the field has received comprehensive research attention. We have selected about 60 references to what we consider as interesting contributions to the field of stochastic integer programming. Our list is almost surely incomplete. Next to the usual disclaimers, we attribute this to the fact that some authors have not been aware of the stochastic integer programming context of their work, and hence did not provide any clues in titles or abstracts. If any reader feels that we have overlooked an important contribution we would be pleased to be informed.

We have divided this bibliography in four sections. In the first section books and surveys are presented. Section 2 exposes theoretical results. General purpose algorithms are presented in Section 3. Algorithmic work on special problems is outlined in Section 4.

1 Books and Surveys

L. Stougie (1987). *Design and analysis of algorithms for stochastic integer programming*, CWI Tract 37, Centre for Mathematics and Computer Science, Amsterdam.

A review is given comprising results on approximation methods for hierarchical planning problems (see [Lenstra, Rinnooy Kan and Stougie 1984] §2.2), and some of the papers in §4.1 and §4.2. Within the same framework a hierarchical bin-packing, a hierarchical location and a hierarchical location-routing problem are studied. Furthermore it contains the dynamic programming approach in [Lageweg, Lenstra, Rinnooy Kan and Stougie 1985] (see §3.4) and the first results on structural properties of objective functions of two-stage stochastic integer programming problems.

A.H.G. Rinnooy Kan, L. Stougie (1988). Stochastic integer programming. Yu. Ermoliev, R.J-B. Wets (eds.) *Numerical Techniques for Stochastic Optimization*, Springer-Verlag, Berlin, 201–213.

This paper is abstracted from [Stougie 1987]. In a few pages it gives a state of the art of theoretical results on stochastic integer programming up to 1985.

G.R. Bitran and D. Tirupati (1993). Hierarchical production planning. S.C. Graves, A.H.G. Rinnooy Kan, P.H. Zipkin (eds.). *Handbooks on Operations Research and Management Science* Vol. 4, North-Holland, Amsterdam, 523–568.

This survey deals with deterministic problems mainly. The last section is devoted to stochastic hierarchical planning problems. It consists of examples that are presented in §4.1.

R. Schultz, L. Stougie, M.H. van der Vlerk (1993). *Two-stage stochastic integer programming: a survey*, Research Memorandum 520, Institute of Economic Research, University of Groningen (to appear in *Statist. Neerl.*).

This survey covers approximation techniques within the framework of [Lenstra, Rinnooy Kan and Stougie 1984] (see §2.2) as well as algorithms and first results on structural properties for (mixed-)integer recourse models.

M.H. van der Vlerk (1995). *Stochastic programming with integer recourse.* Labyrinth Publication, Capelle a/d IJssel.

In addition to the results on simple integer recourse models mentioned in §2.1 and § 3.3, this Ph.D. thesis surveys some of the theoretical and algorithmical results for complete mixed-integer recourse models, treated in §2.1 and §3.1.

R. Schultz (1995). *Discontinuous optimization problems in stochastic integer programming.* Preprint SC 95-20, Konrad-Zuse-Zentrum für Informationstechnik Berlin.

Integer stochastic linear programming is considered from the viewpoint of discontinuous optimization. A concise overview is given of the approaches found in [Ermoliev, Norkin and Wets 1995], [Carøe and Tind 1996] (see §3.4) and [Schultz, Stougie and Van der Vlerk 1995] (see §3.2). The underlying idea of the latter paper is discussed in a different setting.

2 Theory

2.1 Structural properties of the EVF

This subsection contains references to literature on structural properties of *two-stage (mixed-)integer recourse models*, defined as

$$\inf_x \left\{cx + Q(x) : Ax = b, x \in \mathbb{R}^n_+\right\}$$

where

$$\begin{aligned} Q(x) &= E_\omega v(x,\omega), \quad x \in \mathbb{R}^n, \\ v(x,\omega) &= \inf_{y,\bar{y}}\{q(\omega)y + \bar{q}(\omega)\bar{y} : Wy + \bar{W}\bar{y} = p(\omega) - T(\omega)x, \ y \in \mathbb{Z}^d_+, \ \bar{y} \in \mathbb{R}^{\bar{d}}_+\}. \end{aligned}$$

Here, the fixed matrices c, A, b, W, $\bar{W}$ and the random matrices $q(\omega)$, $\bar{q}(\omega)$, $p(\omega)$ and $T(\omega)$ have compatible dimensions, and E_ω denotes the expectation with respect to the random variable ω. The function Q, being the expectation of the value function v of the *second-stage* problem, is called the *expected value function* (EVF).

The first stage variables could also be of (mixed-)integer type. In general, however, this would only introduce complications that are not different from those met in deterministic (mixed-)integer programming. In special cases, one can take advantage of integrality of the first-stage variables in designing algorithms (see §3.1).

Theoretical research has focused on structural properties of the EVF of two special cases of this model. In both cases the second-stage cost coefficients $(q(\omega), \bar{q}(\omega))$ are assumed to be fixed; the right-hand side vector $p(\omega)$ is always random whereas the *technology matrix* $T(\omega)$ is fixed in some studies. The first model is the *complete mixed-integer recourse* (CMIR) model, in which the *recourse matrices* W and $\bar{W}$ are assumed to be such that for every choice of x and every realization of ω the second-stage problem is feasible. The second model, known as *simple integer recourse* (SIR), has a pure integer second-stage problem given by

$$\inf_{y^+,y^-} \{q^+y^+ + q^-y^- : y^+ \geq p(\omega) - T(\omega)x,\ -y^- \leq p(\omega) - T(\omega)x,\ y^+, y^- \in \mathbb{Z}^d_+\}.$$

Note that this is formally not a simple recourse model. In [Van der Vlerk 1995] (see §1) a justification of this formulation is given.

F.V. Louveaux, M.H. van der Vlerk (1993). Stochastic programming with simple integer recourse. *Math. Program.* 61, 301–325.

Analytical as well as computational properties of the EVF of SIR problems with fixed technology matrix are presented. Using the separability of the second-stage problem, properties of the EVF are derived from its one-dimensional version. An explicit expression for the EVF is given, followed by conditions for (Lipschitz) continuity and diferentiability, as well as an approximation formula. First results on the convex hull and piecewise linear approximations of the EVF are given.

W.K. Klein Haneveld, M.H van der Vlerk (1994). On the expected value function of a simple integer recourse problem with random technology matrix. *J. Comput. Appl. Math.* 56, 45–53.

In this paper some of the results found in [Louveaux and Van der Vlerk 1993] are extended to SIR models with random technology matrix.

R. Schultz (1993). Continuity properties of expectation functions in stochastic integer programming. *Math. Oper. Res.* 18, 578–589.

Sufficient conditions for (Lipschitz) continuity of the EVF of the more general CMIR model with fixed technology matrix are presented.

R. Schultz (1992). Continuity and stability in two-stage stochastic integer programming. K. Marti (ed.). *Stochastic optimization: numerical methods and technical applications*, Lecture Notes in Economics and Mathematical Systems 379, Springer-Verlag, Berlin, 81–92.

Continuity of the EVF of CMIR programs jointly in the first-stage decision variables and the distribution function of the random right-hand side parameters is derived. Then, regarding the distribution function as a parameter of the problem, continuity results for locally optimal values and solutions are established. The latter result justifies numerical procedures that rely on approximating the distributions by simpler ones.

R. Schultz (1995). On structure and stability in stochastic programs with random technology matrix and complete integer recourse. *Math. Program.* 70, 73–89.

In this paper the results of the previous two papers are extended to the CMIR model with random technology matrix.

R. Schultz (1993). *Rates of convergence in stochastic programs with complete integer recourse*, Report 483, Schwerpunktprogramm der Deutschen Forschungsgemeinschaft 'Anwendungsbezogene Optimierung und Steuerung', Humboldt-Universität zu Berlin (to appear in *SIAM J. Optimization* 6).

In this paper qualitative as well as quantitative results on the asymptotic convergence of locally optimal values and optimal solutions are presented for the case that the distribution of the parameters is approximated by empirical measures.

2.2 Complexity related results

The computational complexity of stochastic programming problems is studied in the following paper.

M. Dyer, L. Stougie (1996). *Stochastic programming problems: complexity and approximability*, Report AE96/9, Department of Actuarial Sciences and Econometrics, University of Amsterdam.

Some first results on the complexity of multi-stage stochastic programming problems are presented. The evaluation of the expected value function Q appears to be $\sharp$P-complete. This makes two-stage problems $\sharp$P-hard, and in case of integer first stage variables even $\mathrm{NP}^{\sharp\mathrm{P}}$-hard. Increasing the number of stages requires increasingly more powerful oracles. If the number of stages is part of the input the problem is conjectured to be PSPACE-complete. Hardness proofs for problems with more than two stages are still to be established.

As in deterministic combinatorial optimization, approximation by using heuristics is often a good alternative for solving hard stochastic integer programming problems. The performance of the heuristics needs to be analyzed.

J.K. Lenstra, A.H.G. Rinnooy Kan, L. Stougie (1984). A framework for the probabilistic analysis of hierarchical planning systems. *Ann. Oper. Res.* 1, 23–42.

Design and analysis of polynomial time approximation algorithms for two-stage hierarchical planning problems is studied from a general point of view. The main ingredient of approximation methods for problems in this framework is an estimate of the second-stage optimal cost obtained via probabilistic analysis. Using this estimate in the objective instead of the actual expected optimal second-stage cost, the resulting minimization problem is solved for the first stage decision variables. The same procedure but then equipped with a heuristic to solve the second stage problem provides an approximate solution method for the distribution version of the problem. Performance measures for the algorithms for the two models and relations between them are presented. Sufficient conditions are given for asymptotic optimality of the methods as the number of random elements tends to infinity. Applications of this principle are found in various papers described in §4.1 and §4.2. In this paper an example of a hierarchical scheduling problem is worked out.

3 General Purpose Algorithms

3.1 Benders' decomposition methods

R.M. Wollmer (1980). Two-stage linear programming under uncertainty with 0–1 first stage variables. *Math. Program.* 19, 279–288.

This early paper builds on the observation that both 0-1 mixed-integer problems and two-stage stochastic linear programs are solved using Benders' decomposition (in the latter case often called L-shaped algorithm). Combining these basic ideas results in an algorithm for two-stage recourse models with binary first-stage and continuous second-stage variables. This approach is extended in [Laporte and Louveaux 1993].

I.L. Averbakh (1991). An iterative method of solving two-stage discrete stochastic programming problems with additively separable variables. *USSR Comput. Math. Math. Phys.* 31, 21–27.

A Benders type solution method is proposed for a class of problems with integer first-stage, arbitrary second-stage, and continuously distributed random variables. The second-stage problem is to maximize the expected value of the objective function subject to statistical constraints (see [Yudin and Tzoy 1973] §3.4), which is different from the recourse formulation considered in the other papers in this section. It is assumed that the second-stage problems and their Lagrangian duals can be solved analytically. In addition, the case where the distributions of the random parameters are not known in advance is discussed.

G. Laporte, F.V. Louveaux (1993). The integer L-shaped method for stochastic integer programs with complete recourse. *Oper. Res. Lett.* 13, 133–142.

The algorithm presented in this paper applies to recourse problems with binary first-stage and binary or continuous second-stage variables. Embedded in a branch and cut procedure, feasibility and optimality cuts are determined iteratively. The problem of how to approximate the (in general) non-convex EVF is solved by using a class of optimality cuts that provide a lower bound at all binary first-stage solutions (but not necessarily at other points). Moreover, problem dependent improved optimality cuts are discussed.

C.M.A. Leopoldino, M.V.F. Pereira, L.M.V. Pinto, C.C. Ribeiro (1994). A constraint generation scheme to probabilistic linear problems with an application to power system expansion planning. *Ann. Oper. Res.* 50, 367–385.

The recourse model considered here has integer first-stage variables and a continuous second-stage problem. The expected recourse costs do not appear in the objective, but are required to be below a certain threshold. A constraint generation scheme, similar to Benders' decomposition, is presented.

C.C. Carøe and J. Tind (1995). *L-shaped decomposition of two-stage stochastic programs with integer recourse*, Technical report, Institute of Mathematics, University of Copenhagen.

The L-shaped method is generalized to cover two-stage integer recourse models. Using generalized duality theory, it is shown how to determine nonlinear feasibility and optimality cuts depending on the technique used to solve the second-stage problems. The integer L-shaped method in [Laporte and Louveaux 1993] appears as a special case. Finite convergence of the method (disregarding solution of the master problem) is discussed.

3.2 Gröbner basis methods

One of the difficulties met in solving integer recourse problems is that many integer programming problems that differ only in their right-hand side vector need to be solved. In the following two papers Gröbner basis methods, a tool originating in computational algebra, are used to construct a test set to handle this task.

S.R. Tayur, R.R. Thomas, N.R. Natraj (1995). An algebraic geometry algorithm for scheduling in the presence of setups and correlated demands. *Math. Program.* 69, 369–401.

Inspired by a manufacturing problem, an algorithm to solve chance constrained integer programs is presented. In fact, the authors observe that their methodology is not confined to problems of this type, but can be used to solve a very wide class of problems.

R. Schultz, L. Stougie, M.H. van der Vlerk (1995). *Solving stochastic programs with complete integer recourse: a framework using Gröbner Bases*, Discussion Paper 9562, CORE, Louvain-la-Neuve.

A framework for solving stochastic programs with complete integer recourse and discretely distributed right-hand side vector is presented. Using structural properties of the EVF, a finite set containing an optimal solution is constructed, and an enumeration scheme is presented.

3.3 Convex approximations of SIR models

W.K. Klein Haneveld, L. Stougie, M.H. van der Vlerk (1995). On the convex hull of the simple integer recourse objective function. *Ann. Oper. Res.* 56, 209–224.
W.K. Klein Haneveld, L. Stougie, M.H. van der Vlerk (1996). An algorithm for the construction of convex hulls in simple integer recourse programming. *Ann. Oper. Res.* 64, 67–81.

Owing to the existence of the second paper, the main result in the first one is that every reasonable convex approximation of the EVF of a SIR problem with fixed technology matrix is equivalent to the EVF of a simple continuous recourse problem. Consequently, convex approximations of such SIR problems can be solved by existing algorithms for the latter problem type. In the second paper a strongly polynomial time algorithm for computing the convex hull of the EVF of SIR problems with discretely distributed right-hand side vector is presented.

M.H. van der Vlerk (1995). *Stochastic programming with integer recourse*, Chapter 7. Labyrinth Publication, Capelle a/d IJssel.

In Chapter 4 of this Ph.D. thesis a class of continuous distributions is presented that result in a convex EVF for SIR models with random right-hand side vector. Approximation of any distribution by one in this class gives a convex approximation of the EVF. A subclass having piecewise constant densities, called α-approximations, is analyzed. A bound on the resulting error is discussed, and explicit expressions for the distributions of the random parameters in the equivalent continuous simple recourse problems (see [Klein Haneveld, Stougie and Van der Vlerk 1995]) are given.

3.4 Other methods

A. Ettinger, P.L. Hammer (1972). Pseudo-boolean programming with random coefficients. *Cahiers du Centre d'Études de Rec. Oper.* 14, 67–82.

This paper discusses programs with separate chance constraints in terms of random pseudo-Boolean functions, i.e. real valued polynomials in binary variables with independent discrete random coefficients. An algorithm to transform such chance constraints to equivalent deterministic pseudo-Boolean constraints is given. The deterministic equivalent problem is solved by existing algorithms.

D.B. Yudin, E.V. Tzoy (1973). Integer-valued stochastic programming. *Eng. Cybern.* 12, 1–8.

This paper considers maximizing the expected value of nonlinear 0-1 problems with random objective function and constraints in terms of expectations of random functions, where the random variables follow a continuous distribution. The problem is reformulated as finding decision areas (i.e., subsets of realizations resulting in the same optimal solution). This is shown to be equivalent to an infinite-dimensional linear programming problem, which is solved using duality results. For two-stage 0-1 problems with a second stage as described above, a relaxation is formulated and it is shown that there exists a rounding strategy resulting in an optimal first-stage solution.

R.D. Armstrong, J.L. Balintfy (1975). A chance constrained multiple choice programming algorithm. *Oper. Res.* 23, 494–510.

The so-called block pivoting algorithm for multiple choice problems (0-1 programs with precisely one variable equal to one for each predefined subset) is extended to chance-constrained models with a normally distributed constraint matrix. Both separate and joint chance constraints are discussed. A successful application is mentioned, but no numerical details are provided.

B.J. Lageweg, J.K. Lenstra, A.H.G. Rinnooy Kan, L. Stougie (1985). Stochastic integer programming by dynamic programming. *Statist. Neerl.* 39, 97–113.
B.J. Lageweg, J.K. Lenstra, A.H.G. Rinnooy Kan, L. Stougie (1988). Stochastic integer programming by dynamic programming. Yu. Ermoliev, R.J-B. Wets (eds.). *Numerical Techniques for Stochastic Optimization*, Springer-Verlag, Berlin, 403–412.

In these papers, the second being a more concise version of the first, suitable stochastic integer programming problems are solved by a single 'giant' recursion combining the separate dynamic programming computations for all individual realizations of the random parameters. Only problem instances having a small number of variables, constraints and mass points of the distribution are amenable to this approach.

R.L. Carraway, T.L. Morin, H. Moskowitz (1989). Generalized dynamic programming for stochastic combinatorial optimization. *Oper. Res.* 37, 819–829.

A global priority rule on solutions, guided by the stochastics, substitutes the classical real-valued recurrence function in dynamic programming. This recursion, however, may suffer from the loss of the monotonicity property on partial solutions. Therefore, local priority rules are specified related to the states. The drawback of specifying all these rules can be handled via heuristic generation. An application to a stochastic traveling salesman problem is presented.

V. Norkin, Y. Ermoliev, A. Ruszczyński (1994). *On optimal allocation of indivisibles under uncertainty*, Working Paper WP-94-21, IIASA, Laxenburg, Austria (to appear in *Oper. Res.*).

A stochastic version of the branch and bound method applicable to a wide class of stochastic mixed-integer problems is presented. The main idea is to iteratively partition the solution set in increasingly finer subsets, estimate (bounds for) the objective on these subsets, and remove non-promising subsets, respectively focus on promising subsets.

Y.M. Ermoliev, V.I. Norkin, R.J-B. Wets (1995). The minimization of semicontinuous functions: mollifier subgradients. *SIAM J. Control Optim.* 33, 149–167.

In this paper the general problem of minimizing a discontinuous function is tackled by averaging the objective via convolution with smooth kernels (mollifiers).

C.C. Carøe, J. Tind (1996). *A cutting-plane approach to mixed 0-1 stochastic integer programs.* Technical report, Institute of Mathematics, University of Copenhagen.

Working with the equivalent deterministic large scale 0-1 mixed-integer problem, it is shown that a valid inequality for a subproblem (corresponding to one realization of the random parameters) can be made valid for every other subproblem. In general, the quality of the resulting cuts is poor. A possible cutting plane algorithm is discussed.

4 Solution of Specific Problems

4.1 Production planning and scheduling problems

G.R. Bitran, E.A. Haas, H. Matsuo (1986). Production planning of style goods with high setup costs and forecast revisions. *Oper. Res.* 34, 226–236.

A stochastic mixed-integer programming problem is formulated to plan the production of style goods, under uncertain demand and major set-up costs. It is approximated by a deterministic mixed-integer problem. A solution method is tested empirically.

M.A.H. Dempster, M.L. Fisher, L. Jansen, B.J. Lageweg, J.K. Lenstra, A.H.G. Rinnooy Kan (1983). Analysis of heuristics for stochastic programming: results for hierarchical scheduling problems. *Math. Oper. Res.* 8, 525–537.

In a two-stage problem a number of identical machines is to be acquired so as to minimize the sum of acquisition cost and expected cost (makespan) of processing a number of jobs with i.i.d. random processing times. The paper fits in the framework of [Lenstra, Rinnooy Kan and Stougie 1984] (see §2.2), though it precedes the latter paper. An asymptotically correct estimate for the second-stage makespan is used to find an approximate solution to the problem. For the distribution problem a simple list scheduling heuristic is added to solve the second-stage makespan problem. The solution methods are asymptotically optimal for the respective models.

J.B.G. Frenk, A.H.G. Rinnooy Kan, L. Stougie (1984). A hierarchical scheduling problem with a well-solvable second stage. *Ann. Oper. Res.* 1, 43–58.

In the setting of [Lenstra, Rinnooy Kan and Stougie 1984] (see §2.2) a number of identical machines is to be acquired so as to minimize direct cost plus expected average completion time of jobs with i.i.d. processing times. An asymptotically correct estimate of the second stage cost is used to approximate the optimal number of machines. For the distribution version an exact algorithm is used for the well-solved second stage problem. Asymptotically optimal methods result. A rate of convergence to optimality in the form of a law of the iterated logarithm is given.

4.2 Network and routing problems

H. Ishii, S. Shiode, T. Nishida, Y. Namasuya (1981). Stochastic spanning tree problem. *Discr. Appl. Math.* 3, 263–273.
H. Ishii, T. Nishida (1983). Stochastic bottleneck spanning tree problem. *Networks* 13, 443–449.

Chance constrained models for, respectively, the minimum and the bottleneck spanning tree with normally distributed edge costs is transformed into a deterministic equivalent problem. Defining closely related auxiliary problems allows the design of polynomial time parametric type algorithms.

A. Marchetti Spaccamela, A.H.G. Rinnooy Kan, L. Stougie (1984). Hierarchical vehicle routing. *Networks* 14, 571–586.

The paper fits in the framework of [Lenstra, Rinnooy Kan, and Stougie 1984] (see §2.2). In the first stage, a fleet of uniform vehicles with infinite capacity and varying acquisition cost and speed is put together. In the second-stage these vehicles are used to serve a number of i.i.d. points in the Euclidean plane from a single depot. The first-stage costs are acquisition costs and the second-stage costs consist of the maximum route length of any of the vehicles, provided all customer locations are visited. Using probabilistic analysis an asymptotically correct estimate of the second-stage optimal cost is obtained. A region partitioning heuristic is added to solve the distribution problem.

G. Andreatta (1987). Shortest path models in stochastic networks. G. Andreatta, F. Mason, P. Serafini (eds.). *Stochastics in Combinatorial Optimization*, CISM Udine, World Scientific Publishing, Singapore, 178–186.
G. Andreatta, L. Romeo (1988). Stochastic shortest paths with recourse. *Networks*

18, 193–204.

The first paper gives a survey of stochastic shortest path problems. It exposes both chance constrained and recourse formulations. One of them, an interesting recourse model where arcs in the network are failing randomly while traversing the network is studied in the second paper. A stochastic dynamic programming formulation for this problem is given. The claimed complexity of recourse problems in the two papers is false (see [Dyer and Stougie 1996] §2.2 and Papadimitriou (*J. Comput. Syst. Sci.* (1985), 288-301)).

G. Andreatta, G. Romanin-Jacur (1987). Aircraft flow management under congestion. *Transportation Sci.* 21, 249–253.

On a simple star-network the central node represents the destination airport which has random landing capacity. An optimal strategy for ground holding of aircrafts at departure airports minimizing the expected ground holding and airborne delay costs is found via a polynomial time dynamic programming procedure.

G. Laporte, F.V. Louveaux, H. Mercure (1989). Models and exact solutions for a class of stochastic location-routing problems. *European J. Oper. Res.* 39, 71–78.

The objective is to find a single depot location, a number of vehicles of given capacity, and routes for the vehicles, such that the total costs due to location, acquisition and routing are minimized, and random demand at fixed locations is met. A first model uses a bound on the probability of route failure, i.e. the event that total demand on a route exceeds the capacity of the vehicle serving that route. In another model route failure implies that the vehicle must make an intermediate extra visit to the depot. The expected additional route length may not exceed a certain prefixed portion of the length of the route. The simplistic recourse action allows to translate the stochastic constraints directly into feasible routes. A branch-and-bound procedure is used to solve the problems. Test results are provided.

G. Laporte, F.V. Louveaux, H. Mercure (1992). The vehicle routing problem with stochastic travel times. *Transportation Sci.* 26, 161–170.

In this problem customer demands and vehicle capacities are given, but travel times on the edges of a graph are i.i.d random variables. After a choice of routes starting from a central depot a realization of the travel times may cause the total travel time of a vehicle to exceed a certain prefixed threshold. In the probabilistic constrained model the probability of this event is bounded. In the two-stage problem route costs plus expected penalties to be paid at excess of the threshold are minimized. For the latter problem two formulations are presented and solved by a branch-and-cut algorithm from [Laporte and Louveaux 1993] (see §3.1).

The following set of four papers concerns a particular type of two-stage stochastic combinatorial optimization problems, in which the recourse action is a prefixed strategy. Consider, for instance, the probabilistic traveling salesman problem (PTSP) of finding a tour through a random subset of some set V of vertices (points). In the first stage a tour through all vertices in V is determined. Upon a realization of a subset of vertices, the recourse action is to visit these vertices in the order determined by the first-stage tour. The goal is to find the minimum expected length tour, given the recourse rule and the absence-presence distribution of the vertices.

P. Jaillet (1988). A priori solution of a traveling salesman problem in which a random subset of the customers are visited. *Oper. Res.* 36, 929-936.

For any (first stage) tour a closed-form expression for its expected length under any node-invariant distribution and under the above recourse rule is given. This allows its computation in polynomial time and to bound the relative error obtained by approximating the optimal PTSP tour by the optimal TSP tour.

D.J. Bertsimas, P. Jaillet, A.R. Odoni (1990). A priori optimization. *Oper. Res.* 38, 1019-1033.

Next to the PTSP, also a probabilistic vehicle routing problem, a probabilistic minimum spanning tree problem and a probabilistic location-routing problem are considered, all under identically Bernoulli distributed presence of the vertices and simplistic recourse rules. All four are shown to be NP-complete. Through probabilistic analysis the asymptotic characterizations of the optimal values are compared to those of the expected optimal values of the corresponding distribution problems, as the vertices are i.i.d. uniformly distributed in the unit square and their number tends to infinity. Performance behavior of heuristics is studied.

P. Jaillet (1993). Analysis of probabilistic combinatorial optimization problems in Euclidean spaces. *Math. Oper. Res.* 18, 51-70.

The PTSP and the probabilistic minimum spanning tree problem of the previous paper are studied extensively. Upper bounds are derived in terms of the number of vertices and the Bernoulli probability on presence of a vertex. A thorough probabilistic analysis of the asymptotic behavior of the optimal values is the main theme of the paper.

G. Laporte, F.V. Louveaux, H. Mercure (1994). A priori optimization of the probabilistic traveling salesman problem. *Oper. Res.* 42, 543–549.

In this paper a special version of the branch-and-cut algorithm from [Laporte and Louveaux 1993] (see §3.1) is proposed to solve the Euclidean PTSP in its two-stage integer linear programming formulation, wherein second stage costs are minus the expected sum of the shortcuts made by skipping the randomly absent vertices.

S.W. Wallace, R.J-B. Wets (1995). Preprocessing in stochastic programming: the case of capacitated networks. *ORSA J. Comput.* 7, 44–62.

The paper mainly concerns the detection of facets of the capacitated network flow polytope with demand and supply points. The application to solving a two-stage stochastic programming problem having such a second stage is done in a short section.

4.3 Location problems

M.S. Daskin (1983). A maximum expected covering location model: formulation, properties and heuristic solution. *Transportation Sci.* 17, 48–70.

This model for medical emergency service problems has independent Bernoulli variables for the facilities, indicating if they are active or not. A maximum expected coverage problem is formulated. A local search procedure is proposed and tested empirically.

C. ReVelle, K. Hogan (1988). A reliability-constrained siting model with local estimates of busy fractions. *Environment and Planning B: Planning and Design* 15, 143–152.

A chance constrained model for the problem in the previous paper is formulated. It allows more general randomness but no satisfactory solution method can be given.

S. Shiode, H. Ishii, T. Nishida (1985). A chance constrained minimax facility location problem. *Math. Japan.* 30, 783–803.

A single facility location problem with randomly distributed demand points is modeled with a chance constraint. The deterministic equivalent problem is solved analytically.

P.M. Franca, H.P.L. Luna (1982). Solving stochastic transportation-location problems by generalized Benders decomposition. *Transportation Sci.* 16, 113–126.

This paper considers a stochastic facility location problem, with random demands. The first stage involves decisions on which facilities to open and the amounts of product shipped from the open facilities to the demand points. The expected second stage costs emerge from penalties on surplus or shortage that may occur in demand points. A Benders' decomposition algorithm is proposed for solving the resulting stochastic mixed-integer program.

G. Laporte, F.V. Louveaux, L. van Hamme (1994). Exact solution of a stochastic location problem by an integer L-shaped algorithm. *Transportation Sci.* 28, 95–103.

The problem studied here differs only slightly from the one just above. In the first stage the assignment of the demand points to the open facilities is decided, but the actual transportation decision is delayed to the second stage. The branch-and-cut procedure described in [Laporte and Louveaux 1993] (see §3.1) is worked out for this problem.

F.V. Louveaux (1986). Discrete stochastic location models. *Ann. Oper. Res.* 6, 23–34.

This paper gives a clear exposition of two-stage stochastic mixed-integer programming models for facility location and p-median location problems. At the first stage locations and capacities of facilities are to be decided, under random demand at fixed locations, random transportation cost and random prices for unmet demand. The second stage then concerns the transportation decision from the open facilities to the demand points.

F.V. Louveaux, D. Peeters (1992). A dual-based procedure for stochastic facility location. *Oper. Res.* 40, 564–573.

A dual-based approximation algorithm for the two-stage stochastic facility location problem exposed in the previous paper is presented. A critical assumption is that the joint distribution of the random demands, prices for unmet demand, and transportation cost is discrete with only a small number of mass points. Test results are presented.

4.4 Other problems

D. Bienstock, J.F. Shapiro (1988). Optimizing resource acquisition decisions by stochastic programming. *Management Sci.* 34, 215–229.

Resource acquisition problems under uncertainty, represented as probabilities on scenarios, are modeled as large scale deterministic mixed-integer programming problems. Benders' decomposition is proposed for their solution.

K.O. Joernsten (1992). Sequencing offshore oil and gas fields under uncertainty. *European J. Oper. Res.* 58, 191–201.

Exploitation of gas and oil fields is modeled as a mixed-integer programming problem. Uncertainty in oil prices, market developments, and yields are modeled in scenarios, and the scenario aggregation method of Rockafellar and Wets (*Math. Oper. Res.* 16 (1991), 119–147) is proposed to solve the problem.

R.L. Carraway, R.L. Schmidt, L.R. Weatherford (1993). An algorithm for maximizing target achievement in the stochastic knapsack problem with normal returns. *Naval Res. Log. Quarto.* 40, 161–173.

The generalized dynamic programming procedure from [Carraway, Morin and Moskowitz 1989] (see § 3.4) is applied to solve a stochastic knapsack problem, maximizing expected revenue under normally distributed profit coefficients.

C.L. Dert (1995). *Asset Liability Management for Pension Funds, A Multistage Chance Constrained Programming Approach.* Ph.D. Thesis, Erasmus University, Rotterdam.

A multi-stage probabilistic constrained mixed-integer programming problem is formulated for determination of the contribution by participants and portfolio policy of a pension fund. For any time period there are probabilistic constraints on being solvent to make benefit payments and pay salaries . The random state of the world over the planning horizon is approximated by scenarios. The problem is solved approximately by solving a series of suitably linked two-stage problems, one for each time period aggregating the stages up to that period. Ad hoc techniques are used to save computation time for solving the two-stage problems.

10 Randomized Algorithms

Michel X. Goemans
MIT, Cambridge, MA

David Karger
MIT, Cambridge, MA

Jon Kleinberg
Cornell University, Ithaca, NY

Contents

A randomized algorithm is one that chooses randomly among different options at certain points in its computation. There are many examples where randomization leads to algorithms that are very efficient—sometimes more efficient than any deterministic algorithm. Even when they are no more efficient, randomized algorithms are often

Annotated Bibliographies in Combinatorial Optimization, edited by M. Dell'Amico, F. Maffioli and S. Martello

much simpler than their deterministic counterparts.

The use of randomization in algorithms is a vast field, and in this bibliography, we can barely scratch its surface. We have concentrated mostly on techniques and applications that are most relevant to combinatorial optimization. We have also not covered topics such as randomized local search (simulated annealing, etc.) that are covered in other chapters of this book. In several sections, there were too many references to include; in such cases, we have concentrated our attention on the most relevant references of the last decade.

Following an overview of books and surveys on randomized algorithms, the remainder of the chapter is divided into two parts. The first deals with tools and techniques that are often used in the analysis and design of randomized algorithms. The last deals with application areas. Of course, there is much overlap between the parts, and we have tried to avoid too much duplication.

1 Broad Surveys

R. Motwani, P. Raghavan (1995). *Randomized Algorithms*, Cambridge University Press, Cambridge

This recent text is an excellent general reference for the area of randomized algorithms. The structure of this chapter is similar to Motwani and Raghavan's. This is a very nicely written text with many exercises and notes.

L. Lovász (1995). Randomized algorithms in combinatorial optimization. W. Cook, L. Lovász, P. Seymour (eds.). *Combinatorial Optimization*, AMS, Providence, RI.

This survey presents a few successful applications of randomized algorithms in the area of combinatorial optimization.

R. Gupta, S. Smolka, S. Bhaskar (1994). On randomization in sequential and distributed algorithms. *ACM Comput. Surv.* 26, 7–86.

This survey emphasizes applications in distributed computing, and has a long annotated bibliography section.

R. Karp (1991). An introduction to randomized algorithms. *Discr. Appl. Math.* 34, 165–201.

A beautiful survey emphasizing the different uses of randomization in the design of algorithms.

F. Maffioli, M.G. Speranza, C. Vercellis (1985). Randomized algorithms. M. O'hEigeartaigh, J. Lenstra, A. Rinnooy Kan (eds.). *Combinatorial Optimization—Annotated Bibliographies*, John Wiley & Sons, New York, 89–105.

The predecessor of this chapter. It has a large section on complexity models.

D. Welsh (1983). Randomised algorithms. *Discr. Appl. Math.* 5, 133–145.

This paper discusses some of the main uses of randomization as well as various randomized complexity classes.

W. Feller (1968). *An Introduction to Probability Theory and its Applications*, John Wiley & Sons, New York, third edition.

A classic introduction to much of the probability theory and applications used in the analysis of randomized algorithms.

2 Techniques and Tools

The section actually contains several parts. Sections 2.1 and 2.2 are mainly *analytic techniques* that can be used to prove results about the performance of randomized algorithms; Sections 2.3, 2.4, 2.5 and 2.6 discuss *algorithmic techniques* that can be used in the creation of randomized algorithms. Finally, Sections 2.7 and 2.8 discuss *complexity-theoretic* techniques that can be used to prove that certain randomized algorithms are optimal or can be derandomized.

2.1 Tail inequalities

When analyzing randomized algorithms, it is often necessary to bound the probability that a bad event occurs. This is often expressed as the probability that a random variable (such as the sum of independent random variables) differs from its expected value by more than a certain quantity. An upper bound on this probability is referred to as a *tail inequality*. The strongest tail inequalities are known as *Chernoff bounds* and are obtained by applying Markov's inequality to the random variable's generating function. There are many applications of Chernoff bounds; some are mentioned in [Schmidt et al. 1995] below, others are discussed in [Raghavan 1988] (see §2.7.3).

H. Chernoff (1952). A measure of asymptotic efficiency for tests of a hypothesis based on the sum of observations. *Ann. Math. Stat.* 23, 493–509.

The original Chernoff bound.

W. Hoeffding (1963). Probability inequalities for sums of bounded random variables. *J. American Statist. Assoc.* 58, 13–30.

J. Schmidt, A. Siegel, A. Srinivasan (1995). Chernoff–Hoeffding bounds for applications with limited independence. *SIAM J. Discr. Math.* 8, 223–250.

Additional Chernoff bounds and variants. Even though Chernoff bounds typically deal with the sum of independent random variables, [Hoeffding 1963] also derives tail inequalities for martingales, and [Schmidt et al. 1995] only assume k-wise independence.

T. Hagerup, C. Rüb (1990). A guided tour of Chernoff bounds. *Inform. Process. Lett.* 33, 305–308.

A brief summary and derivation of some of the most commonly used Chernoff bounds. There is also a chapter on tail inequalities in [Motwani and Raghavan 1995] (see §1).

R.M. Karp (1991). Probabilistic recurrence relations. *Proc. 23rd Annual ACM Symp. Theory of Comput.*, 190–197.

When a randomized algorithm subdivides a problem into subproblems of random sizes and recurses (in a Divide-and-Conquer or Branch-and-Bound fashion), the running time analysis of the resulting probabilistic recurrence can be a delicate issue, not immediately susceptible to a standard Chernoff bound argument. This reference gives several cookbook schemes to correctly obtain tail inequalities on the running time of such randomized algorithms, and presents some typical applications.

2.2 The probabilistic method

The *probabilistic method* is a technique for providing "non-constructive" proofs of combinatorial statements: one proves the existence of a combinatorial object by defining an appropriate probability space and showing that the object exists with positive probability. This is a large area, and we will not be able to provide anything close to comprehensive coverage here. In addition to tools from elementary probability, work involving the probabilistic method has taken advantage of techniques such as the tail inequalities discussed in §2.1 and the *Lovász Local Lemma* (see §2.2.1).

J. Spencer (1987). *Ten Lectures on the Probabilistic Method*, SIAM, Philadelphia.
N. Alon, J. Spencer (1992). *The Probabilistic Method*, Wiley Interscience, New York.
P. Erdös, J. Spencer (1974). *Probabilistic Methods in Combinatorics*, Academic Press, New York.

Three very good general reference books.

2.2.1 Lovász Local Lemma

The Lovász Local Lemma provides a way of showing that a certain conjunction of events happens with positive probability, provided that the dependences among the events are sufficiently sparse. Applications can be found in §3.6.

P. Erdös, L. Lovász (1975). Problems and results on 3-chromatic hypergraphs and some related questions. A. Hajnal, R. Rado, V. Sos (eds.). *Infinite and Finite Sets*, North-Holland, Amsterdam, 609–628.

This paper represents the first use of the Local Lemma. However, the existence probability guaranteed by the Local Lemma can be exponentially small in the number of events; so there is frequently no straightforward efficient algorithm associated with an existence proof using the Local Lemma.

J. Beck (1991). An algorithmic approach to the Lovász local lemma I. *Rand. Structures and Algorithms* 2, 343–365.

A quite general framework for producing efficient randomized algorithms for finding objects whose existence is guaranteed by the Local Lemma.

2.3 Randomized rounding

A powerful use of the probabilistic method in combinatorial optimization is *randomized rounding.* The term encompasses a variety of randomized methods for obtaining a near-optimal solution to an integer program from an optimal solution to its fractional relaxation. In one sense this is simply a special case of the probabilistic method. Until recently, randomized rounding focused on the linear programming relaxations of integer linear programs. Recently, a new approach has been introduced using *semidefinite programming* relaxations of integer quadratic programs. Since randomized rounding can provide near-optimal solutions to integer programs, it has been used extensively in the design of approximation algorithms (§3.5).

R. Motwani, J. Naor, P. Raghavan (1997). Randomized approximation algorithms in combinatorial optimization. D. Hochbaum (ed.). *Approximation Algorithms for NP-hard Problems*, PWS Publishing Company, Boston.
A survey of randomized rounding.

P. Raghavan, C. Thompson (1987). Randomized rounding. *Combinatorica* 7, 365–374.
The first use of randomized rounding.

J. Spencer (1985). Six standard deviations suffice. *Trans. American Math. Soc.* 289, 679–706.
This paper discusses a related topic: the *discrepancy* of set systems (see also [Alon and Spencer 1992] in §2.2).

A. Srinivasan (1996). An extension of the Lovász local lemma, and its applications to integer programming. *Proc. 7th Annual ACM-SIAM Symp. Discr. Algorithms*, 6–15.
An extension of the Local Lemma (§2.2.1) is developed in order to analyze a randomized rounding algorithm for several classes of integer linear programs.

M.X. Goemans, D.P. Williamson (1995). Improved approximation algorithms for maximum cut and satisfiability problems using semidefinite programming. *J. ACM* 42, 1115–1145.
This paper introduces the randomized rounding of *semidefinite programming relaxations.* Several applications can be found in §3.5.

2.4 Markov chains

In several situations, one would like to sample from a distribution over a set of possible solutions. This can be hard when the solution space is extremely large. One approach is to define an appropriate Markov chain whose state space is the set of solutions and whose stationary distribution is the desired distribution. Starting from an arbitrary state, simulating the Markov chain for a "sufficient" number of steps will converge to the desired distribution.

The key goal is to estimate or bound the number of steps required before approximately reaching the stationary distribution. Somewhat surprisingly, even when the state space is exponential in the natural size of the problem, a polynomial number of steps can often be shown to be sufficient; such Markov chains are referred to as rapidly-mixing.

The two main applications of this approach are randomized local search algorithms and approximate counting (see §3.4). In the popular simulated annealing method, or more generally in randomized local search algorithms for combinatorial optimization problems, the state space is the set of feasible solutions and the transition probabilities are based on the objective function values. We refer the reader to the chapter on local search for references.

A. Sinclair (1992). *Algorithms for Random Generation and Counting: A Markov Chain Approach.* Progress in Theoretical Computer Science, Birkhauser.
M. Jerrum, A. Sinclair (1997). The Markov Chain Monte Carlo method: Analytical techniques and applications. D. Hochbaum (ed.). *Approximation Algorithms for NP-hard Problems*, PWS Publishing Company, Boston.
R. Kannan (1994). Markov chains and polynomial time algorithms. *Proc. 35th Annual IEEE Symp. Found. Comput. Sci.*, 656–671.

Surveys of Markov chain techniques and their applications to approximate counting.

R. Aleliunas, R. Karp, R. Lipton, L. Lovász, C. Rackoff (1979). Random walks, universal traversal sequences, and the complexity of maze problems. *Proc. 20th Annual IEEE Symp. Found. Comput. Sci.*, 218–223.

This paper shows that a random walk will quickly traverse an entire state-space represented by an undirected graph.

A. Sinclair, M. Jerrum (1989). Approximate counting, uniform generation and rapidly mixing Markov chains. *Inf. and Computation* 82, 93–133.
N. Alon (1986). Eigenvalues and expanders. *Combinatorica* 6, 83–96.

These papers introduce *conductance,* a combinatorial quantity that can be used to measure the rate of convergence of a random walk (which depends on the second largest eigenvalue of the transition matrix). The conductance itself is not easy to compute (doing so is NP-hard) but bounds can be obtained using a "canonical path" technique.

M.R. Jerrum, A. Sinclair (1989). Approximating the permanent. *SIAM J. Comput.* 18, 1149–1178.
P. Diaconis, D. Stroock (1991). Geometric bounds for eigenvalues of Markov chains. *Ann. Appl. Prob.* 1, 36–61.
A. Sinclair (1992). Improved bounds for mixing rates of Markov chains and multicommodity flow. *Combinatorics, Probability, and Comp.* 1, 351–370.
N. Kahale (1996). A semidefinite bound for mixing rates of Markov chains. W. Cunningham, S. McCormick, M. Queyranne (eds.). *Proc. 5th Int. IPCO Conf.*, Lecture Notes in Computer Science, 1084. Springer, New York, 190–203.

Development and refinement of the canonical path method.

2.5 Checking identities

If one needs to test whether two functions are equal, say $f(x) = g(x)$ for all x, one can plug in random values for the indeterminate x, and check if the result agrees or not. This very simple technique is known as Freivalds' technique, and has many applications. The key requirement is to show that if the functions are different, a random element (chosen from the right distribution) is reasonably likely to "witness" this fact by evaluating to different results on the two functions. We can then repeat the test to make the error probability arbitrarily and exponentially small.

An important application of this technique is to finding matchings in graphs, which can be reduced to the problem of determining whether the determinant of the *Tutte matrix* of a graph (with indeterminates) is equal to the function which is identically 0 everywhere. See §3.3 and §3.7.

R. Freivalds (1977). Probabilistic machines can use less running time. B. Gilchrist (ed.). *Information Processing 77, Proc. IFIP Congress 77*, North-Holland, Amsterdam, 839–842.

The technique in testing linear equalities is introduced. To test if $AB = C$ where A, B, C are $n \times n$ matrices, one can generate a vector x uniformly at random from $\{0,1\}^n$ and test if $A(Bx) = Cx$, which requires just 3 matrix–vector multiplies instead of a matrix–matrix multiply. The probability that $ABx = Cx$ while $AB \neq C$ can be shown to be at most 0.5.

J. Schwartz (1980). Fast probabilistic algorithms for verification of polynomial identities. *J. ACM* 27, 701–717.

This paper gives the *Schwartz–Zippel Theorem* for multivariate polynomials. This theorem yields an analysis of the random-argument method for comparing two such polynomials.

2.6 Pseudorandom number generators

By definition, randomized algorithms need a source of random numbers. In practice, these numbers are generated by a deterministic computer program, and are not random at all. Such programs are called *pseudorandom number generators.* Since the lack of randomness of these sequences may have much impact on the performance of randomized algorithms, there has been much work devoted to the study of properties of pseudorandom number generators. In this section, we discuss work relevant to practice; in §2.7, we discuss formal definitions of pseudorandom number generators.

D. Knuth (1971). *The Art of Computer Programming*, volume 2, Addison-Wesley, Reading, MA.

A standard and comprehensive reference on the topic which presents many pseudo-random number generators and discusses statistical tests.

P. l'Ecuyer (1994). Uniform random number generator. *Ann. Oper. Res.* 53, 77–120.
P. l'Ecuyer (1996). Random number generators. S. Gass, C. Harris (eds.). *Encyclopedia of Operations Research and Management Science*, Kluwer Academic Publisher, Dordrect.

Two more recent references on the topic.

2.7 Derandomization

Derandomization refers to the process of transforming a randomized algorithm into a deterministic one. Suppose a randomized algorithm is proven to work with nonzero probability. This means that some point in the sample space of algorithmic random choices causes the algorithm to work. Instead of randomly picking a sample point, we can deterministically enumerate all sample points in order to find the one that makes the algorithm work. This gives a deterministic algorithm for the problem. However, the deterministic algorithm is slower than the randomized one by a factor proportional to the size of the sample space. Much attention has therefore been given to generating small sample spaces that "look random" to randomized algorithms.

L. Adleman (1978). Two theorems on random polynomial time. *Proc. 19th Annual IEEE Symp. Found. Comput. Sci.*, 75–83.

This paper proved that for any randomized polynomial time algorithm and any particular problem size, there exists a small sample space that makes that algorithm work on all inputs of that size. However, the small sample space cannot be constructed efficiently, so this reduction is existential and non-uniform.

2.7.1 Pseudorandom number generators

The most general approach to small sample spaces is pseudorandom number generators. We have already described some of the practical issues in §2.6. Theoretical work has defined a pseudorandom number generator to be an algorithm that takes a small input "seed" and generates a large collection of numbers that "look random" to a polynomial-time algorithm. This formalizes some of the ideas that are achieved in practice via the pseudo random number generators described in §2.6.

If a pseudorandom number generator existed for all polynomial time algorithms, then every randomized polynomial time algorithm would be derandomizable. At present, it is not known whether such a pseudorandom number generator exists, although some are known for certain restricted classes of algorithms.

M. Luby (1996). *Pseudorandomness and Cryptographic Applications*, Princeton University Press, Princeton, NJ.

A book surveying much of what is known about pseudorandom generators.

M. Blum, S. Micali (1984). How to generate cryptographically strong sequences of pseudo-random bits. *SIAM J. Comput.* 13, 850–864.
A.C. Yao (1982). Theory and application of trapdoor functions. *Proc. 23rd Annual IEEE Symp. Found. Comput. Sci.*, 80–91.

These papers formalized the concept of a pseudorandom generator. The second established consequences for the derandomization of algorithms.

L. Babai, L. Fortnow, N. Nisan, A. Wigderson (1993). BPP has subexponential time simulations unless EXPTIME has publishable proofs. *Computational Complexity* 3, 307–318.

This paper gives pseudorandom constructions specifically designed for derandomizing algorithms.

2.7.2 Dependent random variables

Many algorithms get their random choices via a collection of independent random variables. One way to reduce the size of this sample space is to create dependence among the variables. This can often be done while preserving the randomized algorithm's performance guarantee. Unlike pseudorandom generators, this approach requires re-analyzing the algorithm to make sure that the restricted independence suffices.

A. Joffe (1974). On a set of almost deterministic k-independent random variables. *Ann. Prob.* 2, 161–162.

This paper shows how to generate small sample spaces for collections of k-wise independent but not fully independent random variables. Perhaps the first application to algorithms is the derandomizations of parallel algorithms for a maximal independent set; see [Luby 1986] and [Alon et al. 1986] in §3.7

B. Chor, O. Goldreich (1989). On the power of two-point based sampling. *J. Complexity* 5, 96–106.

This paper used the previous one to derandomize algorithms.

J. Naor, M. Naor (1993). Small-bias probability spaces: efficient constructions and applications. *SIAM J. Comput.* 22, 838–856.
N. Alon, O. Goldreich, J. Hästad, R. Peralta (1990). Simple construction of almost k-wise independent random variables. *Proc. 31st Annual IEEE Symp. Found. Comput. Sci.*, 544–553.

These papers show that even smaller sample spaces can be obtained if "almost" k-wise independence is sufficient.

A. Wigderson (1994). The amazing power of pairwise independence. *Proc. 26th Annual ACM Symp. Theory of Comput.*, 645–647.

A bibliography of 36 papers applying pairwise independence.

D. Koller, N. Megiddo (1994). Constructing small sample spaces satisfying given constraints. *SIAM J. Comput.* 7, 260–274.

This paper uses linear programming to construct small sample spaces specialized to a given algorithm and input.

2.7.3 Conditional expectations

The method of conditional expectations is a common derandomization technique for optimization problems. Instead of defining a small sample space and examining all of its point, a single "good" sample point is generated incrementally.

Assume that we are considering a minimization problem and we have a randomized algorithm that outputs a solution whose cost is the random variable X. Let $Y_1, \ldots, Y_n$ denote the random choices of the algorithm. Then $E[X] = \sum_{y_1} E[X|Y_1 = y_1]Pr[Y_1 = y_1] \geq \min_{y_1} E[X|Y_1 = y_1]$. As a result, if we can compute $E[X|Y_1 = y_1]$ for all possible realizations y_1 then we can select a value for y_1^* for which $E[X|Y_1 = y_1^*] \leq E[X]$. We fix this value of Y_1 and consider the random choices on all the other variables. We can then fix Y_2 as we did Y_1, and so forth, and eventually find a solution given by the choices $Y_i = y_i^*$ for all i whose cost is at most $E[X]$. We have thus constructed a deterministic algorithm which performs at least as well as the randomized algorithm.

P. Erdös, J. Selfridge (1973). On a combinatorial game. *J. Combin. Theory A* 14, 298–301.

Implicit use of the technique.

P. Raghavan (1988). Probabilistic construction of deterministic algorithms: Approximating packing integer programs. *J. Comput. Syst. Sci.* 37, 130–143.

Explicit use to derandomize the randomized rounding scheme discussed in §2.3. Several of the algorithms described in §3 use the method of conditional expectations. For situations where $E[X|Y_1 = y_1]$ cannot be evaluated exactly, the above paper also introduced the method of *pessimistic estimators*; this is especially useful when combined with Chernoff bounds (§2.1).

2.8 Minimax principle

There is a very nice application of the minimax principle of Von Neumann, or essentially strong duality for linear programming, in the analysis of randomized algorithms. The idea is that in many settings a randomized algorithm can be viewed as a distribution over (or convex combination of) deterministic algorithms. The problem of finding the randomized algorithm minimizing some expected performance (say running time or competitive ratio) can thus often be formulated as a linear program with a constraint for each possible input. Strong duality then shows that the expected performance of the best randomized algorithm is equal to the best performance of any deterministic algorithm on the worst distribution over the possible inputs.

A.C.-C. Yao (1977). Probabilistic computations: towards a unified measure of complexity. *Proc. 18th Annual IEEE Symp. Found. Comput. Sci.*, 222–227.

First description of the principle. For an application to on-line algorithms, see §3.9.

3 Application Areas

3.1 Linear programming

Randomization has led to significant advances in algorithms for linear programming. Two main threads have been apparent. After a first deterministic linear-time (in the number of variables n) algorithm for linear programming was given by Megiddo, randomization was used, typically via *randomized incremental constructions* to dramatically reduce the running time dependence on the dimension d. In a separate vein, randomized simplex-like pivoting rules have been used to give algorithms running in sub-exponential time in arbitrary dimension.

K.L. Clarkson (1988). A Las Vegas algorithm for linear programming when the dimension is small. *Proc. 29th Annual IEEE Symp. Found. Comput. Sci.*, 452–456.
M. Dyer, A. Frieze (1989). A randomized algorithm for fixed-dimensional linear

programming. *Math. Program.* 44, 203–212.
R. Seidel (1991). Small-dimensional linear programming and convex hulls made easy. *Discr. Comput. Geom.* 6, 423–434.
M. Sharir, E. Welzl (1992). A combinatorial bound for linear programming and related problems. *Proc. 9th Symp. on Theoret. Aspects of Comp. Sci.*, Lecture Notes in Computer Science 577. Springer-Verlag, New York, 569–579.

Randomized linear-time algorithms for linear programming in fixed dimension. Seidel's is a particularly simple randomized algorithm with expected running time $O(d!n)$, while the fourth is a simple improvement on it giving an expected running time of $O(d^3 2^d n)$.

G. Kalai (1992). Upper bounds on the diameter of graphs of convex polytopes. *Discr. Comput. Geom.*
G. Kalai, D. Kleitman (1992). A quasi-polynomial bound for diameter of graphs of polyhedra. *Bull. American Math. Soc.* 24, 315–316.

The famous "Hirsch conjecture" posits an upper bound on the diameter of a polytope with n facets in d dimensions that is linear in n and d. The first paper uses randomized arguments to give the first sub-exponential bounds on the diameter of polytopes; the second gives an improved bound.

G. Kalai (1992). A subexponential randomized simplex algorithm. *Proc. 24th Annual ACM Symp. Theory of Comput.*, 475–482.
J. Matoušek, M. Sharir, E. Welzl (1992). A subexponential bound for linear programming. *Proc. 8th Annual ACM Symp. on Comput. Geometry*, 1–8.

Extensions of the previous papers to randomized simplex-like linear programming algorithms with a sub-exponential dependence on d. Combining these with the algorithm of [Clarkson 1988] gives a randomized algorithm with expected running time of $O(d^2 n + e^{\sqrt{d \log d}})$.

B. Gärtner (1992). A subexponential algorithm for abstract optimization problems. *Proc. 33rd Annual IEEE Symp. Found. Comput. Sci.*, 464–472.

This paper extends the above sub-exponential optimization algorithms to a more general class of *abstract optimization problems.*

M. Goldwasser (1995). A survey of linear programming in randomized subexponential time. *SIGACT News* 26, 96–104.

Surveys the relationship between the previous three papers, showing that the first two are essentially dual formulations of the same algorithm.

3.2 Computational geometry

Several powerful and general randomized techniques have been developed for attacking problems in computational geometry. Two of the most prominent techniques are *random sampling* and *randomized incremental constructions.*

The *random sampling* methodology is based on choosing a random sample of the points, lines, or hyperplanes in the problem, and analyzing the sample to extract information about the original problem so that it can be solved quickly. One of its appealing features is that the *selection* of a random sample is typically straightforward; progress has also been made on techniques for deterministic sampling—that is, deterministically selecting a small set of points with the distributional properties of a random sample.

Randomized incremental constructions are based on the idea of randomly ordering the input items, and then bounding the expected amortized cost of producing the desired output when the items are added in this random order. This bounding of the cost can often be achieved by considering the construction in reverse order; this approach is known as *backwards analysis.*

K. Mulmuley (1994). *Computational Geometry: An Introduction Through Randomized Algorithms*, Prentice-Hall, Englewood Cliffs, NJ.
B. Chazelle (1994). Computational geometry: a retrospective. *Proc. 26th Annual ACM Symp. Theory of Comput.*, 75–94.
R. Seidel (1993). Backwards analysis of randomized geometric algorithms. *Algorithms and Comb.* 10, 37–68.

The first of these is a book, the latter two survey articles, covering many of the uses of randomization in geometric algorithms. The third focuses on randomized incremental constructions and backwards analysis.

L.P. Chew (1985). *Building Voronoi diagrams for convex polygons in linear expected time*, Technical report, Dept. Math. Comput. Sci., Dartmouth College, Hanover, NH.

One of the first uses of backwards analysis in computational geometry.

K. Clarkson (1987). New applications of random sampling in computational geometry. *Discr. Comput. Geom.* 2, 195–222.
K.L. Clarkson, P.W. Shor (1989). Applications of random sampling in computational geometry, II. *Discr. Comput. Geom.* 4, 387–421.

These papers give a number of powerful uses of random sampling in the design of fundamental geometric algorithms. Among these, they provide an optimal randomized algorithm for computing the convex hull of n hyperplanes in d dimensions. (Here, "optimal" is meant with respect to the output size of the convex hull, which can be as large as $n^{\lfloor d/2 \rfloor}$.)

B. Chazelle (1993). An optimal convex hull algorithm in any fixed dimension. *Discr. Comput. Geom.* 10, 377–409.

The convex hull algorithm of the previous paper is derandomized.

3.2.1 VC-dimension and ϵ-nets

V. Vapnik, A.Y. Chervonenkis (1971). On the uniform convergence of relative frequencies of events to their probabilities. *Theor. Prob. Appl.* 16, 264–280.
D. Haussler, E. Welzl (1987). Epsilon-nets and simplex range queries. *Discr. Comput. Geom.* 2, 127–151.

The *Vapnik–Chervonenkis dimension* of a set system is developed in the first paper; it has proved to be a powerful way of combinatorializing some of the properties of Euclidean space that underlie the analysis of randomized geometric algorithms. If we have a collection of sets $\mathcal{S}$ over a ground set U, then the VC-dimension of the set system $(U, \mathcal{S})$ is, roughly, the size of the largest subset X of U all of whose subsets can be represented as intersections of X with members of $\mathcal{S}$. When the VC-dimension is bounded (as is common in geometric problems), the two papers above show the existence of good types of finite "samples"—called ϵ-approximations and ϵ-nets respectively. The size of these samples depends only on the VC-dimension and not on the size of the set system.

B. Chazelle, H. Edelsbrunner, M. Grigni, L. Guibas, M. Sharir, E. Welzl (1993). Improved bounds on weak ϵ-nets for convex sets. *Proc. 25th Annual ACM Symp. Theory of Comput.*, 495–504.
B. Chazelle, J. Matoušek (1993). On linear-time deterministic algorithms for optimization problems in fixed dimension. *Proc. 4th Annual ACM-SIAM Symp. Discr. Algorithms*, 281–290.
J. Matoušek (1993). Approximations and optimal geometric divide-and-conquer. *Proc. 25th Annual ACM Symp. Theory of Comput.*, 506–511.
J. Kleinberg (1997). Two algorithms for nearest-neighbor search in high dimensions. *Proc. 29th Annual ACM Symp. Theory of Comput.*, 599–608.
Extensions and applications of this sampling methodology.

3.3 Graph algorithms

Nearly all the techniques discussed in the first section find some application to graph algorithms; no one technique appears dominant. Our focus here is on fast sequential algorithms for tractable problems. For applications to approximating NP-complete graph problems, see §3.5. For the maximal independent set problem, see §3.7 on parallel algorithms. For counting perfect matchings, see §3.4.

R. Seidel (1995). On the all-pairs-shortest-path problem in unweighted undirected graphs. *J. Comput. Syst. Sci.* 51, 400–403.
"Random witness selection" is used to reduce the problem of all pairs shortest paths to that of matrix multiplication.

D. Karger, P. Klein, R. Tarjan (1995). A randomized linear-time algorithm to find minimum spanning trees. *J. ACM* 42, 321–328.
This paper uses random sampling to find minimum spanning trees in linear time.

M. Rauch Henzinger, V. King (1995). Randomized dynamic graph algorithms with polylogarithmic time per operation. *Proc. 27th Annual ACM Symp. Theory of Comput.*, 519–527.
Random sampling is used to update the connected components of a graph as edges are added and removed.

E. Cohen (1994). Estimating the size of the transitive closure in linear time. *Proc. 35th Annual IEEE Symp. Found. Comput. Sci.*, 190–200.
Random sampling is used to estimate various properties of a graph in near linear time, including the number of edges in its transitive closure and the number of vertices

within a given distance of every vertex.

D. Karger, C. Stein (1996). A new approach to the minimum cut problem. *J. ACM* 43, 601–640.
This paper uses random selection to find minimum cuts in undirected graphs.

D. Karger (1996). Minimum cuts in near-linear time. *Proc. 28th Annual ACM Symp. Theory of Comput.*, 56–63.
This paper uses random sampling to find minimum cuts in undirected graphs.

J. Cheriyan, T. Hagerup (1995). A randomized maximum-flow algorithm. *SIAM J. Comput.* 24, 203–226.
Randomization is used to bring the running time of the push-relabel algorithm for maximum flow to slightly below the best currently known deterministic bound.

A. Benczúr, D. Karger (1996). Approximate s–t min-cuts in $\tilde{O}(n^2)$ time. *Proc. 28th Annual ACM Symp. Theory of Comput.*, 47–55.
Random sampling is used to approximate s–t minimum cuts in undirected graphs.

L. Lovász (1979). On determinants, matchings and random algorithms. L. Budach (ed.). *Fundamentals of Computing Theory*, Akademia-Verlag, Berlin.
This paper gives a randomized test to decide if a graph has a perfect matching; it is based on the fact that the determinant of the Tutte matrix of a graph (with indeterminates) is identically 0 if and only if the graph has no perfect matching. This can be detected using Frievald's technique (§2.5). For additional references, see §3.7.

P. Camerini, G. Galbiati, F. Maffioli (1992). Random pseudo-polynomial algorithms for exact matroid problems. *J. Algorithms* 13, 258–273.
This paper develops algebraic methods to obtain randomized pseudo-polynomial algorithms for a class of *exact* optimization problems—in particular, the authors obtain algorithms for finding a base of a fixed value in a class of weighted matroids.

3.4 Approximate counting

A number of well-known combinatorial problems ask for estimates on the size of certain (implicitly defined) sets; often, these are sets of solutions to certain optimization problems. This is essentially the same as determining the probability that a random candidate solution is in fact a solution. Recent work on *approximate counting* has involved the development of efficient algorithms with provable performance guarantees for problems of estimating the sizes of such sets.

M.R. Jerrum, L.G. Valiant, V.V. Vazirani (1986). Random generation of combinatorial structures from a uniform distribution. *Theor. Comput. Sci.* 43, 169–188.
This paper indicates the connection between approximate counting of a set and almost uniform selection of a set element. In particular, subject to some technical conditions, it shows how a randomized algorithm for almost uniform sampling yields an approximation algorithm for the corresponding counting problem.

R.M. Karp, M. Luby, N. Madras (1989). Monte-Carlo approximation algorithms for enumeration problems. *J. Algorithms* 10, 429–448.

Biased Monte-Carlo simulation is used to estimate the number of satisfying assignments to a Boolean formula in disjunctive normal form.

A. Broder (1986). How hard is it to marry at random? *Proc. 18th Annual ACM Symp. Theory of Comput.*, 50–58. Erratum in *Proc. 20th Annual ACM Symp. Theory of Comput.*, 1988, pp. 51.

Along with [Jerrum and Sinclair 1989] (see §2.4), these papers show how to approximately count the number of perfect matchings in a graph by considering a Markov chain defined on the set of perfect and near-perfect matchings. [Jerrum and Sinclair 1989] gives a correct analysis of the rate of convergence. This is a fully polynomial randomized approximation scheme only if the ratio of the numbers of near-perfect matchings to perfect matchings is polynomial.

3.4.1 Volume computation

One application of approximate counting is to estimate the volume of a convex body.

M. Dyer, A. Frieze, R. Kannan (1991). A random polynomial time algorithm for approximating the value of convex bodies. *J. ACM* 38, 1–17.

The authors use (rapidly mixing) Markov chains (§2.4) to randomly sample points from a convex body in an arbitrary dimensional space. This leads to fully polynomial randomized approximation schemes for the volume of such bodies (exact computation is #P-hard).

I. Bárány, Z. Füredi (1986). Computing the volume is difficult. *Proc. 18th Annual ACM Symp. Theory of Comput.*, 442–447.
G. Elekes (1986). A geometric inequality and the complexity of computing the volume. *Discr. Comput. Geom.* 1, 289–292.

These two papers show that if the convex body is specified by a membership oracle, no deterministic algorithm can approximate the volume to within a polynomial factor (the first paper shows a lower bound that is exponential). This proves that the use of randomization in the previous papers is crucial.

D. Applegate, R. Kannan (1991). Sampling and integration of near log-concave functions. *Proc. 23rd Annual ACM Symp. Theory of Comput.*, 156–163.
M. Dyer, A. Frieze (1991). Computing the volume of convex bodies: A case where randomness provably helps. *Proc. AMS Symposia in Applied Mathematics*, Probabilistic Combinatorics and its Applications 44, 123–170.
L. Lovász, M. Simonovits (1993). Random walks in a convex body and an improved volume algorithm. *Rand. Structures and Algorithms* 4, 359–412.

These papers significantly improve the running times of the randomized volume approximation schemes.

3.5 Integer programming

Randomized rounding (see §2.3) has had many applications in the approximate solution of integer programs. A randomized rounding algorithm solves the linear programming relaxation (or some other type of relaxation) and then "rounds" in a randomized fashion the fractional solution to an integral solution.

[Raghavan and Thompson 1987] (see §2.3) give the first use of randomized rounding. One of the problems they consider is the integral multicommodity flow problem. They give a randomized constant-factor approximation for such problems when the edge capacities are superlogarithmic. Their results exploit Chernoff bounds (see §2.1), and the algorithms can be derandomized using the method of pessimistic estimators (see §2.7).

M.X. Goemans, D.P. Williamson (1994). New 3/4-approximation algorithms for the maximum satisfiability problem. *SIAM J. Discr. Math.* 7, 656–666.

Randomized rounding is used to derive a 3/4-approximation algorithm for the maximum satisfiability problem. Here each literal is set to TRUE or FALSE independently with a probability depending non-linearly on the fractional LP solution.

[Goemans and Williamson 1995] (see §2.3) proposed a different rounding procedure in which the different variables are set in a dependent fashion. The technique is based on solving a semidefinite programming relaxation, viewing the solution in terms of vectors, and generating a random hyperplane in this vector space. This lead to a randomized 0.878-approximation algorithm for MAX CUT and improved algorithms for other problems.

D. Karger, R. Motwani, M. Sudan (1994). Approximate graph coloring by semidefinite programming. *Proc. 35th Annual IEEE Symp. Found. Comput. Sci.*, 2–13. To appear in *Journal of the ACM.*

U. Feige, M.X. Goemans (1995). Approximating the value of two prover proof systems, with applications to MAX 2SAT and MAX DICUT. *Proc. of 3rd Israel Symp. on Theory of Comput. and Syst.*, 182–189.

A. Frieze, M. Jerrum (1997). Improved approximation algorithms for MAX k-CUT and MAX BISECTION. *Algorithmica*, 18, 67–81.

These three papers use semidefinite programming techniques to derive improved approximation algorithms for vertex coloring, maximum satisfiability, and maximum k-cut problems.

S. Mahajan, H. Ramesh (1995). Derandomizing semidefinite programming based approximation algorithms. *Proc. 36th Annual IEEE Symp. Found. Comput. Sci.*, 162–169.

It is shown how the method of conditional expectations (see §2.7) can be used to derandomize these semidefinite-based approximation algorithms.

3.6 Routing and scheduling

A. Broder, A. Frieze, E. Upfal (1994). Existence and construction of edge-disjoint paths on expander graphs. *SIAM J. Comput.* 23, 603–612.

The Lovász Local Lemma (see §2.2) is used to show the existence and efficient construction of large sets of disjoint paths in any graph with a "sufficiently strong" expansion parameter.

F. Leighton, B. Maggs, S. Rao (1994). Packet routing and job-shop scheduling in O(congestion + dilation) steps. *Combinatorica* 14, 167–186.
F. Leighton, B. Maggs (1995). Fast algorithms for finding O(congestion + dilation) packet-routing schedules. *Proc. 28th Hawaii Int. Conf. on Syst. Sci.* 2, 555–563.

Applications of the Lovász Local Lemma to packet-routing and scheduling problems. In the basic problem, one is given a set of packets in a network, each of which has a simple path that it must traverse. At most one packet can traverse any edge in any time step. The first paper shows that there always exists a *schedule*, giving the time at which each packet crosses each edge, which completes in time $O(c + d)$, where d is the maximum path length and c is the maximum number of paths containing any single edge. The second paper shows how to construct the schedule efficiently. There is a natural generalization of this to job-shop scheduling problems with unit-length jobs and no "cycles" in the sequence of machines required by a job.

3.7 Parallel algorithms

Randomization has been important in developing RNC algorithms (randomized parallel algorithms running in polylogarithmic time on a polynomial number of processors) for many problems that are quite easy to solve sequentially. Perhaps its greatest power has been to "break symmetry" allowing processors that start with the same input to divide the work on it easily. In addition, the use of small sample spaces (see §2.7) often allows one to convert such randomized algorithms into deterministic NC algorithms (by assigning a processor to each sample point). A classic problem tackled this way was the maximal independent set problem.

R. Karp, A. Wigderson (1985). A fast parallel algorithm for the maximal independent set problem. *J. ACM* 32, 762–773.

The first RNC algorithm (and its derandomization) for the problem of finding a maximal independent set in a graph.

M. Luby (1986). A simple parallel algorithm for the maximal independent set problem. *SIAM J. Comput.* 15, 1036–1053.
N. Alon, L. Babai, A. Itai (1986). A fast and simple randomized algorithm for the maximal independent set problem. *J. Algorithms* 7, 567–583.
M. Luby (1993). Removing randomness in parallel computation without a processor penalty. *J. Comput. Syst. Sci.* 47, 250–86.

The first two papers give faster RNC and NC algorithms for the maximal independent set problem. The derandomizations exploit pairwise independence. The last paper adds the method of conditional expectations to the previous papers in order to reduce the processor cost of the derandomization.

R. Karp, E. Upfal, A. Wigderson (1986). Constructing a perfect matching in random NC. *Combinatorica* 6, 35–48.

K. Mulmuley, U. Vazirani, V. Vazirani (1987). Matching is as easy as matrix inversion. *Combinatorica* 7, 105–113.
H. Karloff (1986). A Las Vegas RNC algorithm for maximum matching. *Combinatorica* 6, 387–391.

These papers show that maximum matchings can be found efficiently in parallel using randomization. An efficient deterministic parallel algorithm for this problem is not known. The first paper gives the first RNC algorithm for constructing a maximum matching. The second gives a simpler solution based on an elegant randomized argument for selecting costs which make the minimum cost maximum matching unique (the so-called isolating lemma). The third paper shows that by using the Edmonds–Gallai structure, one can make sure that the matching returned (if any) is a maximum matching.

H. Narayanan, H. Saran, V. Vazirani (1994). Randomized parallel algorithms for matroid union and intersection, with applications to arboresences and edge-disjoint spanning trees. *SIAM J. Comput.* 23, 387–397.

The matching approach is generalized to matroid intersection problems.

A. Aggarwal, R. Anderson (1988). A random NC algorithm for depth first search. *Combinatorica* 8, 1–12.

Depth first search trees are found via a reduction to the (randomized) algorithm for maximum matching.

H. Gazit (1991). An optimal randomized parallel algorithm for finding connected components in a graph. *SIAM J. Comput.* 20, 1046–1067.

This paper gives an algorithm with best-possible parallel running time and processor count.

3.8 Data structures

A randomized data structure for maintaining the connected components of a graph is referenced in §3.3.

The most common use of randomization in data structures is in several implementations of *dictionaries* that allow insertion, deletion, and looking up of keys drawn from a large set. One dictionary implementation is hash tables. Hashing is often argued to have good expected time performance when the set of input keys is random. By introducing randomization in the choice of the hash function, we can ensure that good performance is likely even when the input keys are not chosen randomly.

J. Carter, M. Wegman (1979). Universal classes of hash functions. *J. Comput. Syst. Sci.* 18, 143–154.
M. Wegman, J. Carter (1981). New hash functions and their use in authentication and set equality. *J. Comput. Syst. Sci.* 22, 265–279.

Universal hash functions are introduced; choosing a random universal hash function makes good performance likely on any input.

A. Yao (1981). Should tables be sorted? *J. ACM* 28, 615–628.

M. Fredman, J. Komlós, E. Szemerédi (1984). Storing a sparse table with $o(1)$ worst-case access time. *J. ACM* 31, 538–544.

These papers develop hash tables with linear storage that support constant-time query access.

M. Dietzfelbinger, A. Karlin, K. Mehlhorn, F. Meyer auf der Heide, H. Rohnert, R. Tarjan (1994). Dynamic perfect hashing: Upper and lower bounds. *SIAM J. Comput.* 23, 738–761.

Universal hash functions are used to build extended hash tables that support insertions, deletions, and queries in constant amortized time.

C. Aragon, R. Seidel (1989). Randomized search trees. *Proc. 30th Annual IEEE Symp. Found. Comput. Sci.*, 540–545.

A simple randomized implementation of balanced binary search trees is given.

W. Pugh (1990). Skip lists: a probabilistic alternative to balanced trees. *Commun. ACM* 33, 668–676.

An alternative randomized implementation of dictionaries.

3.9 On-line algorithms

Many algorithms respond to series of "requests" that arrive over time. Such *on-line* algorithms must allocate resources at request time, without knowledge of the future. *Competitive Analysis* was invented to study the behaviour of such algorithms; the competitive ratio of an algorithm measures its performance against that which could be achieved with perfect knowledge of the future (or, equivalently, in hindsight).

D. Sleator, R. Tarjan (1985). Amortized efficiency of list update and paging rules. *Commun. ACM* 28, 202–208.

This paper proposes the competitive analysis framework as a tool to study the problem of virtual memory management (paging). A paging algorithm is c-competitive if the number of pages it evicts is, in every scenario, asymptotically within a factor of c of the optimal number. The paper shows that if the memory contains k pages, then no deterministic on-line algorithm can be better than k-competitive; and it identifies a number of k-competitive on-line algorithms.

A. Fiat, R.M. Karp, M. Luby, L.A. McGeoch, D.D. Sleator, N. Young (1991). Competitive paging algorithms. *J. Algorithms* 12, 685–699.
L. McGeoch, D. Sleator (1991). A strongly competitive randomized paging algorithm. *Algorithmica* 6, 816–825.

In contrast to the preceding paper on deterministic algorithms, these papers study the use of randomization for on-line paging. The first paper develops a simple $2H_k$-competitive randomized paging algorithm, when memory has k pages, and shows that no randomized on-line algorithm can achieve a ratio better than H_k. The second presents an H_k-competitive randomized algorithm.

S. Ben-David, A. Borodin, R. Karp, G. Tardos, A. Wigderson (1990). On the power of randomization in on-line algorithms. *Algorithmica* 11, 2–14.

This paper formalizes various models for randomized on-line algorithms; it also develops non-trivial relationships among these models, including a method for inferring the existence of good deterministic on-line algorithms from good randomized ones.

P. Raghavan, M. Snir (1989). Memory versus randomization in on-line algorithms. *Proc. 16th Int. Coll. Automata, Lang. and Program.*, Lecture Notes in Computer Science 372. Springer-Verlag, Berlin, 687–703.
E. Grove (1991). The harmonic online k-server algorithm is competitive. *Proc. 23rd Annual ACM Symp. Theory of Comput.*, 260–266.

For the on-line *k-server problem*, a generalization of paging, the first paper proposes HARMONIC, a simple randomized algorithm, and conjectures that it is $O(k^2)$-competitive. The second paper establishes the first bound on its competitive ratio, exponential in k.

A. Blum, H. Karloff, Y. Rabani, M. Saks (1992). A decomposition theorem and bounds for randomized server problems. *Proc. 33rd Annual IEEE Symp. Found. Comput. Sci.*, 197–207.

This paper establishes the strongest lower bound known for randomized k-server algorithms: $\Omega\sqrt{\log k / \log\log k}$.

A. Borodin, N. Linial, M. Saks (1992). An optimal online algorithm for metrical task systems. *J. ACM* 39, 745–763.

An application of Yao's minimax principle (see §2.8) to the analysis of randomized on-line algorithms is presented.

S. Vishwanathan (1990). Randomized online coloring of graphs. *Proc. 31st Annual IEEE Symp. Found. Comput. Sci.*, 464–469.

A randomized on-line algorithm for graph coloring is provided.

B. Awerbuch, R. Gawlick, F. Leighton, Y. Rabani (1994). On-line admission control and circuit routing for high performance computing and communication. *Proc. 35th Annual IEEE Symp. Found. Comput. Sci.*, 412–423.
J. Kleinberg, E. Tardos (1995). Disjoint paths in densely embedded graphs. *Proc. 36th Annual IEEE Symp. Found. Comput. Sci.*, 52–61.
Y. Bartal, A. Fiat, A. Rosén (1996). Lower bounds for on-line graph problems with applications to on-line circuit and optical routing. *Proc. 28th Annual ACM Symp. Theory of Comput.*, 531–560.

These papers consider randomized on-line algorithms for the disjoint paths problem. The third gives a lower bound for such algorithms.

11 Local Search

Emile Aarts
Philips Research Laboratories, Eindhoven and
Eindhoven University of Technology

Marco Verhoeven
Philips Research Laboratories, Eindhoven

CONTENTS

Local search algorithms constitute a widely used, general approach to hard combinatorial optimization problems. They are are typically instantiations of various general search schemes, but all have the same feature of an underlying neighborhood function, which is used to guide the search for a good solution.

The use of a local search algorithm presupposes the definition of a problem and a neighborhood. An *instance* of a combinatorial optimization problem consists of a set $\mathcal{S}$ of feasible solutions and a cost function f. The problem is to find a *globally optimal solution*, i.e., a solution with minimal cost in the case of minimization and maximal cost in the case of maximization. A *neighborhood function* is a mapping $\mathcal{N} : \mathcal{S} \rightarrow 2^{\mathcal{S}}$, which defines for each solution $i \in \mathcal{S}$ a set $\mathcal{N}(i) \subseteq \mathcal{S}$ of solutions that are in some sense close to i. The set $\mathcal{N}(i)$ is called the *neighborhood* of solution i, and each $j \in \mathcal{N}(i)$ is called a *neighbor* of i.

A basic version of local search is *iterative improvement*, which starts with some initial solution and searches its neighborhood for a solution of lower cost (in the case of minimization). If such a solution is found, it replaces the current solution, and the search continues. Otherwise, the algorithm returns the current solution, which is then called *locally optimal.* A local optimum obtained by iterative improvement may be bad. This has led to the development of a number of more sophisticated local search algorithms which aim at finding better solutions by introducing mechanism that allow the search to escape from locally optimal solutions.

In general, local search can be viewed as a walk in a *neighborhood graph*, whose node set is given by the set of solutions and there is an arc from node i to node j if j is a neighbor of i. The sequence of nodes visited by the search process defines the walk.

Roughly speaking, the two main issues of a local search algorithm are the choice of

Annotated Bibliographies in Combinatorial Optimization, edited by M. Dell'Amico, F. Maffioli and S. Martello

the neighborhoods and the search strategy that is used. Good neighborhoods often take advantage of the combinatorial structure of the problem at hand, and are therefore typically problem dependent. Most search strategies are generally applicable, and in this bibliography we classify the existing ones in the following four categories: simulated annealing, tabu search, genetic algorithms, and neural networks. We do not present the classical iterative improvement algorithms as a separate class, since their application is rather elementary and straightforward. They are mentioned in the section on history and can be found in most references mentioned in the section on best practice.

1 Books and Surveys

C.R. Reeves, (ed.) (1993). *Modern Heuristic Techniques for Combinatorial Problems*, Blackwell Scientific Publications, Oxford.

This book presents an introductory overview of the standard search strategies, i.e., simulated annealing, tabu search, genetic algorithms, and neural networks. It also includes a chapter on lower bound calculations.

I.H. Osman, J.P. Kelley (eds.) (1996). *Metaheuristics: Theory and Applications*, Kluwer Academic Publishers, Hingham, MA.

G. Laporte, I.H. Osman (eds.) (1996). Meta-Heuristics in Combinatorial Optimization. *Ann. Oper. Res.* 63.

V.J. Rayward-Smith, I.H. Osman, C.R. Reeves, G.D. Smith (eds.) (1996). *Modern Heuristic Search Methods*, Wiley, Chichester.

These three books contain an extensive collection of papers that present a state of the art overview of techniques and applications of local search. The books present a good overview of the recent development in the field. The nature of most of the contributions is aimed at practical use of techniques, rather than on theoretical aspects.

E.H.L. Aarts, J.K. Lenstra (eds.) (1997). *Local Search in Combinatorial Optimization*, Wiley, Chichester.

This book reviews presents theoretical aspects of local search, including the performance of local search for well-structured problems and a complexity theory for local search. It discusses four basic local search methodologies, i.e., simulated annealing, tabu search, genetic algorithms, and neural networks. Furthermore, it reviews the best practice of applications in sequencing, scheduling, vehicle routing, VLSI design, and coding theory; see also Section 8.

M. Yannakakis (1990). The analysis of local search problems and their heuristics. *Lecture Notes Comput. Sci.* 415, 298–311.

This paper presents an introductory review of classical local search concepts and briefly outlines the complexity theory of local search.

M. Pirlot (1992). General local search heuristics in combinatorial optimization: a tutorial. *Belgian J. Oper. Res. Stat. Comp. Sci.* 32, 8–67.

This survey presents a review of the standard local search techniques and their application.

E. Pesch, S. Voss (eds.) (1995). Applied local search. *Oper. Res. Spekt.* 17, 2–3.
This special journal issue presents a survey of recent developments in the application and analysis of local search.

I.H. Osman, G. Laporte (1996). Meta-heuristics: a bibliography. *Ann. Oper. Res.* 63, 513–623.
This bibliography provides a classification of a list of more than 1400 references both on the theory and application of meta-heuristics.

2 History

The use of local search in combinatorial optimization reaches back to the late fifties and early sixties of the twentieth century.

F. Bock (1958). An algorithm for solving 'Travelling-Salesman' and related network optimization problems (abstract). *Bull. 14th Nat. Meeting Oper. Res. Soc. America*, 897.
G.A. Croes (1958). A method for solving traveling salesman problems. *Oper. Res.* 6, 791–812.
S. Reiter, G. Sherman (1965). Discrete optimizing. *J. Soc. Ind. App. Math.* 13, 864–889.
These three papers are among the first that mention edge-exchange algorithms for the traveling salesman problem.

S. Lin (1965). Computer solutions of the traveling salesman problem. *Bell Syst. Tech. J.* 44, 2245–2269.
This paper present further refinements of the classical edge-exchange algorithms for the traveling salesman problem, including 3-exchange and Or-exchange neighborhood functions.

S. Lin, B.W. Kernighan (1973). An effective heuristic algorithm for the traveling-salesman problem. *Oper. Res.* 21, 498–516.
This is the classical paper on *variable-depth search*, which is a sophisticated edge-exchange algorithms for the TSP. It presents the notorious Lin-Kernighan algorithm and shows its performance on the "large" 318 cities instance.

B.W. Kernighan, S. Lin (1970). An efficient heuristic procedure for partitioning graphs. *Bell Syst. Tech. J.* 49, 291–307.
This paper presents a variable-depth search algorithm for uniform graph partitioning.

T.A.J. Nicholson (1971). A method for optimizing permutation problems and its industrial applications. D.J.A. Welsh (ed.). *Combinatorial Mathematics and its Applications*, Academic Press, London, 201–217.
The author extends the concept of edge-exchange strategies for the traveling sales-

man problem to a more general class of permutation problems including network layout design, scheduling, vehicle routing, and cutting stock problems.

Despite the early successes, local search was not considered a mature technique for a long time. This was mainly because the success could be attributed to its practical usefulness only, and no major conceptual progress was made at that time. The past decade shows a strong renewed interest in the subject for the following three reasons.

Appeal. Processes in nature related to statistical physics, biological evolution, and neurophysiology are shown to be quite appealing in the design of new local search paradigms. Well-known examples are *simulated annealing, genetic algorithms* and *neural networks.*

Theory. Some of the newly proposed local search algorithms, as for instance simulated annealing, could be mathematically modeled in such a way that theoretical results on their performance could be obtained. Furthermore, a complexity theory of local search was introduced, which provided more theoretical insight, not only into the complexity of local search but also into the combinatorial structure of discrete optimization problems.

Practice. The flexibility and ease of implementation of local search algorithms in combination with the availability of powerful computational resources have made local search algorithms strong competitors within the class of algorithms designed to handle large practical problem instances.

3 Computational Complexity

The theoretical performance analysis of local search shows a number of "bad" examples, which prove that the performance of local search algorithms may be arbitrarily bad, both with respect to running time and solution quality.

C.H. Papadimitriou, K. Steiglitz (1977). On the complexity of local search for the traveling salesman problem. *SIAM J. Comp.* 6, 76–83.

The authors prove for the traveling salesman problem that no polynomially searchable exact neighborhoods can exist unless P = NP. Furthermore, examples of instances are presented with a single optimal solution and exponentially many second best solutions that are locally optimal with respect to k-exchange neighborhoods, for $k < 3/8n$, while their costs are arbitrarily far from optimal.

G. Lueker (1976). Two NP-complete problems in nonnegative integer programming. Manuscript, Princeton University.

This frequently cited unpublished manuscript presents instances and start solutions of the traveling salesman problem for which 2-exchange algorithms require an exponential number of steps to find a local optimum.

V. Rodl, C.A. Tovey (1987). Multiple optima in local search. *J. Algorithms* 8, 250–259.

This paper presents examples of a bad worst-case behavior for the independent set problem.

W. Kern (1989). A probabilistic analysis of the switching algorithm for the Euclidean

TSP. *Math. Program.* 44, 213–219.

This paper addresses the theoretical probabilistic behavior of local search, and shows that the 2-exchange algorithm for Euclidean instances of the traveling salesman problem with n points in the unit square converges with high probability within $O(n^{18})$ steps.

C.A. Tovey (1985). Hill climbing with multiple local optima. *SIAM J. Algebraic Discr. Meth.* 6, 384–393.
C.A. Tovey (1986). Low order polynomial bounds on the expected performance of local improvement algorithms. *Math. Program.* 35, 193–224.

In these two papers, the author considers classes of neighborhood graphs with a regular structure, e.g., the hypercube. For several types of random cost functions he proves low-order polynomial average-case running times. The worst-case results are invariably pessimistic.

C.A. Tovey (1996). Local improvement on discrete structures. E.H.L. Aarts, J.K. Lenstra (eds.). *Local Search in Combinatorial Optimization*, Wiley, Chichester.

This book chapter presents a review of the theoretical results obtained in the study of local search on regularly shaped combinatorial problems.

A major achievement was the introduction of the complexity class PLS of *Polynomial-time Local Search* problems. Informally speaking, PLS contains those local search problems for which local optimality can be verified in polynomial time. Furthermore, just as the NP-complete problems were defined as the hardest decision problems in NP, *PLS-complete* problems are the hardest local search problems in PLS. None of the PLS-complete problems is known to be solvable in polynomial time, and if any of them can, then so can all problems in PLS. There is a "root" problem called Flip, which is as hard as any other problem in PLS. The notion of a PLS-*reduction* has been formalized which transforms instances of one problem in PLS into instances of another problem in PLS, and several local search problems have been proved PLS-complete using such PLS-reductions.

M. Yannakakis (1996). Computational complexity of local search. E.H.L. Aarts, J.K. Lenstra (eds.). *Local Search in Combinatorial Optimization*, Wiley, Chichester.

This book chapter presents a comprehensive survey of the complexity theory of local search.

D.S. Johnson, C.H. Papadimitriou, M. Yannakakis (1988). How easy is local search? *J. Comput. Syst. Sci.* 37, 79–100.

This is the seminal paper of the complexity theory for local search. It introduces the class PLS and the concepts of PLS-completeness and PLS-reductions. The "root" problem Flip is introduced and it is proved that Flip is PLS-complete. Furthermore, it is proved that uniform graph partitioning with the Kernighan-Lin variable-depth neighborhood function is PLS-complete.

M.W. Krentel (1989). Structure in locally optimal solutions. *Proc. IEEE 30th Ann. Symp. Found. Comput. Sci.*, 216–222.

The author proves PLS-completeness of the traveling salesman problem with the k-exchange neighborhood function for fixed k.

M.W. Krentel (1990). On finding and verifying locally optimal solutions. *SIAM J. Comp.* 19, 742–751.
The author proves PLS-completeness of Max-2Sat and Pos NAE Max-3Sat with the Flip neighborhood function.

A.A. Schäffer, M. Yannakakis (1991). Simple local search problems that are hard to solve. *SIAM J. Comp.* 20, 56–87.
The authors prove PLS-completeness of graph partitioning with the Swap, the Fiduccia-Mattheyses, and Fiduccia-Mattheyses-Swap neighborhood functions, of Max-Cut with the Flip neighborhood function, and of the stable configuration for neural networks problem with the Flip neighborhood function.

C.H. Papadimitriou (1992). The complexity of the Lin-Kernighan heuristic for the traveling salesman problem. *SIAM J. Comp.* 21, 450–465.
The author proves PLS-completeness of the traveling salesman problem with the Lin-Kernighan variable-depth neighborhood function.

G. Ausiello, M. Protasi (1995). Local search, reducibility and approximability of NP-optimization problems. *Inform. Process. Lett.* 54, 73–79.
This paper discusses optimization problems that can be approximated within a guaranteed constant factor using (polynomial) local search for a class of neighborhood functions with a bounded Hamming distance, as in the k-exchange neighborhood function.

4 Simulated Annealing

Simulated annealing is based on an analogy with the physical process of annealing, in which a pure lattice structure of a solid is made by heating up the solid in a heat bath until it melts, and next cooling it slowly until it solidifies into a low-energy state. From the point of view of combinatorial optimization, simulated annealing is a randomized neighborhood search algorithm. In addition to better-cost neighbors, which are always accepted if they are selected, worse-cost neighbors are also accepted, although with a probability that is gradually decreased in the course of the algorithm's execution. Lowering the *acceptance probability* is controlled by a set of parameters whose values are determined by a *cooling schedule*.

E.H.L. Aarts, J.H.M. Korst (1989). *Simulated Annealing and Boltzmann Machines*, Wiley, Chichester.
P.J.M. van Laarhoven, E.H.L. Aarts (1987). *Simulated Annealing: Theory and Applications*. Reidel, Dordrecht.
These are introductory textbooks describing both theoretical and practical issues of simulated annealing.

F. Romeo, A. Sangiovanni-Vincentelli (1991). A theoretical framework for simulated annealing. *Algorithmica* 6, 302–345.

This review paper presents a comprehensive survey of the theory of simulated annealing.

R.V.V. Vidal (ed.) (1993). Applied Simulated Annealing. *Lecture Notes Econ. Math. Syst.* 396.

This book presents an edited collection of papers on practical aspects of simulated annealing, ranging from empirical studies of cooling schedules up to implementation issues of simulated annealing for problems in engineering and planning.

R.W. Eglese (1990). Simulated annealing: a tool for operational research. *European J. Oper. Res.* 46, 271–281.

This review paper presents a survey of the application of simulated annealing to problems in operations research.

N.E. Collins, R.W. Eglese, B.L. Golden (1988). Simulated annealing – an annotated bibliography. *American J. Math. Management Sci.* 8, 209–307.

This annotated bibliography presents more than a thousand references to papers on simulated annealing. It is organized in two parts; one on the theory, and the other on applications. The applications range from graph theoretic problems up to problems in engineering, biology, and chemistry.

S. Kirkpatrick, C.D. Gelatt, Jr., M.P. Vecchi (1983). Optimization by simulated annealing. *Science* 220, 671–680.

This is a seminal paper on simulated annealing explaining the basic concept of the physical annealing process and the analogy with combinatorial optimization.

V. Černy (1985). A thermodynamical approach to the travelling salesman problem: an efficient simulation algorithm. *J. Optim. Theory Appl.* 45, 41–51.

This paper can be viewed as an independent introduction of the physical annealing concept into the area of combinatorial optimization.

W. Metropolis, A. Rosenbluth, M. Rosenbluth, A. Teller, E. Teller (1953). Equation of state calculations by fast computing machines. *J. Chem. Phys.* 21, 1087–1092.

This paper presents the first results on computer simulations of the Boltzmann distribution for physical systems. Although this is the central issue of simulated annealing, no reference is made to its use for combinatorial optimization problems.

B. Hajek (1988). Cooling schedules for optimal annealing. *Math. Oper. Res.* 13, 311–329.

This paper presents necessary and sufficient conditions for asymptotic convergence of simulated annealing.

S. Anily, A. Federgruen (1987). Simulated annealing methods with general acceptance probabilities. *J. Appl. Probab.* 24, 657–667.

This paper presents theoretical results on the convergence of simulated annealing for a set of acceptance probabilities that are much more general than the classical Metropolis acceptance probabilities.

5 Tabu Search

Tabu search combines deterministic iterative improvement with a possibility to accept cost increasing solutions. In this way the search is directed away from local minima, such that other parts of the search space can be explored. The next solution visited is chosen to be a *legal neighbor* of the current solution with the best cost, even if that cost is worse than that of the current solution. The set of legal neighbors is restricted by a *tabu list* designed to prevent us going back to recently-visited solutions. The tabu list is dynamically updated in the course of the algorithm's execution, and defines solutions that are not acceptable in the next few iterations. However, a solution on the tabu list may be accepted if its quality is in some sense good enough, in which case it is said to attain a certain *aspiration level.*

F. Glover, M. Laguna, E. Taillard, D. de Werra (eds.) (1993). Tabu search. *Ann. Oper. Res.* 41.
This special issue presents the theory and many practical applications of tabu search.

D. de Werra, A. Hertz (1989). Tabu search techniques: a tutorial and an application to neural networks. *Oper. Res. Spekt.* 11, 131–141.
This survey reviews many practical issues related to the use of tabu search for combinatorial optimization problems.

F. Glover, M. Laguna (1993). Tabu search. C. Reeves (ed.). *Modern Heuristic Techniques for Combinatorial Problems*, Blackwell Scientific Publishing, Oxford, 70–141.
This book chapter presents a comprehensive overview of the existing tabu search variants.

F. Glover (1989). Tabu search – Part I. *ORSA J. Comp.* 1, 190–206.
This is the first paper of a diptych that introduces the basic concepts of tabu search. It presents the basic concepts of short-term and long-term memory, and data structures for implementing the tabu conditions. It makes a distinction between *simple tabu search*, applying only short term memory, and more sophisticated versions with various types of long term memory.

F. Glover (1990). Tabu search – Part II. *ORSA J. Comp.* 2, 4–32.
This is the second part of the diptych, which discusses more advanced dynamic elements of tabu search. It also presents tabu search methods for solving mixed integer programming problems, and the relation with neural networks.

F. Glover (1986). Future paths for integer programming and links to artificial intelligence. *Computers Oper. Res.* 13, 533–549.
This paper is one of the first papers that mention tabu search explicitly. It can be

viewed as the seminal paper on the subject. The motivating themes of tabu search go back to earlier developments, especially to those that grew out of surrogate constraint relaxations and heuristics used to take advantage of them in mixed integer linear programming.

F. Glover (1968). Surrogate constraints. *Oper. Res.* 16, 741–749.

This is one of the early papers that present heuristics similar to tabu search, without mentioning it explicitly.

P. Hansen, B. Jaumard (1990). Algorithms for the maximum satisfyability problem. *Computing* 44, 279–303.

This paper presents an approach called *steepest descent mildest ascent*, which can be viewed as a tabu search approach that has been developed independently from Glover's version.

U. Faigle, W. Kern (1992). Some convergence results for probabilistic tabu search. *ORSA J. Comput.* 4, 32–37.

The authors present convergence results for a stochastic version of tabu search. The analysis is based on Markov models and shows a resemblance with the approach used for simulated annealing.

R. Battiti, G. Tecchiolli (1994). The reactive tabu search, *ORSA J. Comput.* 6, 126–140.

This paper presents a variant of tabu search in which the length of the tabu list is computed from an explicit check of the repetition of solutions found by the algorithm. They show that cycle detection can be efficiently implemented by hashing or digital tree techniques.

6 Genetic Algorithms

Genetic algorithms apply concepts from population genetics and evolution theory to construct algorithms that try to optimize the *fitness* of a *population* of elements through *recombination* and *mutation* of their genes. The recombination and mutation can be viewed as neighbor selection in terms of local search. There are many variations known in the literature of algorithms that follow these concepts. *"Evolutionary computing"* is also often used to denote this field.

D.E. Goldberg (1989). *Genetic Algorithms in Search, Optimization, and Machine Learning.* Addison-Wesley, Reading, MA.

This is the first comprehensive text on genetic algorithms. It describes the basic algorithmic concepts and their application to global optimization and classifier systems.

Z. Michalewicz (1992). *Genetic Algorithms + Data Structures = Evolutionary Programs.* Springer, Berlin.

This textbook presents in a comprehensive fashion the concepts and applications of genetic algorithms at a graduate level. It explains the fundamentals of genetic algorithms and their implementation. It also presents in-depth discussions of many applications in combinatorial optimizations.

G. Rawlins (ed.) (1991). *Foundations of Genetic Algorithms.* Morgan Kaufmann, San Mateo, CA.
L.D. Whitley (ed.) (1993). *Foundations of Genetic Algorithms 2.* Morgan Kaufmann, San Mateo, CA.
These two textbooks present a selection of papers describing results of the ongoing studies of genetic algorithms.

Th. Bäck, F. Hoffmeister, H.-P. Schwefel (1991). A survey of evolution strategies. R.K. Belew, L.B. Booker (eds.). *Proc. 4th Int. Conf. Genetic Algorithms*, 2–9.
This is a well-written short survey that classifies most of the existing approaches to genetic programming. It also mentions some applications.

I. Rechenberg (1973). *Evolutionsstrategie - Optimierung Technischer Systeme nach Prinzipien der Biologischen Information*, Fromman, Freiburg.
This is one of the early works presenting results on the application of evolutionary strategies to function optimization.

J.H. Holland (1975). *Adaptation in Natural and Artificial Systems*, University of Michigan Press, Ann Arbor, MI.
This textbook is often viewed as the seminal work on genetic algorithms. Most of the work presented in later stages follows the lines of the *simple genetic algorithm* and the *parallel genetic algorithms* introduced in this text.

R.M. Brady (1985). Optimization strategies gleaned from biological evolution. *Nature* 317, 804–806.
This is a short popular text which has stimulated the research on genetic algorithms and their use in combinatorial optimization.

H. Mühlenbein, M. Gorges-Schleuter, O. Krämer (1988). Evolution algorithms in combinatorial optimization. *Parallel Comput.* 7, 65–85.
This paper introduces the concept of *genetic local search*, which combines the evolution of populations of solutions through recombination, with the improvement of individual solutions in a population through simple neighborhood search. In population genetics jargon this is called Lamarquian evolution.

A. Kolen, E. Pesch (1994). Genetic local search in combinatorial optimization. *Discr. Appl. Math.* 48, 273–284.
This review paper discusses approaches of genetic local search algorithms to the traveling salesman problem, with the emphasis on situations where severe time constraints are imposed on the running time of the algorithm.

D.B. Fogel, L.J. Fogel (1996). An introduction to evolutionary programming. J.-M. Alliot, E. Lutton, E. Ronald, M. Schoenauer, D. Snyers (eds). *Proc. European Conf.*

Artificial Evolution, 21–33.
This paper reviews recent efforts of evolutionary programming approaches in combinatorial and continuous optimization. Some areas of current investigation are mentioned, including empirical assessment of the optimization performance of the technique and extensions of the method to include mechanisms to self-adapt to the topology of the search space.

J.J. Grefenstette (1986). Optimization of control parameters for genetic algorithms. *IEEE Trans. Syst. Man Cybern.* 16, 122–128.
The author presents a generic approach to the problem of determining the parameter values of genetic algorithms.

A.E. Eiben, E.H.L. Aarts, K.M. van Hee (1991). Global convergence of genetic algorithms: a Markov chain analysis. H.-P. Schwefel, R. Männer (eds.). Parallel Problem Solving from Nature, *Lecture Notes Comput. Sci.* 496, Springer, Berlin, 4–12.
This paper presents a Markov model for stochastic variants of genetic algorithms and discusses sufficient conditions for convergence to optimal populations.

7 Neural Networks

A neural network consists of a network of elementary nodes (neurons) that are linked by weighted connections. The nodes represent computational units, which are capable of performing a simple computation, consisting of a summation of the weighted inputs, followed by the addition of a constant called the threshold or bias, and the application of a non-linear response function. The result of the computation of a unit constitutes its output. This output is used as an input for the nodes to which it is linked through an outgoing connection. The overall task of the network is to achieve a certain network configuration, for instance a required input-output relation by means of the collective computation of the nodes. This process is often called *self-organization*. The literature on neural networks is abundant. Below we only mention two introductory textbooks, a few seminal papers and surveys of the application of neural networks to combinatorial optimization.

S. Haykin (1994). *Neural Networks: A Comprehensive Foundation*, Macmillan, New York, NY.
This is a general text at the graduate level presenting the theory and application of neural networks.

J. Hertz, A. Krogh, R.G. Palmer (1991). *Introduction to the Theory of Neural Computation*, Addison-Wesley, Redwood City, CA.
This introductory textbook discusses most of the existing neural network methods and their application to combinatorial optimization.

I. Parberry (1990). A primer on the complexity theory of neural networks. R.B. Banerji (ed.). *A Sourcebook on Formal Techniques in Artificial Intelligence*, North-Holland, Amsterdam, 217–268.

This book chapter presents an overview of complexity issues related to neural networks. Special attention is paid to the relation of neural networks with parallel models of computation.

R. Rosenblatt (1962). *Principles of Neurodynamics*, Spartan Books, New York.
This book introduces perceptrons as artificial neural computing elements and discusses learning in networks consisting of perceptrons organized in a single layer. This book has greatly influenced the research on neural networks and is often viewed as the seminal work in this area.

M. Minsky, S. Papert (1969). *Perceptrons*, MIT Press, Cambridge, MA.
In this book the authors discuss the limitations of networks of perceptrons. They show that for some learning problems Rosenblatt's learning procedure can take exponential time, and that there exist learning problems that cannot be solved by a single-layered perceptron network at all.

J.J. Hopfield (1982). Neural networks and physical systems with emergent collective computational abilities. *Proc. Nat. Acad. Sci. Unit. Stat. America* 79, 2554–2558.
This is the seminal paper on Hopfield networks which are recurrent symmetric neural networks which can apply both continuous and discrete state values. The Hopfield networks are of great interest for their applicability in combinatorial optimization.

J.J. Hopfield, D.W. Tank (1985). 'Neural' computation of decisions in optimization problems. *Biol. Cybern.* 52, 141–152.
In this paper the authors discuss the application of Hopfield networks to combinatorial optimization problems. Especially their treatment of the traveling salesman problem has been of great influence in the literature.

J. Bruck, J.W. Goodman (1987). A generalized convergence theorem for neural networks and its applications in combinatorial optimization. *Proc. IEEE 1st Int. Conf. Neural Networks*, 649–56.
The authors prove that binary Hopfield networks converge in pseudo-polynomial time to a stable state when operating in a serial mode and to a cycle of length at most two when operating in a fully parallel mode. The authors also present a general convergence theorem that unifies the two known cases. Some applications for combinatorial optimization are included.

J. Bruck, J.W. Goodman (1990). On the power of neural networks for solving hard problems. *J. Complexity* 34, 1089–1092.
This paper presents a complexity result that proves that no polynomial-size neural network can exist that solves a given arbitrary NP-hard problem, unless NP = coNP.

E.B. Baum (1986). Towards practical 'neural' computation for combinatorial optimization problems. J.S. Denker (ed.). *Neural Networks for Computing*, Proc. AIP Conference 151, American Institute of Physics, 53–58.
This is one of the first papers that discusses a generic approach of neural networks to combinatorial optimization. The treatment is basic but the elementary concepts

are well-explained.

C. Peterson, B. Söderberg (1989). A new method for mapping optimization problems onto neural networks. *Int. J. Neural Systems* 1, 3–22.
This paper presents a neural network approach to combinatorial optimization based in mean field annealing techniques and Potts-glass models. The authors show that these techniques can effectively handle traveling salesman and graph partitioning problems.

E.H.L. Aarts, J.H.M. Korst (1991). Boltzmann machines as a model for massively parallel annealing. *Algorithmica* 6, 437–465.
This paper presents a general framework for mapping combinatorial optimization problems onto a stochastic variant of Hopfield networks that are called Boltzmann machines.

Y. Takefuji (1992). *Neural Network Parallel Computing*, Kluwer Academic Publishers, Dordrecht.
This textbook presents a broad overview of applications of neural networks to combinatorial optimization problems with an emphasis on scheduling problems.

C.-K. Looi (1992). Neural network methods in combinatorial optimization. *Computers Oper. Res.* 19, 191–208.
This survey presents a comprehensive overview of neural networks to combinatorial optimization problems with pointers to the literature on more than thirty well-known problems. The author, furthermore, presents a discussion on the use of neural networks in combinatorial optimization.

J. Ramanujam, P. Sadayappan (1995). Mapping combinatorial optimization problems onto neural networks. *Inform. Sci.* 3-4, 239–255.
This paper reviews the general problem of mapping of combinatorial optimization problems onto neural networks. Examples are discussed from graph theory, VLSI placement, and operations research.

8 Best Practice

Local search algorithms have been applied to a wide range of combinatorial (optimization) problems. Over the past three decades several thousands of papers have been published scattered all over the literature. Here we confine ourselves to a listing of a limited number of papers that are selected because they present thorough computational studies that aim at comparing several approaches for well-known combinatorial problems. Annotations are omitted because the titles are self explanatory.

D.S. Johnson, C.R. Aragon, L.A. McGeoch, C. Schevon (1989). Optimization by simulated annealing: an experimental evaluation, Part I (graph partitioning). *Oper. Res.* 37, 865–892.
D.S. Johnson, C.R. Aragon, L.A. McGeoch, C. Schevon (1991). Optimization by sim-

ulated annealing: an experimental evaluation, Part II (graph coloring and number partitioning). *Oper. Res.* 39, 378–406.
C. Peterson (1990). Parallel distributed approaches to combinatorial optimization problems – benchmark studies on TSP. *Neural Comput.* 2, 261–269.
P.M. Pardalos, K.A. Murty, T.P. Harrison (1993). A computational comparison of local search heuristics for solving quadratic assignment problems. *Informatica* 4, 172–187.
B.S. Cooper (1994). Selected applications of neural networks in telecommunication systems: A review. *Australian Telecomm. Res.* 28, 9–29.
R. Batitti, G. Tecchiolli (1994). Simulated annealing and tabu search in the long run: a comparison on QAP tasks. *Comput. Math. Appl.* 28, 1–8.
G. Reinelt (1994). The Traveling Salesman: computational solutions for TSP applications. *Lecture Notes Comput. Sci.* 840.
C.A. Glass, C.N. Potts (1996). A comparison of local search methods for flow shop scheduling, *Ann. Oper. Res.* 63, 489-512.
R.J.M. Vaessens, E.H.L. Aarts, J.K. Lenstra (1996). Local Search for Job Shop Scheduling. *INFORMS J. Comp.* 8, 302–317.

The following six references are book chapters from E.H.L. Aarts, J.K. Lenstra (eds.), *Local Search in Combinatorial Optimization*, Wiley, Chichester.

D.S. Johnson, L.A. McGeoch (1996). The traveling salesman problem: a case study. 215–310.
M. Gendreau, G. Laporte, J.-Y. Potvin (1996). Vehicle routing: modern search heuristics. 311–336.
G.A.P. Kindervater, M.W.P. Savelsbergh (1996). Vehicle routing: handling edge exchanges. 337–360.
E.J. Anderson, C.A. Glass, C.N. Potts (1996). Machine scheduling. 361–414.
E.H.L. Aarts, P.J.M. van Laarhoven, C.L. Liu, P. Pan (1996). VLSI layout synthesis. 415–440.
I.S. Honkala, P.R.J. Ostergard (1996). Code design. 441–456.

9 Parallel Algorithms

Local search algorithms often find high-quality solutions, but at the cost of substantial amounts of running time for the larger instances. To cope with this drawback several approaches have been proposed to implement these algorithms on parallel machines. Early approaches concentrate on parallel simulated annealing algorithms. More recently, also parallel approaches to other local search methods have been reported.

We classify parallel local search algorithms as follows. First, we distinguish between tailored and general approaches. A *tailored approach* requires parallel execution of problem specific parts of the algorithm, such as cost computations or manipulations of the data structures. A *general approach* typically applies problem independent parallel calculations.

Next, we distinguish between single-walk and multiple-walk algorithms. A *multiple-*

walk algorithm is based on the parallel evaluation of multiple walks in the neighborhood graph. These walks may interact or may be independent. The latter can be considered as the most straightforward approach to parallel local search. A *single-walk* algorithm evaluates a single walk in the neighborhood graph in parallel. This requires the distribution of a solution over different processors.

In the class of single-walk algorithms, we distinguish between multiple-step and single-step parallelism. In *single-step parallelism*, neighbors are simultaneously evaluated to subsequently make a single step. To this end neighborhoods need to be partitioned into subsets that are searched in parallel. In *multiple-step parallelism* a sequence of consecutive steps in the neighborhood graph is made simultaneously.

Finally, in both single-walk and multiple-walk parallel local search we distinguish between synchronous and asynchronous algorithms. *Synchronous parallelism* requires a global clocking scheme which guarantees that communication occurs at fixed points in time. *Asynchronous parallelism* does not require a central clocking scheme and communication may take place at arbitrary moments in time.

R. Azencott (1992). *Simulated Annealing: Parallelization Techniques*, Wiley, New York.

This book presents a collection of more-or-less independent chapters on theoretical and experimental aspects of parallel simulated annealing. Both fine and coarse grained parallel approaches are discussed, and a number of applications are discussed including a comparative study of implementations on different hardware architectures.

D. Greening (1990). Parallel simulated annealing techniques. *Physica D* 42, 293–306.

This paper reviews parallel simulated annealing techniques, using a classification that distinguishes between correct and incorrect cost evaluations, and between algorithms that maintain the sequential convergence properties or not.

E.H.L. Aarts, F.M.J. de Bont, E.H.A. Habers (1986). Parallel implementations of the statistical cooling algorithm. *Integration* 3, 209–238.

This is one the first examples of parallel simulated annealing applying multiple walk parallelism. It presents a *clustered algorithm* that shows a linear speed-up on the traveling salesman problem for a shared memory machine with eight processors.

S.A. Kravitz, R.A. Rutenbar (1987). Placement by simulated annealing on a multiprocessor. *IEEE Trans. Comput. Aid. Des.* 6, 534–549.

This is one the first examples of single-step parallelism in the literature. The authors apply asynchronous single-step parallelism to simulated annealing for the cell placement problem in combination with a tailored approach to compute the cost of solution in parallel. They report a speed-up of 2.5 on a shared memory machine with four processors.

M.G.A. Verhoeven, E.H.L. Aarts (1995). Parallel local search. *J. Heuristics* 1, 43–65.

This paper presents a general overview of parallel local search algorithms. It discusses *tailored* and *general* approaches, and introduces a classification of existing approaches by distinguishing between *single-walk* and *multiple-walk parallelism*, and between asynchronous and synchronous algorithms. In the class of single-walk algorithms, the authors distinguish between multiple-step and single-step parallelism.

R. Battiti, G Tecchiolli (1992). Parallel biased search for combinatorial optimization: genetic algorithms and tabu search. *Microproc. Microsyst.* 16, 351–367.

The authors empirically investigate the behavior of tabu search for randomly generated instances of the quadratic assignment problem. They observe that the probability of finding a (locally) optimal solution with a tabu search algorithm is distributed exponentially, provided the search is started with a local minimum. This indicates that good efficiencies can be obtained with multiple independent parallel walks in tabu search, if the time needed to find the first local minimum is relatively small compared to the time spent in the remainder of the search process.

T.G. Crainic, M. Toulouse, M. Gendreau (1995). Synchronous tabu search parallelization strategies for multicommodity location-allocation with balancing requirements. *Oper. Res. Spekt.* 17.

The authors present various synchronous multiple-walk tabu search algorithms. They propose two forms of interaction: one in which the best solution in the entire population is selected by all processors, and one in which each of the P processors selects a different solution among the P best solutions encountered since the previous interaction. Different walks are obtained by choosing a different parameter setting for processors. The best solutions are found by multiple independent walks.

J. Chakrapani, J. Skorin-Kapov (1993). Massively parallel tabu search for the quadratic assignment problem. *Ann. Oper. Res.* 41, 327–342.

The authors use synchronous single-step parallelism in combination with best improvement for the traveling salesman problem and the quadratic assignment problem, respectively. Their algorithm is implemented on a massively parallel Connection Machine with 16,384 processors. They report that 55% of the time is spent on communication.

J.R.A. Allwright, D.B. Carpenter (1989). A distributed implementation of simulated annealing for the traveling salesman problem. *Parallel Comput.* 10, 335–338.

In this paper a parallel simulated annealing algorithm for the traveling salesman problem is presented, based on a distribution of a tour over a linear array of processors. A partial solution in the distribution consists of two non-adjacent paths in a tour. Edges are randomly reassigned to processors to obtain a new distribution.

M.G.A. Verhoeven, E.H.L. Aarts, P.C.J. Swinkels (1995). A parallel 2-opt algorithm for the traveling salesman problem. *Future Gener. Comput. Syst.* 11, 175–182.

The authors present a distributed neighborhood for the traveling salesman problem, based on partitioning a tour into several subtours. They prove that the neighborhood is isomorphic with the well-known 2-exchange and 3-exchange neighborhoods.

J.S. Rose, W.M. Snelgrove, Z.G. Vranesic (1988). Parallel standard cell placement algorithms with quality equivalent to simulated annealing. *IEEE Trans. Comput. Aid. Des.* 7, 387–396.

The authors present a hybrid algorithm that uses both multiple-walk parallelism and multiple-step parallelism is presented. The authors apply multiple-walk parallelism at large values of control parameter of the simulated annealing algorithm and multiple-step parallelism at low values.

J.E. Savage, M.G. Wloka (1991). Parallelism in graph-partitioning. *J. Parallel Distrib. Comput.* 13, 257–272.
An algorithm is presented for the graph partitioning problem based on splitting a solution. In order to deal with erroneous cost computations, either all proposed exchanges are accepted if the resulting global solution has lower cost, or all proposed exchanges are rejected if the resulting global solution has higher cost.

P.M. Pardalos, L. Pitsoulis, T. Mavridou, M.G.C. Resende, M.G.C. (1995). Parallel search for combinatorial optimization: genetic algorithms, simulated annealing, tabu search and GRASP. A. Ferreira, J. Rolim (eds). *Parallel Algorithms for Irregularly Structured Problems*, Springer, Berlin, 317–331.
This paper reviews parallel search techniques for the approximate solution of combinatorial optimization problems. It discusses recent developments on parallel implementations of genetic algorithms, simulated annealing, tabu search, and greedy randomized adaptive search procedures (GRASP); see also Section 10.

10 Local Search Variants

The literature on local search contains many variants of the approaches discussed in the previous sections. Many of these are rather specific and tailored to the problem under consideration. Their specialistic nature is not suited for the general presentation we had in mind for this bibliography and we therefore will not attempt to review them. The interested reader is referred to the bibliography by Laporte and Osman mentioned in Section 1. There are, however, a few variants that are more widely applicable, and among these we mention the following ones.

T.A. Feo, M.G.C. Resende (1995). Greedy randomized adaptive search procedures. *J. Global Optim.* 6, 109.
GRASP (greedy randomized adaptive search procedure) is a polynomial-time constructive approach to generate start solutions for a local search procedure. In each construction step, a number of candidate elements is specified using an adaptive function to qualify the benefits of adding a particular element to the partial solution constructed so far. From the set of candidates one element is randomly selected and added to the partial solution until a complete solution is obtained.

G. Dueck and T. Scheuer (1990). Threshold accepting: A general purpose optimization algorithm. *J. Comput. Phys.* 90, 92–98.
Threshold accepting is a deterministic variant of simulated annealing in which a neighboring solution replaces the current solution if their cost difference does not exceed a certain threshold. In this way cost deteriorations not exceeding the threshold are accepted. As in simulated annealing the threshold value is lowered in the course of the search.

F. Glover (1995). Tabu thresholding: improved search by non-monotonic trajectories. *ORSA J. Comp.* 7, 426.

Tabu thresholding is a form of tabu search in which memory is replaced by thresholds that determine whether neighboring solutions are accepted. The selection of neighbors is governed by candidate lists, and the thresholds for accepting neighbors are randomly varied within a given interval.

F. Glover (1996). Ejection chains, reference structures and alternating path methods for traveling salesman problems. *Discr. Appl. Math.* 65, 223–253.

Ejection chains are neighbor construction procedures in which neighbors are generated using a sequences of exchanges leading from one solution to another via several intermediate partial solutions which are not necessarily feasible. Ejection chains were first proposed within the scope of tabu search, but their general nature also enable their use in other local search algorithms.

R.H. Storer, S.D. Wu, R. Vaccari (1992). New search spaces for sequencing problems with applications to job shop scheduling. *Management Sci.* 38, 1495.

Space search methods iterate over a problem-instance space rather than over neighboring solutions in a solution set. The method applies a fast heuristic to construct feasible solutions for a given problem instance with an automatically computed cost criterion that is derived from the cost function of the original problem instance. Problem-instance neighbors are constructed by means of a parametric perturbation mechanism.

A. Colorni, M. Dorigo, V. Maniezzo (1992). An investigation of some properties of an ant algorithm. R. Männer, B. Manderick (eds.). *Parallel Problem Solving from Nature, 2*, North-Holland, Amsterdam, 509–520.

Ant algorithms use ants as simple *agents* that perform a local search and colonies of ants that perform a collective search. This approach can be viewed as a parallel bi-level local search strategy; one level corresponds to the search of the individual ants and the other to the search of the ant colonies.

12 Sequencing and Scheduling

Johannes A. Hoogeveen
Eindhoven University of Technology, Eindhoven

Jan Karel Lenstra
Eindhoven University of Technology, Eindhoven

Steef L. van de Velde
University of Twente, Enschede

CONTENTS

E.L. Lawler, J.K. Lenstra, A.H.G. Rinnooy Kan, D.B. Shmoys (1993). Sequencing and scheduling: algorithms and complexity. S.C. Graves, P.H. Zipkin, A.H.G. Rinnooy Kan (eds.). *Logistics of Production and Inventory; Handbooks in Operations Research and Management Science, Vol. 4*, North-Holland, Amsterdam, 445-522.

Sequencing and scheduling is concerned with the optimal allocation of scarce resources to activities over time. The area is motivated by questions that arise in production planning, project scheduling, and computer control, and to a lesser extent in routing, personnel scheduling, maintenance scheduling, and materials handling. The models in this area are highly standardized: they concern the scheduling of n *jobs* (the activities) on m *machines* (the resources), which can process no more than one activity at a time, so as to optimize some function of the job completion times.

Annotated Bibliographies in Combinatorial Optimization, edited by M. Dell'Amico, F. Maffioli and S. Martello

The specification of a machine scheduling problem requires the description of a *machine environment*, *job characteristics*, and an *optimality criterion*. The simplest environment is the *single machine*, on which jobs j, each consisting of a single operation, have to spend given processing times p_j ($j = 1, \ldots, n$). In a *parallel machine* environment, job j has to spend a given time on any of m machines. These can be *identical*, in which case the machines operate at the same speed; *uniform*, in which case each machine has its own speed; and *unrelated*, in which case the speed of a machine is job-dependent. Open shops, flow shops, and job shops are m-machine environments in which each job consists of several operations, each of which has to be executed on a designated machine; no job can undergo more than one operation at a time. In *job shops*, the order in which the operations of a job have to be executed is fixed; in *flow shops*, the order is fixed and the same for all jobs; and in *open shops*, the order is free and hence up to the scheduler. A generalization of these multi-stage models is the situation in which there are parallel machines available at one or more stages.

The job characteristics include the possibility of allowing preemption, and of specifying precedence constraints, release dates and deadlines. If *preemption* is allowed, then an operation may be interrupted and resumed later on; otherwise, an operation, once started, must be processed till completion without interruption. A *precedence constraint* stipulates that a certain job cannot start before another one has been completed. Job availability may be restricted by imposing *release dates* r_j, before which the jobs j cannot be started, and *deadlines* $\bar{d}_j$, by which they have to be completed.

A *feasible schedule* is an allocation of the operations to time intervals on the machines such that all restrictions are met. The optimality criterion is usually a function of the job completion times $C_1, \ldots, C_n$. Common criteria are *maximum completion time* or *makespan* $C_{\max} = \max_j C_j$ and *total completion time* $\sum C_j$. If job j has a *due date* d_j, we can compute its *lateness* $L_j = C_j - d_j$, its *tardiness* $T_j = \max\{0, C_j - d_j\}$, and its *earliness* $E_j = \max\{0, d_j - C_j\}$, for any given schedule. Important criteria involving due dates are *maximum lateness* $L_{\max} = \max_j L_j$, *total tardiness* $\sum T_j$, *total earliness* $\sum E_j$, and the *number of tardy jobs* (i.e., with $C_j > d_j$). If each job j has a *weight* w_j, then we can also have weighted versions of these criteria.

Machine scheduling has been an active field for many years. In this chapter, we review the main contributions to the area after the completion of the chapter by Lawler et al. (1993), which offers a comprehensive survey up to November 1990, providing 378 references.

Compiling this bibliography, we have focused on *deterministic* scheduling problems, where all parameters involved are given; such problems belong to the area of *combinatorial optimization*. We have put emphasis on classical models, new and relevant models such as on-line scheduling and scheduling with communication delays, and new and interesting techniques like polyhedral combinatorics. To save space, we have clustered papers and referred to surveys when possible and appropriate.

We do not cover purely experimental comparisons of heuristic rules. Models that, in our evaluation, are too specific or too artificial have been deleted, along with models that have not quite caught on or that are already dying off. We have also deleted models that are methodologically too different, such as stochastic scheduling models. Of the many papers surviving, we offer a survey of the most noteworthy results.

Acknowledgements. We are grateful for the comments of David Shmoys and Gerhard Woeginger. This research was partially supported by NSF grant CCR-9307391.

1 Books

J. Błazewicz, K. Ecker, G. Schmidt, J. Węglarz (1993). *Scheduling in Computer and Manufacturing Systems*, Springer, Berlin.
T.E. Morton, D.W. Pentico (1993). *Heuristic Scheduling Systems*, Wiley, New York.
V.S. Tanaev, V.S. Gordon, Y.M. Shafransky (1994). *Scheduling Theory: Single-Stage Systems*, Kluwer, Dordrecht.
V.S. Tanaev, Y.N. Sotskov, V.A. Strusevich (1994). *Scheduling Theory: Multi-Stage Systems*, Kluwer, Dordrecht.
P. Brucker (1995). *Scheduling Algorithms*, Springer, Berlin.
P. Chrétienne, E.G. Coffman, Jr., J.K. Lenstra, Z. Liu (eds.) (1995). *Scheduling Theory and Its Applications*, Wiley, Chichester.
M.L. Pinedo (1995). *Scheduling: Theory, Algorithms, and Systems*, Prentice Hall, Englewood Cliffs, NJ.

This is a selection of the scheduling books that have appeared in the 1990's. Morton and Pentico convey an engineering approach to scheduling manufacturing systems. Tanaev et al. integrate the Soviet and Western literature. Chrétienne et al. collected 17 tutorials and surveys on a broad range of topics. Pinedo's book is an attractive undergraduate text, also covering stochastic models and practical applications.

2 Single Machine

2.1 Regular criteria

S.K. Gupta, J. Kyparisis (1987). Single machine scheduling research. *Omega* 15, 207-227.

A notable survey, containing 204 references.

M.C. Fields, G.N. Frederickson (1990). A faster algorithm for the maximum weighted tardiness problem. *Inform. Process. Lett.* 36, 39-44.

An $O(k + n \log n)$ algorithm for the problem with deadlines and k precedence constraints.

L.A. Hall, D.B. Shmoys (1992). Jackson's rule for single-machine scheduling: making a good heuristic better. *Math. Oper. Res.* 17, 22-35.
L.A. Hall, D.B. Shmoys (1990). Near-optimal sequencing with precedence constraints. R. Kannan, W.R. Pulleyblank (eds.). *Proc. 1st MPS Conf. IPCO*, University of Waterloo Press, Waterloo, 249-260.

A 4/3-approximation algorithm and two approximation schemes for minimizing maximum lateness in case of release dates and negative due dates that run in time $O(n^2 \log n)$, $O(16^{1/\epsilon}(n/\epsilon)^{3+4/\epsilon})$, and $O(n \log n + n(4/\epsilon)^{8/\epsilon^2+8/\epsilon+2})$ —and a polynomial approximation scheme for the precedence-constrained case.

E. Balas, J.K. Lenstra, A. Vazacopoulos (1995). The one-machine problem with delayed precedence constraints and its use in job shop scheduling. *Management Sci.* 41, 94-109.

Carlier's (1982) algorithm for minimizing $L_{\max}$ subject to release dates is adapted to handle *precedence delays*, i.e., minimum delays between precedence-related jobs. The method is used in the *shifting bottleneck* heuristic for job shop scheduling.

R. Ahmadi, U. Bagchi (1990). Lower bounds for single-machine scheduling problems. *Naval Res. Log. Quart.* 37, 967-979.

Proof that the preemptive bound dominates all proposed polynomial-time lower bounds for the problem of minimizing total completion time subject to release dates.

J. Du, J.Y.-T. Leung (1993). Minimizing mean flow time with release time and deadline constraints. *J. Algorithms* 14, 45-68.

An NP-hardness proof for the preemptive problem, along with the identification of some well-solvable cases.

J.A. Hoogeveen, S.L. van de Velde (1995). Stronger Lagrangian bounds by use of slack variables: applications to machine scheduling problems. *Math. Program.* 70, 173-190.

A generic method to improve Lagrangian lower bounds; applications to the problems of minimizing total weighted completion time subject to precedence constraints, total weighted tardiness, and total completion time in a two-machine flow shop.

H. Kellerer, T. Tautenhahn, G.J. Woeginger (1996). Approximability and nonapproximability results for minimizing total flow time on a single machine. *Proc. 28th Annual ACM Symp. Theory of Comput.*, 418-426.

A polynomial $O(n^{1/2})$-approximation algorithm for minimizing $\sum(C_j - r_j)$, and a lower bound of $O(n^{1/2-\epsilon})$ on the performance of any polynomial-time algorithm.

T.S. Abdul-Razaq, C.N. Potts, L.N. Van Wassenhove (1990). A survey of algorithms for the single machine total weighted tardiness scheduling problem. *Discr. Appl. Math.* 26, 235-253.

An insightful survey of dynamic programming and branch-and-bound algorithms.

B.C. Tansel, I. Sabuncuoglu (1994). *Geometry based analysis of single machine tardiness problem and implications on solvability*, Report IEOR-9405, Bilkent University.
B.C. Tansel, B. Kara-Yetis, I. Sabuncuoglu (1995). *Advances in solvability of the single machine total tardiness scheduling problem*, Report IEOR-9517, Bilkent University.

A method to distinguish between presumably easy and hard instances of the total tardiness problem, and the best existing code for this problem.

D.S. Hochbaum, R. Shamir (1991). Strongly polynomial algorithms for the high multiplicity scheduling problem. *Oper. Res.* 39, 648-653.
D.S. Hochbaum, R. Shamir (1990). Minimizing the number of tardy job units under release time constraints. *Discr. Appl. Math.* 28, 45-57.

If there are g groups of identical unit-time jobs, one only needs to specify one copy of the job data per group and the size of each group. The weighted number of late jobs can be minimized in $O(g \log g)$ time, in $O(g^2)$ time in case of release dates, and in $O(g \log g)$ time in case of release dates and equal weights.

C.N. Potts, L.N. Van Wassenhove (1992). Single machine scheduling to minimize total late work. *Oper. Res.* 40, 586-595.
A.M.A. Hariri, C.N. Potts, L.N. Van Wassenhove (1995). Single machine scheduling to minimize total weighted late work. *ORSA J. Comput.* 7, 232-242.
M.Y. Kovalyov, C.N. Potts, L.N. Van Wassenhove (1995). A fully polynomial approximation scheme for scheduling a single machine to minimize total weighted late work. *Math. Oper. Res.* 19, 86-93.

The *late work* of job j is the amount of processing done after d_j. The preemptive weighted problem is solvable in $O(n \log n)$ time. The nonpreemptive problems, with or without weights, are both NP-hard but solvable in pseudopolynomial time. For the weighted version, a branch-and-bound algorithm is given, and a fully polynomial approximation scheme that runs in time $O(n^3 \log n + n^3/\epsilon)$.

2.2 Regular criteria, polyhedral techniques

M. Queyranne, A.S. Schulz (1997). Polyhedral approaches to machine scheduling. *Math. Program. (B)*, to appear.

A comprehensive survey of the area, which was independently initiated by Balas, Queyranne, and Wolsey in the mid-1980's. Originally meant to provide strong lower bounds, polyhedral methods are recently being used as a tool to develop polynomial approximation algorithms with excellent performance guarantees. Depending upon the variables used, several approaches can be distinguished.

M. Queyranne, Y. Wang (1991). Single-machine scheduling polyhedra with precedence constraints. *Math. Oper. Res.* 16, 1-20. Erratum. *Math. Oper. Res.* 20 (1995), 768.

A formulation based on job completion times for the problem of minimizing total weighted completion time subject to precedence constraints, and a complete polyhedral characterization for the case of series-parallel constraints by two classes of inequalities, which are polynomially separable.

M.E. Dyer, L.A. Wolsey (1990). Formulating the single machine sequencing problem with release dates as a mixed integer program. *Discr. Appl. Math.* 26, 255-270.
J.P. Sousa, L.A. Wolsey (1992). A time indexed formulation of non-preemptive single machine scheduling problems. *Math. Program.* 54, 353-367.
J.M. van den Akker (1994). *LP-based solution methods for single-machine scheduling problems*, PhD Thesis, Eindhoven University of Technology. See also Reports COSOR 93-27, 95-24, 96-14, Dept. Math. & Comp. Sci., Eindhoven University of Technology.

Dyer & Wolsey analyze a time-indexed formulation, in which binary variables x_{jt} indicate if job j is started in time period t. It is stronger than other formulations but of pseudopolynomial size. Sousa & Wolsey derive a class of valid inequalities with right-hand side 1 and solve the separation problem in polynomial time. Van den Akker characterizes all facet-inducing inequalities with right-hand side 1 or 2 and solves the corresponding separation problem in polynomial time. She uses column generation to solve the LP-relaxation and presents a branch-and-cut algorithm.

J.B. Lasserre, M. Queyranne (1992). Generic scheduling polyhedra and a new mixed-integer formulation for single-machine scheduling. E. Balas, G. Cornuéjols, R. Kannan (eds.). *Proc. 2nd MPS Conf. IPCO*, University Printing and Publications, Carnegie Mellon University, Pittsburgh, 136-149.

A formulation in which the variables denote the start time of the jth job in the sequence, and its application to the problem of minimizing makespan in case of release dates and deadlines. The lower bound obtained is computationally competitive.

L.A. Hall, A.S. Schulz, D.B. Shmoys, J. Wein (1997). Scheduling to minimize average completion time: off-line and on-line approximation algorithms. *Math. Oper. Res.* 22, to appear.

A number of polynomial approximation algorithms with constant performance ratios based on the solution to an LP-relaxation. Completion time formulations are used for the problems of minimizing total weighted completion time subject to release dates and/or precedence constraints on a single machine and on identical parallel machines. The resulting bounds on the quality of the LP-relaxations are also valid for some other formulations, as these dominate the ones based on completion times in strength. A variant of the time-indexed formulation, using binary variables to indicate if a certain job completes in an interval of the form $(2^{k-1}, 2^k]$, is applied to the problem of minimizing $\sum w_j C_j$ on unrelated parallel machines subject to release dates.

2.3 Regular criteria, setup times

C.N. Potts, L.N. Van Wassenhove (1992). Integrating scheduling with batching and lot-sizing: a review of algorithms and complexity. *J. Oper. Res. Soc.* 43, 395-406.
S. Webster, K.R. Baker (1995). Scheduling groups of jobs on a single machine. *Oper. Res.* 43, 692-703.

In case of job setup times, the issue is to batch similar jobs to save setups, without delaying other jobs too much. Not surprisingly, most types of batching models are intractable. The above papers survey more of these than we consider here.

J.M.J. Schutten, S.L. van de Velde, W.H.M. Zijm (1996). Single-machine scheduling with release dates, due dates, and family setup times. *Management Science* 42, 1165-1174.
H.A.J. Crauwels, A.M.A. Hariri, C.N. Potts, L.N. Van Wassenhove (1997). Branch-and-bound algorithms for single machine scheduling with batch set-up times to minimize total weighted completion time. *Ann. Oper. Res.*, to appear.
A.M.A. Hariri, C.N. Potts (1997). Single machine scheduling with batch set-up times to minimize maximum lateness. *Ann. Oper. Res.*, 79, 75–92.
S. Zdrzałka (1992). *Analysis of an approximation algorithm for single-machine scheduling with delivery times and sequence independent batch setup times*, Manuscript, Inst. Eng. Cybernetics, Technical University of Wrocław.

Branch-and-bound algorithms for three single-machine problems complicated by setup times. The crux lies in the computation of good lower bounds. Schutten et al. handle setups as additional jobs, Crauwels et al. use Lagrangian bounds, while Hariri & Potts apply combinatorial bounds. For minimizing maximum lateness in case of negative due dates, the latter also present an $O(n \log n)$ time 5/3-approximation algorithm, while Zdrzałka gives an $O(n^2)$ time 3/2-approximation algorithm.

2.4 Irregular criteria

K.R. Baker, G.D. Scudder (1990). Sequencing with earliness and tardiness penalties. *Oper. Res.* 38, 22-37.

Irregular criteria are not monotone in the job completion times. Objective functions of the form $\sum(\alpha_j E_j + \beta_j T_j)$ have attracted much attention over the last ten years. Due to the intractability of problems with general due dates, most work assumes a due date d that is common to all jobs. The due date d is called *large* if $d \geq \sum p_j$: it does not restrict the decision to schedule a job early or tardy; otherwise, d is called *small.* Baker & Scudder give a comprehensive survey of the early work, including Kanet's (1981) $O(n \log n)$ algorithm for the problem with large d and $\alpha_j = \beta_j = 1$ for all j.

N.G. Hall, M.E. Posner (1991). Earliness-tardiness scheduling problems, I: weighted deviation of completion times about a common due date. *Oper. Res.* 39, 836-846.
N.G. Hall, W. Kubiak, S.P. Sethi (1991). Earliness-tardiness scheduling problems, II: deviation of completion times about a restrictive common due date. *Oper. Res.* 39, 847-856.
J.A. Hoogeveen, S.L. van de Velde (1991). Scheduling around a small common due date. *European J. Oper. Res.* 55, 237-242.
J.A. Hoogeveen, H. Oosterhout, S.L. van de Velde (1994). New lower and upper bounds for scheduling around a small common due date. *Oper. Res.* 42, 102-110.

Hall & Posner prove that minimizing $\sum w_j|C_j - d|$ is NP-hard if d is large. Hall et al. and Hoogeveen & Van de Velde show that minimizing $\sum |C_j - d|$ becomes NP-hard for small d; the former authors also give an $O(n \sum p_j)$ time algorithm, while the latter give an $O(n^2 \sum p_j)$ algorithm for the weighted problem. For the unweighted problem, Hoogeveen et al. provide Lagrangean lower and upper bounds that always seem to concur in practice and present an $O(n \log n)$ 4/3-approximation algorithm.

M.R. Garey, R.E. Tarjan, G.T. Wilfong (1988). One-processor scheduling with symmetric earliness and tardiness penalties. *Math. Oper. Res.* 13, 330-348.
J.A. Hoogeveen, S.L. van de Velde (1997). Earliness-tardiness scheduling around almost equal due dates. *INFORMS J. Comput.*, 9, 92–99.
S. Verma, M. Dessouky (1998). Single-machine scheduling of unit-time jobs with earliness and tardiness penalties. *Math. Oper. Res.*, to appear.

Few memorable results are known for the case of unequal due dates. Garey et al. are the first to present a polynomial algorithm to find an optimal schedule for a *given* job sequence. Hoogeveen & Van de Velde present an $O(n^2)$ algorithm for the case that all intervals $[d_j - p_j, d_j]$ overlap, where the due dates are assumed to be large. Verma & Dessouky show that the problem is solvable as a linear program in case of unit processing times and identically ordered weights α_j and β_j.

H.G. Kahlbacher (1989). SWEAT – a program for a scheduling problem with earliness and tardiness penalties. *European J. Oper. Res.* 43, 111-112.
W. Kubiak (1993). Completion time variance minimization on a single machine is difficult. *Oper. Res. Lett.* 14, 49-59.

For small d, an $O(n \sum p_j)$ algorithm for minimizing $\sum(\alpha E_j^c + \beta T_j^c)$ with arbitrary positive c, and an NP-hardness proof for $\alpha = \beta$ and $c = 2$.

2.5 Multiple criteria

T.D. Fry, R.D. Armstrong, H. Lewis (1989). A framework for single machine multiple objective sequencing research. *Omega* 17, 595-607.

An early survey of this vast area, providing 43 references. We only highlight the few polynomial-time algorithms below.

J.A. Hoogeveen, S.L. van de Velde (1995). Minimizing total completion time and maximum cost simultaneously is solvable in polynomial time. *Oper. Res. Lett.* 17, 205-208.

Determining all $O(n^2)$ Pareto-optimal points takes $O(n^3 \min\{n, \log(\sum p_j)\})$ time.

J.A. Hoogeveen (1996). Minimizing maximum promptness and maximum lateness on a single machine. *Math. Oper. Res.* 21, 100-114.

An $O(n^2 \log n)$ algorithm to determine the trade-off curve for maximum lateness and maximum promptness, where the *promptness* of a job is the difference between its start time and its target start time. It can also be used to minimize $L_{\max}$ subject to release dates if $d_j - p_j - A \leq r_j \leq d_j - A$ $(j = 1, \ldots, n)$ for some constant A.

J.A. Hoogeveen (1996). Single-machine scheduling to minimize a function of two or three maximum cost criteria. *J. Algorithms* 21, 415-433.

Determining all Pareto-optimal points in $O(n^4)$ and $O(n^8)$ time, respectively.

3 Parallel Machines

T.C.E. Cheng, C.C.S. Sin (1990). A state-of-the-art review of parallel-machine scheduling research. *European J. Oper. Res.* 47, 271-292.

A useful survey, providing 113 references.

3.1 Independent jobs, regular criteria

M. Dell'Amico, S. Martello (1995). Optimal scheduling of tasks on identical parallel processors. *ORSA J. Comput.* 7, 191-200.

A very effective branch-and-bound algorithm for minimizing makespan.

S.L. van de Velde (1993). Duality-based algorithms for scheduling unrelated parallel machines. *ORSA J. Comput.* 5, 192-205.
S. Martello, F. Soumis, P. Toth (1997). An exact algorithm for makespan minimization on unrelated parallel machines. *Discr. Appl. Math.*, to appear.

Optimization and approximation algorithms for the makespan problem.

A. Marchetti Spaccamela, W.S. Rhee, L. Stougie, S. van de Geer (1992). Probabilistic analysis of the minimum weighted flowtime scheduling problem. *Oper. Res. Lett.* 11, 67-71.

In a standard probabilistic model, the authors analyze the solution value obtained by the ratio rule for the single-machine problem and prove asymptotic optimality of this rule for the variant with identical parallel machines.

H. Belouadah, C.N. Potts (1994). Scheduling identical parallel machines to minimize total weighted completion time. *Discr. Appl. Math.* 48, 201-218.

A branch-and-bound algorithm with a polynomial Lagrangian lower bound obtained from a time-indexed formulation with a pseudopolynomial number of variables.

S. Webster (1995). Weighted flow time bounds for scheduling identical processors. *European J. Oper. Res.* 80, 103-111.

Latest paper in a series on obtaining lower bounds by job splitting.

L.M.A. Chan, P. Kaminsky, A. Muriel, D. Simchi-Levi (1995). *Machine scheduling, linear programming and list scheduling heuristics*, Manuscript, Dept. IE, Northwestern University, Evanston.
J.M. van den Akker, J.A. Hoogeveen, S.L. van de Velde (1995). *Parallel machine scheduling by column generation*, Report COSOR 95-35, Dept. Math. Comp. Sci., Eindhoven University of Technology.
Z.L. Chen, W.B. Powell (1995). *Solving parallel machine total weighted completion time problems by column generation*, Manuscript, Dept. Civil Eng. & Oper. Res., Princeton University.

Parallel machine problems can be formulated as set partitioning problems with an exponential number of variables. For minimizing $\sum w_j C_j$, Chan et al. prove that the optimal solution value is at most $(1+\sqrt{2})/2$ times the value of the LP-relaxation and that the latter is, under mild conditions, asymptotically optimal. The other authors show that the LP-bound is effective. They give different column-generation algorithms to solve the LP-relaxation along with different branching strategies to close the gap.

J.Y.-T. Leung, V.K.M. Yu (1994). Heuristic for minimizing the number of late jobs on two processors. *Int. J. Foundations Comput. Sci. 5,* 261-279.

An $O(n \log n)$ 4/3-approximation algorithm for the case of two identical machines.

3.2 Independent jobs, multiple criteria

J.Y.-T. Leung, G.H. Young (1989). Minimizing schedule length subject to minimum flow time. *SIAM J. Comput.* 18, 314-326.

An $O(n \log n)$ algorithm for the preemptive minimization of makespan on identical parallel machines subject to minimum total flow time.

B.T. Eck, M. Pinedo (1993). On the minimization of the makespan subject to flowtime optimality. *Oper. Res.* 41, 797-801.

An $O(n \log n)$ 28/27-approximation algorithm for two identical parallel machines.

D.B. Shmoys, É. Tardos (1993). An approximation algorithm for the generalized assignment problem. *Math. Program.* 62, 461-474.

Among related results, a polynomial 2-approximation algorithm for minimizing a weighted combination of makespan and total cost on unrelated parallel machines.

3.3 Independent jobs, on-line models

J. Sgall (1997). On-line scheduling. A. Fiat, G.J. Woeginger (eds.). *On-Line Algorithms: the State of the Art*, Lecture Notes in Computer Science, Springer, Berlin, to appear.

A survey of the area. There are two types of on-line models. The first stems from on-line bin-packing: the jobs arrive in a list, and the next job in the list is only revealed after all previous jobs have been assigned irrevocably to a time slot on a machine. In the second type of model, the jobs arrive over time.

B. Chen, A. van Vliet, G.J. Woeginger (1994). New lower and upper bounds for on-line scheduling. *Oper. Res. Lett.* 16, 221-230.
Y. Bartal, A. Fiat, H. Karloff, R. Vohra (1995). New algorithms for an ancient scheduling problem. *J. Comput. Syst. Sci.* 51, 359-366.
D.R. Karger, S.J. Phillips, E. Torng (1996). A better algorithm for an ancient scheduling problem. *J. Algorithms* 20, 400-430.
J. Aspnes, Y. Azar, A. Fiat, S. Plotkin, O. Waarts (1993). On-line load balancing with applications to machine scheduling and virtual circuit routing. *Proc. 25th Annual ACM Symp. Theory of Comput.*, 623-631.

The papers in the first branch deal with the makespan problem on m identical parallel machines. Graham's (1966) list scheduling rule has worst-case ratio $2 - 1/m$. More than 25 years later, the problem got renewed attention, where the emphasis is on the questions 'Does there exist an on-line algorithm with a better performance ratio?' and 'How closely can any on-line algorithm approach the optimum off-line solution?' Chen et al. determine lower and upper bounds on the on-line performance for problems with $m \leq 10$. Bartal et al. present an algorithm with performance ratio smaller than $2 - 1/70$ for any m; Karger et al. achieve a ratio of 1.945 for $m \geq 6$. Extending this analysis to uniform and unrelated machines, Aspnes et al. give algorithms with worst-case ratios 8 and $\log m$, respectively.

For models in the second branch, the schedule needs to be built over time. At time t, the jobs released before or at t are fully known, but all other jobs and their characteristics are unknown. The questions that have to be answered are the same, though.

K.S. Hong, Y.-T. Leung (1992). On-line scheduling of real-time tasks. *IEEE Trans. Comput.* 41, 1326-1331.

In case of identical parallel machines, makespan can be minimized on-line when preemption is allowed, but maximum lateness cannot.

A.P.A. Vestjens (1997). Scheduling uniform machines on-line requires nondecreasing speed ratios. *Math. Program. (B)*, to appear.

Necessary and sufficient conditions on the machine speeds to allow the existence of an on-line algorithm that minimizes makespan on uniform parallel machines when preemption is allowed but processor-sharing is not.

G.J. Woeginger (1994). On-line scheduling of jobs with fixed start and end times. *Theor. Comput. Sci.* 130, 5-16.

Necessary and sufficient conditions for the existence of an on-line 1/4-approximation algorithm for the single-machine preemptive scheduling problem of maximizing the total weight of the fully processed jobs, where job j needs to be processed during the interval $[r_j, r_j + p_j]$. The algorithm is shown to be best possible.

B. Chen, A.P.A. Vestjens (1996). *Scheduling on identical machines: how good is LPT in an on-line setting?*, Report COSOR 96-11, Dept. Math. & Comp. Sci., Eindhoven University of Technology.

On-line 'longest processing time first' approximates the minimum makespan within a factor of 3/2 in case of identical parallel machines. No on-line algorithm can do better than 1.3820 for $m = 2$ and better than 1.3473 for $m \geq 3$.

J.A. Hoogeveen, A.P.A. Vestjens (1996). Optimal on-line algorithms for single-machine scheduling. M. Queyranne, W.H. Cunningham (eds.). *Proc. 5th MPS Conf. IPCO*, Lecture Notes in Computer Science 1084, Springer, Berlin, 404-414.

Best possible on-line algorithms for the respective minimization of total completion time and maximum lateness (in case of negative due dates) on a single machine.

C. Phillips, C. Stein, J. Wein (1997). Minimizing average completion time in the presence of release dates. *Math. Program. (B)*, to appear.

An algorithm that converts a preemptive schedule to a nonpreemptive one while increasing the completion times by a factor of at most 2 on a single machine or 3 on identical machines. On-line 2- and $(8 + \epsilon)$-approximation algorithms for the preemptive minimization of total completion time on identical machines and of total weighted completion time on unrelated machines, respectively.

D.B. Shmoys, J. Wein, D.P. Williamson (1995). Scheduling parallel machines on-line. *SIAM J. Comput.* 24, 1313-1331.

General techniques for adjusting off-line algorithms to cope with jobs with unknown release dates and jobs with unknown processing times.

L.A. Hall, A.S. Schulz, D.B. Shmoys, J. Wein (1997) (see Section 2.2).

A framework to design on-line algorithms for minimizing total weighted completion time. The performance guarantees are $3 + \epsilon$, $4 + \epsilon$, and 8 in case of a single machine, identical, and unrelated machines, respectively.

S. Chakrabarti, C.A. Phillips, A.S. Schulz, D.B. Shmoys, C. Stein, J. Wein (1997). Improved scheduling algorithms for minsum criteria. Lecture Notes in Computer Science, Springer, Berlin, to appear.

Improvements of some of the bounds in the previous two papers. An on-line algorithm that approximates both $C_{\max}$ and $\sum w_j C_j$ within constant factors.

3.4 Independent multiprocessor jobs

Nonmalleable multiprocessor jobs need a given number of machines during a given processing time. For *malleable* jobs, the processing time is a nonincreasing function of the number of machines put on the job.

J. Błazewicz, M. Drozdowksi, J. Węglarz (1994). Scheduling multiprocessor tasks: a survey. *Int. J. Microcomputer Applications* 13, 89-97.

A survey of results for problems with nonmalleable jobs. See also Veltman et al. (1990) (Section 3.5).

W. Ludwig, P. Tiwari (1994). Scheduling malleable and nonmalleable parallel tasks. *Proc. 5th Annual ACM–SIAM Symp. Discr. Algorithms*, 167-176.

The most recent of a number of papers on approximating makespan. Performance bounds of 2 and 2.5 for malleable and nonmalleable jobs, respectively.

J. Turek, W. Ludwig, J.L. Wolf, L. Fleischer, P. Tiwari, J. Glasgow, U. Schwiegelshohn, P.S. Yu (1994). Scheduling parallelizable tasks to minimize average response time. *Proc. 6th Annual Symp. Parallel Algorithms and Architectures*, 200-209.

An $O(n^3 + mn)$ 2-approximation algorithm for minimizing total completion time in case of malleable jobs under mild conditions on the processing times.

U. Schwiegelshohn, W. Ludwig, J.L. Wolf, J. Turek, P.S. Yu (1997). Smart SMART bounds for weighted response time scheduling. *SIAM J. Comput.*, to appear.

An $O(n \log n)$ 8-(8.53-)approximation algorithm for minimizing total (weighted) completion time in case of nonmalleable jobs.

3.5 Precedence-constrained jobs

E.G. Coffman, Jr., M.R. Garey (1993). Proof of the 4/3 conjecture for preemptive vs. nonpreemptive two-processor scheduling. *J. ACM 40*, 991-1018.

In case of two identical machines and precedence constraints, the nonpreemptive minimum makespan is at most 4/3 times as large as the preemptive one.

B. Berger, L. Cowen (1995). Scheduling with concurrency constraints. *J. Algorithms* 18, 98-123.

The paper addresses the problem of scheduling unit-time jobs on identical parallel machines to minimize makespan with 'before', 'no later than' and 'at the same time' constraints. The problem is solvable in linear time for $m = 2$ but NP-hard for $m \geq 3$.

B. Veltman, B.J. Lageweg, J.K. Lenstra (1990). Multiprocessor scheduling with communication delays. *Parallel Comput.* 16, 173-182.
P. Chrétienne, C. Picouleau (1995). Scheduling with communication delays: a survey. P. Chrétienne, E.G. Coffman, Jr., J.K. Lenstra, Z. Liu (eds.) (1995). *Scheduling Theory and Its Applications*, 65-90, Wiley, Chichester.

Two surveys of an active area. *Communication delays* occur between precedence-related jobs assigned to different machines. Some of the basic results: for the case that the number of identical machines is unrestricted and jobs may be duplicated, Papadimitriou & Yannakakis (1990) proved that minimizing makespan is NP-hard and gave a polynomial 2-approximation algorithm. For m part of the input, no job duplication, and unit-time delays and jobs, the problem is NP-hard, even for trees; for $m = 2$, the case of tree-type constraints is solvable in linear time, but the general case is open.

J.A. Hoogeveen, J.K. Lenstra, B. Veltman (1994). Three, four, five, six, or the complexity of scheduling with communication delays. *Oper. Res. Lett.* 16, 129-137.

The complexity of finding short schedules for unit-time jobs subject to unit-time communication delays on a restricted or unrestricted number of identical machines, and lower bounds on the polynomial-time approximability of the makespan.

A. Munier, J.-C. König (1997). A heuristic for a scheduling problem with communication delays. *Oper. Res.*, 45, 145–147.
C. Hanen, A. Munier (1995). An approximation algorithm for scheduling dependent tasks on m processors with small communication delays. *IEEE Symp. Emerging Technologies and Factory Automation*, Paris, 167-189.

A polynomial 4/3-approximation algorithm for the case of unrestricted m and unit-time delays and jobs, based on rounding the solution to an LP-relaxation—and an extension to the case of 'small' delays, both for unrestricted and restricted m.

4 Open, Flow and Job Shops

R.A. Dudek, S.S. Panwalkar, M.L. Smith (1992). The lessons of flowshop scheduling research. *Oper. Res.* 40, 7-13.

A critical appraisal of the practical relevance of a substantial body of research.

4.1 Complexity

W. Kubiak, C. Sriskandarajah, K. Zaras (1991). A note on the complexity of openshop scheduling problems. *INFOR* 29, 284-294.

A detailed survey of complexity results for open shop scheduling.

P. Brucker, B. Jurisch, M. Jurisch (1993). Open shop problems with unit time operations. *Z. Oper. Res.* 37, 59-73.

This paper ties unit-time open shop problems to preemptive identical parallel machine problems and offers a survey as well as new results.

J. Du, J.Y.-T. Leung (1993). Minimizing mean flow time in two-machine open shops and flow shops. *J. Algorithms* 14, 24-44.
C. Sriskandarajah, E. Wagneur (1994). On the complexity of preemptive openshop scheduling problems. *European J. Oper. Res.* 77, 404-414.

Minimizing total completion time in a preemptive two-machine shop is NP-hard in case of an open shop, strongly NP-hard for a flow shop—and also strongly NP-hard for an open shop with release dates.

M.M. Dessouky, M.I. Dessouky, S.K. Verma (1996). *Flowshop scheduling with identical jobs and uniform parallel machines*, Manuscript, Dept. Industrial & Systems Eng., University of Southern California, Los Angeles.

Minimizing makespan in a unit-time flow shop with uniform machines at each stage is easy for two stages and strongly NP-hard for three. For the latter case, a polynomial 7/4-approximation algorithm and a branch-and-bound method are given.

J.A. Hoogeveen, J.K. Lenstra, B. Veltman (1996). Minimizing makespan in a multiprocessor flow shop is strongly NP-hard. *European J. Oper. Res.* 89, 172-175.

Even the preemptive two-stage flow shop problem with two identical machines at one stage and a single machine at the other is strongly NP-hard.

W. Kubiak, S. Sethi, C. Sriskandarajah (1995). An efficient algorithm for a job shop problem. *Ann. Oper. Res.* 57, 203-216.

Various authors have observed that instances of the unit-time two-machine job shop can be encoded more compactly than in the 'standard' way. Here: an algorithm for minimizing maximum lateness that is polynomial with respect to such an encoding.

P. Brucker, R. Schlie (1990). Job-shop scheduling with multi-purpose machines. *Computer* 45, 369-375.

A polynomial algorithm for the two-job case of a generalized job shop, in which an operation can be processed by any machine in its associated machine set.

P. Brucker (1994). A polynomial algorithm for the two machine job-shop scheduling problem with a fixed number of jobs. *Oper. Res. Spekt.* 16, 5-7.
Y.N. Sotskov, N.V. Shakhlevich (1995). NP-hardness of shop-scheduling problems with three jobs. *Discr. Appl. Math.* 59, 237-266.

Job shop scheduling is easy for $m = 2$, n fixed, and NP-hard for $m = n = 3$.

P. Brucker, S.A. Kravchenko, Y.N. Sotskov (1997). Preemptive job-shop scheduling problems with a fixed number of jobs. *Math. Methods Oper. Res.*, to appear.

Preemptive job shop scheduling is NP-hard for $m = 2, n = 3$. Note that the non-preemptive problem is easy!

N.G. Hall, C. Sriskandarajah (1996). A survey of machine scheduling problems with blocking and no-wait in process. *Oper. Res.* 44, 510-525.

An overview of shop models with the additional feature of *no wait in process* or *no intermediate storage*.

4.2 Approximability

B. Chen, V.A. Strusevich (1993). Approximation algorithms for three-machine open shop scheduling. *ORSA J. Comput.* 5, 321-326.
B. Chen, V.A. Strusevich (1993). Worst-case analysis of heuristics for open shops with parallel machines. *European J. Oper. Res.* 70, 379-390.

Rácsmány proved that *dense* open shop schedules are shorter than twice the optimum. For three machines, the ratio becomes 5/3; a linear-time heuristic improves this to 3/2. The two-stage case with identical machines at each stage has a polynomial 2-approximation algorithm.

B. Chen, C.A. Glass, C.N. Potts, V.A. Strusevich (1996). A new heuristic for three-machine flow shop scheduling. *Oper. Res.* 44, 891-898.

An $O(n \log n)$ time 5/3-approximation algorithm for the three-machine flow shop.

C.-Y. Lee, G.L. Vairaktarakis (1994). Minimizing makespan in hybrid flowshops. *Oper.*

Res. Lett. 16, 149-158.
B. Chen (1994). Scheduling multiprocessor flow shops. D.-Z. Du, J. Sun (eds.). *Advances in Optimization and Approximation*, Kluwer, Dordrecht, 1-8.

A polynomial algorithm with performance ratio $2 - 1/\max\{m_1, m_2\}$ for the two-stage flow shop with m_i identical machines at stage i—and the same algorithm along with a literature survey of the area.

C.N. Potts, D.B. Shmoys, D.P. Williamson (1991). Permutation vs. non-permutation flow shop schedules. *Oper. Res. Lett.* 10, 281-284.

The restriction to permutation schedules may cost a factor of more than $\sqrt{m}/2$.

E. Nowicki, C. Smutnicki (1994). A note on worst-case analysis of approximation algorithms for a scheduling problem. *European J. Oper. Res.* 74, 128-134.

The fourth paper in a series on performance bounds for permutation flow shops, which also summarizes the other three.

D.P. Williamson, L.A. Hall, J.A. Hoogeveen, C.A.J. Hurkens, J.K. Lenstra, S.V. Sevast'janov, D.B. Shmoys (1997). Short shop schedules. *Oper. Res.*, 45, 288–294.

Given an open, flow or job shop instance with integral processing times, does there exist a schedule of length at most c? This question is easy for $c = 3$ but NP-complete for $c = 4$. Hence, finding a schedule shorter than 5/4 times the optimum is NP-hard ...

L.A. Hall (1997). Approximability of flow shop scheduling. *Math. Program. (B)*, to appear.
S.V. Sevast'janov, G.J. Woeginger (1997). A polynomial approximation scheme for the open shop problem. *Math. Program. (B)*, to appear.

... but for minimizing makespan in flow shops and open shops with a *fixed* number of machines, there exist polynomial approximation schemes.

S.V. Sevast'janov (1994). On some geometric methods in scheduling theory: a survey. *Discr. Appl. Math.* 55, 59-82.
S.V. Sevast'janov (1995). Vector summation in Banach space and polynomial algorithms for flow shops and open shops. *Math. Oper. Res.* 20, 90-103.

The first paper surveys work by the author and others on using vector summation in obtaining absolute and relative performance bounds for shop scheduling and other problems. The second presents new results along the same lines.

D.B. Shmoys, C. Stein, J. Wein (1994). Improved approximation algorithms for shop scheduling problems. *SIAM J. Comput.* 23, 617-632.

Polylogarithmic performance bounds for job shops and extensions with parallel machines at each stage or with jobs corresponding to a partial order of their operations.

4.3 Branch-and-bound

F. Della Croce, V. Narayan, R. Tadei (1996). The two-machine total completion time flow shop problem. *European J. Oper. Res.* 90, 227-237.

A survey and empirical comparison of old and new lower bounds.

A.M.G. Vandevelde, J.A. Hoogeveen, C.A.J. Hurkens, J.K. Lenstra (1997). *Lower bounds for the multiprocessor flow shop*, Manuscript, Dept. Math. & Comp. Sci., Eindhoven University of Technology.

A theoretical and computational investigation of lower bounds for the m-stage flow shop with identical parallel machines at each stage.

J. Carlier, E. Pinson (1990). A practical use of Jackson's preemptive schedule for solving the job shop problem. *Ann. Oper. Res.* 26, 269-287.
J. Carlier, E. Pinson (1994). Adjustment of heads and tails for the job-shop problem. *European J. Oper. Res.* 78, 146-161.
D. Applegate, W. Cook (1991). A computational study of the job-shop scheduling problem. *ORSA J. Comput.* 3, 149-156.
P. Brucker, B. Jurisch (1993). A new lower bound for the job-shop scheduling problem. *European J. Oper. Res.* 64, 156-167.
P. Brucker, B. Jurisch, B. Sievers (1994). A branch and bound algorithm for the job-shop scheduling problem. *Discr. Appl. Math.* 49, 107-127.
P. Brucker, B. Jurisch, A. Krämer (1994). The job-shop problem and immediate selection. *Ann. Oper. Res.* 50, 73-114.
P. Martin, D.B. Shmoys (1996). A new approach to computing optimal schedules for the job shop scheduling problem. M. Queyranne, W.H. Cunningham (eds.). *Proc. 5th MPS Conf. IPCO*, Lecture Notes in Computer Science 1084, Springer, Berlin, 389-403.

These papers represent the continuing efforts to develop optimization algorithms for job shop scheduling. The most effective lower bound is the classical single-machine bound, enhanced by techniques to increase heads and tails and to fix the order between operations on the same machine. The two-job bound of Brucker & Jurisch may work well when m/n is large. LP-based bounds are not yet competitive, as shown by Applegate & Cook, who tested cutting-plane methods for standard disjunctive and mixed-integer formulations, and by Martin & Shmoys, who computed the linear relaxation of a giant packing formulation. The branching strategies used settle an essential conflict between two operations or choose an operation to precede (or follow) a larger set of operations. The infamous 10×10 problem is now within reach, but 15×15 instances may still pose a computational challenge.

4.4 Local search

E. Taillard (1990). Some efficient heuristic methods for the flow shop sequencing problem. *European J. Oper. Res.* 47, 65-74.
E. Nowicki, C. Smutnicki (1996). A fast tabu search algorithm for the permutation flow-shop problem. *European J. Oper. Res.* 91, 160-175.
H. Ishibuchi, S. Misaki, H. Tanaka (1995). Modified simulated annealing algorithms for the flow shop sequencing problem. *European J. Oper. Res.* 81, 388-398.
C.R. Reeves (1995). A genetic algorithm for flowshop sequencing. *Computers Oper. Res.* 22, 5-13.

Tabu search (twice), simulated annealing, and a genetic algorithm for flow shop scheduling.

E. Nowicki, C. Smutnicki (1996). A fast taboo search algorithm for the job shop problem. *Management Sci.* 42, 797-813.

E. Balas, A. Vazacopoulos (1994). *Guided local search with shifting bottleneck for job shop scheduling*, Report MSRR-609, GSIA, Carnegie Mellon University, Pittsburgh.
R.J.M. Vaessens, E.H.L. Aarts, J.K. Lenstra (1996). Job shop scheduling by local search. *INFORMS J. Comput.* 8, 302-317.

The champions in job shop scheduling by local search, based on variants of tabu search and variable-depth search, and a survey covering 71 references.

5 Miscellaneous

P. Serafini, W. Ukovich (1989). A mathematical model for periodic scheduling problems. *SIAM J. Discr. Math.* 2, 550-581.
C. Hanen, A. Munier (1995). Cyclic scheduling on parallel processors: an overview. Chrétienne, Coffman, Lenstra & Liu (eds.) (see Section 1), 193-226.

Two surveys. In periodic scheduling, operations have to be periodically executed at a constant rate over time. In cyclic scheduling, a set of generic tasks must be performed infinitely often.

E. Nowicki, S. Zdrzałka (1990). A survey of results for sequencing problems with controllable processing times. *Discr. Appl. Math.* 26, 271-287.

A survey of results on another variation of the standard model, in which the processing times may vary at some cost.

N.G. Hall, S.P. Sethi, C. Sriskandarajah (1991). On the complexity of generalized due date scheduling problems. *European J. Oper. Res.* 51, 100-109.

Complexity categorization of problems with positional due dates.

L. Özdamar, G. Ulusoy (1995). A survey on the resource-constrained project scheduling problem. *IIE Trans.* 27, 574-586.
E. Demeulemeester (1995). Minimizing resource availability costs in time-limited project networks. *Management Sci.* 41, 1590-1598.

Resource-constrained project scheduling is a practical and classical area. The focus is on branch-and-bound and local search. The first paper surveys the area, the second gives a state-of-the-art branch-and-bound algorithm.

Y. Crama (1997). Combinatorial optimization models for production scheduling in automated manufacturing systems. *European J. Oper. Res.*, to appear.

A survey of results on models involving tool management, part selection, robotic cells, and automated guided vehicles.

13 The Traveling Salesman Problem

Michael Jünger
Institut für Informatik der Universität zu Köln, Köln

Gerhard Reinelt
Institut für Angewandte Mathematik, Universität Heidelberg, Heidelberg

Giovanni Rinaldi
Istituto di Analisi dei Sistemi ed Informatica, CNR, Roma

CONTENTS

The *Traveling Salesman Problem* (TSP) is the problem of finding a shortest *Hamiltonian cycle* (or *tour*), i.e., a shortest roundtrip through a given set of cities. This problem has model character in many branches of Mathematics, Computer Science, and Operations Research.

When scanning through the literature one observes that probably every approach for attacking hard optimization problems has also been tested for the TSP or has even been first formulated for the TSP. And, therefore, there are several hundreds of papers available. We refrain from trying to give a complete list which would be outdated soon anyway, since about every week a new paper is published on this subject. Rather we will classify the main contributions and give pointers to survey articles, to recent papers, or to papers that are historically relevant.

Annotated Bibliographies in Combinatorial Optimization, edited by M. Dell'Amico, F. Maffioli and S. Martello

Depending on whether the distance between two cities depends on the direction of travel or not, we have two versions of the TSP, namely the *asymmetric* (ATSP) and the *symmetric* one (STSP), respectively, which will be both considered in the following. In all issues that pertain to both versions of the problem we use the acronym TSP.

1 Books and Surveys

In this section, we list some books and surveys that will serve as general introduction to the area.

E.L. Lawler, J.K. Lenstra, A.H.G. Rinnooy Kan, D.B. Shmoys (eds.) (1985). *The Traveling Salesman Problem*, John Wiley & Sons, Chichester.

This book was the first one devoted entirely to one combinatorial optimization problem. Starting with the history of the TSP it addresses all of its theoretical and computational aspects.

M. Jünger, G. Reinelt, G. Rinaldi (1995). The traveling salesman problem. M.O. Ball, T.L. Magnanti, C.L. Monma, G.L. Nemhauser (eds.). *Network Models*, Handbooks in Operations Research and Management Science, 7, Elsevier Publisher B.V., Amsterdam, 225–330.

This survey article provides a self-contained introduction into algorithmic and computational aspects of the STSP as seen from the point of view of an operations researcher who wants to solve practical problem instances. The survey also contains many computational results, partly collected from literature, partly presented for the first time. In addition to the performance of several heuristics and implementations of the branch-and-cut algorithm on TSPLIB instances (see below), also the related rural postman problem, which arises in printed circuit board mask production, is treated.

G. Reinelt (1994). *Contributions to Practical Traveling Salesman Problem Solving*, Lecture Notes Comp. Science 840, Springer-Verlag, Berlin.
D.S. Johnson, L.A. Mc Geoch (1996). The traveling salesman problem: A case study in local optimization. E.H.L. Aarts, J.K. Lenstra (eds.). *Local Search in Combinatorial Optimization*, John Wiley & Sons, New York. to appear.

These two recent monographs cover almost all heuristic approaches to the STSP including extensive computational results.

R.E. Burkard, V.G. Deineko, R. van Dal, J.A.A. van der Veen, G.J. Woeginger (1995). *Well-solvable special cases of the TSP: a survey*, Technical Report 52, Karl-Franzenz-Universität & Technische Universität Graz.

Since the TSP is NP-hard, it is interesting to identify classes of problem instances that are efficiently solvable. This is a survey on such special cases.

M. Grötschel, M.W. Padberg (1993). *Ulysses 2000: in search of optimal solutions to hard combinatorial problems*, Technical Report 93-34, Konrad-Zuse-Zentrum für Informationstechnik, Berlin.

This paper is the first successful attempt to relate the TSP to Greek mythology and nicely introduces the novice to the TSP world.

G. Reinelt (1991). TSPLIB – A traveling salesman problem library. *ORSA J. Comput.* 3, 376–384. Internet: `http://www.iwr.uni-heidelberg.de/iwr/comopt/soft/TSPLIB95/TSPLIB.html`.

It is now common to report computational results for the STSP and the ATSP on the electronically available collection TSPLIB of problem instances presented in this paper. The collection is maintained and periodically updated by the author and makes it possible to compare the different approaches on a common test-bed.

2 Applications

Applications of the pure TSP are relatively rare. Usually additional side constraints limiting the set of feasible tours are present in real world applications. Nevertheless, TSP methodology can be adopted for some variations of the problem.

J.K. Lenstra, A.H.G. Rinnooy Kan (1974). *Some simple applications of the travelling salesman problem*, Technical Report BW 38/74, Stichting Mathematisch Centrum, Amsterdam.

R.E. Bland, D.F. Shallcross (1989). Large traveling salesman problems arising from experiments in X-ray crystallography: a preliminary report on computation. *Oper. Res. Lett.* 8, 125–128.

M. Grötschel, M. Jünger, G. Reinelt (1991). Optimal control of plotting and drilling machines: a case study. *Z. Oper. Res.—Methods and Models of Operations Research* 35, 61–84.

N. Ascheuer (1995). *Hamiltonian Path Problems in the On-line Optimization of Flexible Manufacturing Systems*, Technische Universität, Berlin. TR-96-03.

All these papers address some interesting applications.

3 Approximation Algorithms

Because the TSP is a hard optimization problem, there is a need for developing heuristics, i.e., algorithms that find good tours quickly. In the following we will proceed according to a rough classification. Many heuristics, however, are hybrid in the sense that they contain several of the components addressed below. We reference papers usually in the subsection corresponding to the approach which is predominant.

No matter which approach is favored for computing near-optimal TSP solutions, it is an important issue to design appropriate data structures and develop preprocessing routines for obtaining efficient implementations that are able to handle problem instances with up to 10,000 or even 100,000 cities.

J.L. Bentley (1990). Experiments on traveling salesman heuristics. *Proc. 1st Ann. ACM-SIAM Symp. Discr. Algorithms*, 91–99.

M.L. Fredman, D.S. Johnson, L.A. McGeoch, G. Ostheimer (1995). Data structures for traveling salesmen. *J. Algorithms* 18, 432–479.

These two papers focus on this topic, which is also addressed by many papers of the following sections.

3.1 Construction

By the term *construction heuristic* we denote procedures that successively build a tour where parts already constructed basically remain unchanged. Many of these approaches can be considered folklore as, e.g., *nearest neighbor*, *greedy* or *insertion* heuristics.

G. Clarke, J.W. Wright (1964). Scheduling of vehicles from a central depot to a number of delivery points. *Oper. Res.* 12, 568–581.
N. Christofides (1976). *Worst case analysis of a new heuristic for the travelling salesman problem*, Technical Report 388, GSIA, Carnegie-Mellon University, Pittsburgh.
The savings heuristic of Clarke/Wright and the Christofides heuristic are of particular interest. Both perform very well in practice and the latter also has good theoretical worst-case performance for the *metric TSP* (for details see section 4).

A lot of effort has been spent in analyzing worst-case performance of construction heuristics.

D.J. Rosenkrantz, R.E. Stearns, P.M. Lewis (1977). An analysis of several heuristics for the traveling salesman problem. *SIAM J. Comput.* 6, 563–581.
G. Conuéjols, G.L. Nemhauser (1978). Tight bounds for Christofides' traveling salesman heuristic. *Math. Program.* 14, 116–121.
A.M. Frieze (1979). Worst-case analysis of algorithms for travelling salesman problems. *Methods of Oper. Res.* 32, 97–112.
A.M. Frieze, G. Galbiati, F. Maffioli (1982). On the worst-case performance of some algorithms for the asymmetric traveling salesman problem. *Networks* 12, 23–39.
H.L. Ong, J.B. Moore (1984). Worst-case analysis of two travelling salesman heuristics. *Oper. Res. Lett.* 2, 273–277.
These five papers discuss worst-case performance of several heuristics for the symmetric and asymmetric TSP including spanning tree, nearest neighbor, nearest addition, greedy and Clarke/Wright heuristics.

B.L. Golden, W.R. Stewart (1985). Empirical analysis of heuristics. E.L. Lawler, J.K. Lenstra, A.H.G. Rinnooy Kan, D.B. Shmoys (eds.). *The Traveling Salesman Problem*, John Wiley & Sons, Chichester, 207–249.
This is a further reference reporting about empirical tests and also discussing testing methodology.

R.M. Karp, J.M. Steele (1985). Probabilistic analysis of heuristics. E.L. Lawler, J.K. Lenstra, A.H.G. Rinnooy Kan, D.B. Shmoys (eds.). *The Traveling Salesman Problem*, John Wiley & Sons, Chichester, 181–205.
A probabilistic analysis of the TSP exhibits some interesting properties and is discussed in this paper. In particular it is shown that a certain partitioning heuristic is

asymptotically optimal for a class of random problem instances.

Special attention has been given to geometric instances where cities correspond to points in 2- or 3-space and distances are evaluated according to some metric.

L.K. Platzman, J.J. Bartholdi III (1989). Spacefilling curves and the planar travelling salesman problem. *J. ACM* 36, 719–737.
J.L. Bentley (1992). Fast algorithms for geometric traveling salesman problems. *ORSA J. Comput.* 4, 387–411.
G. Reinelt (1992). Fast heuristics for large geometric traveling salesman problems. *ORSA J. Comput.* 2, 206–217.
Speed-up possibilities for the heuristics mentioned above and special approaches are discussed in these three papers.

3.2 Deterministic improvement

Usually the tour provided by a construction heuristic is taken as input for a subsequent improvement heuristic which tries to obtain a shorter tour by making local alterations of the sequence of cities. Such heuristics are characterized by the type of local moves they are based on.

Straightforward approaches are k-opt exchanges where k edges are removed from the current tour and the resulting paths are then reconnected in the best possible way.

M.M. Flood (1956). The traveling-salesman problem. *Oper. Res.* 4, 61–75.
G.A. Croes (1958). A method for solving traveling salesman problems. *Oper. Res.* 6, 791–812.
These are the first references addressing the case $k = 2$.

S. Lin (1965). Computer solutions of the traveling salesman problem. *Bell Syst. Tech. J.* 44, 2245–2269.
The case $k = 3$ is treated here.

For larger values of k, only a subset of all possibilities to reconnect the paths should be taken into account. In any case, for obtaining efficient implementations, care has to be given to choosing data structures and selection of exchange moves.

F. Margot (1992). Quick updates for p-opt TSP heuristics. *Oper. Res. Lett.* 11, 45–46.
This paper considers the efficient implementation of k-opt.

B. Chandra, H. Karloff, Tovey C. (1994). New results on the old k-opt algorithm for the TSP. *Proc. 5th Ann. ACM-SIAM Symp. on Discrete Algorithms*, 150–159.
This paper addresses worst-case analysis for the k-opt approach.

S. Lin, B.W. Kernighan (1973). An effective heuristic algorithm for the traveling-salesman problem. *Oper. Res.* 21, 498–516.
The most famous improvement approach is the Lin-Kernighan heuristic described

in this paper, where each length decreasing move consists of a sequence of simpler moves where some of these moves may also increase the tour length. The advantage of this fundamental idea is that local minima can be escaped from, which is not possible when using pure k-opt. Very many improvement heuristics for the TSP are variants of the Lin-Kernighan principle.

K.-T. Mak, A.J. Morton (1993). A modified Lin-Kernighan traveling salesman heuristic. *Oper. Res. Lett.* 13, 127–132.

This paper gives examples of how the Lin-Kernighan heuristic can be modified and accelerated.

I. Or (1976). *Traveling salesman-type combinatorial problems and their relation to the logistics of regional blood banking*, Technical report, Northwestern University, Evanston, IL.

M. Gendreau, A. Hertz, G. Laporte (1992). New insertion and postoptimization procedures for the traveling salesman problem. *Oper. Res.* 40, 1086–1094.

C. Potts, S. van De Velde (1995). *Dynasearch – iterative local improvement by dynamic programming: Part I, The traveling salesman problem*, Technical report.

In these three papers some further improvement approaches are described.

It appears that the Lin-Kernighan heuristic is unrivalled among the deterministic improvement methods. On the other hand, implementation effort is substantial.

So far all heuristics cited were deterministic. For several years, however, there has been considerable research in introducing randomness into the heuristics in a systematic way. We will discuss four generic approaches in the next subsections. For general introductory references we refer to chapter 11 of this book where also some further references concerning the TSP can be found.

3.3 Simulated annealing

The approach of simulated annealing originates from theoretical physics where Monte-Carlo methods are employed to simulate phenomena in statistical mechanics.

E. Bonomi, J.-L. Lutton (1984). The N-city travelling salesman problem: statistical mechanics and the Metropolis algorithm. *SIAM Rev.* 26, 551–568.

S. Kirkpatrick (1984). Optimization by simulated annealing: quantitative studies. *J. Statist. Physics* 34, 975–986.

V. Cerny (1985). A thermodynamical approach to the travelling salesman problem: an efficient simulation algorithm. *J. Optim. Theory Appl.* 45, 41–51.

Y. Rossier, M. Troyon, T.M. Liebling (1986). Probabilistic exchange algorithms and Euclidean traveling salesman problems. *Oper. Res. Spekt.* 8, 151–164.

E.H.L. Aarts, J. Korst, P.J.M. van Laarhoven (1988). A quantitative analysis of the simulated annealing algorithm. *J. Statist. Physics* 50, 189–206.

All these papers provide examples of application of this methodology to the TSP.

In general, it can be concluded that simulated annealing has the advantage of being easily implementable and has the capability to find very good tours. On the other hand, parameter tuning is necessary and a substantial amount of CPU time has to be spent.

G. Dueck, T. Scheuer (1990). Threshold accepting: a general purpose algorithm appearing superior to simulated annealing. *J. Comp. Phys.* 90, 161–175.
G. Dueck (1993). New optimization heuristics: the great deluge algorithm and record-to-record travel. *J. Comp. Phys.* 104, 86–92.
J. Lee, M.Y. Choi (1994). Optimization by multicanonical annealing and the traveling salesman problem. *Physical Review E* 50, 651–654.

Several variations of the annealing principle have been studied for speeding-up or simplifying the approach. Names are, e.g., *threshold accepting*, *great deluge algorithm*, *record-to-record travel* or *multicanonical annealing* addressed in these three papers, respectively.

3.4 Tabu search

For finding good tours it is important that heuristics can leave local minima. In some sense this is already accomplished by the Lin-Kernighan heuristic or simulated annealing. *Tabu search* also allows the leaving of local optima, but uses in addition the basic concept of *forbidden moves* to avoid visiting the same local minimum several times. Besides this, a tabu search algorithm contains further components to guide the search.

M. Troyon (1988). *Quelques heuristiques et résultats asymptotiques pour trois problémes d'optimisation combinatoire*, Technical report, Ecole Polytechnique Fédérale, Lausanne, Switzerland.
M. Malek, M. Guruswamy, M. Pandya (1989). Serial and parallel search simulated annealing and tabu search for the traveling salesman problem. *Ann. Oper. Res.* 21, 59–84.
C.-N. Fiechter (1994). A parallel tabu search algorithm for large traveling salesman problems. *Discr. Appl. Math.* 51, 243–267.
J. Knox (1994). Tabu search performance on the symmetric traveling salesman problem. *Computers Oper. Res.* 21, 867–876.
C. Rego, C. Roucairol (1996). *Relaxed tours and path ejections for the travelling salesman problem*, Technical report, Université de Versailles.

Here examples of applications to the TSP are given.

Up to now, the conclusion is that tabu search is not yet competitive with the best TSP heuristics, but is worth further studies. In particular the idea of combining efficient heuristics with a tabu mechanism seems interesting.

3.5 Genetic algorithms

Genetic algorithms try to mimic evolution processes in nature. They have algorithmic analogues of *mutation*, *selection* and *genetic recombination*. A basic operation is the *mating* of two tours using a crossover operator to form a new tour. Various possibilities exist to follow this approach.

H. Mühlenbein, M. Gorges-Schleuter, O. Krämer (1988). Evolution algorithms in combinatorial optimization. *Parallel Comput.* 7, 65–85.
N.L.J. Ulder, E.H.L. Aarts, H.-J. Bandelt, P.J.M. van Laarhoven, E. Pesch (1991). Genetic local search algorithms for the traveling salesman problem. H.P. Schwefel, R. Männer (eds.). *Proc. 1st Int. Workshop on Problem Solving from Nature.* Springer, Berlin, 109–116.
O. Martin, S.W. Otto, E.W. Felten (1992). Large-step Markov chains for the TSP incorporating local search heuristics. *Oper. Res. Lett.* 11, 219–224.
These are some of the many reports about applications to the TSP.

Combined with other improvement techniques genetic algorithms bear the potential of finding very good tours significantly faster than simulated annealing or tabu search, but not yet competitive with the best Lin-Kernighan implementations if CPU time is limited.

3.6 Neural networks

Neural networks model a set of neurons connected by some interconnection network. Based on the input that a neuron receives, an output is computed which is propagated to other neurons. A variety of models addresses activation status of neurons, determinations of outputs and propagation of signals. The basic goal is to realize some kind of learning mechanism and thereby compute the solution of optimization problems. The solution then either appears explicitly as output or can be read from the final state of the network.

J.-Y. Potvin (1993). The traveling salesman problem: a neural network perspective. *ORSA J. Comput.* 5, 328–347.
This is a comprehensive survey on such approaches focusing on the TSP.

Basically, neural net approaches can be classified into three classes.

J.J. Hopfield, D.W. Tank (1985). 'Neural' computation of decisions in optimization problems. *Biol. Cybern.* 52, 141–152.
Y.P.S. Foo, H. Szu (1989). Solving large-scale optimization problems by divide-and-conquer neural networks. *Proc. Int. Joint Conference on Neural Networks.* IEEE, Piscataway, NJ, 507–511.
Approaches based on integer programming formulations are described in these papers.

R. Durbin, D. Willshaw (1987). An analogue approach to the travelling salesman problem using an elastic net method. *Nature* 326, 689–691.
M.C.S. Boeres, L.A.V. De Carvalho, V.C. Barbosa (1992). A faster elastic-net algorithm for the traveling salesman problem. *Proc. Int. Joint Conference on Neural Networks.* IEEE, Piscataway, NJ, 215–220.
These two deal with the concept of elastic nets.

B. Fritzke, P. Wilke (1991). FLEXMAP—a neural network for the traveling salesman

problem with linear time and space complexity. *Proc. Int. Joint Conference on Neural Networks*. IEEE, Piscataway, NJ, 929–934.
S. Amin (1994). A self-organized travelling salesman. *Neural Comput. & Applic.* 2, 129–133.

Self-organizing maps are addressed in these two articles.

In spite of the enormous effort spent for adopting the neural net approach to the TSP, its competitiveness with the approaches above has not been established so far.

Heuristics for the ATSP can be designed analogously to the heuristics for the symmetric case. However, because the direction of connections has to be observed, updates are more complicated and lead to substantially longer running times.

4 Theoretical Approximability

Are there heuristic algorithms for the TSP having a performance that can be guaranteed a priori?

A *polynomial-time approximation algorithm* (PTAA) is a polynomial-time algorithm that is able to produce an approximate solution to a problem whose objective function is, in the worst case, within a constant factor of the optimal value. A classical result of Sahni and Gonzalez states that no PTAAs exist for the general TSP. However, for the metric TSP, [Christofides 1976] (see above) gave a PTAA with a factor 1.5.

A *polynomial-time approximation scheme* (PTAS) is a polynomial-time algorithm (or a family of such algorithms) that can approximate a problem within a factor $1+\epsilon$ for each fixed $\epsilon > 0$. The running time may depend upon ϵ, but for each fixed ϵ it has to be polynomial in the input size.

S. Arora, L. Carsten, R. Motwani, M. Sudan, M. Szegedy (1992). Proof verification and hardness of approximation problems. *Proc. 33rd Annual IEEE Symp. Found. Comput. Sci.*, 14–23.

It is shown that metric TSP, among other NP-hard problems, is MAX SNP-hard and that a MAX SNP-hard problem does not have polynomial time approximation schemes unless P = NP. This result leaves open the problem of whether a factor better than 1.5 can be achieved by a PTAA for this class of TSP instances.

C.H. Papadimitriou, M. Yannakakis (1993). The traveling salesman problem with distance one and two. *Math. Oper. Res.* 18, 1–11.

A polynomial-time approximation algorithm with worst-case factor $\frac{7}{6}$ is proposed for the TSP with objective function coefficients in the set $\{1, 2\}$. This special case of TSP is also shown to be MAX SNP-hard.

M. Grigni, E. Koutsoupias, C.H. Papadimitriou (1995). An approximation scheme for planar graph TSP. *Proc. 36th Annual IEEE Symp. Found. Comput. Sci.*, 640–645.

The *graph TSP* is a special case where the distance function is obtained by associating with each city a node of a given graph G. The distance between two cities is then given by the number of edges of the shortest path in G between the corresponding two nodes. In this paper a PTAS is given for the case where G is a planar graph and it is conjectured that a PTAS exists for the *Euclidean TSP*, which is a special case of the metric TSP, where each city is associated with a point in the 2-dimensional plane and the distance between two cities is given by the Euclidean distance between the corresponding two points.

S. Arora (1996). *Polynomial time approximation schemes for euclidean TSP and other geometric problems. Ann. IEEE Symp. Found. Comp. Sci.*, 2–11.

Whether the Euclidean TSP is MAX SNP-hard was a long standing open problem. A solution to this problem is given here, where a PTAS is described for such special case of the TSP.

A.I. Barvinok (1996). Two algorithmic results for the traveling salesman problem. *Math. Oper. Res.* 21, 64–84.

A PTAS is given for the "maximum" length TSP in a d-dimensional Euclidian space, based on algebraic methods (a variation on the notion of permanent).

5 Polyhedral Relaxations

The convex hull P_S^n of the incidence vectors of all the Hamiltonian cycles of a complete undirected graph with n nodes is called the *symmetric traveling salesman polytope.* Analogously, for the directed Hamiltonian cycles of a directed graph, P_A^n is called the *asymmetric traveling salesman polytope.* The knowledge of the linear inequalities that define these polytopes provides a powerful tool for solving the TSP exactly with a cutting-plane algorithm.

M. Grötschel, M.W. Padberg (1985). Polyhedral theory. E.L. Lawler, J.K. Lenstra, A.H.G. Rinnooy Kan, D.B. Shmoys (eds.). *The Traveling Salesman Problem*, John Wiley & Sons, Chichester, 251–305.

This is an excellent introduction to the study of the TSP polytopes.

5.1 Small polytopes

The polytopes associated with the TSP are very complex and are defined by sets of linear inequalities whose sizes grow extremely fast with n. The following papers contain a full description of the polytopes for small values of n.

R.Z. Norman (1955). On the convex polyhedra of the symmetric traveling salesman problem (abstract). *Bull. American Math. Soc.* 61, 559.
H.W. Kuhn (1955). On certain convex polyhedra. *Bull. American Math. Soc.* 61, 557–558.

These are among the first attempts to describe TSP polytopes for a small value of n. In the first paper a complete description of a minimal set of linear inequalities defining P_S^6 is given in terms of 6 equations and 100 inequalities. A description of P_A^5 in terms of 9 equations and 390 inequalities is given in the second one.

S.C. Boyd, W.H. Cunningham (1991). Small travelling salesman polytopes. *Math.*

Oper. Res. 16, 259–271.

A mathematical proof is given that 3 437 inequalities, falling into only four distinct classes, are sufficient to describe P_S^7. The authors believe that it is a difficult task to extrapolate their proof technique to larger values of n.

T. Christof, M. Jünger, G. Reinelt (1991). A complete description of the traveling salesman polytope on 8 nodes. *Oper. Res. Lett.* 10, 497–500.
T. Christof, G. Reinelt (1996). Combinatorial optimization and small polytopes. *TOP* 4, 1–53.

It is shown how the complete description of P_S^n for a few larger values can be accomplished by means of a computer program, a variant of the *double-description method* (that produces all the inequalities defining a polytope given as convex hull of a set of points) implemented by the authors. In the first paper P_S^8 is characterized in terms of 194 187 inequalities belonging to 24 classes, three of which were unknown. In the second paper P_S^9 is completely described with 42 104 442 inequalities (192 classes) and a (possibly incomplete) set of 51 043 900 866 inequalities in 15 379 classes is given for P_S^{10}. The classes of facets for P_S^n with $n \leq 10$ are available via `http://www.iwr.uni-heidelberg.de/iwr/comopt/soft/SMAPO/tsp/tsp.html`.

R. Euler, H. Le Verge (1995). Complete linear descriptions of small asymmetric traveling salesman polytopes. *Discr. Appl. Math.* 62, 193–208.

Using their own implementation of Chernikova's algorithm, the authors have computed the inequalities defining all the facets of P_A^6 and of the lower monotonization of P_A^5. For the first polytope they found 319 015 inequalities grouped into 287 different classes, all described in the paper. For the latter they found 7 615 that can be classified into 51 distinct classes, also completely described.

5.2 The STSP polytope

Due to the complexity of the TSP polytopes, a considerable amount of research work has been devoted to characterizing or describing classes of facet defining inequalities for the TSP polytopes for arbitrary values of n.

G.B. Dantzig, D.R. Fulkerson, S.M. Johnson (1954). Solution of a large scale traveling-salesman problem. *Oper. Res.* 2, 393–410.

For the first time some valid inequalities for P_S^n are introduced, the *subtour elimination inequalities*, which make infeasible the incidence vector of any collection of two or more cycles that cover all the nodes of the graph (subtours). The issue whether these inequalities are facet defining for the polytope is not addressed here.

V. Chvátal (1973). Edmonds polytopes and weakly Hamiltonian graphs. *Math. Program.* 5, 29–40.

A class of valid inequalities, the *comb* inequalities, is defined. These inequalities are a generalization of the *2-matching* inequalities defined by Edmonds in 1965 to provide a complete description of the polytope associated with the *2-matching problem.* These inequalities are today called *Chvátal combs.* It is also shown that an inequality defined by the Petersen graph is facet defining for P_S^{10}.

J.F. Maurras (1975). Some results on the convex hull of Hamiltonian cycles of symmetric complete graphs. B. Roy (ed.). *Combinatorial Programming: Methods and Applications*, Reidel, Dordrecht, 179–190.

A generalization of the inequality based on the Petersen graph is described, which is facet defining for $n > 10$.

M. Grötschel, M.W. Padberg (1979). On the symmetric traveling salesman problem I: inequalities. *Math. Program.* 16, 265–280.
M. Grötschel, M.W. Padberg (1979). On the symmetric traveling salesman problem II: lifting theorems and facets. *Math. Program.* 16, 281–302.

The first systematic study of the polyhedral structure of P_S^n was done by the authors of these two papers, which actually summarize the results published in many other papers and in Grötschel's doctoral dissertation. The major results are that all the valid inequalities for P_S^n known at the time of writing, i.e., the *trivial* (variable upper and lower bounds), the *subtour elimination*, and a generalization of the *Chvátal's comb* inequalities (called *comb* by the authors) define facets of P_S^n. These results were obtained with a new proof technique based on two main steps: the first was to prove that an inequality of "small size" is facet defining for a small value of n and the second (called *lifting*) was to "extend" the inequality for higher values of n, while preserving the property of being facet defining.

M. Grötschel, W.R. Pulleyblank (1986). Clique tree inequalities and the symmetric traveling salesman problem. *Math. Oper. Res.* 11, 537–569.

A new class of inequalities is given that generalize the comb inequalities. The new inequalities are called *clique-tree* inequalities and shown to define facets of the polytope. This is the first example of an inequality facet defining for P_S^n whose *support graph* (the graph whose edges correspond to the nonzero coefficients of the inequality) is obtained by composing the support graphs of other facet defining inequalities, the comb inequalities.

M.W. Padberg, S. Hong (1980). On the symmetric traveling salesman problem: a computational study. *Math. Program. Study* 12, 78–107.

Another generalization of the comb inequalities is described that produces the *chain* inequalities. Only a proof of validity is given.

Due to the complexity of the facial structure of P_S^n, several approaches have been followed for finding new valid and/or faced defining inequalities. Subtour elimination, comb, clique-tree, and chain inequalities are all described by a collection subsets (*handles* and *teeth*) of the node set, satisfying some simple conditions. All the edges of the support graph of the inequality have both the endpoints belonging to some set of the collection. One approach to describe more inequalities is to define collections of node sets satisfying more complex conditions.

B. Fleischmann (1988). A new class of cutting planes for the symmetric travelling salesman problem. *Math. Program.* 40, 225–246.

The *star* inequalities are defined as a generalization of the comb inequalities, by allowing the subsets called *handles* to be nested. Only a proof of the validity of these inequalities is given.

B. Fleischmann (1987). *Cutting planes for the symmetric traveling salesman problem*, Technical report, Universität Hamburg.

Further generalizations lead to the *hyperstar* inequalities that are proved here to be valid for P_S^n.

A generalization of the clique-tree inequalities obtained by relaxing some conditions on the defining node subsets leads to the *bipartition inequalities* introduced in [Boyd and Cunningham 1991] (see above), where their validity for P_S^n is proved.

D. Naddef (1992). The binested inequalities for the symmetric traveling salesman polytope. *Math. Oper. Res.* 17, 882–900.

The *binested* inequalities obtained by generalizations of the same kind are defined and shown to be valid for P_S^n. These inequalities properly generalize the hyperstar and a large subclass of the bipartition inequalities.

D. Naddef (1990). Handles and teeth in the symmetric traveling salesman polytope. W. Cook, P.D. Seymour (eds.). *Polyhedral Combinatorics*, AMS, Baltimore.

The above classes of inequalities, defined by collections of subsets, are surveyed in this paper.

The question whether these inequalities define facets is never addressed systematically in the above papers. Actually it is known that some of them do not define facets of P_S^n. However, being valid, they can always be used to tighten an LP relaxation of P_S^n.

Other approaches to study P_S^n are based on some relaxations of TSP. The idea is that it might be easier to describe facet defining inequalities for the polytope associated with a TSP relaxation. If this is the case, the study of the facets of P_S^n can proceed in two steps. First, one provides a (partial) description of the inequalities defining the polyhedron of the relaxation. Second one gives some (manageable) conditions under which an inequality facet defining for the relaxation is such also for P_S^n. Two are the relaxations that have been considered. The first relaxation is given by the set of all the Hamiltonian cycles and all their subsets; the convex hull of the corresponding incidence vectors is the *monotone TSP polytope*. Although many facet defining inequalities have been found for this polytope, which are surveyed in [Grötschel and Padberg 1985] (see above), it is not easy to give manageable conditions for some of them to be facet defining for P_S^n. A more fruitful approach uses a second relaxation, the *graphical traveling salesman problem* (GTSP). A feasible solution for this problem is any closed walk in a graph that visits each city at least once and may travel an edge more than once. The convex hull of the incidence vectors of all the feasible solutions for GTSP is the *graphical traveling salesman polyhedron* G_S^n.

G. Cornuéjols, J. Fonlupt, D. Naddef (1985). The traveling salesman problem on a graph and some related polyhedra. *Math. Program.* 33, 1–27.

Many new classes of inequalities for G_S^n are defined. All these inequalities, among which we mention the *path* inequalities, the *wheelbarrow* inequalities and the *bicycle* inequalities are shown to be facet defining for the polyhedron. Since P_S^n is a face of G_S^n all these inequalities are also valid for P_S^n. These inequalities belong all to the class of the star inequalities and generalize the comb inequalities.

D. Naddef, G. Rinaldi (1991). The symmetric traveling salesman polytope and its graphical relaxation: composition of valid inequalities. *Math. Program. (A)* 51, 359–400.

A composition operation (the s-sum) is defined that permits to produce new facet defining inequalities by combining inequalities that are known to define facets of G_S^n. Applying a special case of s-sum to some of the path, wheelbarrow and bicycle inequalities the *regular parity path-tree* inequalities are defined and shown to be facet defining for G_S^n. These inequalities generalize and include (with marginal exceptions) the clique-tree inequalities.

D. Naddef, G. Rinaldi (1993). The graphical relaxation: a new framework for the symmetric traveling salesman polytope. *Math. Program. (A)* 58, 53–88.

The relationship between P_S^n and G_S^n is studied and exploited to derive conditions for a facet defining inequality for G_S^n to be facet defining for P_S^n as well. A special case of s-sum is defined that yields facet defining inequalities for P_S^n by composing known facet defining inequalities. Moreover the operations of *node lifting* and *edge cloning* are defined that extend facet defining inequalities, while preserving the property of being facet defining, by adding one node and a pair of nodes, respectively, to their support graphs.

D. Naddef, G. Rinaldi (1988). *The symmetric traveling salesman polytope: new facets from the graphical relaxation*, Technical Report 248, IASI-CNR, Rome.

By using the above results it is shown that the path, the wheelbarrow, the bicycle (*PWB inequalities*, for short), and the regular parity path-tree inequalities are facet defining also for P_S^n. By applying the edge-cloning operation to the PWB inequalities it is shown that all the inequalities of a large class (*extended PWB*), which includes the chain inequalities, are facet defining for P_S^n.

D. Naddef, G. Rinaldi (1992). The crown inequalities for the symmetric traveling salesman polytope. *Math. Oper. Res.* 17, 308–326.

The *crown* inequalities are introduced and shown to be facet defining for P_S^n along with their extensions obtained by node lifting and edge cloning.

The inequalities described in the last two papers include most of those that are known today to define facets of P_S^n. A relevant exception is given by the *ladder inequalities* introduced in [Boyd and Cunningham 1991] (see above).

S.C. Boyd, W.H. Cunningham, M. Queyranne, Y. Wang (1995). Ladders for travelling salesmen. *SIAM J. Optim.* 5, 408–420.

The ladder inequalities and their extensions obtained by node lifting are shown to define facets of P_S^n.

Another way to study the facets of STSP is by exploiting the relationship between P_S^n and the polytope associated with the *Hamiltonian path problem*, i.e., the problem of finding a shortest simple path connecting any two nodes and including any other node of the graph.

M. Queyranne, Y. Wang (1993). Hamiltonian path and symmetric travelling salesman polytopes. *Math. Program. (A)* 58, 89–110.

The *projective approach* for the study of P_S^n via the above relationship is introduced and an extension of the lifting operation described in [Grötschel and Padberg 1979] (see above) is given.

M. Queyranne, Y. Wang (1991). *Composing facets of symmetric travelling salesman polytopes*, Technical report, Faculty of Commerce and Business Administration, University of British Columbia.
M. Queyranne, Y. Wang (1990). *Facet-tree composition for symmetric travelling salesman polytopes*, Technical Report 90-MSC-001, Faculty of Commerce and Business Administration, University of British Columbia.

Two new operations *tooth fusing*, and *ring composition* are described that preserve the property of being facet defining if the components satisfy simple structural properties; these compositions can be repeated to create complex facets, including clique trees and regular parity path trees as special cases.

5.3 The ATSP polytope

The first systematic study of P_A^n, the polytope associated with ATSP, and of the connections with its relaxation the *monotone ATSP polytope*, was carried out by Grötschel and Padberg and later by Grötschel and Wakabayashi, and it is surveyed in [Grötschel and Padberg 1985] (see above). Among the results, there is the definition of several inequalities valid for P_A^n, like the *subtour elimination*, the *comb*, the *C2*, the T_k, the *C3*, the D_k^+, and the D_k^- inequalities, and the proof that some of them are facet defining.

E. Balas (1989). The asymmetric assignment problem and some new facets of the traveling salesman polytope. *SIAM J. Discr. Math.* 2, 425–451.

The polytope associated with a relaxation of ATSP, the *asymmetric assignment problem* is studied. The feasible solutions of this relaxation are assignments with no 2-arc directed cycles. Based on this relaxation the class of *odd CAT inequalities* is introduced. These inequalities are shown to define facets of P_A^n.

M. Fischetti (1991). Facets of the asymmetric traveling salesman polytope. *Math. Oper. Res.* 16, 42–56.
M. Fischetti (1995). Clique tree inequalities define facets of the asymmetric traveling salesman polytope. *Discr. Appl. Math.* 56, 9–18.

It is shown that the D_k^+, the D_k^-, the $C3$, the comb, the $C2$, and the clique tree inequalities define facets of P_A^n.

E. Balas, M. Fischetti (1992). The fixed out-degree 1-arborescence polytope. *Math. Oper. Res.* 17, 1001–1018.

Another relaxation of ATSP is defined, the *fixed outdegree 1-arborescence*. The feasible solutions of this relaxation are connected spanning directed subgraphs with n arcs, where all the nodes have indegree 1 and the nodes of a given subset of V have outdegree 1. Based on this relaxation the *FDA inequalities* are introduced and shown to define facets of P_A^n.

E. Balas, M. Fischetti (1993). A lifting procedure for the asymmetric traveling salesman polytope and a large new class of facets. *Math. Program. (A)* 58, 325–352.

The translation of the node lifting of the STSP into the ATSP case is described (here the operation is called *node cloning*). Since node cloning always preserves the facet defining property of an inequality, the study of P_A^n can be restricted to the *primitive inequalities*, since all the other inequalities are produced from those by node cloning. The class of *SD inequalities* is defined. Except for a few simple cases, all the SD inequalities define facets for P_A^n. The T_k, some of the FDA (called *simple*), the comb, the $C2$, and the odd CAT inequalities belong to this class.

E. Balas, M. Fischetti (1993). On the monotonization of polyhedra. G. Rinaldi, L.A. Wolsey (eds.). *Proc. 3rd Math. Progr. Conf. Int. Progr. Comb. Opt.*, Erice. CORE, Louvain-la-Neuve, 23–38.

The connections of a polyhedron with its monotonizations are studied. As a byproduct of the techniques developed, it is shown that the *lifted cycle inequalities* define facets of P_A^n.

M. Fischetti (1992). Three lifting theorems for the asymmetric traveling salesman polytope. E. Balas, G. Cornuéjols, R. Kannan (eds.). *Proc. 2nd Math. Progr. Conf. Int. Progr. Comb. Opt.*, Carnegie Mellon University, Pittsburgh. 260–273.

Three lifting operations are described that can be applied to known facet defining inequalities to produce new and more complex inequalities defining facets of P_A^n (e.g., generalizations of the clique tree that also include the chain inequalities).

An inequality of the linear description of P_A^n is called *symmetric* if for each pair i, j, the coefficients associated with the arcs (i, j) and (j, i) coincide; otherwise it is called *asymmetric*.

S. Chopra, G. Rinaldi (1990). The graphical asymmetric traveling salesman polyhedron. R. Kannan, W.R. Pulleyblank (eds.). *Proc. 1st Math. Progr. Conf. Int. Progr. Comb. Opt.*, University of Waterloo Press, Waterloo. 129–145.

The *graphical asymmetric traveling salesman problem*, a relaxation of ATSP is introduced. Like for the undirected case, a feasible solution for this problem is any directed closed walk in a directed graph that visits each city at least once and may travel an arc more than once. The polyhedron G_A^n associated with this problem is studied. Several classes of asymmetric inequalities are introduced, namely the *a-path* the *a-wheelbarrow*, and the *a-bicycle* inequalities that are related to the path, wheelbarrow, and the bicycle inequalities of G_S^n and define facets of G_A^n.

S. Chopra, G. Rinaldi (1996). The graphical asymmetric traveling salesman polyhedron: symmetric inequalities. *SIAM J. Discr. Math.*, 9, 602–624.

A general simple condition is given under which a facet defining inequality for G_S^n

yields a symmetric facet defining inequality for G_A^n. It is shown that the condition applies to most of the known facet defining inequalities for P_S^n (PWBs, crowns, regular parity path-trees, chains).

M. Queyranne, Y. Wang (1995). Symmetric inequalities and their composition for asymmetric travelling salesman polytopes. *Math. Oper. Res.* 20, 838–863. (Minor errata at URL: `http://acme.commerce.ubc.ca:80/quey/atsp-errata.html`).

An elegant proof technique is developed that is used to prove that many facet defining inequalities for P_S^n yield symmetric inequalities defining facets of P_A^n. In particular the path, the wheelbarrow, the chain, and the ladder inequalities are shown to be facet defining for P_A^n. A composition operation is described that, under some conditions, produces facet defining inequalities if the components are also facet defining. Complex inequalities can be constructed using this operation in a tree-like fashion. The produced inequalities are called *facet tree inequalities* and, by construction, define facets of P_A^n.

5.4 Additional work on TSP polytopes

Although not aimed at chasing facet defining inequalities, there are some more papers that provide further insights on the structure of the TSP polytopes.

M.X. Goemans (1995). Worst-case comparison of valid inequalities for the TSP. *Math. Program. (A)* 69, 335–349.

How do different classes of facet defining inequalities perform in providing good lower bounds to the problem? Is there a formal notion of effectiveness for ranking such inequalities? The *strength* of a class of inequalities is formally defined here for GTSP and a ranking of most of the known facet inducing inequalities is provided. The path inequalities hold the top position of this ranking.

When the TSP is formulated on a general graph, rather than on a complete graph, an interesting question to ask is for which graphs the problem is easily solvable.

G. Conuéjols, D. Naddef, W.R. Pulleyblank (1983). The traveling salesman problem in Halin graphs. *Math. Program.* 26, 287–294.
G. Conuéjols, D. Naddef, W.R. Pulleyblank (1985). The traveling salesman problem in graphs with 3-edge cutsets. *J. ACM* 32, 383–410.

A decomposition operation, based on the notion of *3-edge cutset* is described that, if applicable to the graph on which the problem is defined, yields a polynomial-time solution algorithm and a complete description of the polytope associated with the problem. A composition operation based on the 3-edge cutset is also used to construct graphs on which the problem is polynomially solvable, like the *Halin graphs*.

J. Fonlupt, D. Naddef (1992). The traveling salesman problem in graphs with excluded minors. *Math. Program. (A)* 53, 147–172.

It is shown that the subtour elimination inequalities along with the nonnegativity constraints completely describe the polyhedron G_S^n on graphs that do not contain three special graphs as minors.

M. Padberg, T.-Y. Sung (1988). A polynomial-time solution to Papadimitriou and Steiglitz's 'traps'. *Oper. Res. Lett.* 7, 117–125.

Are there classes of objective function for which some simple polyhedral relaxations are sufficient to produce always an optimal solution? Here it is shown that optimizing over the subtour polytope (the relaxation of P_S^n where trivial and subtour elimination are the only inequalities) gives always a Hamiltonian cycle for all the objective functions that come from certain instances of TSP that, conversely, are very hard for heuristic improvement algorithms.

M. Queyranne, Y. Wang (1994). *On the Chvátal rank of certain inequalities*, Technical report, Faculty of Commerce and Business Administration, University of British Columbia.

A common way to express the complexity of a facet defining inequality is the Chvátal rank. Here it is proven that a popular method for showing that the Chvátal rank is at least two, is incorrect. The paper provides a correct result and method. The method is applied to several classes of STSP and ATSP inequalities.

5.5 Separation

Given a point $\overline{x}$ in the space of a polyhedron P, the *separation problem* for a class $\mathcal{I}$ of valid inequalities for P is to produce an inequality of $\mathcal{I}$ violated by $\overline{x}$ or to prove that all the inequalities $\mathcal{I}$ are satisfied by $\overline{x}$. The separation is the key to exploiting the knowledge on a huge system of linear inequalities describing P in a practical cutting plane algorithm. Moreover, due to a classical result of Grötschel, Lovász, and Schrijver, and of Padberg and Rao, the polynomial-time solvability of a problem is equivalent to the polynomial-time solvability of the separation for the inequalities of the associated polyhedron.

Only for two classes of facet defining inequalities for P_S^n and P_A^n the separation problem is known to be solvable in polynomial time. These classes are the subtour elimination and the 2-matching inequalities, for which the problem is solvable with the Gomory-Hu algorithm and with the Padberg-Rao algorithm, respectively. The Padberg-Rao algorithm amounts to performing a simple processing of the *cut-tree* produced by the Gomory-Hu algorithm. A separation algorithm is called *heuristic* if its answer "no inequalities are violated" can be wrong; it is called *exact* otherwise. Heuristic algorithms have been proposed also for the separation of inequalities for which an exact polynomial-time algorithm is known: usually these algorithms have the advantage of being faster, and so they may be useful in a practical computation.

M.W. Padberg, M. Grötschel (1985). Polyhedral computations. E.L. Lawler, J.K. Lenstra, A.H.G. Rinnooy Kan, D.B. Shmoys (eds.). *The Traveling Salesman Problem*, John Wiley & Sons, Chichester, 307–360.

This is an excellent introduction to the cutting plane algorithms based on the (partial) knowledge of the polyhedron associated with the problem. It surveys all the results concerning the polynomial-time separation of TSP inequalities and gives some simple heuristic separation algorithms for the comb inequalities (for which no exact polynomial-time separation algorithm is known at present).

The separation of the subtour elimination inequalities amounts to finding a minimum cut in a weighted graph. The Gomory-Hu algorithm accomplishes this task by solving $n - 1$ maximum flow problems. Several more efficient algorithms have been proposed in the last ten years. Due to space limitations, rather than providing a complete list of the papers describing them, we refer to the survey paper [Jünger et al. 1995] (see above).

D. Gusfield (1990). Very simple algorithms and programs for all pairs network flow analysis. *SIAM J. Comput.* 19, 143–155.

Gives a very simple implementation of the Gomory-Hu algorithm that can be used to produce the cut-tree necessary for the exact separation of the 2-matching inequalities.

M.W. Padberg, G. Rinaldi (1990). Facet identification for the symmetric traveling salesman polytope. *Math. Program.* 47, 219–257.

Several heuristic algorithms are described for the separation of the 2-matching, of the comb, and of the clique tree inequalities along with some preprocessing procedures that speed up the exact separation for the subtour elimination and the 2-matching inequalities. A contraction operation is defined on the support graph associated with the point $\overline{x}$ (the graph whose edges correspond to the positive components of $\overline{x}$). Conditions are given under which an inequality violated in the contracted graph yields, by node lifting, a violated inequality in the original graph. Some of the heuristic for the separation of the comb inequalities are based on these conditions.

M. Grötschel, O. Holland (1987). A cutting plane algorithm for minimum perfect 2-matching. *Computing* 39, 327–344.

A heuristic algorithm for the separation of the 2-matching inequalities is described as well as an effective implementation of the exact Padberg-Rao algorithm.

M. Grötschel, O. Holland (1991). Solution of large-scale symmetric traveling salesman problems. *Math. Program.* 51, 141–202.

Heuristic algorithms are described for the separation of subtour elimination, 2-matching, and comb inequalities. The comb separation is based on the contraction of suitable node sets in the support graph of $\overline{x}$.

J.M. Clochard, D. Naddef (1993). Using path inequalities in a branch and cut code for the symmetric traveling salesman problem. G. Rinaldi, L.A. Wolsey (eds.). *Proc. 3rd Math. Progr. Conf. Int. Progr. Comb. Opt.*, CORE, Louvain-la-Neuve. 291–311.

The authors point out how a separation algorithm for the path inequalities can improve the performances of a cutting plane algorithm on a selected set on TSP instances. This claim is also supported by computational tests.

D. Naddef, J.M. Clochard (1994). *Some fast and efficient heuristics for comb separation in the symmetric traveling salesman problem*, Technical Report RR 941, ARTEMIS, Grenoble.

Several improvements on heuristic separation algorithms for comb and clique-tree inequalities proposed in the literature are described. An effective heuristic separation is described for the comb inequalities having three teeth.

R.D. Carr (1995). Separating clique tree and bipartition inequalities in polynomial time. E. Balas, J. Clausen (eds.). *Integer Programming and Combinatorial Optimization 4*, Lecture Notes in Computer Science, 920, Springer-Verlag, Berlin. 40–49.

An exact separation algorithm is described for the class of bipartition inequalities (which include clique-tree and comb inequalities as special cases) when the total number of handles and teeth is fixed. Let $\mathcal{C}$ be the set of all such inequalities having a given number h and t of handles and of teeth, respectively. A *backbone* is a set of k nodes of the graph (where k polynomially depends on h and t) that is shared by a subset of the inequalities in $\mathcal{C}$. For a fixed backbone the most violated inequality of the corresponding set is found in polynomial-time. The separation for the whole family $\mathcal{C}$ can then be done by enumerating all the backbones, which are polynomially many (since k is fixed). Due to its enumerative nature this procedure is not practically usable but can give useful hints for a heuristic algorithm.

R.D. Carr (1996). Separating over classes of TSP inequalities defined by 0 node-lifting in polynomial time. W.H. Cunningham, S.T. McCormick, M. Queyranne (eds.). *Integer Programming and Combinatorial Optimization 5*, Lecture Notes in Computer Science, 1084, Springer-Verlag, Berlin. 460–474.

The *cycle-shrink* relaxation for STSP is introduced. The number of variables of the relaxation is polynomial in n but it is larger than the number of edges of the graph. The number of inequalities is polynomial in n as well. A projection of this relaxation yields the subtour polytope. Based on this relaxation a polynomial-time separation algorithm is described for a class of valid inequalities for P_S^n obtained by node lifting of any fixed inequality defined on a subset of V having k nodes. Once the k nodes are fixed the separation amounts to solving a polynomially sized linear program, which can be done in polynomial-time. The separation is completed by repeating the procedure on all possible subsets of k nodes, which are a numbers polynomial in n, since k is fixed.

When a point $\overline{x}$ satisfies all the subtour elimination inequalities, a comb inequality (expressed as, e.g., in [Padberg and Grötschel 1979] (see above)) is violated by at most 0.5. A *maximally violated* comb inequality is violated by $\overline{x}$ by exactly 0.5. A *tight set* is a subset of V that defines a subtour elimination satisfied by $\overline{x}$ at equality.

D. Applegate, R.E. Bixby, V. Chvátal, W. Cook (1995). *Finding cuts in the TSP (a preliminary report)*, Technical Report 95–05, DIMACS, Rutgers University, New Brunswick, NJ.

A heuristic procedure is described to find maximally violated comb inequalities when a point $\overline{x}$ satisfying all the subtour elimination inequalities is given. It is shown that the teeth of these inequalities are *dominos*, i.e., tight subsets that can be partitioned into two tight subsets. A PQ-tree is used as data structure to efficiently represent the tight sets, while the candidate teeth for a maximally violated comb are the solution of a system of congruential equations. Unfortunately, the solution space of such a system has a size exponential in n. For this reason a heuristic procedure is proposed to select a polynomial number of candidates. Variations of the algorithm are described to find inequalities which may not be maximally violated. In a cutting plane algorithm for TSP several comb inequalities are generated at each iteration. At a given iteration the handles and the teeth of all the comb inequalities generated at the previous iterations can be used and suitably modified to generate a new inequality violated at the current iteration. Several techniques to efficiently store these special sets and to use them in the generation of a new violated inequality are described.

L. Fleischer, É. Tardos (1996). Separating maximally violated comb inequalities in planar graphs. W.H. Cunningham, S.T. McCormick, M. Queyranne (eds.). *Integer Programming and Combinatorial Optimization 5*, Lecture Notes in Computer Science, 1084, Springer-Verlag, Berlin. 475–489.

It is shown that, when the support graph of $\overline{x}$ is planar, the congruential system of the previous paper has a solution space whose size is polynomial in n. Consequently, a polynomial-time separation algorithm is derived for finding a maximally violated comb in this case. Since in practical computation it is not so rare to have a planar support graph for $\overline{x}$, this is the first polynomial-time algorithm that solves a special case of the comb separation and that is also efficient enough to be of practical use. Some variations of the algorithm are described that may produce also violated comb inequalities which are not maximally violated.

A. Caprara, M. Fischetti, A.N. Letchford (1997). *On the separation of maximally violated mod-k cuts*, Technical Report 3/97, DMI, Università di Udine.

The problem of separating the maximally violated comb inequalities in polynomial time is finally solved in this paper in a simple and elegant way. The proposed algorithm actually finds a class of maximally violated inequalities (called *mod-k cuts*) that includes, beside the comb, also the extended comb defined in [Naddef and Rinaldi 1988] (see above) and other (possibly non defining facets) inequalities.

M. Fischetti, P. Toth (1996). *A polyhedral approach for the exact solution of hard ATSP instances*, Technical report, DEIS, University of Bologna.

This is the only paper we could find that explicitly deals with the separation problem for P_A^n. Heuristic separation algorithms are described for symmetric inequalities, D_k^+ and D_k^- inequalities, odd CAT inequalities and node lifted inequalities.

6 Algorithms for Finding Optimal and Provably Good Solutions

The references of the previous sections provide a variety of tools for computing undirected or directed Hamiltonian cycles and lower bounds on their lengths. Such lower bounds allow qualifying a given solution as, say, at most $p\%$ longer than an optimum solution. In this section we give references to articles that describe algorithms that can achieve any prescribed quality, including $p = 0$, i.e., optimality. Due to the NP-hardness of the problem, it is not surprising that all such algorithms use enumerative methods like branch-and-bound or branch-and-cut. They differ on the branching strategy, and most importantly, on the bounding method. The latter consists of choosing appropriate relaxations and (efficient) algorithms for optimizing over them.

E. Balas, P. Toth (1985). Branch and bound methods. E.L. Lawler, J.K. Lenstra, A.H.G. Rinnooy Kan, D.B. Shmoys (eds.). *The Traveling Salesman Problem*, John Wiley & Sons, Chichester, 361–401.
G. Laporte (1992). The traveling salesman problem: an overview of exact and approximate algorithms. *European J. Oper. Res.* 59, 231–247.
Both survey papers are mainly concerned with several relaxations such as n-path, 1-tree or assignment relaxation, and with branching strategies.

J.F. Pekny, D.L Miller (1992). A parallel branch and bound algorithm for solving large asymmetric traveling salesman problems. *Math. Program.* 55, 17–33.
G. Carpaneto, M. Dell'Amico, P. Toth (1995). Exact solution of large-scale, asymmetric traveling salesman problems. *ACM Trans. Math. Software* 21, 394–409.
The first paper contains a parallel, the second a sequential branch-and-bound approach to the ATSP, based on the assignment relaxation. The algorithms perform well, as long the lengths of antiparallel arcs are uncorrelated, but they fail on (almost) symmetric instances.

G. Carpaneto, M. Fischetti, P. Toth (1989). New lower bounds for the symmetric travelling salesman problem. *Math. Program. (B)* 45, 233–254.
M. Fischetti, P. Toth (1992). An additive bounding procedure for the asymmetric travelling salesman problem. *Math. Program. (A)* 53, 173–197.
Here the so-called "additive bounding procedure" is introduced in order to improve bounds obtained from solving the assignment relaxation, first in the symmetric, second in the asymmetric case.

M. Held, R.M. Karp (1970). The traveling salesman problem and minimum spanning trees. *Oper. Res.* 18, 1138–1162.
M. Held, R.M. Karp (1971). The traveling salesman problem and minimum spanning trees: Part II. *Math. Program.* 1, 6–25.
The Held-Karp approach of approximating the optimum value of the Lagrangian dual based on the 1-tree relaxation of the TSP has been analyzed and refined by several authors. Because the 1-tree bound is the same as the bound obtained by optimizing over the subtour polytope, implementations of this approach in a branch-and-bound environment belong to the most successful methods for solving reasonably sized TSPs to optimality. However, such methods are outperformed by far by branch-and-cut methods to be surveyed next.

The paper [Dantzig, Fulkerson, and Johnson 1954] (see §5.2) is one of the corner-stones on which much of the modern methodology of using heuristics, linear programming, and separation, to attack combinatorial optimization problems is founded. The authors solved a 48-city instance of the TSP to optimality, this was truly large scale in the fifties. Today's most successful approaches to both TSP and ATSP are refinements of the methods introduced in this paper.

M.W. Padberg, G. Rinaldi (1991). A branch and cut algorithm for the resolution of large-scale symmetric traveling salesman problems. *SIAM Rev.* 33, 60–100.
The authors describe the first branch-and-cut algorithm for the TSP. At each node

of the enumeration tree, relaxations involving subtour elimination, 2-matching, comb, and clique-tree constraints, are solved by the simplex method. The instances solved to optimality with the new method are of about one order of magnitude larger than those solved in other previous computational studies. This paper was preceded by some less general attempts and followed by some refinements.

M. Jünger, G. Reinelt, S. Thienel (1994). Provably good solutions for the traveling salesman problem. *Z. Oper. Res.—Methods and Models of Operations Research* 40, 183–217.

This paper, as well as [Clochard and Naddef 1993] (see §5.5), and [Applegate et al. 1995] (see §5.5) describe several enhancements of the Padberg-Rinaldi branch-and-cut algorithm, together with which they represent the practically most successful algorithms that can solve most of the instances of TSPLIB, with sizes up to a few thousand cities.

In [Fischetti and Toth 1996] (see §5.5) the authors describe a branch-and-cut algorithm for the ATSP. They give computational results for hard ATSP instances, comparing them with their own implementation of a branch-and-bound algorithm based on additive bounding of four years earlier. Among other problem instances, ATSP instances of TSPLIB are solved to optimality.

7 Parallel Algorithms

When problem instances become large or when approaches with need for CPU time are employed, it is reasonable to think about parallelization options to make use of multi-processor hardware. Parallelization can either be performed on problem level where, for example, an instance is partitioned into smaller parts whose partial solutions are then combined, or on algorithm level where parts of an algorithm are parallelized.

J.R.A. Allwright, D.B. Carpenter (1989). A distributed implementation of simulated annealing for the travelling salesman problem. *Parallel Comput.* 10, 335–338.
P. Jog, J.Y. Suh, D. van Gucht (1991). Parallel genetic algorithms applied to the traveling salesman problem. *SIAM J. Optim.* 1, 515–529.
M.G.A. Verhoeven (1996). *Parallel Local Search*, Technical University, Eindhoven.

Very many experiments with parallelization techniques for TSP heuristics have been developed. These papers, as well as some of those cited in preceding section, report on computational experience on algorithms of this kind.

T. Christof, G. Reinelt (1995). Parallel cutting plane generation for the TSP. P. Fritzson, L. Finmo (eds.). *Parallel Programming and Applications*, IOS Press, Amsterdam, 163–169.

This is a paper reporting about the parallelization of the separation for facets obtained from polytopes associated with small instances.

14 Vehicle Routing

Gilbert Laporte
GERAD, École des Hautes Études Commerciales, Montreal and
Centre for Research on Transportation, Montreal

CONTENTS

The classical *Vehicle Routing Problem* (VRP) is defined on a graph $G = (V, A)$ where $V = \{v_0, v_1, \ldots, v_n\}$ is the vertex set and v_0 denotes a depot, and $A = \{(v_i, v_j) : v_i, v_j \in A, i \neq j\}$ is the arc set. Each vertex v_i has a demand q_i and a service duration s_i. A travel time matrix $c = (c_{ij})$ is defined on A. In some contexts q_i must be interpreted as a supply, and c_{ij} can also represent a distance or a travel cost. There are m identical vehicles of capacity Q based at the depot. The VRP consists of constructing a set of at most m vehicle routes of least total duration *(i)* starting and ending at the depot; *(ii)* such that each remaining vertex is visited exactly once; *(iii)* such that the total demand of each route does not exceed Q and its total duration does not exceed a preset constant L.

A number of variants of the basic VRP have also been studied. The most common is the VRP with time windows (VRPTW) in which each vertex must be visited within a time window $[a_i, b_i]$. Typically waiting is allowed at a vertex if the vehicle arrives before a_i. In this context, several objectives can be considered, sometimes hierarchically: minimize the number of vehicles, minimize the total route durations, minimize the total distance traveled.

Annotated Bibliographies in Combinatorial Optimization, edited by M. Dell'Amico, F. Maffioli and S. Martello

In the stochastic VRP, one of several components of the problem, e.g., demands, travel times, customer presence, can be stochastic. In this case, one first constructs a *first stage solution*, but it may prove impossible to follow this solution as planned (e.g., in a collection context, vehicle capacity may become exceeded at some point along the route). Then, a *recourse action* must be implemented, such as returning to the depot to unload and resuming collections at the point of failure. Such an action usually generates an extra cost. The problem is then to design a solution of *least expected cost.*

Several other variants of the VRP have also been studied over the years, such as the multi-depot VRP, the VRP with backhaul, the multi-period VRP, the inventory VRP, etc.

An important class of routing problems arises in contexts where all arcs or edges of a graph must be traversed by a single vehicle. This gives rises to the *Chinese Postman Problem.* When only a subset of arcs or edges must be traversed, one obtains the *Rural Postman Problem.* Constrained problems such as the *Capacitated Arc Routing Problem* also occur in a variety of practical contexts.

Over the last ten years, there have been important advances in the development of exact and approximate algorithms for the VRP and its variants. In the first case, the most significant progress has been made in the design of branch-and-cut and of column generation algorithms. In the second case, most advances have taken place in the area of so-called metaheuristics, including simulated anealing, tabu search and genetic search. In what follows, we summarize some of the most representative or significant work published since 1985, with a particular emphasis on survey papers.

1 The Classical VRP with Capacity and Distance Restrictions

1.1 Surveys and general references

N. Christofides (1985). Vehicle routing. E.L. Lawler, J.K. Lenstra, A.H.G. Rinnooy Kan, D.B. Shmoys (eds). *The Traveling Salesman Problem*, Wiley, Chichester, 431–448.

This survey paper presents formulations, exact algorithms and heuristics for the VRP. Exact methods include the Fisher-Jaikumar algorithm, and algorithms based on set covering and on branch-and-bound. Heuristics include constructive methods, two-phase methods, and incomplete optimization methods. Computational comments are offered.

G. Laporte, Y. Nobert (1987). Exact algorithms for the vehicle routing problem. S. Martello, G. Laporte, M. Minoux, C. Ribeiro (eds). *Surveys in Combinatorial Optimization*, Ann. Discr. Math. 31, North-Holland, Amsterdam, 147–184.

A survey of exact algorithms for the vehicle routing problem is presented. Algorithms are classified under three main headings: direct tree search methods, dynamic programming, integer linear programming. The latter category includes set partitionning formulations, vehicle flow formulations, and commodity flow formulations.

M. Desrochers, J.K, Lenstra, M.W.P. Savelsbergh (1990). A classification scheme for vehicle routing and scheduling problems. *European J. Oper. Res.* 46, 322–332.

A multi-field classification scheme is proposed for various types of vehicle routing and scheduling problems. Fourteen examples are provided.

C.F. Daganzo (1991). *Logistics Systems Analysis.* Lecture Notes in Economics and Mathematical Systems, Springer-Verlag, Berlin.

This monograph describes several techniques for solving large-scale complex distribution systems. These techniques typically describe general policies rather than specific solutions. This work is rooted in previous models developed by the author. It contains an extensive bibliography.

G. Laporte (1992). The vehicle routing problem: an overview of exact and approximate algorithms. *European J. Oper. Res.* 59, 345–358.

This survey paper describes some of the main algorithms for the vehicle routing problem. Exact algorithms include: assignment based branch-and-bound procedures, algorithms based on trees, dynamic programming, set partitioning and column generation, two- and three-index vehicle flow formulations. Heuristics include: the Clarke and Wright algorithm, the sweep algorithm, the Christofides-Mingozzi-Toth two-phase algorithm, tabu search.

M.L. Fisher (1995). Vehicle routing. M.O. Ball, T.L. Magnanti, C.L. Monma, G.L. Nemhauser (eds.). *Network Routing*, Handbooks in Operations Research and Management Science 8, North-Holland, Amsterdam, 1–33.

In this survey paper, algorithms for the vehicle routing problem are classified into three sections. *1)* Simple heuristics, including the Clarke and Wright method and its variants, the sweep algorithm, algorithms by Christofides and Eilon, and by Russell. *2)* Mathematical programming based heuristics, including generalized assignment based methods and set partitioning based methods. *3)* New algorithms, including simulated annealing, tabu search, and branch-and-cut optimization methods. Several computational comparisons are presented.

G. Laporte, I.H. Osman (1995). Routing problems: a bibliography. G. Laporte, M. Gendreau (eds.). *Freight Transportation*, Ann. Discr. Math. 61, North-Holland, Amsterdam, 227–262.

This bibliography contains 500 titles, mostly articles published since 1985, and some classical papers. Articles are classified according to four main headings, with several subdivisions: the Traveling Salesman Problem, the Vehicle Routing Problem, the Chinese Postman Problem, and the Rural Postman Problem.

1.2 Exact algorithms

G. Laporte, Y. Nobert, M. Desrochers (1985). Optimal routing under capacity and distance restrictions. *Oper. Res.* 33, 1050–1073.

A branch-and-cut algorithm is developed for the vehicle routing problem with capacity and distance restrictions. The algorithm uses a two-index vehicle flow formulation including generalized subtour elimination constraints. Computational results are presented for randomly generated instances involving up to 60 customers.

A.P. de Lucena Filho (1986). *Exact Solution Approaches to the Vehicle Routing Problem*, Ph.D. Thesis, Imperial College, London.

An exact algorithm is developed for the capacitated vehicle routing problem. The procedure first eliminates several suboptimal routes through the use of bounding and dominance conditions derived from state-space relaxation techniques. A set of routes guaranteed to include an optimal solution is identified and a set partitioning is solved to reach the optimum. Instances involving up to 45 customers are solved exactly. An approach based on a two-commodity flow formulation is also described.

M. Labbé, G. Laporte, H. Mercure (1991). Capacitated vehicle routing problem on trees. *Oper. Res.* 39, 616–622.

An exact branch-and-bound algorithm is proposed for the capacitated vehicle routing problem on trees. Lower bounds are based on the solution of bin packing problems. A heuristic with worst-case performance ratio of 2 is described. Instances involving up to 140 vertices are solved to optimality.

G. Laporte, M. Desrochers (1991). Improvements and extensions to the Miller-Tucker-Zemlin subtour elimination constraints. *Oper. Res. Lett.* 10, 27–36.

The Miller-Tucker-Zemlin subtour elimination constraints are lifted and extended to three types of vehicle routing problems: the capacitated VRP, the distance constrained VRP, and the VRP with time windows. Computational results are reported on the quality of the linear relaxation at the root of the search tree.

G. Laporte, H. Mercure, Y. Nobert (1992). A branch-and-bound algorithm for a class of asymmetrical vehicle routeing problems. *J. Oper. Res. Soc.* 43, 469–481.

This paper describes a branch-and-bound algorithm for asymmetric vehicle routing problems containing specified and unspecified vertices, capacity, and distance restrictions; in addition, vertices are grouped into clusters and a minimum number of vertices per cluster must be visited. The algorithm exploits the assignment relaxation of these problems. Instances involving up to 40 vertices are solved to optimality.

G. Cornuéjols, F. Harche (1993). Polyhedral study of the capacitated vehicle routing problem. *Math. Program.* 60, 21–52.

Polyhedral results are presented for the capacitated vehicle routing problem. In particular, extensions of subtour elimination constraints and of comb inequalities for the traveling salesman problem are described. A branch-and-cut algorithm is developed and applied to two instances involving 18 and 50 customers.

J.R. Araque, G. Kudva, T.L. Morin, J.F. Pekny (1994). A branch-and-cut algorithm for vehicle routing problems. *Ann. Oper. Res.* 50, 37–59.

Facet inducing inequalities are presented for the capacitated vehicle routing problem with unit demands. These include multistars, partial multistars, and generalized subtour elimination constraints. A branch-and-cut algorithm is developed and sample instances involving up to 60 customers are solved to optimality.

M. Fischetti, P. Toth, D. Vigo (1994). A branch-and-bound algorithm for the capacitated vehicle routing problem on directed graphs. *Oper. Res.* 42, 846–859.

Two new lower bounding procedures are described for the asymmetric capacitated vehicle routing problem. These are based on the so-called additive approach. Each procedure computes a sequence of non-decreasing lower bounds obtained by solving different relaxations of the problem, which are then embedded within a branch-and-bound procedure. Extensive tests are presented on randomly generated instances involving up to 300 vertices, and on eight real-life problems containing up to 70 customers.

M.L. Fisher (1994). Optimal solution of vehicle routing problems using minimum K-trees. *Oper. Res.* 42, 626–642.

An exact branch-and-bound algorithm is proposed for the vehicle routing with capacities in which single customer trips are disallowed. The problem is modeled as that of determining a minimum cost tree with $2K$ edges incident to the depot, subject to capacity constraints and to the requirement that each customer is visited exactly once. Lower bounds are obtained by dualizing the side constraints in a Lagrangean fashion. Computational results are reported for instances involving between 25 and 100 customers.

E. Hadjiconstantinou, N. Christofides, A. Mingozzi (1995). A new exact algorithm for the vehicle routing problem based on q-paths and k-shortest paths relaxations. G. Laporte, M. Gendreau (eds.). *Freight Transportation*, Ann. Oper. Res. 61, 21–44.

New lower bounds for the capacitated vehicle routing problem are developed. These are based on the q-path and k-shortest path relaxations of the problem. Tests are applied to reduce the size of the problem, using information provided by these bounds. A branch-and-bound algorithm is developed. Tight bounds are obtained for instances involving 150 vertices. The largest size problem solved to optimality contains 50 vertices.

S.P. Hill (1995). *Branch-and-Cut Methods for the Symmetric Capacitated Vehicle Routing Problem*, Ph.D. Thesis, Curtin University of Technology, Australia.

This thesis presents a new form of subtour elimination constraint and new cutting planes for the symmetric capacitated vehicle routing problems. A branch-and-cut algorithm is developed and extensive computational results are presented. The largest problem solved to optimality includes 134 customers.

D.L. Miller (1995). A matching based exact algorithm for capacitated vehicle routing problems. *ORSA J. Comput.* 7, 1–9.

A branch-and-bound algorithm is developed for the capacitated vehicle routing problem. The algorithm uses a two-index vehicle flow formulation and relaxes subtour elimination constraints. The relaxed problem is a b-matching problem. A procedure for efficiently computing good lower bounds is described and computational results are reported for instances containing up to 51 vertices.

1.3 Heuristics

M.D. Nelson, K.E. Nygard, J.H. Griffin, W.E. Shreve (1985). Implementation techniques for the vehicle routing problem. *Computers Oper. Res.* 273–283.

Six implementations of the Clarke and Wright savings algorithm are described and compared on 55 test problems. Various way to efficiently exploit data structures are presented.

M. Haimovich, A.H.G. Rinnooy Kan, L. Stougie (1988). Analysis of heuristics for vehicle routing problems. B.L. Golden and A.A. Assad (eds), *Vehicle Routing: Methods and Studies*, North-Holland, Amsterdam, 47–61.

This article analyzes the probabilistic and worst-case performance behaviour of several heuristics for the capacitated vehicle routing problem.

H. Paessens (1988). The savings algorithm for the vehicle routing problem. *European J. Oper. Res.* 34, 336–344.

Several implementations of the savings algorithm for the vehicle routing problem are described and compared. It is shown how CPU times and storage requirements can be reduced.

J. Bramel, E.G. Coffman Jr., P.W. Shor, D. Simchi-Levi (1992). Probabilistic analysis of the capacitated vehicle routing problem with unsplit demands. *Oper. Res.* 40, 1095–1106.

Capacitated vehicle routing problems with customers independently and identically distributed over the plane are considered. The asymptotic optimal solution value for these problems is determined and polynomial time heuristics are developed. It is shown that under some circumstances, the heuristic solutions are asymptotically optimal.

M.W.P. Savelsbergh (1992). *Computer Aided Routing*, CWI Tract 75, Center for Mathematics, Amsterdam.

This monograph reproduces the doctoral thesis of M.W.P. Savelsbergh. Part I contains the description of several heuristics for routing problems with time windows, with mixed collections and deliveries, with precedence constraints and with fixed paths. Part II described CAR (Computer Aided Routing), an interactive graphic system for the above problems. Part III reports on efforts and ideas to design a model and algorithm management system for vehicle routing and scheduling problems.

I.H. Osman (1993). Metastrategy simulated annealing and tabu search algorithms for the vehicle routing problem. F. Glover, M. Laguna, É. Taillard, D. de Werra (eds.). *Tabu Search*, Ann. Oper. Res. 41, Baltzer, Amsterdam, 421–451.

Several implementations of simulated annealing and tabu search heuristics are described for the vehicle routing problem. Hybrid approaches are also described. The various implementations are compared on 17 benchmark problems and several new best solutions are obtained.

É.D. Taillard (1993). Parallel iterative search methods for vehicle routing problems. *Networks* 23, 661–676.

A tabu search algorithm is developed for the vehicle routing problem with capacity and distance restrictions. The algorithm exploits a partitioning of the customer geographical space. As the algorithm evolves, so does the partitioning. Parallel computing is used to accelerate the search. Best known solutions are reported on fourteen benchmark problems.

M. Gendreau, A. Hertz, G. Laporte (1994). A tabu search heuristic for the vehicle routing problem. *Management Sci.* 40, 1276–1290.

A tabu search algorithm is developed for the vehicle routing problem with capacity and distance constraints. The algorithm considers a sequence of adjacent solutions obtained by repeatedly moving a vertex from its current route into another route. This is done by means of generalized insertions. Intermediate infeasible solutions are allowed and penalized. Computational results are presented on fourteen benchmark problems.

Y. Rochat, É.D. Taillard (1995). Probabilistic diversification and intensification in local search for vehicle routing. *Journal of Heuristics* 1, 147–167.

This paper describes an adaptive memory procedure used on conjunction with a tabu search algorithm. A population of good solutions is kept in a memory, and new solutions are generated based on these. The algorithm is applied to the capacitated vehicle routing problem and to the vehicle routing problem with time windows. Several new best solutions are reported.

D. Vigo (1996). A heuristic algorithm for the asymmetric capacitated vehicle routing problem. *European J. Oper. Res.* 89, 108–126.

This paper describes heuristics for the asymmetric VRP. These include extensions of the well-known Clarke-Wright and Fisher-Jaikumar algorithms, and a new heuristic combining insertions and vertex exchanges. Computational results are presented.

C. Rego, C. Roucairol (1996). A parallel tabu search algorithm using ejection chains for the vehicle routing problem. I.H. Osman, J.P. Kelly (eds.). *Metaheuristics: theory and Applications*, Kluwers, Boston, 661–675.

This paper describes a tabu search algorithm for the capacity and distance constrained vehicle routing problem. The algorithm uses ejection chains that produce compound moves from one solution to another. Crossing several feasible solutions is allowed. The algorithm is shown to produce very good feasible solutions in short computing times.

M. Gendreau, G. Laporte, J.-Y. Potvin (1997). Vehicle routing: modern heuristics. E.H.L. Aarts, J.K. Lenstra (eds.). *Local Search in Combinatorial Optimization*, Wiley, Chichester, 311–336.

Modern local search algorithms for the vehicle routing problems are classified and described under four headings: simulated annealing, tabu search, genetic algorithms and neural networks. Computational comparisons are presented.

2 The VRP with Time Windows

2.1 Surveys and general references

M.M. Solomon, J. Desrosiers (1988). Time window constrained routing and scheduling problems. *Transportation Sci.* 22, 1–13.

This is a survey of several families of vehicle routing problems with time windows: the multiple traveling salesman problem, the shortest path problem, the minimum spanning tree problem, the vehicle routing problem, the pick up and delivery problem, the multi-period vehicle routing problem, the shoreline problem.

J. Desrosiers, Y. Dumas, M.M. Solomon, F. Soumis (1995). Time constrained routing and scheduling. M.O. Ball, T.L. Magnanti, C.L. Monma, G.L. Nemhauser (eds.). *Network Routing*, Handbooks in Operations Research and Management Science 8, North-Holland, Amsterdam, 35–139.

This is an extensive survey of formulations and algorithms for time constrained, routing and scheduling problems. It is organized as follows: fixed schedule problems, the traveling salesman problem with time windows, constrained shortest path problems, the vehicle routing problem with time windows, pick-up and delivery problems with time windows, fleet and crew scheduling problems. Semi-structured algorithms and computational comparisons are provided.

2.2 Exact algorithms

M. Desrochers, J. Desrosiers, M.M. Solomon (1992). A new optimization algorithm for the vehicle routing problem with time windows. *Oper. Res.* 40, 342–354.

The vehicle routing problem with time windows is formulated as a set partitioning problem and solved by means of GENCOL, a column generation software. Exact solutions are obtained on benchmark problems containing up to 100 vertices. Problems with tight time windows are easier to solve.

K. Halse (1992). *Modeling and Solving Complex Vehicle Routing Problems*, Ph.D. Thesis, IMSOR, Technical University of Denmark.

Methods for solving complex vehicle routing problems are developed. Emphasis is placed on two problems: the VRP with time windows and the VRP with simultaneous pick-ups and deliveries. Formulations, exact algorithms and heuristics are proposed for these two problems and some of their extensions. Computational results are reported.

N. Kohl (1995). *Exact Methods for Time Constrained Routing and Related Scheduling Problems*, Ph.D. Thesis, IMM-DTU, Technical University of Denmark.

This thesis presents a number of optimization methods for constrained shortest path problems, vehicle routing with time windows, as well as some generalizations. A theoretic framework is developed, formulations, valid inequalities, and cutting planes methods are described, and branching strategies are compared. Computational results on bechmark problems are presented. These include solutions to a large number of previously unsolved instances.

2.3 Heuristics

M.M. Solomon (1987). Algorithms for the vehicle routing and scheduling problem with time window constraints. *Oper. Res.* 35, 254–265.

Several heuristics for the vehicle routing problem with time windows are designed and compared on several sets of benchmark problems. Extensive computational results are presented. Insertion-type heuristics perform consistently well.

L.F. Frantzeskakis, W.B. Powell (1990). A successive linear approximation procedure for stochastic, dynamic vehicle allocation problems. *Transportation Sci.* 24, 40–57.

The stochastic dynamic vehicle allocation problem is formulated as a stochastic mathematical program which is solved by means of a heuristic. Results indicate the superiority of the proposed approach over previous methods.

Y.A. Koskosidis, W.B. Powell, M.M. Solomon (1992). An optimization-based heuristic for vehicle routing and scheduling with soft time window constraints. *Transportation Sci.* 26, 69–85.

The VRPTW is formulated and solved as a mixed integer program by means of an optimization-based heuristics that extends the Fisher-Jaikumar algorithm. A new formulation enables the treatment of soft time windows and a new decomposition heuristic is presented. Computational results are reported.

P.M. Thompson, H.N. Psaraftis (1993). Cyclic transfer algorithms for multivehicle routing and scheduling problems. *Oper. Res.* 41, 935–946.

A new class of neighbourhood search algorithms – cyclic transfers –is described and applied to routing and scheduling problems. Some worst-case results are derived and computational results are presented for three classes of problems: the VRP, the precedence constrained vehicle routing and scheduling problem, and the VRPTW.

J. Bramel, C.-L. Li, D. Simchi-Levi (1993). Probabilistic analysis of a vehicle routing problem with time windows. *American J. Math. and Management Sci.* 13, 267–322.

The following version of the vehicle routing problem with time windows is considered. A working day is divided into time windows of equal length, customers select with some probability a time window during which they would like to be serviced, and the number of vehicles to satisfy all requests is always sufficient. The aim is to minimize the expected total distance traveled by all vehicles. A polynomial-time heuristic is developed and it is shown to be asymptotically optimal assuming customers are independently and identically distributed in the plane. Computational results are reported.

S.R. Thangiah (1993). *Vehicle routing with time windows using genetic algorithms*, Technical Report SRU-CpSc-TR-93-23, Slippery Rock University, PA.

A genetic search heuristic, called GIDEON, is developed for the vehicle routing problem with time windows. This heuristic consists of a customer clustering phase, followed by a local improvement phase. Computational results on 56 benchmark problems indicate that the proposed heuristic identifies a new best solution in 41 cases. Improvements are found both for the fleet size (average: 3.9%) and for the total distance traveled (average: 4.4%).

G. Kontoravdis, J.F. Bard (1995). A GRASP for the vehicle routing problem with time windows. *ORSA J. Comput.* 7, 10–23.

This paper considers the problem of determining the minimum number of vehicles for vehicle routing problems with time windows. A secondary objective is the total distance traveled. A greedy randomized adaptive search procedure (GRASP) is used to obtain feasible solutions. Computational results are reported on benchmark problems containing 100 vertices, and on some real-life instances involving up to 417 customers.

É.D. Taillard, P. Badeau, M. Gendreau, F. Guertin, J.-Y. Potvin (1995). *A New Neighborhood Structure for the Vehicle Routing Problem with Time Windows*, Publication 95-66, Centre for Research in Transportation, Montreal.

This article describes a new neighborhood structure for a tabu search mechanism applied to the vehicle routing with soft time windows. The search uses an adaptive memory that contains the routes of the best previously visited solutions. These routes are used as starting points for new solutions. The algorithm is tested on a set of benchmark problems and several new best solutions are identified.

R.A. Russell (1995). Hybrid heuristics for the vehicle routing problem with time windows. *Transportation Sci.* 29, 156–166.

This paper describes efficient heuristics for the TSPTW. These combine construction and local search improvement techniques. Local improvements are sought during the construction phase. This procedure is effective in reducing the number of vehicles.

J.-Y. Potvin, T. Kervahut, B.L. Garcia, J.-M. Rousseau (1996). The vehicle routing problem with time windows – Part I: tabu search. *INFORMS J. Comput.* 8, 158–164.

This paper describes a tabu search heuristic for the vehicle routing with time windows. It uses specialized exchanges that maintain feasibility. Computational tests and comparisons are carried out on a set of benchmark problems.

J.-Y. Potvin, S. Bengio (1996). The vehicle routing problem with time windows – Part II: genetic search. *INFORMS J. Comput.* 8, 165–172.

A genetic algorithm, called GENEROUS (GENEtic ROUting System), is developed for the vehicle routing with time windows. A methodology is devised for the merging of two solutions into a single solution that is likely to be feasible with respect to the time window constraints. Computational comparisons with alternative heuristics are reported.

3 The Stochastic VRP

3.1 Surveys and general references

P. Jaillet, A.R. Odoni (1988). The probabilistic vehicle routing problem. B.L. Golden, A.A. Assad (eds.). *Vehicle Routing: Methods and Studies*, North-Holland, Amsterdam, 293–318.

This article reviews properties, bounds, and algorithmic results related with probabilistic traveling salesman problems and probabilistic vehicle routing problems. These are problems in which customers are present with some probability and an a priori solution of least expected cost must be determined.

W.B. Powell (1988). A comparative review of alternative algorithms for the dynamic vehicle allocation problem. B.L. Golden, A.A. Assad (eds.). *Vehicle Routing: Methods and Studies*, North-Holland, Amsterdam, 249–291.

This paper reviews the dynamic vehicle allocation problem in the context of truckload trucking, with attention given to dispatching and repositioning trucks in anticipation of forecasted future demand. Four approaches are reviewed: deterministic transshipment networks, stochastic and non-linear networks, Markov decision processes, and stochastic programming.

H.N. Psaraftis (1988). Dynamic vehicle routing problems. B.L. Golden, A.A. Assad (eds.). *Vehicle Routing: Methods and Studies*, North-Holland, Amsterdam, 223–248.

In this paper, dynamic routing is defined and compared with the static and stochastic cases. Methodological and design features issues are examined. An algorithm for dynamic routing is described in the context of cargo ships in an emergency situation.

M. Dror, G. Laporte, P. Trudeau (1989). Vehicle routing with stochastic demands: properties and solution frameworks. *Transportation Sci.* 23, 166–176.

This is an overview of the vehicle routing problem with stochastic demands. The most common operating and service policies are first described. Solution properties are investigated and various formulations are examined, including chance constrained and recourse model. A new Markov decision process representation is also described

W.B. Powell, P. Jaillet, A.R. Odoni (1995). Stochastic and dynamic network and routing. M.O. Ball, T.L. Magnanti, C.L. Monma, G.L. Nemhauser (eds.). *Network Routing*, Handbooks in Operations Research and Management Science 8, North-Holland, Amsterdam, 141–295.

This survey covers the field of dynamic and stochastic routing problems on networks. It contains an introduction to stochastic models and discusses modeling issues associated with dynamic problems. Formulations, mathematical properties and algorithms are provided both for stochastic and dynamic routing problems. A full section is devoted to the evaluation of dynamic models. Algorithmic issues associated with several of the models are discussed in some detail.

M. Gendreau, G. Laporte, R. Séguin (1996). Stochastic vehicle routing. *European J. Oper. Res.* 88, 3–12.

This survey paper on the stochastic vehicle routing problem describes the main types of problems and summarizes the most important contributions in table form. The main headings are: the traveling salesman problem with stochastic customers, the traveling salesman problem with stochastic travel times, the m-traveling salesman problem with stochastic travel times, the vehicle routing problem with stochastic demands, the vehicle routing problem with stochastic customers, the vehicle routing problem with stochastic customers and demands.

3.2 Exact algorithms

G. Laporte, F.V. Louveaux, H. Mercure (1992). The vehicle routing problem with stochastic travel times. *Transportation Sci.* 26, 161–170.

Several formulations are proposed for the vehicle routing problem with stochastic travel times. An integer L-shaped algorithm is then developed and applied to one of these formulations. Exact solutions are obtained for instances containing up to 20 customers.

M. Gendreau, G. Laporte, R. Séguin (1995). An exact algorithm for the vehicle routing problem with stochastic demands and customers. *Transportation Sci.* 29, 143–155.

The vehicle routing problem in which customers are present with a certain probability and have a random demand is considered. The recourse version of the problem is formulated and the problem is solved to optimality by means of an integer L-shaped branch-and-cut algorithm. Depending on problem parameters and characteristics, instances involving up to 70 vertices can be solved to optimality.

3.3 Heuristics

D.J. Bertsimas, G.J. van Ryzin (1991). A stochastic and dynamic vehicle routing problem in the euclidean plane. *Oper. Res.* 39, 601–615.

The problem considered in this paper arises in contexts where points in the plane generate demands whose time of arrival, location and on-site service are stochastic. The objective is to find a policy to service demands over an infinite time horizon in order to minimize the expected system time. Several policies are developed and analyzed. Optimal policies can be obtained under some circumstances.

D.J. Bertsimas (1992). A vehicle routing problem with stochastic demands. *Oper. Res.* 40, 574–585.

The vehicle routing problem with stochastic demands is considered. Closed form expressions and algorithms are developed to compute the expected length of an a priori solution. Upper and lower bounds are developed for the probabilistic VRP and the VRP with re-optimization strategy in which an optimal route is found at every instance. Heuristics are developed and their worst-case performance ratio is computed.

P. Trudeau, M. Dror (1992). Stochastic inventory routing: route design with stockouts and route failures. *Transportation Sci.* 26, 171–184.

The stochastic inventory routing problem consists of planning delivery routes to customers maintaining a low inventory of a commodity consumed at a daily rate. The problem is analyzed in the framework of stochastic programming. Several solution procedures are proposed, compared and tested on real-life data.

D.J. Bertsimas, G.J. van Ryzin (1993). Stochastic and dynamic vehicle routing in the euclidean plane with multiple capacitated vehicles. *Oper. Res.* 41, 60–76.

This paper extends the work of Bertsimas and van Ryzin (1991) in several directions. The problem of m identical vehicles of identical capacities is considered. Then the case where any vehicle can serve at most q customers is analyzed. Policies for these two cases are developed, analyzed and compared. Extensions to mixed travel cost and system time objectives are discussed.

D.J. Bertsimas, P. Chervi, M. Peterson (1995). Computational approaches to stochastic vehicle routing. *Transportation Sci.* 29, 342–352.

Several graph-based a priori heuristics for the probabilistic traveling salesman problem and the probabilistic vehicle routing problem are described. Computational comparisons are made with sample averages of a posteriori solutions for these problems. Depending on the type of implementation, gaps of between 1% and 5% are obtained.

M. Gendreau, G. Laporte, R. Séguin (1996). A tabu search heuristic for the vehicle routing problem with stochastic demands and customers. *Oper. Res.* 44, 469–477.

This paper considers a version of the stochastic vehicle routing problem where customers are present with given probabilities and have random demands. A tabu search algorithm is developed. It uses proxy approximation for the objective function values of intermediate solutions. Tests on instances involving up to 46 customers indicate that optimal solutions are identified in 89.45% of all cases, and the average deviation from optimality is 0.38%.

4 Other Variants of the VRP

4.1 Surveys and general references

B.L. Golden, A.A. Assad (eds.) (1988). *Vehicle Routing: Methods and Studies*, North-Holland, Amsterdam.

This book contains twenty-one chapters classified as follows: Overview, Algorithmic techniques for vehicle routing, Models for complex routing environments, Practical applications, and Development of vehicle routing systems. The book contains a good overview of the field, as well as several articles on variants of the vehicle routing problem: the VRP with time windows, the VRP with backhauls, the VRP with site dependencies, location-routing problems, allocation-routing problems, dynamic VRPs, probabilistic VRPs, and the prize-collecting traveling salesman problem.

G. Laporte (1988). Location-routing problems. B.L. Golden, A.A. Assad (eds). *Vehicle Routing: Methods and Studies*, North-Holland, Amsterdam, 163–197.

In location-routing problems, location and routing problems must be solved simultaneously. This article provides an overview of the field. It contains a taxonomy, as well as a description of the most common heuristics and exact algorithms for these classes of problems.

A. Federgruen, D. Simchi-Levi (1995). Analysis of vehicle routing and inventory-routing problems. M.O. Ball, T.L. Magnanti, C.L. Monma, G.L. Nemhauser (eds.). *Network Routing*, Handbooks in Operations Research and Management Science 8, North-Holland, Amsterdam, 297–373.

This paper analyzes the worst-case performance and asymptotic properties of a number of heuristics for the capacitated vehicle routing problem with split or unequal demands, for inventory-routing models, for the multi-depot vehicle routing problem, and for a number of extensions and generalizations of these problems.

4.2 Exact algorithms

J. Desrosiers, G. Laporte, M. Sauvé, F. Soumis, S. Taillefer (1988). Vehicle routing with full loads. *Computers Oper. Res.* 15, 219–226.

Several requests with known origins and destinations have to be satisfied by a fleet of identical vehicles. Side constraints can be imposed on the vehicle routes. This problem is transformed into an equivalent directed vehicle routing problem in which each vertex corresponds to a request. An exact branch-and-bound algorithm is developed and instances containing up to 104 requests are solved to optimality.

G. Laporte, Y. Nobert, S. Taillefer (1988). Solving a family of multi-depot vehicle routing and location-routing problems. *Transportation Sci.* 22, 161–172.

Multi-depot vehicle routing and location-routing problems on directed graphs are considered. A graph transformation is first applied to the problems in order to formulate them as constrained assignment problems. A branch-and-bound algorithm that exploits the assignment relaxation of the problems is then developed. Instances involving up to 80 vertices for each of the two problems are solved to optimality.

G. Carpaneto, M. Dell'Amico, M. Fischetti, P. Toth (1989). A branch-and-bound algorithm for the multiple depot vehicle scheduling problem. *Networks* 19, 531–548.

The vehicle scheduling problem consists of assigning a set of pre-timetabled trips to vehicles in order to minimize a cost function. This paper considers a version of the problem with several depots. Lower bounds based on assignment and connectivity constraints are developed and embedded within a branch-and-bound algorithm. Instances containing up to 70 vertices are solved to optimality.

M. Dror, G. Laporte, P. Trudeau (1994). Vehicle routing with split deliveries. *Discr. Appl. Math.* 50, 239–254.

A relaxation of the classical capacitated vehicle routing problem in which split deliveries are allowed is considered. The problem is formulated as an integer linear program and several valid inequalities are derived. A branch-and-bound procedure is proposed. The algorithm is applied to the root node of the search tree on 10, 15 and 20 vertex instances. The quality of the lower bound is measured by comparing the solution values to those produced by a heuristic.

C.C. Ribeiro, F. Soumis (1994). A column generation approach to the multi-depot vehicle scheduling problem. *Oper. Res.* 42, 41–52.

The multi-depot vehicle scheduling problem is formulated as a set partitioning problem and solved by column generation. Various properties of the formulation are presented and exploited in the solution process. Several instances involving up 300 vertices are solved to optimality.

4.3 Heuristics

D.O. Casco, B.L. Golden, E.A. Wasil (1988). Vehicle routing with backhauls: models, algorithms, and case studies. B.L. Golden, A.A. Assad (eds.). *Vehicle Routing: Methods and Studies*, North-Holland, Amsterdam, 127–147.

The vehicle routing with backhauls is a variant of the VRP in which vehicles first make deliveries, then collections. This article reviews several heuristics for this problem and then presents some case studies.

M. Goetschalckx, C. Jacobs-Blecha (1989). The vehicle routing with backhauls. *European J. Oper. Res.* 42, 39–51.

A two-phase heuristic is presented for the vehicle routing problem with backhauls. An initial solution is first obtained by means of a spacefilling curve. Then an improved solution is sought by optimizing subproblems. Several computational results are reported to compare various algorithmic options.

R.A. Russell, D. Gribbin (1991). A multiphase approach to the period routing problem. *Networks* 21, 747–765.

A four-phase heuristic for the period routing problem is described. In the first phase, an initial solution heuristic is determined. Two arc interchange phases are then applied, finally further improvements are sought by solving a 0-1 model. Computational experiments indicate that the proposed heuristic improves upon previous methods.

C. Malandraki, M.S. Daskin (1992). Time dependent vehicle routing problems: formulations, properties and heuristic algorithms. *Transportation Sci.* 26, 185–200.

The problem considered in this paper is a vehicle routing problem in which travel times vary throughout the day. Several heuristics are proposed and compared in randomly generated instances.

I M. Chao (1993). *Algorithms and Solutions to Multi-level Vehicle Routing Problems*, Ph.D. Dissertation, University of Maryland at College Park.

New heuristics for various extensions of the vehicle routing problem are developed. The problems include: the orienteering problem, the team orienteering problems, the multi-depot VRP, the period VRP, the period traveling salesman problem. New test problems are introduced and extensive computational results are reported.

M. Dell'Amico, M. Fischetti, P. Toth (1993). Heuristic algorithms for the multiple depot vehicle scheduling problem. *Management Sci.* 39, 115–125.

The authors propose a heuristic for the multi-depot vehicle scheduling problem in which a set of time-tabled trips have to be assigned to vehicles based at different depots. A polynomial time heuristic guarantees a solution containing the minimum number of vehicles, and attempts to minimize routing costs. Computational results are reported for instances involving 1000 trips and 10 depots.

A. Van Breedam (1994). *An Analysis of the Behavior of Heuristics for the Vehicle Routing Problem for a Selection of Problems with Vehicle-Related, Customer-Related and Time-Related Constraints*, Ph.D. Thesis, University of Antwerp.

This thesis describes and compares several classical heuristics for the VRP with vehicle-related, customer-related and time-related constraints. The heuristics considered include sequential and parallel versions of the following procedures: nearest neighbour, savings, insertion, assignment based insertions, generalized assignment heuristics, two-phase heuristics, and the sweep method. Improvements of these heuristics are also presented. The thesis includes several computational comparisons.

I.M. Chao, B.L. Golden, E.A. Wasil (1995). An improved heuristic for the period vehicle routing problem. *Networks* 26, 25–44.

A new heuristic is presented for the period vehicle routing problem. Integer programming is used to initially allocate visit combinations to customers. A vehicle routing problem is then solved for each day by means of a savings heuristic. Local improvements are then sought. Computational experiments are performed and several new best solutions are identified.

J. Renaud, G. Laporte, F.F. Boctor (1996). A tabu search heuristic for the multi-depot vehicle routing problem. *Computers Oper. Res.* 23, 229–235.

This article describes a tabu search heuristic for the multi-depot vehicle routing problem with capacity and distance restrictions. The algorithm is made up of three phases: fast improvement, intensification, and diversification. Computational results are reported in a set of 23 benchmark instances and several new best solutions are identified. Good results are often obtained even if only the first phase is applied.

J.-F. Cordeau, M. Gendreau, G. Laporte (1997). A tabu search heuristic for periodic and multi-depot vehicle routing problems, *Networks*. Forthcoming.

A unified tabu search algorithm is developed for three extensions of the classical vehicle routing problem: the periodic VRP, the periodic traveling salesman problem, and the multi-depot VRP. The latter problem is shown to be a special case of the periodic VRP. Computational results on benchmark instances indicate that the algorithm yields several new best solutions for each of the three problems.

5 Arc Routing Problems

5.1 Surveys and general references

H. Fleischner (1990,1991). *Eulerian Graphs and Related Topics*, Ann. Discr. Math., 45 and 50, North-Holland, Amsterdam.

These two volumes are a compendium of Eulerian graph theory which constitutes one of the foundations of arc routing. Algorithms for the determination of Eulerian walks are described for undirected, directed, mixed and windy postman problems. The first of these books includes the original Latin version of Euler's (1736) manuscript on the Königsberg bridges problems, and the German version of Hierholzer's (1873) work on the determination of Eulerian cycles in graphs.

A.A. Assad and B.L. Golden (1995). Arc routing methods and applications. M.O. Ball, T.L. Magnanti, C.L. Monma, G.L. Nemhauser (eds.). *Network Routing*, Handbooks in Operations Research and Management Science 8, North-Holland, Amsterdam, 375–483.

This is a review of the field of arc routing. The following problems are described: the Chinese postman problem, the rural postman problem, the capacitated arc routing problem. There is a strong emphasis on applications such as sanitation services, postal delivery, meter reading, snow control, and manufacturing. Several computational comparisons of bounds and algorithms are presented.

H.A. Eiselt, M. Gendreau, G. Laporte (1995). Arc routing problems, part I: the chinese postman problem. *Oper. Res.* 43, 231–242.

A survey of the main known results on the Chinese postman problem (CPP) is presented. The paper is organized as follows: applications, the undirected CPP, the directed CPP, the windy CPP, the mixed CPP, the hierarchical CPP. The emphasis is placed on integer linear programming formulations and on algorithms.

H.A. Eiselt, M. Gendreau, G. Laporte (1995). Arc routing problems, part II: the rural postman problem. *Oper. Res.* 43, 399–414.

A survey of the main known results on the rural postman problem (RPP) is presented. The paper is organized as follows: applications, the undirected RPP, the directed RPP, the stacker crane problem, the capacitated arc routing problem. Various exact and approximate algorithms are described, together with comparative computational results.

5.2 Exact algorithms

N. Christofides, V. Campos, A. Corberán, E. Mota (1986). An algorithm for the rural postman problem on a directed graph. *Math. Program. Study* 26, 155–166.

A heuristic and an exact algorithm are presented for the directed rural postman problem. The heuristic solves in sequence a shortest spanning arborescence problem, a transportation problem and a Eulerian circuit. The exact algorithm applies branch-and-bound to an integer linear programming formulation. Twenty-three instances involving up to 80 vertices, 180 arcs, and 74 required arcs are solved to optimality.

Z. Win (1987). *Contributions to Routing Problems*, Doctoral Dissertation, Universität Augsburg, Germany.

This thesis investigates two NP-hard variants of postman problems: the windy postman problem and the capacitated postman problem. It contains integer linear programming formulations, polyhedral results, branch-and-cut algorithms, as well as heuristics. Computational results are presented.

E. Benavent, V. Campos, A. Corberán, E. Mota (1992). The capacitated arc routing problem: lower bounds. *Networks* 22, 669–690.

Lower bounds are developed for the capacitated arc routing problem. Three are based on the solution of a minimum cost perfect matching problem, while the fourth one uses dynamic programming. Computational results indicate that the new bounds are superior to several previous bounds for the same problem.

M. Grötschel, Z. Win (1992). A cutting plane algorithm for the windy postman problem. *Math. Program.* 55, 339–358.

The windy postman problem is defined on an undirected graph, but the cost of traversing an edge depends on the direction of travel. A branch-and-cut algorithm is proposed for this problem. The algorithm was applied to instances involving up to 264 vertices and 489 edges. The linear relaxation provided a feasible solution 31 times out of 36.

A. Corberán, J.M. Sanchis (1994). A polyhedral approach to the rural postman problem. *European J. Oper. Res.* 79, 95–114.

The polyhedral structure of the rural postman problem is investigated and several facet inducing inequalities are developed. These are used in a cutting plane algorithm. Benchmark problems involving up to 50 vertices and 184 edges, as well as two real-life instances are solved to optimality.

Y. Nobert, J.-C. Picard (1996). An optimal algorithm for the mixed chinese postman problem. *Networks* 27, 95–108.

A branch-and-cut algorithm is developed for the mixed Chinese postman problem. The problem of determining a Eulerian graph is formulated as an integer linear program, incorporating the necessary and sufficient conditions stated by Ford and Fulkerson. Gomory cuts may be generated to help gain integrality. The algorithm was tested on 440 problems including up to 225 vertices, 5569 arcs and 4455 edges. Of these problems, 313 were solved to optimality without branching.

5.3 Heuristics

W.-L. Pearn (1989). Approximate solutions for the capacitated arc routing problem. *Computers Oper. Res.* 6, 589–600.

Two new heuristics are developed for the capacitated arc routing problem: the modified construct-strike and path-scanning algorithms. On a large series of test problems, these two heuristics are shown to outperform three previous heuristics. On dense graphs (70% to 100% density), the modified construct-strike algorithm often produces an optimal solution.

W.-L. Pearn (1991). Augment-insert algorithms for the capacitated arc routing problem. *Computers Oper. Res.* 18, 189–198.

Two versions of the so-called augment-insert procedure for the capacitated arc routing problem are described. Extensive comparisons with several alternative heuristics indicate that these are particularly efficient on sparse graphs having a density not exceeding 30%.

A. Hertz, G. Laporte, P. Nanchen (1996). *Improvement Procedures for the Undirected Rural Postman Problem*, Publication 96-30, Center for Research on Transportation, Montreal.

Several new procedures are developed and integrated into powerful heuristics for the undirected rural postman problem. Computational results are presented.

15 Max-Cut Problem

Monique Laurent
CNRS, DMI, Ecole Normale Supérieure, Paris

I dedicate this paper to the memory of Svata Poljak, who would have been the perfect coauthor for it. I thank his wife Jana for giving me access to his files about max-cut.

CONTENTS

Cuts in graphs play an ubiquitous role in combinatorial optimization since the early days of network flow theory. Indeed, the celebrated max-flow min-cut result of Ford and Fulkerson shows how to find in polynomial time a minimum capacity cut separating two nodes in a graph using flow techniques. flow The minimum cut problem is treated in detail in Chapter 18 of this Annotated Bibliography, as well as the role of cuts in multicommodity flow problems. We consider here the *max-cut problem*, which, given a graph $G = (V, E)$ and edge weights $w = (w_e)_{e \in E} \in \Re^E$, consists of finding a cut $\delta_G(S)$ ($S \subseteq V$) whose weight $\sum_{e \in \delta_G(S)} w_e$ is maximum (the *cut* $\delta_G(S)$ consisting of the edges of G having one end node in S and the other one in $V \backslash S$). (Note that the optimum value is 0 and attained at the empty cut $\delta_G(\emptyset) = \delta_G(V)$ when all edge weights are nonpositive.) Variations of this problem arise when considering partitions of the node set into more than two sets and/or asking conditions on the cardinalities of the sets in the partition; some of them will be mentioned in §6. We provide here a number of references about max-cut relevant to complexity issues and various methods for solving the problem, based on polyhedral, spectral, and semidefinite approaches.

Annotated Bibliographies in Combinatorial Optimization, edited by M. Dell'Amico, F. Maffioli and S. Martello

1 Formulations and Applications

1.1 Formulations

The max-cut problem has several equivalent formulations. One of them is as a quadratic ± 1 optimization problem:

$$\max(\frac{1}{2}\sum_{ij\in E} w_{ij}(1 - x_i x_j) \mid x \in \{\pm 1\}^n). \tag{P1}$$

Another well-known formulation is in terms of *unconstrained quadratic* 0,1 *programming*, or UBQP, for short. This is the problem:

$$\min(\sum_{1\le i\le j\le n} q_{ij} x_i x_j \mid x \in \{0,1\}^n). \tag{P2}$$

Let H denote the graph with node set $V := \{1, \ldots, n\}$ and edges the pairs ij with $q_{ij} \neq 0$ $(i < j)$. It is known since long that (P2) is equivalent to a max-cut problem on the graph $G := H + u$ (obtained by adding a new node u to H adjacent to all nodes in H) with suitable edge weights.

J-C. Picard and H.D. Ratliff (1973). A graph-theoretic equivalence for integer programs. *Oper. Res.* 21, 261–269.

The above transformation is made explicit, e.g., in this paper.

J-C. Picard and H.D. Ratliff (1975). Minimum cuts and related problems. *Networks* 5, 357–370.

These authors show that problem (P2) can be solved efficiently when $q_{ij} \le 0$ for all $i < j$ as it can then be formulated as a minimum cut problem in a network with $n+2$ vertices. Further solvable instances will be presented in §2.2.

P.L. Hammer and I.G. Rosenberg (1972). Equivalent forms of zero-one programs. S.K. Zaremba (ed.). *Applications of Number Theory to Numerical Analysis*, Academic Press, New York, 453–463.

In this paper it is explained how general 0-1 programs of the form: $\min(f(x) \mid x \in \{0,1\}^n)$ where $f : \{0,1\}^n \longrightarrow \Re$ (a *pseudo-boolean function*) can, in fact, be brought in the form (P2) (at the cost of possibly introducing exponentially many new variables).

P.L. Hammer and S. Rudeanu (1968). *Boolean Methods in Operations Research and Related areas*, Springer-Verlag, Berlin (and Dunod, Paris, 1970).

This is an early textbook on the topic of pseudo-boolean programming.

P. Hansen (1979). Methods of nonlinear 0-1 programming. P.L. Hammer et al. (eds.). *Ann. Discr. Math.* 5, 53–70.

A survey of results obtained before 1980 in this area can be found in this paper.

Finally, we point out that the maximum stable set problem in a graph $G = (V, E)$

with node weights $c = (c_i)_{i \in V}$ can be formulated as the following instance of UBQP:

$$\max(\sum_{i \in V} c_i x_i - M \sum_{ij \in E} x_i x_j \mid x \in \{0,1\}^V)$$

where $M := \max_i c_i$. Conversely, UBQP can be formulated as a stable set problem in a slightly larger graph; cf. [Hammer, Hansen and Simeone 1984] in §3.2.

1.2 Surveys

A number of survey papers and PhD dissertations related to max-cut (or UBQP) have been written in recent years. Some of them are listed below.

E. Boros and P.L. Hammer (1991). The max-cut problem and quadratic 0-1 optimization: polyhedral aspects, relaxations and bounds. *Ann. Oper. Res.* 33, 151–180.
M. Deza and M. Laurent (1994). Applications of cut polyhedra - I, II. *J. Comput. & Appl. Math.* 55, 191–216 and 217–247.
S. Poljak and Z. Tuza (1995). Maximum cuts and largest bipartite subgraphs. W. Cook, L. Lovász, and P. Seymour (eds.). *Combinatorial Optimization*, DIMACS Ser. in Discr. Math. and Theoretical Comput. Sci. 20, AMS, 181–244.

J.-M. Bourjolly (1986). *Integral and Fractional Node-Packings and Pseudo-Boolean Programs.*, PhD thesis, University of Waterloo, Canada.
G.P. Rodgers (1989). *Algorithms for Unconstrained Quadratic 0-1 Programming and Related Problems on Contemporary Computer Architectures*, PhD thesis, The Pennsylvania State University.
C. De Simone (1992). *The Max Cut Problem.* PhD thesis, Rutgers University, New Brunswick.
C.C. De Souza (1993). *The Graph Equipartition Problem: Optimal Solutions, Extensions and Applications*, PhD thesis, Université Catholique de Louvain.
N. Boissin (1994). *Optimisation des Fonctions Quadratiques en Variables Bivalentes.* PhD thesis, Centre National d'Etudes des Télécommunications, Issy Les Moulineaux.
A. Deza (1994). *Graphes et Faces de Polyèdres Combinatoires*, PhD thesis, Université de Paris-Sud, Orsay.
C. Helmberg (1994). *An Interior Point Method for Semidefinite Programming and Max-Cut Bounds.*, Doctoral dissertation, Graz University of Technology.

The survey of [Deza and Laurent 1994] emphasizes the relevance of several polyhedra associated with cuts to other areas of mathematics and its applications. The thesis of [Deza 1994] considers geometric questions for these polyhedra, in particular, for small graphs. The other works treat max-cut and, in particular, polyhedral theory, approximation and computational aspects.

M. Deza and M. Laurent (1997). *Geometry of Cuts and Metrics.* Vol. 15 of *Algorithms and Combinatorics*, Springer-Verlag, Berlin.
A global overview of various aspects of cuts and metrics can be found in this book.

1.3 Applications

The max-cut problem arises in a natural way in several applications. A first important application is in statistical physics for the determination of ground states of spin glasses.

F. Barahona, M. Grötschel, M. Jünger, and G. Reinelt (1988). An application of combinatorial optimization to statistical physics and circuit layout design. *Oper. Res.* 36, 493–513.

The authors describe this application in detail and give an extended bibliography. In the (short range) Ising model, the problem can be formulated as $\min(-\sum_{ij\in E} J_{ij}s_is_j - h\sum_{i=1}^n s_i \mid s \in \{\pm 1\}^n)$, where J_{ij} represents the interaction between two magnetic impurities i and j, and h the exterior magnetic field. The impurities are assumed to be distributed regularly on a 2- or 3-dimensional grid and the interactions are supposed to be nonzero only on the edge set E of the grid. Thus, we find an instance of the max-cut problem (P1) on a 2D or 3D grid plus a universal node. This problem can be solved efficiently in the 2D case when there is no exterior magnetic field by the reduction of [Hadlock 1975] (cf. §2.2), as in that case the graph is planar.

G. Toulouse (1977). Theory of the frustration effect in spin glasses: I. *Commun. Phys.* 2, 115–119.
I. Bieche, R. Maynard, R. Rammal, and J-P. Uhry (1982). On the ground states of the frustration model of a spin glass by a matching method of graph theory. *J. Physics A Math. Gen.* 13, 2553–2576.

This reduction as well as a polynomial algorithm were discovered independently in the field of physics by the above authors.

The max-cut problem also comes up in the design of printed circuit boards and VLSI design. The design of a chip is usually broken into several phases: placement, routing, and layer assignment. In the placement phase, after placing the modules on the chip one tries to reduce the total net length by flipping some of the modules (exchanging left and right, or top and bottom). In the last phase arises the problem of minimizing the number of vias (holes in a printed circuit board or contacts on a chip) that are needed so that crossing wire segments are assigned to different layers.

C.K. Cheng, S.Z. Yao, and T.C. Hu (1991). The orientation of modules based on graph decomposition. *IEEE Trans. Comput.* 40, 774–780.
printed circuit boards. K.C. Chang and D. H-C. Du (1987). Efficient algorithms for layer assignment problem. *IEEE Trans. Computer-Aided Design* CAD-6, 67–78.

It is shown in these papers that the two above mentioned problems from VLSI design can both be formulated as instances of max-cut. See also [Boros and Hammer 1991] in §1.2 and [Barahona et al. 1988] (cited above), as well as Chapter 23 in this volume. Computational results on the above instances of max-cut are reported in §5.

Many other problems involve max-cut or UBQP instances. Among them, the problems of balancing signed graphs, or generating tests detecting logical faults in combinational networks, or identifying vowels/consonants while trying to decode some cryptograms, or determining differential errors in digital-to-analogue convertors, or finding the radius of nonsingularity of a matrix considered, respectively, in [Barahona and Mahjoub 1986] (cf §3.1) and in the papers listed below.

S.G. Papaioannou (1977). Optimal test generation in combinational networks by pseudo-boolean programming. *IEEE Trans. Comput.* C-26, 553–560.
C. Moler and D. Morrison (1983). Singular value analysis of cryptograms. *American Math. Monthly* 90, 78–87.
J. Nešetřil and S. Poljak (1986). A remark on max-cut problem with an application to digital-analogue convertors. *Oper. Res. Lett.* 4, 289–291.
S. Poljak and J. Rohn (1993). Checking robust nonsingularity is NP-hard. *Math. of Control, Signal & Systems* 6, 1–9.

A problem in quantum mechanics (the representability problem for density matrices of order 2) is directly relevant to UBQP and the associated polytope $Q(K_n)$ (cf. §3.2 for a definition of this polytope). There is a large literature on this topic, including the following two basic papers.

M.L. Yoseloff and H.W. Kuhn (1969). Combinatorial approach to the N-representability of p-density matrices. *J. Math. Phys.* 10, 703–706.
W.B. MacRae and E.R. Davidson (1972). Linear inequalities for density matrices II. *J. Math. Phys.* 13, 1527–1538.

The problem of determining linear conditions that are satisfied by the joint probabilities $\mu_{ij} := \mu(A_i \cap A_j)$ of a set of n events in a probability space is also directly relevant to the polytope $Q(K_n)$ (as $Q(K_n)$ consists precisely of such vectors μ). This question arises in various settings, such as the study of pair distributions of particles in lattice sites (cf. [MacRae and Davidson 1972]) or the Boole problem.

S. Kounias and J. Marin (1976). Best linear Bonferroni bounds. *SIAM J. Appl. Math.* 30, 307–323.
I. Pitowsky (1986). The range of quantum probability. *J. Math. Phys.* 27, 1556–1565.

The Boole problem is considered in these papers as well as in [Pitowsky 1991] (cf. §3.2).

P. Assouad (1979/80). Plongements isométriques dans L^1: aspect analytique. *Séminaire d'Initiation à l'Analyse (G. Choquet, H. Fakhoury, J. Saint-Raymond)* 14, Université Paris VI.

It is explained there how the analogue problem in the context of cuts leads to the study of ℓ_1-embeddable metrics. A class of valid inequalities for cuts turns out to be of special interest for the study of ℓ_1-metrics, as well as for the notion of Delaunay polytopes in lattices; they are the hypermetric inequalities.

M. Deza, V.P. Grishukhin, and M. Laurent (1995). Hypermetrics in geometry of numbers. W. Cook, L. Lovász, and P.D. Seymour (eds.). *Combinatorial Optimization,* DIMACS Ser. in Discr. Math. and Theoretical Comput. Sci. 20, AMS, 1–109.

This survey treats hypermetric inequalities in detail.

Further bibliography on the above applications can be found in [Deza and Laurent 1994,1997] (cited in §1.2).

2 Complexity Results

The max-cut problem belongs to the most basic combinatorial problems; its complexity was already studied in the fundamental paper of [Karp 1972] (mentioned in §2.1). In what follows, we refer for short to 'max-cut' when the problem applies to a graph with arbitrary edge weights and to 'simple max-cut' when all edge weights are assumed to be equal to 1, which corresponds then to the problem of finding a cut of maximum cardinality. The complexity results apply to the problem in its decision version.

2.1 Hard instances

R.M. Karp (1972). Reducibility among combinatorial problems. R.E. Miller and J.W. Thatcher (eds.). *Complexity of Computer Computations*, Plenum Press, New York, 85–103.

It is shown in this paper that max-cut is NP-complete (by a simple reduction from the partition problem, for weights of the form $w_{ij} := b_i b_j$ on the complete graph K_n where $b_1, \ldots, b_n$ are integers).

M.R. Garey, D.S. Johnson, and L. Stockmeyer (1976). Some simplified NP-complete graph problems. *Theoretical Comput. Sci.* 1, 237–267.

The authors show that, in fact, the unweighted problem remains NP-complete.

M. Yannakakis (1978). Node- and edge-deletion NP-complete problems. *Proc. 10th Annual ACM Symp. Theory of Comput.*, 253–264.
F. Barahona (1982). On the computational complexity of Ising spin glass models. *J. Physics A Math. Gen.* 15, 3241–3253.

Moreover, these authors show, respectively, that simple max-cut is still NP-complete when restricted to cubic graphs or to those graphs having a node whose deletion results in a planar graph. The latter paper also demonstrates NP-completeness of max-cut in a three-dimensional grid with two levels in one direction and with edge weights $0, \pm 1$.

D.S. Johnson (1985). The NP-completeness column: an ongoing guide. *J. Algorithms* 6, 434–451.

This paper gives an update on the complexity status of max-cut.

H.L. Bodlaender and K. Jansen (1994). On the complexity of the maximum cut problem. P. Enjalbert, E.W. Mayr and K.W. Wagner (eds) *Proc. 11th Annual Symp. Theoretical Aspects of Comput. Sci.*, Lecture Notes Comput. Sci. 775, Springer-Verlag, Berlin, 769–780.

The authors solve some of the open cases posed in Johnson's paper. They show, in particular, that simple max-cut is NP-complete for chordal graphs, tripartite graphs, and complements of bipartite graphs.

S. Arora, C. Lund, R. Motwani, M. Sudan, and M. Szegedy (1992). Proof verification and hardness of approximation problems. *Proc. 33rd Annual IEEE Symp. Found. Comput. Sci.*, 14–23.

Approximating max-cut is also hard. These authors show that there is a constant

$0 < \epsilon < 1$ for which no (polynomial time) ϵ-approximation algorithm exists for max-cut unless P=NP. Moreover, in [Ngoc and Tuza 1993] (cf. §4.1) it is shown that deciding whether there exists a cut of cardinality at least $m(\frac{1}{2} + \epsilon)$ is an NP-complete problem for any $0 < \epsilon < \frac{1}{2}$ ($\frac{m}{2}$ is an obvious lower bound for the maximum cardinality of a cut if m is the total number of edges).

2.2 Polynomial instances

On the other hand, there exist several classes of graphs for which max-cut can be solved in polynomial time. A first fundamental such instance is the class of planar graphs.

G.I. Orlova and Y.G. Dorfman (1972). Finding the maximum cut in a graph. *Engrg. Cybernetics* 10, 502–506.
F. Hadlock (1975). Finding a maximum cut of a planar graph in polynomial time. *SIAM J. Comput.* 4, 221–225.

These authors show (independently) that max-cut in a planar graph can be solved in polynomial time (for arbitrary edge weights) by a reduction to the Chinese postman problem in the dual graph.

J. Edmonds and E.L. Johnson (1973). Matching, Euler tours and the Chinese postman. *Math. Program.* 5, 88–124.

In this basic paper the authors present a polynomial-time algorithm for solving the Chinese postman problem using matching theory.

F. Barahona (1990). Planar multiflows, max cut, and the Chinese postman problem. W. Cook and P.D. Seymour (eds.). *Polyhedral Combinatorics*, DIMACS Ser. in Discr. Math. and Theoretical Comput. Sci. Ser. in Discr. Math. and Theoretical Comput. Sci. 1, AMS, New York, 189–202.

The author gives an algorithm with running time $O(n^{3/2} \log n)$ for solving max-cut in a planar graph with n nodes.

problem F. Barahona (1983). The max-cut problem on graphs not contractible to K_5. *Oper. Res. Lett.* 2, 107–111.
F. Barahona (1981). *Balancing signed graphs of fixed genus in polynomial time*, Technical Report, Departamento de Matematicas, Universidad de Chile, Santiago.

As extensions of the planar case it is shown, respectively, that max-cut can be solved in polynomial time for graphs with no K_5-minor, and for graphs of fixed genus with weights $0, \pm 1$.

B. Monien and I.H. Sudborough (1981). Bandwidth-constrained NP-complete problems. *Proc. 13th Annual ACM Symp. Theory of Comput.*, 207–217.

Dynamic programming is used there to solve simple max-cut over graphs with fixed bandwidth in polynomial time.

T.V. Wimer (1987). *Linear Algorithms on k-Terminal Graphs*, PhD thesis, Department of Computer Science, Clemson University.

E. Wanke (to appear). k-NLC graphs and polynomial algorithms. *Ann. Discr. Math.*.

In Wimer's thesis it is shown that simple max-cut can be solved in linear time for graphs with bounded tree-width. The authors in [Bodlaender and Jansen 1994] (cf. §2.1) give an algorithm for simple max-cut over cographs (graphs with no induced P_4) with running time $O(n^2)$. Wanke presents a common generalization of cographs and graphs with bounded tree-width for which simple max-cut remains polynomial.

M. Grötschel and G.L. Nemhauser (1984). A polynomial algorithm for the max-cut problem on graphs without long odd cycles. *Math. Program.* 29, 28–40.
C. Arbib (1987/88). A polynomial characterization of some graph partitioning problems. *Inform. Process. Lett.* 26, 223–230.

Further polynomial cases are treated there, with line-graphs being considered in the latter paper. Max-cut is also polynomially solvable over the class of weakly bipartite graphs when restricted to nonnegative edge weights (see §3.3).

2.3 Local search

D.S. Johnson, C.H. Papadimitriou and M. Yannakakis (1988). How easy is local search ? *J. Comput. Syst. Sci.* 37, 79–100.

The authors raise the question of determining the complexity of finding locally optimum solutions for hard optimization problems. They introduce the complexity class PLS (polynomial-time local search) and show that it contains *complete* problems. In particular, they show that the graph bisection problem (which asks for a partition of the nodes of a graph into two sets of equal sizes so as to minimize the total weight of the edges cut by the partition) is PLS-complete under the Kernighan-Lin neighborhood.

A.A. Schäffer and M. Yannakakis (1991). Simple local search problems that are hard to solve. *SIAM J. Comput.* 20, 56–87.

This paper shows that the graph bisection problem remains PLS-complete under the simpler SWAP heuristic (which consists of exchanging two nodes beween the two sides of the partition). It also shows that max-cut is PLS-complete with the FLIP neighbourhood (consisting now of moving a single vertex to the opposite class of the partition). Hence, finding a cut which is locally optimum with respect to flipping may require an exponential number of steps in the worst case.

S. Poljak (1995). Integer linear programs and local search for max-cut. *SIAM J. Comput.* 24, 822–839.

On the other hand, it is proved there that the flip local search terminates in $O(n^2)$ steps for a weighted cubic graph on n nodes.

3 Polyhedral Results

3.1 The cut polytope

The classical polyhedral approach for the max-cut problem leads to the study of the cut polytope $P_C(G)$, which is defined as the convex hull of the incidence vectors of the cuts in G.

F. Barahona and A.R. Mahjoub (1986). On the cut polytope. *Math. Program.* 36, 157–173.
F. Barahona (1993). On cuts and matchings in planar graphs. *Math. Program.* 60, 53–68.

The cut polytope has been studied, in particular, in these papers and in [Barahona 1983] (cf. §2.2). One of the main results is a complete linear description of $P_C(G)$ when G has no K_5-minor. One way to derive it is from the 'sums of circuits property'.

P.D. Seymour (1981). Matroids and multicommodity flows. *European J. Combinatorics* 2, 257–290.

The sums of circuits property is studied there.

P.D. Seymour (1979). Sums of circuits. J.A. Bondy and U.S.R. Murty (eds.). *Graph Theory and Related Topics*, Academic Press, New York, 341–355.

This paper gives the linear description of the Eulerian subgraph cone, which permits to derive the description of $P_C(G)$ for a planar graph G.

F. Barahona and M. Grötschel (1986). On the cycle polytope of a binary matroid. *J. Combin. Theory B* 40, 40–62.
M. Grötschel and K. Truemper (1989). Decomposition and optimization over cycles in binary matroids. *J. Combin. Theory B* 46, 306–337.
M. Grötschel and K. Truemper (1989). Master polytopes for cycles of binary matroids. *Linear Algebra Appl.* 114/115, 523–540.

These authors carry out the study of the cut polytope in the more general setting of binary matroids.

M. Deza and M. Laurent (1992). Facets for the cut cone I, II. *Math. Program.* 56, 121–160 and 161–188.
M. Deza and M. Laurent (1992). New results on facets of the cut cone. *J. Comb., Inf. & Syst. Sci.* 17, 19–38.
S. Poljak and D. Turzik (1992). Max-cut in circulant graphs. *Discr. Math.* 108, 379–392.
C.C. De Souza and M. Laurent (1995). Some new classes of facets for the equicut polytope. *Discr. Appl. Math.* 62, 167–191.

These papers provide an intensive study of the facial structure of the cut polytope in the case of the complete graph. Further references can be found in [Deza and Laurent 1994,1997] (cf. §1.2). Large classes of facets for $P_C(K_n)$ are described in the above papers. Among them, the class of hypermetric inequalities is of special interest, in particular, in view of its many applications (see §1.3). A complete linear description of the polytope $P_C(K_n)$ is known for $n \leq 8$; it is reported in [Deza and Laurent 1992] for $n \leq 7$. The computation for $n = 8$ has been carried out by T. Christof and G. Reinelt, whose results are available on the following WWW site:
http://www.iwr.uni-heidelberg.de/iwr/comopt/soft/SMAPO/SMAPO.html.

A.M.H. Gerards (1985). Testing the bicycle odd wheel inequalities for the bipartite subgraph polytope. *Math. Oper. Res.* 10, 359–360.

The author gives a polynomial algorithm that solves the separation problem for the class of bicycle odd wheel inequalities. Separation heuristics for further classes of inequalities valid for the cut polytope are presented in the theses of [De Simone 1992] and [De Souza 1993] (cf. §1.2).

M. Deza, M. Laurent, and S. Poljak (1993). The cut cone III: on the role of triangle facets. *Graphs & Combin.* 9, 135–152.
A. Deza, M. Deza and K. Fukuda (1996). On skeletons, diameters and volumes of metric polyhedra. M. Deza, R. Euler and Y. Manoussakis (eds.). *Combinatorics and Computer Science*, Lecture Notes Comput. Sci. 1120, Springer-Verlag, Berlin, 112–128.
These papers study the geometry of $P_C(K_n)$. They consider, respectively, properties of $P_C(K_n)$ with respect to its linear relaxation by triangle inequalities and various geometric properties (adjacency, diameter, volumes, etc.) of metric polyhedra of K_n for small values of n.

3.2 The boolean quadric polytope

Another polytope which has received a lot of attention in the literature is the polytope $Q(G)$, defined for graph $G = (V, E)$ as the convex hull of the vectors $(x, y) \in \{0,1\}^V \times \{0,1\}^E$ with $y_{ij} = x_i x_j$ for all edges $ij \in E$. This polytope permits to model the UBQP problem (P2).

M.W. Padberg (1989). The boolean quadric polytope: some characteristics, facets and relatives. *Math. Program.* 45, 139–172.
I. Pitowsky (1991). Correlation polytopes: their geometry and complexity. *Math. Program.* 50, 395–414.
E. Boros and P.L. Hammer (1993). Cut polytopes, boolean quadratic polytopes and nonnegative quadratic pseudo-boolean functions. *Math. Oper. Res.* 18, 245–253.
These authors study the polytope $Q(G)$, under the names of boolean quadric polytope and correlation polytope.

C. De Simone (1989/90). The cut polytope and the boolean quadric polytope. *Discr. Math.* 79, 71–75.
This paper exposes the well-known connection existing between the polytope $Q(G)$ and the cut polytope $P_C(G + u)$. Therefore, any result concerning the polytope $Q(G)$ has a direct counterpart for the cut polytope.

P.L. Hammer, P. Hansen, and B. Simeone (1984). Roof duality, complementation and persistency in quadratic 0-1 optimization. *Math. Program.* 28, 121–155.
E. Boros, Y. Crama, and P.L. Hammer (1990). Upper-bounds for quadratic 0-1 maximization. *Oper. Res. Lett.* 9, 73–79.
The authors formulate some lower bounds for the problem (P2). Among other formulations, these bounds can be obtained by optimizing over successive linear relaxations of the boolean quadric polytope.

P.L. Hammer, P. Hansen, and B. Simeone (1992). Chvátal cuts and odd cycle inequalities in quadratic 0-1 optimization. *SIAM J. Discr. Math.* 5, 163–177.
Properties of some of the above mentioned linear relaxations are studied there.

3.3 The bipartite subgraph polytope

When all edge weights are nonnegative, the max-cut problem can be reformulated as the problem of finding a bipartite subgraph of maximum weight and, thus, as an optimization problem over the bipartite subgraph polytope $P_B(G)$.

M. Grötschel and W.R. Pulleyblank (1981). Weakly bipartite graphs and the max-cut problem. *Oper. Res. Lett.* 1, 23–27.
F. Barahona, M. Grötschel, and A.R. Mahjoub (1985). Facets of the bipartite subgraph polytope. *Math. Oper. Res.* 10, 340–358.
J. Fonlupt, A.R. Mahjoub and J.P. Uhry (1992). Compositions in the bipartite subgraph polytope. *Discr. Math.* 105, 73–91.

These papers study the bipartite subgraph polytope $P_B(G)$; they contain, in particular, results about facets with respect to operations and graph compositions.

An obvious linear relaxation for the polytope $P_B(G)$ is provided by the inequalities (i) $0 \leq x \leq 1$ and (ii) $x(C) \leq |C| - 1$ for C odd cycle in G. The graph G is said to be *weakly bipartite* if the inequalities (i) and (ii) constitute the full linear description of $P_B(G)$. In [Grötschel and Pulleyblank 1981] it is shown that the separation problem for the system of inequalities (i), (ii) can be solved in polynomial time. Therefore, using the ellipsoid method, one can find a maximum cut in a weakly bipartite graph with nonnegative edge weights in polynomial time.

Several classes of weakly bipartite graphs are known, including planar graphs and graphs with no K_5-minor (as shown in [Fonlupt, Mahjoub and Uhry 1992] or [Seymour 1981] (cf. §3.1)).

F. Barahona (1983). On some weakly bipartite graphs. *Oper. Res. Lett.* 2, 239–242.
A.M.H. Gerards (1992). *Odd paths and odd circuits in planar graphs with two odd faces*, Report BS-R9218, CWI, Amsterdam.

Two more classes of weakly bipartite graphs are presented there; namely, those graphs having two nodes covering all odd cycles and those graphs G having a node u such that $G\backslash u$ has at most two odd faces.

A.M.H. Gerards. Odd paths and odd circuits in graphs with no odd-K_4. Unpublished.

As a common extension of these two classes, this author shows that the graphs G having a node u such that $G\backslash u$ contains no odd K_4 are weakly bipartite.

P.D. Seymour (1977). The matroids with the max-flow min-cut property. *J. Combin. Theory B* 23, 189–222.

There is stated a conjecture concerning a characterization of the weakly bipartite graphs. This conjecture is formulated, in fact, in the more general setting of binary matroids. However, it remains unsolved for graphs. This is certainly one of the current major open problems in the field of combinatorial optimization.

4 Approximation Results

4.1 Lower bounds

As max-cut is NP-hard it is of interest to find good lower bounds for the maximum weight of a cut as well as efficient algorithms for constructing cuts achieving the guaranteed lower bounds. (In what follows $w(G)$ denotes the sum of all edge weights in graph G, m is the number of edges in G and n its number of nodes.)

P. Erdös (1967). On bipartite subgraphs of graphs. *Math. Lapok* 18, 283–288.
C.S. Edwards (1973). Some extremal properties of bipartite graphs. *Canadian J. Math.* 25, 475–485.

These are early papers on this question; they establish the lower bounds: $\frac{w(G)}{2}(1+\frac{1}{n-1})$ (n even) and $\frac{m}{2}+\frac{n-1}{4}$ (in the unweighted case), thus improving on the obvious lower bound $\frac{w(G)}{2}$. Their proofs are nonconstructive.

T. Hofmeister and H. Lefmann (1996). A combinatorial design approach to MAXCUT. C. Puech und R. Reischuk (eds.). *Proc. 13th Annual Symp. Theoretical Aspects of Comput. Sci.*, Lecture Notes Comput. Sci. 1046, Springer-Verlag, Berlin, 441–452.
N.V. Ngoc and Z. Tuza (1993). Linear-time approximation algorithms for the max cut problem. *Combin., Probab. & Comput.* 2, 201–210.

These authors propose linear-time algorithms for finding cuts achieving the above mentioned bounds.

S. Poljak and D. Turzik (1986). A polynomial time heuristic for certain subgraph optimization problems with guaranteed worst case bound. *Discr. Math.* 58, 99–104.

There is given an extension of Edwards' bound to the weighted case and a polynomial algorithm. Better bounds have been obtained for special classes of graphs; e.g., $\frac{m}{2}+\frac{3}{8}(n-1)$ for triangle-free graphs (in [Ngoc and Tuza 1993]), $\frac{3}{4}m-\frac{1}{4}$ for graphs with max. degree 3, or $\frac{4}{5}m$ for triangle-free graphs with max. degree 3 in the 7th and 3rd papers of the following list of works in the area.

G. Hopkins and W. Staton (1982). Maximal bipartite subgraphs. *Ars Combin.* 13, 223–226.
S.C. Locke (1982). Maximum k-colorable subgraphs. *J. Graph Theory* 6, 123–132.
J.A. Bondy and S.C. Locke (1986). Largest bipartite subgraphs in triangle-free graphs with maximum degree three. *J. Graph Theory* 10, 477–504.
P. Erdös, R. Faudree, J. Pach, and J. Spencer (1988). How to make a graph bipartite ? *J. Combin. Theory B* 45, 86–98.
S.C. Locke (1990). A note on bipartite subgraphs of triangle-free regular graphs. *J. Graph Theory* 14, 181–185.
J.B. Shearer (1990). A note on bipartite subgraphs of triangle-free graphs. *Random Struct. & Alg.* 3, 223–226.
S. Bylka, A. Idzik, and J. Komar (1993). *Bipartite subgraphs of graphs with maximum degree three*, Report 727, Institute of Computer Science, Polish Academy of Sciences.
O. Zýka (1990). On the bipartite density of regular graphs with large girth. *J. Graph Theory* 14, 631–634.

S. Poljak and Z. Tuza (1994). Bipartite subgraphs of triangle-free graphs. *SIAM J. Discr. Math.* 7, 307–313.

An ϵ-approximation algorithm is a polynomial time algorithm that returns a solution of value at least ϵ times the optimum value; this definition applies for max-cut when all weights are assumed to be nonnegative.

S. Sahni and T. Gonzalez (1976). P-complete approximations problems. *J. ACM* 23, 555–565.

This paper gives the first approximation algorithm for max-cut, which is a 1/2-approximation algorithm. As $w(G)$ is an obvious upper bound for the optimum value, some of the results mentioned above yield approximation algorithms whose performance guarantee (e.g., $\frac{1}{2} + \frac{1}{2(n-1)}$, $\frac{1}{2} + \frac{n-1}{4m}$) is a slight improvement over $\frac{1}{2}$.

M.X. Goemans and D.P. Williamson (1994). 0.878-approximation algorithms for MAX CUT and MAX 2SAT. *Proc. 26th Annual ACM Symp. Theory of Comput.*, 422–431. (Updated version as: Improved approximation algorithms for maximum cut and satisfiability problems using semidefinite programming. *J. ACM* 42, 1115–1145, 1995.)

These authors achieve a substantial improvement over $\frac{1}{2}$; they show indeed a performance guarantee $\alpha - \eta$ for any $\eta > 0$, where $\alpha > 0.87856$.

H. Karloff (1996). How good is the Goemans-Williamson MAX CUT algorithm ? *Proc. 28th Annual ACM Symp. Theory of Comput.*, 427–434.

The author analyzes the exact performance guarantee of the Goemans-Williamson algorithm as well as its behaviour under adding linear constraints.

W. Fernandez de la Vega (1996). Max-cut has a randomized approximation scheme in dense graphs. *Random Struct. & Alg.* 8, 187–198.

A randomized approximation scheme for dense graphs is given there, which has a performance guarantee arbitrarily close to 1.

4.2 Upper bounds

Three types of upper bounds for max-cut have been considered in the literature: the polyhedral bound $\sigma(G, w)$, the eigenvalue bound $\varphi(G, w)$ and the positive semidefinite bound $\psi(G, w)$.

S. Poljak (1991). Polyhedral and eigenvalue approximations of the max-cut problem. D. Miklós et al. (eds.). *Sets, Graphs and Numbers*, Colloq. Math. Societatis János Bolyai 60, North-Holland, Amsterdam, 569–581.
S. Poljak and Z. Tuza (1994). The expected relative error of the polyhedral relative approximation of the max-cut problem. *Oper. Res. Lett.* 16, 191–198.

These authors study the bound $\sigma(G, w)$, which is obtained by optimizing over a linear relaxation of the cut polytope (defined by projections of triangle inequalities).

B. Mohar and S. Poljak (1990). Eigenvalues and the max-cut problem. *Czec. Math. J.* 40, 343–352.
C. Delorme and S. Poljak (1993). Laplacian eigenvalues and the maximum cut

problem. *Math. Program.* 62, 557–574.
C. Delorme and S. Poljak (1993). Combinatorial properties and the complexity of a max-cut approximation. *European J. Combinatorics* 14, 313–333.

The bound $\varphi(G, w)$, which is defined in terms of the maximum eigenvalue of the Laplacian matrix of the weighted graph G, is introduced in the first paper and further studied in the other two papers.

The third bound $\psi(G, w)$ is obtained by optimizing over a nonpolyhedral relaxation of the cut polytope (consisting of the positive semidefinite matrices with diagonal entries 1) (recall formulation (P1)). It forms the basis of the approximation algorithm of [Goemans and Williamson 1994].

S. Poljak and F. Rendl (1995). Nonpolyhedral relaxations of graph-bisection problems. *SIAM J. Optim.* 5, 467–487.

These authors show, as an application of cone LP-duality, that the two bounds $\varphi(G, w)$ and $\psi(G, w)$ in fact coincide.

S. Poljak and H. Wolkowicz (1995). Convex relaxations of 0-1 quadratic programming. *Math. Oper. Res.* 20, 550–561.
S. Poljak, F. Rendl, and H. Wolkowicz (1995). A recipe for semidefinite relaxation for (0,1) quadratic programming. *J. Global Optim.* 7, 51–73.

These papers present further equivalent formulations for the quantity $\varphi(G, w)$.

M. Laurent and S. Poljak (1995). On a positive semidefinite relaxation of the cut polytope. *Linear Algebra Appl.* 223/224, 439–461.
M. Laurent and S. Poljak (1996). On the facial structure of the set of correlation matrices. *SIAM J. Matrix Analysis & Appl.* 17, 530–547.

In these papers are investigated geometric properties of this semidefinite relaxation of the cut polytope.

M. Laurent, S. Poljak, and F. Rendl (1997). Connections between semidefinite relaxations of the max-cut and stable set problems. *Math. Program.*, 77, 225–246.
M. Laurent (1997). The real positive semidefinite completion problem for series-parallel graphs. *Linear Algebra Appl.*, 252, 347–366.
M. Laurent (1997). A tour d'horizon on positive semidefinite and Euclidean distance matrix completion problems. P. Pardalos and H. Wolkowicz (eds.). *Topics in Semidefinite and Interior-Points Methods, The Fields Inst. Res. Math. Sci., Commun. Ser.*, Providence, to appear.

These papers mention links with other known semidefinite relaxations for maximum stable set problem and with matrix completion problems in linear algebra.

A. Kamath and N. Karmarkar (1992). A continuous method for computing bounds in integer quadratic optimization problems. *J. Global Optim.* 2, 229–241.

The authors investigate a continuous approach for obtaining further bounds for max-cut.

5 Computational Results

We present here papers dealing with computational studies on max-cut or UBQP problems in the recent years. See, e.g., [Hansen 1979] (cf. §1.1) for information about earlier results. The methods used can be roughly divided into the following groups: branch and bound methods, linear programming based methods (branch and cut algorithms), eigenvalue based methods and approaches via semidefinite programming. This subdivision matches, in fact, quite well the chronological developments.

M.W. Carter (1984). The indefinite zero-one quadratic problem. *Discr. Appl. Math.* 7, 23–44.
P.M. Pardalos and G.P. Rodgers (1990). Computational aspects of a branch and bound algorithm for quadratic zero-one programming. *Computing* 45, 131–144.
B. Kalantari and A. Bagchi (1990). An algorithm for quadratic zero-ones programs. *Naval Res. Log. Quart.* 37, 527–538.

These authors, as well as [Rodgers 1989] (cf. §1.2), propose algorithms based on branch and bounds (with possibly further techniques such as preprocessing for fixing variables). Roughly, these methods work well for sparse problems or under some sort of diagonal dominance, but they seem to fail on dense instances with more than 50 nodes.

Cutting plane algorithms constitute a widely used technique in Combinatorial Optimization for attacking hard problems such as max-cut; we refer to Chapter 4 in this volume for a detailed treatment on these techniques.

A.C. Williams (1989). *Quadratic 0-1 programming using the roof dual with computational results*, RUTCOR Research Report #8-85, The State University of New Jersey, New Brunswick.
F. Barahona, M. Jünger, and G. Reinelt (1989). Experiments in quadratic 0-1 programming. *Math. Program.* 44, 127–137.
F. Barahona and H. Titan (1991). Max mean cuts and max cuts. In *Combinatorial Optimization in Science and Technology*, DIMACS Technical Report 91-18, Rutgers University, New Brunswick, 30–45.
C. De Simone and G. Rinaldi (1994). A cutting plane algorithm for the max-cut problem. *Opt. Methods & Software* 3, 195–214.
C. De Simone, M. Diehl, M. Jünger, P. Mutzel, G. Reinelt, and G. Rinaldi (1995). Exact ground states of Ising spin glasses: new experimental results with a branch-and-cut algorithm. *J. Statistical Phys.* 80 (1-2), 487–496.
C. De Simone, M. Diehl, M. Jünger, P. Mutzel, G. Reinelt, and G. Rinaldi (1996). Exact ground states of two-dimensional $\pm J$ Ising spin glasses. *J. Statistical Phys.* 84 (5-6), 1363–1371.

In these papers, as well as in [Barahona et al. 1988] (cf. §1.3), cutting plane algorithms have been tested on a variety of instances of max-cut and UBQP problems. With one exception these papers use essentially cycle inequalities (projections of triangle inequalities) in the separation phase. Several of them study instances arising from spin glasses in statistical physics and mainly the case when the graph is a toroidal grid with a universal node. The largest graphs considered have up to 10,000 nodes (solved at optimality in [De Simone et al. 1995]); they are typically rather sparse. On the other hand, the authors in [De Simone and Rinaldi 1994] solve instances on dense graphs with up to 25 nodes, using further classes of inequalities (hypermetric inequalities).

S. Poljak and F. Rendl (1994). Node and edge relaxations of the max-cut problem. *Computing* 52, 123–137.
S. Poljak and F. Rendl (1995). Solving the max-cut problem using eigenvalues. *Discr. Appl. Math.* 62, 249–278.

Recently, algorithms have been designed, based on the upper bounds for max-cut obtained by any of the following two dual methods: minimization of the largest eigenvalue of the Laplacian matrix (with diagonal shift) and maximization over the positive semidefinite relaxation of the cut polytope (cf. §4.2). The above papers report on computational experiments on various instances; e.g., dense graphs up to 1000 nodes, or sparse graphs up to 50,000 nodes.

C. Helmberg, S. Poljak, F. Rendl, and H. Wolkowicz (1995). Combining semidefinite and polyhedral relaxations for integer programs. E. Balas and J. Clausen (eds.). *Integer Programming and Combinatorial Optimization*, Lecture Notes Comput. Sci. 920, Springer-Verlag, Berlin, 124–134.
C. Helmberg and F. Rendl (1995). Solving quadratic (0,1)-problems by semidefinite programs and cutting planes. Preprint. To appear in *Math. Program.*

These papers, as well as [Helmberg 1994] (cf. §1.2), investigate algorithms combining the semidefinite approach with cutting planes.

6 Graph Partitioning Problems

Graph partitioning problems are variations of the max-cut problem in which one considers partitions of the node set into more than two sets and/or one requires additional conditions on the cardinality of the partition classes. Among them, the *bisection problem* (or *equicut problem*) asks for a cut $\delta_G(S)$ of minimum weight for which both sets S and $V \backslash S$ in the partition have sizes that differ by at most one. The *max k-cut problem* asks for a partition of the node set into k nonempty parts so as to maximize the total weight of the edges cut by the partition. One may further require that the sets in the partition have prescribed cardinalities.

T. Lengauer (1990). *Combinatorial Algorithms for Integrated Circuit Layout.* John Wiley & Sons, Chichester.

A good overview on graph partitioning problems and their applications can be found in Chap. 6 of this book.

T.N. Bui, S. Chaudhuri, T. Leighton, and M. Sipser (1987). Graph bisection algorithms with good average case behavior. *Combinatorica* 7, 171–191.
R.M. MacGregor. *On Partitioning a Graph: A Theoretical and Empirical Method.* PhD thesis, Electronics Research Lab., University of California, Berkeley.

The various graph partitioning problems are all NP-hard. For instance, the graph bisection problem is shown to be NP-hard in [Garey, Johnson and Stockmeyer 1976] (cf. §2.1). The authors of the first above mentioned paper show that the graph bisection problem remains NP-hard in the unweighted case when restricted to regular graphs of fixed degree $d \geq 3$. On the other hand, the problem becomes polynomial for line-graphs (cf. [Arbib 1987/88] in §2.2). Moreover, MacGregor shows that the graph bisection problem is also polynomial for trees (or partial k-trees). However, the complexity for planar graphs is not known.

O. Goldschmidt and D.S. Hochbaum (1988). A polynomial algorithm for the k-cut problem. *Proc. 29th Annual IEEE Symp. Found. Comput. Sci.*, 444--451. (Updated version in *Math. Oper. Res.* 19, 24–37, 1994.)
D.S. Hochbaum and D.B. Shmoys (1985). An $O(|V|^2)$ algorithm for the planar 3-cut problem. *SIAM J. Algebraic. Disc. Meth.* 6, 707–712.

The minimum k-cut problem minimum weight of cut edges) is NP-hard for arbitrary k, but it can be solved in polynomial time for any fixed k by the algorithm given in the paper by Goldschmidt and Hochbaum. Hochbaum and Shmoys give a faster algorithm for planar graphs in the case $k = 3$.

E. Dahlhaus, D.S. Johnson, C.H. Papadimitriou, P.D. Seymour, and M. Yannakakis (1992). The complexity of the multiway cuts. *Proc. 24th Annual ACM Symp. Theory of Comput.*, 241–251.

These authors show that the minimum k-cut problem becomes NP-hard for any fixed $k \geq 3$ if one asks for a partition whose classes contain precribed terminal nodes (even in the unweighted case) (this problem is also called the *k-way cut problem*); the case $k = 2$ is the well-known polynomially solvable minimum st-cut problem. They also show that the problem with arbitrary k remains NP-hard over planar graphs, and that the problem can be solved in polynomial time for any fixed $k \geq 3$ over planar graphs. trees and 2-trees

O. Goldschmidt and D.S. Hochbaum (1990). Asymptotically optimal linear algorithm for the minimum k-cut in a random graph. *SIAM J. Discr. Math.* 3, 58–73.
P.L. Erdös and L.A. Székely (1994). On weighted multiway cuts in trees. *Math. Program.* 65, 93–105.

The first paper gives a fast algorithm for the k-cut problem in random graphs and the second paper considers variations of the multiway cut problem (involving colouring).

H. Saran and V.V. Vazirani (1991). Finding k-cuts within twice the optimal. *Proc. 32nd Annual IEEE Symp. Found. Comput. Sci.*, 743–751.

Approximation algorithms for the k-way cut and the minimum k-cut problems with a performance guarantee of $2 - \frac{2}{k}$ are given in the above paper and in [Dahlhaus et al. 1992]. An $(1 - \frac{1}{k})$-approximation algorithm for max k-cut is presented in [Sahni and Gonzalez 1976] (cf. §4.1).

A. Frieze and M. Jerrum (1995). Improved approximation algorithms for MAX k-CUT and MAX BISECTION. E. Balas and J. Clausen (eds.). *Integer Programming and Combinatorial Optimization*, Lecture Notes Comput. Sci. 920, Springer-Verlag, Berlin, 1–13.

These authors give further approximation algorithms for the bisection and max k-

cut problems, that are based on positive semidefinite programming, thus extending the results for max-cut presented in §4.2.

Some bounds are proposed in the literature for the various graph partitioning problems, that are based mainly on eigenvalue techniques and continuous optimization; see, in particular, [Poljak and Rendl 1995], [Poljak, Rendl and Wolkowicz 1995] in §4.2 and the following papers:

W.E. Donath and A.J. Hoffman (1973). Lower bounds for the partitioning of graphs. *IBM J. Res. Develop.* 17, 420–425.
R.B. Boppana (1987). Eigenvalues and graph bisection: an average case analysis. *Proc. 28th Annual IEEE Symp. Found. Comput. Sci.*, 280–285.
F. Rendl and H. Wolkowicz (1995). A projection technique for partitioning the nodes of a graph. *Ann. Oper. Res.* 58, 155–179.

Further combinatorial lower bounds for max k-cut can be found in [Locke 1982], [Ngoc and Tuza 1993], [Hofmeister and Lefmann 1996] (cf. §4.1).

Y. Wakabayashi (1986). *Aggregation of Binary Relations: Algorithmic and Polyhedral Investigations*, PhD thesis, University of Augsburg, Germany.
M. Grötschel and Y. Wakabayashi (1990). Facets of the clique partitioning polytope. *Math. Program.* 47, 367–387.
M. Conforti, M.R. Rao, and A. Sassano (1990). The equipartition polytope. *Math. Program.* 49, 49–70 and 71–90.
M. Deza, M. Grötschel, and M. Laurent (1992). Clique-web facets for multicut polytopes. *Math. Oper. Res.* 17, 981–1000.
S. Chopra and M.R. Rao (1991). On the multiway cut polyhedron. *Networks* 21, 51–89.
S. Chopra and M.R. Rao (1993). The partition problem. *Math. Program.* 59, 87–115.
S. Chopra and M.R. Rao (1995). Facets of the k-partition polytope. *Discr. Appl. Math.* 61, 27–48.
C.E. Ferreira, A. Martin, C.C. De Souza, R. Weismantel, and L.A. Wolsey (1996). Formulations and valid inequalities for the node capacitated graph partitioning problem. *Math. Program.* 74, 247–267.

These papers as well as [De Souza 1993] (cf. §1.2) and [De Souza and Laurent 1995] (cf. §3.1) study the polyhedral approach for graph partitioning.

A. Kamath and N. Karmarkar (1991). A continuous approach to compute upper bounds in quadratic maximization problems with integer constraints. C.A. Floudas and P.M. Pardalos (eds.). *Recent Advances in Global Optimization*, Princeton University Press, 125–140.
J. Falkner, F. Rendl, and H. Wolkowicz (1994). A computational study of graph partitioning. *Math. Program.* 66, 211–239.
S.E. Karisch (1995). *Nonlinear Approaches for Quadratic Assignment and Graph Partition Problems*, PhD thesis, Graz University of Technology.
S.E. Karisch and F. Rendl (1997). Semidefinite programming and graph equipartition. P. Pardalos and H. Wolkowicz (eds.). *Topics in Semidefinite and Interior-Points*

Methods, The Fields Inst. Res. Math. Sci., Commun. Ser., Providence, to appear.

L. Brunetta, M. Conforti, and G. Rinaldi (1997). A branch-and-cut algorithm for the equicut problem. *Math. Program. (B)*, to appear.

Computational studies on graph partitioning problems have been made by the above authors, following the same approaches as for max-cut; namely, continuous, spectral, and polyhedral approaches.

16 Location Problems

Martine Labbé
Université Libre de Bruxelles, Brussels

François V. Louveaux[1]
Facultés Universitaires Notre Dame de la Paix, Namur

CONTENTS

The uncapacitated facility location problem is defined as

$$(UFLP) \ \min \sum_{i=1}^{n}\sum_{j=1}^{m} c_{ij}x_{ij} + \sum_{j=1}^{m} f_j y_j \qquad (Z)$$

$$\text{s.t} \quad \sum_{j=1}^{m} x_{ij} = 1 \qquad i = 1,\ldots,n \qquad (D)$$

$$x_{ij} \le y_i \qquad i = 1,\ldots,n, j = 1,\ldots,m \qquad (B)$$

$$x_{ij} \ge 0 \qquad i = 1,\ldots,n, j = 1,\ldots,m \qquad (N)$$

$$y_j \in \{0,1\} \qquad j = 1,\ldots,m. \qquad (I)$$

[1]This author wishes to thank the NFSR of Belgium for financial support. Part of this author's research has been done while visiting CORE

In this model, n customers must be supplied from a facility (plant, warehouse, retailer shop) where a commodity is made available (produced, stored, sold). There is a discrete number of potential facility sites, which may come out of some selection process or may correspond to the vertices of a graph (as optimal location at the vertices of a graph is a frequent feature in location on networks).

In the UFLP model, the y_j variables correspond to the decision of opening facility j, which is a binary yes-no decision (see constraint (I)) with associated fixed cost f_j, the x_{ij} variables correspond to the fraction of i's demand satisfied from facility j, with c_{ij} the cost of satisfying all of i's demand from j. Constraint (D) represents demand satisfaction for all customers. Constraint (B) states that demand can only be satisfied from open plants. This particular formulation is known as the strong formulation of UFLP.

In the capacitated facility location problem (CFLP), an additional constraint is placed on the total supply available from a single facility:

$$\sum_{i=1}^{n} d_i x_{ij} \leq s_j y_j \tag{C}$$

where d_i is the demand of customer i and s_j the maximum supply at facility j.

In the p-median problem, the number of facilities to be open is fixed:

$$\sum_{j=1}^{m} y_j = p$$

and fixed costs are omitted ($f_j = 0, j = 1, \ldots, m$).

In a covering location model, a customer or demand area is said to be covered if the distance t_{ij} (or transportation cost, or travel time) from an open facility to the customer or demand area is below some given treshold t_i. The covering location model consists of finding a set of locations which minimize the total set-up costs while covering all customers.

$$(CLP) \min \sum_{j=1}^{m} f_j \cdot x_j$$

$$s.t. \sum_{j \in N_i} x_j \geq 1 \quad i = 1, \ldots, n$$

where $N_i = \{j : t_{ij} \leq t_i\}$ is sometimes called the eligible set.

This selective bibliography is mainly concerned with combinatorial location problems, in which a finite number of potential facility locations are considered. Thus, continuous location problems are not covered here. Similarly, papers on location on networks are covered only if their major concern is of combinatorial nature. We have chosen to make a somewhat severe (and perhaps arbitrary) selection as recent surveys on location on networks are available.

1 Books and Surveys

1.1 Books

R.F. Love, J.G. Morris, G.O. Wesolowsky (1988). *Facilities Location. Models & Methods*, North Holland, New York.
A.P. Hurter, J.S. Martinich (1989). *Facility Location and the Theory of Production*, Kluwer Academic Publishers, Boston.
P.B. Mirchandani, R.L. Francis (eds.) (1990). *Discrete Location Theory*, Wiley Interscience, New York.
R.L. Francis, L.F. McGinnis Jr., J.A. White (1992). 2nd edition, *Facility Layout and Location: An Analytical Approach*, Prentice Hall, Englewood Cliffs.
M.S. Daskin (1995). *Network and Discrete Location, Models, Algorithms and Applications*, Wiley Interscience, New York.
Z. Drezner, (ed.) (1995). *Facility Location, A Survey of Applications and Methods*, Springer Series in Operations Research, New York.

1.2 Surveys

Surveys relating to one particular section of this bibliography are included in the corresponding section.

P. Hansen, M. Labbé, D. Peeters, J.-F. Thisse (1987). Single facility location on networks. *Ann. Discr. Math.* 31, 113-146.
P. Hansen, M. Labbé, D. Peeters, J.-F. Thisse (1987). Facility location analysis. *Fundam. Pure Appl. Econ.* 22, 1-70.
T.L. Friesz, T. Miller, R.L. Tobin (1988). Competitive network facility location models: A Survey. *Papers of the Regional Science Assoc.* 65, 47-57.
E. Erkut, S. Neuman (1989). Analytical models for locating undesirable facilities. *European J. Oper. Res.* 40, 275-291.
M.L. Brandeau, S.S. Chiu (1989). An overview of representative problems in location research. *Management Sci.* 35, 645-674.
D. Chhajed, R.L. Francis, T.J. Lowe (1993). Contributions of operations research to location analysis. *Location Sci.* 1, 263-287.
H.A. Eiselt, G. Laporte, J.-F. Thisse (1993). Competitive location models: a framework and bibliography. *Transportation Sci.* 27, 44-54.
A. Ghosh, F. Harche (1993). Location-allocation models in the private sector: progress, problems and prospects. *Location Sci.* 1, 81-106.
F.V. Louveaux (1993). Stochastic location analysis. *Location Sci.* 1, 127-154.
M. Labbé, D. Peeters, J.-F. Thisse (1995). Location on networks. M.O. Ball, T.L. Magnanti, C.L. Monma, G.L. Nemhauser (eds.). *Handbooks in Operations Research and Management Science* 8, North Holland, Amsterdam, 551-624.

1.3 Special issues

J.P. Osleeb, S.J. Ratick (eds.) (1986). Locational decisions: methodology and applications. *Ann. Oper. Res.* 6.
F.V. Louveaux, M. Labbé, J.-F. Thisse (eds.) (1989). Facility location analysis: theory

and applications. *Ann. Oper. Res.* 18.
J.R. Current, D.A. Schilling (eds.) (1991). Special issue on location analysis. *Infor* 29, 65-183.
Z. Drezner (ed.) (1992). Locational decisions. *Ann. Oper. Res.* 40.

2 UFLP

2.1 Properties, general methods and applications of UFLP

M. Körkel (1989). On the exact solution of large-scale simple plant location problems. *European J. Oper. Res.* 39, 157-173.

This paper presents refinements to the DUALOC procedure (Erlenkotter 1978), a dual-ascent heuristic procedure to solve the dual of the LP-relaxation of the UFLP combined with the use of the complementarity slackness conditions to construct primal solutions. Körkel's experiments show that large-sized problems can be solved.

H.P. Simão, J.M. Thizy (1989). A dual simplex algorithm for the canonical representation of the uncapacitated facility location problem. *Oper. Res. Lett.* 8, 279-286.

The use of LP algorithms provide distinct advantages, in particular direct amenability tocutting plane techniques. A streamlined dual simplex algorithm is designed, based on a covering formulation of the problem. Computational results are presented.

G. Cornuéjols, G.L. Nemhauser, L.A. Wolsey (1990). The uncapacitated facility location. P.B. Mirchandani, R.L. Francis (eds.). *Discrete Location Theory*, Wiley, New York, 119-171.

This survey presents a detailed and up to date account of the structural properties of the UFLP.

A.R. Conn, G. Cornuéjols (1990). A projection method for the uncapacitated facility location problem. *Math. Program.* 46, 273-298.

This paper considers the dual of the strong linear programming relaxation of the UFLP. Based on the observation that the condensed form of this dual is an unconstrained minimization problem with a piecewise linear convex objective function, this paper develops an exact solution of the condensed dual via orthogonal projections. Interestingly enough, this approach can handle side constraints. In particular, valid inequalities can be added when a duality gap exists between the exact solution of the dual and the primal solution recovered from the complementarity slackness conditions. Numerical results for some classical test problems are included.

P.D. Domich, K.L. Hoffman, R.H.F. Jackson, M.A. McClain (1991). Locating tax facilities: a graphics-based micro computer optimization model. *Management Sci.* 37, 960-979.

The problem of optimal selection of locations for Internal Revenue Service Posts-of-Duty is formulated as an uncapacitated, fixed charge location allocation model to minimize travel and facility costs. It is solved using a greedy-interchange heuristic coupled with a Lagrangean relaxation technique.

P.M. Dearing, P.L. Hammer, B. Simeone (1992). Boolean and graph theoretic formulations of the simple plant location problem. *Transportation Sci.* 26, 138-148.

The paper gives a pseudo-Boolean formulation of the UFLP, then transforms this formulation into both a set covering problem as well as a vertex packing problem.

N. Xu, T.J. Lowe (1993). On the equivalence of dual methods for two location problems. *Transportation Sci.* 27, 194-199.

This paper shows the equivalence between the dual-ascent procedure for the uncapacitated facility location problem and a standard greedy for the minimum cost operating problem, or extended covering problem of minimizing the sum of the set-up costs for facility locations plus the penalty costs of not covering some demands.

V. Verter, M.C. Dincer (1995). Facility location and capacity acquisition: an integrated approach. *Naval Res. Log. Quart.* 42, 1141-1160.

This paper considers the situation where, in addition to a fixed cost for opening a plant, a concave cost of acquisition of capacity is incurred. Due to concavity, each customer is served by only one facility. An algorithm is designed, using successively refined piecewise linear concave approximations.

M. Desrochers, P. Marcotte, M. Stan (1995). The congested facility location problem. *Location Sci.* 3, 9-23.

In this model, a congestion factor is added to the objective function. It corresponds to an increasing convex cost related to the delay incurred by each customer at facility j. The problem is solved using a columm generation technique within a branch and bound scheme.

2.2 Extensions, related models

H.N. Psaraftis, G.G. Tharakan, A. Ceder (1986). Optimal response to oil spills. The strategic decision case. *Oper. Res.* 34, 203-217.

The paper considers the optimal location and size of equipment to fight oil spills. A detailed description of the cost to recover damages and of the environmental cost of the unrecovered damages is provided. Random elements include the arrival of oil spills (independent and non-concomitant Poisson processes) and the sizes of the spills (small, medium, large).

J.G. Klincewicz, H. Luss (1987). A dual-based algorithm for multi product uncapacitated facility location. *Transportation Sci.* 21, 198-206.

In the multiproduct uncapacitated facility location problem (MUFLP), different products are required by the customers. The problem consists of minimizing the total of fixed costs for opening facilities, fixed costs for handling a particular product at a given location and assignment costs for satisfying demands of various products.

This paper extends Erlenkotter's dual ascent and dual adjustment procedures to generate a good feasible solution to the dual of the MUFLP. Primal feasible solutions are constructed from the dual solutions and the procedure is embedded in a branch and bound algorithm to reach optimality.

J.G. Klincewicz (1990). Solving a freight transport problem using facility location techniques. *Oper. Res.* 38, 99-109.

The problem consists of finding a minimum pattern of direct and indirect shipments. A typical example of indirect shipment is the use of consolidation terminals which allow advantage to be taken of economies of scale in transportation costs. If either the source-to-terminal or the terminal-to-destination shipping costs are linear, the freight transport problems decompose into a concave cost uncapacitated facility location problem.

A. Wagelmans, S. van Hoesel, A. Kolen (1992). Economic lot sizing: an $O(n \log n)$ algorithm that runs in linear time in the Wagner-Within case. *Oper. Res.* 40, S 145-S 156.

The paper presents an $O(n \log n)$ algorithm for the economic lot-sizing problem, that runs in $O(n)$ time in some special cases, including the Wagner-Within case. This result is of relevance to location theory, as the economic lot-sizing problem can be reformulated as an uncapacitated facility location, and the proposed algorithm directly relates to solving the dual of the LP relaxation of the UFLP.

F.V. Louveaux, D. Peeters (1992). A dual-based procedure for stochastic facility location. *Oper. Res.* 40, 564-573.

The uncapacitated facility location is transformed into a two-stage stochastic program with recourse to take uncertainty into account. Demand, selling prices, production and transportation costs may be considered as random elements. A dual-ascent procedure is presented, and the monotone improvements available in the static case are proved to be also achievable in the stochastic case.

G. Laporte, F.V. Louveaux, L. Van hamme (1994). Exact solution of a stochastic location problem by an integer L-shaped algorithm. *Transportation Sci.* 28, 95-103.

This paper considers the optimal location and sizing of facilities given that future demand is uncertain. When demand is known, the available capacity is allocated to clients, which may result in lost sales. An exact algorithm is presented and tested on medium size problems.

P.C. Jones, T.J. Lowe, G. Muller, N. Xu, Y. Ye, J.L. Zydiak (1995). Specially structured uncapacitated facility location problems. *Oper. Res.* 43, 661-669.

The paper considers the uncapacitated facility location problem with two structural assumptions on the model, one on the network structure related to the set of customers a given plant can supply, one on the monotonicity of the variable costs. The authors develop an $O(mn)$ algorithm for solving this specially structured problem. They also show that several other problems are instances of the specially structured UFLP. In particular, this is the case for the stochastic UFLP, the economic lot sizing problem and the capacity expansion problem.

A. Caprara, J.J. Salazar Gonzalez (1996). A branch-and-cut algorithm for a generalization of the uncapacitated facility location problem. *TOP* 4, 135–163.

This paper presents a generalization of the UFLP in which a gain is realized when a client is served by a set of facilities (called a configuration) instead of being served by a single facility. The problem is inspired by the Index Selection Problem in physical data base design. The problem is formulated as a Set Packing Problem, and solved through Branch and Cut, with an exact separation procedure for odd-hole inequalities. Computation results are given.

Price effects

R. Logendran, M.P. Terrell (1988). Uncapacitated plant location-allocation with price sensitive stochastic demands. *Computers Oper. Res.* 15, 189-198.

Demands are assumed stochastic and price-sensitive. First stage decisions consist of selecting facilities to open as well as the allocation of clients to facilities. Quantities to be transported are optimized for each plant-customer combination.

P. Hanjoul, P. Hansen, D. Peeters, J.-F. Thisse (1990). Uncapacitated plant location under alternative spatial price policies. *Management Sci.* 36, 41-57.

The paper presents solution methods to determine the price, number, locations, sizes and market areas of the plants supplying the clients to maximize the profit of the firm. Alternative spatial price policies are considered (uniform mill pricing, uniform delivered pricing, spatial discriminatory pricing). The solution methods consist of branch and bound algorithms based on a new upper bounding function.

Two-level models

L.L. Gao, E.P. Robinson Jr. (1992). A dual-based optimization procedure for the two-echelon uncapacitated facility location problem. *Naval Res. Log. Quart.* 39, 191-212.

An uncapacited facility location problem with two levels in which the products are shipped from one level, echelon 1, to the demand points via an intermediate second level of facilities, echelon 2. A fixed cost is associated to each open facility of echelon 1 and to each pair of echelon 1 and echelon 2 facilities which serve at least one client. A dual ascent heuristic is proposed and embedded into a branch and bound algorithm. Computational experiments are reported for problems with up to 25 possibles locations at each level and 35 clients.

A.I. Barros, M. Labbé (1994). A general model for the uncapacitated facility and depot location problem. *Location Sci.* 3, 173-191.

A two level uncapacitated facility location model involving fixed costs for opening facilities of both types and for having pairs of facilities of different type operating together is considered.

A Lagrangean dual approach is used within a branch and bound method for solving the problem. Computational experiments are reported for problems with up to 10 or 15 possible locations for each level and 50 clients.

A.I. Barros, M. Labbé (1994). The multi-level uncapacitated facility location problem is not submodular. *European J. Oper. Res.* 72, 607-609.

Unlike the classical UFLP, the objective function of the corresponding multi-level problem is not submodular.

K. Aardal, M. Labbé, J. Leung, M. Queyranne (1996). On the two-level uncapacitated facility location problem. *INFORMS J. Comput.* 8, 289-301.

The polyhedral structure of the convex hull of feasible solutions of the two-level uncapacitated facility location problem is studied. In particular, it is shown that all nontrivial facets for the UFLP define facets for the two-level problem, and conditions when facets of the two-level problem are also facets for the CFLP are derived. New families of facets and valid inequalities are also introduced.

Public facility location

F.V. Louveaux (1986). Discrete stochastic location models. *Ann. Oper. Res.* 6, 23-34.

This paper presents a stochastic version of the p-median problem where demands and service costs may be random. Alternative ways to impose the budget constraint and the links to the stochastic UFLP are provided.

S.R. Gregg, J.M. Mulvey, J. Wolpert (1988). A stochastic planning system for siting and closing public service facilities. *Environment and Planning A* 20, 83-98.

The paper considers optimal siting and closing of public service facilities. The level of service and the allocation of service to demand regions are decided in the first-stage. When demands actually occur, overage and underage penalties are paid. A detailed application to the public libraries in Queens Borough is provided.

J. Perl, P.-K. Ho (1990). Public facilities location under elastic demand. *Transportation Sci.* 24, 117-136.

The paper considers the location problem of public emergency facilities (e.g. schools, post office, day-care centers). Assuming demands for such services are elastic, an integer programing formulation is presented to maximize the consumer's surplus, under different demand functions. Comparisons of location behaviors under elastic and inelastic demands are provided.

Multimode multicommodity

T.G. Crainic, P.J. Dejax, L. Delorme (1989). Models for multimode multicommodity location with interdepot balancing requirements. *Ann. Oper. Res.* 18, 279-302.
T.G. Crainic, M. Gendreau, P. Soriano, M. Toulouse (1992). A tabu search procedure for multicommodity location/allocation with balancing requirements. *Ann. Oper. Res.* 41, 359-383.
T.G. Crainic, L. Delorme (1993). Dual-ascent procedures for multicommodity location-allocation problems with balancing requirements. *Transportation Sci.* 27, 90-101.
T.G. Crainic, L. Delorme, P.J. Dejax (1993). A branch-and-bound method for multi-commodity location with balancing requirements. *European J. Oper. Res.* 65, 368-382.
B. Gendron, T.G. Crainic (1995). A branch and bound algorithm for depot location and container fleet management. *Location Sci.* 3, 39-53.

A series of papers for the multicommodity location-allocation problem with balancing requirements, a problem related to the management of a fleet of vehicles over a medium to long-term planning horizon.

The first paper presents the modelling issues, the four others various solution techniques.

2.3 Capacitated facility location problem

B.M. Baker (1986). A partial dual algorithm for the capacitated warehouse location problem. *European J. Oper. Res.* 23, 48-56.

In order to solve the LP-relaxation of the strong formulation (P) of the CFLP, the variable upper bound constraints (B) are dualized in a Lagrangean way to obtain a transportation problem as subproblem and the associated multipliers are updated heuristically. Then, this lower bounding procedure is embedded into a branch and bound method. Computational results are provided for problems with up to 40 potential facilities and 80 clients.

J.G. Klincewicz, H. Luss (1986). A Lagrangean relaxation heuristic for capacitated facility location with single source constraints. *J. Oper. Res. Soc.* 37, 495-500.

This papers studies a special version of the CFLP in which each client must be served by a single facility, i.e. $x_{ij} \in \{0,1\}$, $i = 1,\ldots,n$, $j = 1,\ldots,m$. A heuristic procedure based on a Lagrangean relaxation obtained by dualizing the capacity constraints (C) is proposed. Computational results are reported for the 12 tests problems of Kuehn and Hamburger with two different capacity levels. For all problems but one, the heuristic finds a feasible solution within 12% of the lower bound.

T.J. Van Roy (1986). A cross decomposition algorithm for capacitated facility location. *Oper. Res.* 34, 145-163.

The Cross Decomposition method exploits the primal and dual structures of a problem simultaneously. Applied to formulation (P) of the CFLP, it amounts to solving successively some transportation problem, corresponding to the allocation problem of the clients to some given facilities, and some UFLP, emerging as subproblem in the Lagrangean relaxation obtained by dualizing the capacity constraints (C). Computational experiments on problems with up to 100 potential facility locations and 200 clients show that the Cross Decomposition algorithm finds near optimal Lagrangean multipliers within just a few iterations and is about 10 times faster than the alternative algorithms using the same Lagrangean relaxation.

H. Pirkul (1987). Efficient algorithms for the capacitated concentrator location problem. *Computers Oper. Res.* 14, 197-208.

The variant of the CFLP in which each demand must be allocated to one single facility is considered. The Lagrangean relaxation obtained by dualizing the demand constraints (D) is used to derive lower and upper bounds, which in turn are used in a branch and bound procedure. Computational experience is reported for problems with up to 20 possible locations and 100 demand points.

J.E. Beasley (1988). An algorithm for solving large capacitated warehouse location problems. *European J. Oper. Res.* 33, 314-325.

A more general version of the CFLP involving lower and upper bounds on the used facility capacities is considered. A lower bound is proposed by considering a Lagrangean relaxation obtained by dualizing the demand constraints (D), the new capacity constraints, and some feasible solution exclusion constraints.

By incorporating this lower bound and some problem reduction tests into a branch and bound procedure, problems involving up to 500 potential facility locations and 1000 clients are solved on a Cray-1S computer.

J.M.Y. Leung, T.L. Magnanti (1989). Valid inequalities and facets of the capacitated plant location problems. *Math. Program.* 44, 271-291.

This paper studies the polyhedral structure of several variants of the CFLP. A first version where all facilities have the same capacity and customers' demand does not need to be fully met is considered. The so-called residual capacity inequalities are proved to be valid for that model and conditions under which they are facets are identified. Then, it is investigated whether these inequalities remain facets for other variants of the CFLP in which the demand of each client must be fully met, is indivisible, and if the facilities have varying capacities.

M. Guignard, K. Opaswongkarn (1990). Lagrangean dual ascent algorithms for computing bounds in capacitated plant location problems. *European J. Oper. Res.* 46, 73-83.

This paper considers a Lagrangean relaxation of the strong formulation (P) obtained by dualizing constraints (B) and (C) and strengthened by adding the total demand constraint (T) and some simple capacity constraints $\sum_{i=1}^{n} x_{ij} \leq s_j$, $j = 1, \ldots, n$. A dual ascent procedure is proposed to update the Lagrangean multipliers. Computational experiments on problems with up to 70 possible locations and clients show that the gap between the best lower bound obtained and some heuristic primal solution value is generaly not larger than 3% .

G. Cornuéjols, R. Sridharan, J.M. Thizy (1991). A comparison of heuristics and relaxations for the capacitated plant location problem. *European J. Oper. Res.* 50, 280-297.

A systematic study of the relative quality of the bounds used in Branch and Bound algorithms for the CFLP is performed. Specifically, 41 relaxations are reviewed and yield only seven genuinely different bounds for which dominance relations are identified. A computational study is also performed to compare the relaxations and heuristic feasible solutions they generate in terms of various characteristics of the test problems.

J. Barcelo, E. Fernandez, K.O. Jörnsten (1991). Computational results from a new Lagrangean relaxation algorithm for the capacitated plant location problem. *European J. Oper. Res.* 53, 38-45.

Lagrangean decomposition is applied to the weak formulation of the CFLP plus the total demand constraint (T). The so-obtained lower bound as well as an associated upper bound are embedded into a Branch and Bound procedure. Computational results for problems involving up to 20 possible locations and 50 customers are presented.

A. Shulman (1991). An algorithm for solving dynamic capacitated plant location problems with discrete expansion sizes. *Oper. Res.* 39, 423-436.

This paper considers the special class of Dynamic Capacitated Facility Location problems in which the facility capacities can only take a finite number of given values and several facilities can be opened at the same location. The problem is modeled as a MILP and a heuristic resolution method based on the Lagrangean relaxation in which the demand constraints are dualized is proposed. Two versions of the algorithm are developed depending on whether facilities of different capacities are allowed at the same location or not. Computational results for problems with up to 62 facility locations, 62 demands and 10 planning periods are reported: the method produces solutions within 3% of the optimum except when transportation costs are very low.

Q. Deng, D. Simchi-Levi (1992). *Valid Inequalities Facets and Computational Results for the Capacitated Concentrator Location Problem*, Research Report, Department of Industrial Engineering and Operations Research, Columbia University, New York.

This report studies the polyhedral structure of a variant of the CFLP in which each customer must be assigned to exactly one facility. Families of valid inequalities are identified and some of them are proved to be facets. Computational experiments with a cutting plane algorithm based on the identified valid inequalities are presented. The test problems involve 50 customers and 25 or 26 possible locations.

P. Hansen, E. de Luna Pedrosa Filho, C. Carneiro Ribeiro (1992). Location and sizing of offshore platforms for oil exploration. *European J. Oper. Res.* 58, 202-214.

An extension of the CFLP called the multicapacitated Facility Location Problem and in which each facility capacity must be chosen among a given set of possible sizes is considered. Exact algorithms based on an integer programming code and using either a weak or a strong formulation are compared to a tabu search heuristic. Computational results for problems with up to 100 clients and 40 possible locations are reported.

K. Aardal, Y. Pochet, L.A. Wolsey (1995). Capacitated facility location: valid inequalities and facets. *Math. Oper. Res.* 20, 562-582.

The polyhedral structure of the convex hull of feasible solutions of the CFLP is examined. Two new families of valid inequalities are introduced: the family of effective capacity inequalities which generalizes the flow cover inequalities and the family of submodular inequalities which in turn generalizes the effective capacity inequalities. Two additional families of inequalities are considered: the class of combinatorial inequalities previously introduced by Cho et al. (1983) for the UFLP and for which sufficient conditions for them to be facet defining are provided and the class of (k, ℓ, S, I) inequalities proposed by Pochet and Wolsey (1993) for the lot-sizing problem with constant batch sizes.

K. Aardal (1997). Capacitated facility location: separation algorithms and computational experience. *Math. Program.* Forthcoming.

This companion paper of Aardal, Pochet and Wolsey (1995) considers the cutting plane approach for solving the CFLP. A subfamily of the submodular inequalities and the combinatorial inequalities are used and heuristics are proposed for solving the associated separation problems. Computational experiments are reported for 60 problems with up to 100 customers and 75 potential facility locations.

3 P-Facility Location Problems

3.1 P-median

The following survey presents different variants and extensions of the p-median problem.

P.B. Mirchandani (1990). The p-median problem and generalizations. P.B. Mirchandani, R.L. Francis (eds.). *Discrete Location Theory*, Wiley, New York, 55-117.

B.L. Golden and C.C. Skiscin (1986). Using simulated annealing to solve routing and location problems. *Naval Res. Log. Quart.* 33, 261-279.

This article presents a computational study of the simulated annealing approach for the p-median problem. Results concerning instances with up to 100 clients and possible locations and $p = 20$ are reported.

S. Ahn, C. Cooper, G. Cornuéjols, A. Frieze (1988). Probabilistic analysis of a relaxation for the k-median problem. *Math. Oper. Res.* 13, 1-30.

A probabilistic analysis of the strong linear relaxation of the p-median problem is performed when the number of points (clients) tends to infinity. Four underlying metric space models are considered: Euclidean, network, tree and uniform cost models. Computational experiments are also reported for problems with $n = 50$ and $n = 100$.

M.E. Captivo (1991). Fast primal and dual heuristics for the p-median location problem. *European J. Oper. Res.* 52, 65-74.

Two heuristics are proposed for the p-median: a greedy-type one with local improvement and a dual based one. Computational experiments compare these solutions to others previously presented in the literature on problems with $6 \leq n \leq 400$ and $2 \leq p \leq 133$.

R. Hassin, A. Tamir (1991). Improved complexity bounds for location problems on the real line. *Oper. Res. Lett.* 10, 395-402.

The paper considers the optimal location of p facilities on a real line. The model extends the p-median, p-coverage and simple uncapacitated plant location on the line. Applying recent results in dynamic programing, improved complexity bounds are obtained for these problems.

B. Gavish and S. Sridhar (1995). Computing the 2-median on tree networks in $O(n \log n)$. *Networks* 26, 305-317.

An $O(n \log n)$ algorithm is proposed for the 2-median when the underlying metric space is a tree.

A. Tamir (1996). An $O(pn^2)$ Algorithm for the p-median and related problems on tree graphs. *Oper. Res. Lett.* 19, 59–64.

It is shown that the complexity of the "leaves to root" dynamic programming algorithm for the p-median is $O(pn^2)$.

3.2 P-center

A. Tamir (1993). The least element property of center location on tree networks with applications to distance and precedence constrained problems. *Math. Program.* 62, 475-496.

In the classical p-center location model on a network, the problem consists of finding p service centers that minimize the maximum distance of a customer to a closest center. Here, a secondary objective is to locate the centers that optimize the primary objective as close as possible to the central depot. For two p-center models on tree networks, the set of optimal solutions to the primary objective is shown to have a semi lattice structure w.r.t. some natural ordering. It is thus proved that there is a p-center solution to the primary objective that simulteanously minimizes every secondary objective function which is monotone nondecreasing in the distance of the p centers from the central depot.

M. Jaeger, J. Goldberg (1994). A polynomial algorithm for the equal capacity p-center problem on trees. *Transportation Sci.* 28, 167-175.

A polynomial time algorithm is provided for the capacitated version of the p-center problem where all centers have equal capacity and the underlying metric space is a tree.

3.3 P-facility location problems with mutual communication

D. Chhajed, T.J. Lowe (1992). M-median and m-center problems with mutual communication: solvable special cases. *Oper. Res.* 40, Supp.1, S56-S66.

Polynomial time algorithms are provided for the p-median and p-center problems with mutual communication when the graph representing the interactions between pairs of new facilities has a special structure.

A. Tamir (1993). Complexity results for the p-median problem with mutual communication. *Oper. Res. Lett.* 14, 79-84.

The p-median problem with mutual communication is shown to be strongly NP-hard even when $n = 3$ and p is variable. When the metric space is a tree, an $O(p^3 \log n + pn + n \log n)$ algorithm is proposed.

D. Chhajed, T.J. Lowe (1994). Solving structured multifacility location problems efficiently. *Transportation Sci.* 28, 104-115.

A generic p-facility location model which generalizes the p-median and p-center problems with mutual communication is considered. A polynomial time algorithm is presented when the dependency graph is a k-tree.

A. Tamir (1994). A distance constrained p-facility location problem on the real line. *Math. Program.* 66, 201-204.

The problem considered is to locate p new facilities while satisfying distance constraints between pairs of existing and new facilities and between pairs of new facilities and in order to minimize the total cost for setting the new facilities. An $O(p^3 n^2 \log n)$ algorithm is presented for the special case of a linear metric space.

4 Covering Models

Surveys on covering include:
A.W.J. Kolen, A. Tamir (1990). Covering problems. P.B. Mirchandani, R.L. Francis (eds.). *Discrete Location Theory*, Wiley, New York, 263-304.
D.A. Schilling, V. Jayaraman, R. Barkhi (1993). A review of covering problems in facility location. *Location Sci.* 1, 25-55.

In this section, we only select papers explicitly dealing with location aspects.

4.1 Properties, general methods and applications

A. Tamir (1992). On the core of cost allocation games defined on location problems. *Transportation Sci.* 27, 81-86.

Most of the location literature has concentrated on where to establish facilities in order to serve given demand points. Interestingly, this paper is devoted to the problem of allocating the costs for opening facilities among the users. The underlying location model is a covering one with minimum total cost. An allocation of this total cost is sought for such that no group of users has an incentive to separate from the others and to build its own set of facilities. Such an allocation concept originates from game theory and is called a core allocation.

M.W. Broin, T.J. Lowe (1994). Coverage location problems and totally balanced relaxations: simulation results. *Location Sci.* 2, 241-257.

Several types of covering location models are easily solved if the underlying covering matrix is totally balanced. This paper gives a heuristic algorithm for finding maximal totally balanced submatrices. Simulation results imply that planar covering location problems have much more structure than comparable randomly generated problems.

B.T. Downs, J.D. Camm (1996). An exact algorithm for the maximal covering problem. *Naval Res. Log. Quart.* 43, 435-461.

Using a known primal heuristic, a dual solution of the dual-relaxation of the LP is obtained. This dual solution is improved by subgradient optimization to produce an upper bound on the optimal solution value, as well as a possible improved primal solution. Finiteness of the method is obtained through Branch-and-Bound. Results of extensive computational experiments are reported.

O. Fujiwara, K. Kachenchai, T. Makjamroen, K.K. Gupta (1988). An efficient scheme for deployment of ambulances in metropolitan Bangkok. G.K. Rand (ed.). *Operational research* 87, North Holland, Amsterdam, 730-742.

The paper presents an application of covering models to the location of ambulances in metropolitan Bangkok, with 879 demand points and 46 potential ambulance locations.

A.J. Swersey, L.S. Thakur (1995). An integer programming model for locating vehicle emissions testing stations. *Management Sci.* 41, 496-512.

The problem is to determine the number, size and locations of stations given constraints on the maximal travel distance as well as the maximal average waiting time at the facility. A simulation model is used to find the maximum allowable arrival rates at stations of different sizes. The problem is then formulated and solved as a set covering location model.

Discretionary service facilities

A. Tamir (1987). Totally balanced and totally unimodular matrices defined by center location problems. *Discr. Appl. Math.* 16, 245-263.

Consider the case of customers visiting facilities, such as shopping centers, called servers. Any customer v is assumed to be willing to visit any server in a set $S(v)$, which is e.g. those within distance r_v of v. The question is to locate secondary stations (e.g. recycling depots for empty bottles and paper products). To give the customers an incentive to visit the secondary stations, one wishes to locate them on the path from the customers to the servers. For each customer, let $G(S(v))$ be the subgraph of all shortest paths connecting v to vertices in $S(v)$. The location model consists of finding a minimum cost set of secondary stations such that every customer will find at least one station in $G(S(v))$. This paper studies the properties of the subgraphs $G(S(v))$ when the underlying graph is a tree. It is proved that a large collection of such subgraphs is totally balanced, and in some cases totally unimodular. These properties imply polynomial solvability of the location problem, and some of its generalizations.

O. Berman, R.C. Larson, N. Fouska (1992). Optimal location of discretionary service facilities. *Transportation Sci.* 26, 201-211.

The motivation for the so-called discretionary service facilities is based on the perceived behavior on the part of the customers. It is argued that many customers carry out several purchases as part of routine preplanned trips (instead of a return trip from home or workplace). An optimal set of facility locations (which maximizes the flow of potential customers who pass at least one discretionary service facility along their preselected travel paths) is proved to exist on the nodes of the network. Exact and heuristic algorithms are developed to solve this problem.

O. Berman, D. Bertsimas, R.C. Larson (1995). Locating discretionary service facilities. Maximizing market size, minimizing inconvenience. *Oper. Res.* 43, 623-632.

The previous model is generalized in three directions: first in allowing the customers to make some detour from the preplanned route to visit a discretionary service facility; second in considering maximizing the expected number of potential customers who become actual customers, assuming the probability of becoming customer is a convex decreasing function of the extra distance to a facility, third in locating the service facilities to minimize the total deviation distance per unit of time for all customers, assuming they all have to purchase at a service facility. Those formulations yield to a greedy heuristic solution and for one of the formulations the greedy performance bound is proved to be tight.

4.2 Workload capacities, backup and multiple service

K. Hogan, C. Revelle (1986). Concepts and applications of backup coverage. *Management Sci.* 32, 1434-1444.
M.S. Daskin, K. Hogan, C. Revelle (1988). Integration of multiple, excess, backup and expected covering models. *Environmental and Planning B* 15, 15-35.
These two papers introduce the concept and applications of backup coverage and study alternative ways to require multiple coverage.

H. Pirkul, D.A. Schilling (1988). The siting of emergency service facilities with workload capacities and backup service. *Management Sci.* 34, 896-908.
This paper presents a deterministic model formulation for emergency service systems where facility workload is limited and some or all demand points require a backup service. A solution procedure is presented which uses Lagrangean relaxation. Results of computational experiments as well as of an example using real data are provided.

R. Batta, N.R. Mannur (1990). Covering location models for emergency situations that require multiple response units. *Management Sci.* 36, 16-23.
This paper presents a model formulation of the set covering location and the maximal covering location problem when demands require a response from multiple units (units of different types or many units) instead of a response of at least one unit. The paper studies some dominance properties to reduce the size of the IP, which is solved by an implicit enumeration branch and bound algorithm.

H. Pirkul, D.A. Schilling (1991). The maximal covering location problem with capacities on the total workload. *Management Sci.* 37, 233-248.
This paper presents a model formulation of the maximal covering location problem with capacities on the total workload, discusses some difficulties related to the assignment of uncovered demand to facilities and presents a Lagrangean relaxation method to solve the problem.
(see also J. Current, J. Storbeck (1986). Capacitated covering models. *Environment and Planning B* 153-164 and B. Gavish, H. Pirkul (1986). Computer and data base location in distributed computer systems. *IEEE Trans. Comput.* C-35, 583-590.)

C. Revelle, V. Marianov (1991). The probabilistic FLEET model with individual vehicle reliability requirements. *European J. Oper. Res.* 93-105.
The paper presents a probabilistic model for the case where two types of servers are needed to cover a call. The number and location of both types of servers have to be found under uncertainty over their availability. Applications come from fire protection systems.

P. Mirchandani, R. Kohli and A. Tamir (1996). Capacitated location problems on a line. *Transportation Sci.* 30, 75-80.
Several capacitated covering location problems are considered where customers lie on a straight line. The objective is either to serve all the demand with a set of facilities of minimum total fixed cost or to maximize the profit by locating up to q facilities that serve some or all demands with idiosyncratic returns and penalties. Polynomial-

time dynamic programming algorithm are proposed when additional assumptions are satisfied.

4.3 Probabilistic and stochastic models

C. Revelle, K. Hogan (1988). A reliability-constrained siting model with local estimates of busy fractions. *Environment and Planning B* 15, 143-152.

This paper considers the maximal expected covering location model when the probability of a server being busy is made local, i.e. client or sector dependent. The major difficulty here is that this probability is a function of the decisions on the locations to be opened.

R. Batta, J. Dolan, N.N. Krisnamurthy (1989). The maximal expected covering location problem revisited. *Transportation Sci.* 277-287.

In the maximal expected covering problem (Daskin 1983), the probability that a server is busy is assumed to be the same for every server, independent of its location and the servers are assume to operate independently. To remove those assumptions, the authors use a node substitution heuristic combined with Larson's hypercube model (Larson 1974), a descriptive model that provides statistics about the servers and the demand.

C. Revelle, K. Hogan (1989). The maximum reliability location problem and α-reliable p-center problem: Derivatives of the probabilistic location set covering problem. *Ann. Oper. Res.* 18, 155-174.

C. Revelle, K. Hogan (1989). The maximum availability location problem. *Transportation Sci.* 23, 192-200.

Two papers which study a variety of probabilistic constraint models (locate p facilities to either maximize the minimum probability of service, or maximize the population served with confidence level α e.g.).

R. Batta (1989). The stochastic queue median over a finite discrete set. *Oper. Res.* 37, 648-652.

The Stochastic Queue Median model of Berman, Larson and Chiu concerns the location of a single facility on a network operating as an M/G/1 queue. This note considers the case where the location of the server is restricted to a finite discrete set of points.

J. Goldberg, L. Paz (1991). Locating emergency vehicle bases when service time depends on call location. *Transportation Sci.* 264-280.

This paper considers a maximal expected coverage model where the probability that a call is reached within a given threshold does not only depend on the probability that an idle server is available at a given facility but also on the stochastic travel times.

H.D. Sherali, S. Kim, E.L. Parrish (1991). Probabilistic partial set covering problems. *Naval Res. Log. Quart.* 38, 41-51.

Due to operational reliability or limited availability at any given point of time, a facility j is able to serve an assigned customer i with fixed probability p_{ij}. The problem then consists of minimizing the sum over all customers of the probability of non coverage, subject to a budget constraint on the cost of opening facilities. A branch and bound algorithm is developed and tested.

M. Ball, F.L. Lin (1993). A reliability model applied to emergency service vehicle location. *Oper. Res.* 41, 18-36.

A system failure is said to occur when an emergency service cannot respond to a call within a specified time limit. An upper bound on this failure probability is derived and use to formulate an I.P. with constraints that guarantee that each demand point (such as a residential subdivision) achieves a certain level of reliability. By incorporating valid inequalities as a preprocessing technique, drastic improvement in solution times is obtained.

5 Miscelleanous

5.1 Path and tree-shaped facility location problems

J.R. Current, C.S. Revelle, J.L. Cohon (1987). The median shortest path problem: a multiobjective approach to analyse cost vs. accessibility in the design of transportation networks. *Transportation Sci.* 21, 188-197.

The problem of locating a path on a network in order to minimize the total path length and the total travel time required for reaching a node on the path from all nodes of the network is considered. An ILP formulation of the problem and an exact algorithm for generating non dominated solutions are provided.

V.A. Hutson, C.S. Revelle (1989). Maximal direct covering tree problems. *Transportation Sci.* 23, 288-299.

The problem of locating a subtree of a tree in order to maximize the total demand covered and to minimize the subtree length is considered. Non dominated solutions are sought for.

A. Tamir, T.J. Lowe (1992). The generalized p-forest problem on a tree network. *Networks* 22, 217-230.

A polynomial dynamic programming algorithm is provided for locating p tree-shaped facilities on a tree network. The location model considered captures other models previously proposed: the p-center tree-shaped facility location problem and the minimal cost/maximal covering forest problem.

A. Tamir (1992). On the complexity of some classes of location problems. *Transportation Sci.* 26, 352-354.

R. Church, J. Current (1993). Maximal covering tree problems. *Naval Res. Log. Quart.* 40, 129-142.

In both papers, it is shown that the weighting method used by [Hutson and Revelle 1989] (see above) for the maximum direct cover tree problem can be implemented in $O(n^2)$.

S.L. Hakimi, E.F. Schmeichel, M. Labbé (1993). On locating path- or tree-shaped facilities on networks. *Networks* 23, 543-555.
In this paper, 64 different path- or tree-shaped facility location problems are identified and 32 of them consist of discrete problems. The algorithmic complexity of those problems is systematically identified.

A. Tamir (1993). A unifying location model on tree graphs based on submodularity properties. *Discr. Appl. Math.* 47, 275-283.
A general model for locating a subtree of a tree is proposed. Most common objective functions used in facility location theory are shown to have the submodularity property, which allows one to solve the problem in polynomial time.

E.H. Aghezzaf, T.L. Magnanti, L.A. Wolsey (1995). Optimizing constrained subtrees of trees. *Math. Program.* 71, 113-126.
This paper studies complexity results for the rooted subtree problem and the subtree packing problem. In particular, the subtree packing problem, which relates to location problems on trees, is shown to be polynomial if and only if each rooted subtree problem is polynomial.

T.U. Kim, T.J. Lowe, A. Tamir, J.E. Ward (1996). On the location of a tree-shaped facility. *Networks.* 28, 167–175.
A general model is proposed for locating a subtree of a tree. The objective is to minimize the total fixed cost for building the subtree facility plus the total transportation cost. An $O(n^2)$ algorithm is proposed for the problem in general and algorithms with subquadratic complexity are discussed for special cases such as the subtree, the direct covering subtree and the undirect covering subtree location problems.

5.2 Location-routing

Most available results on location-routing are known as the traveling salesman facility location or the a priori optimization of location-routing. Many papers in this area are concerned with location-routing on networks or in the Euclidean plane. As indicated in the introduction, these topics are not covered here. The following two surveys are available.

M.S. Daskin (1987). Location, dispatching and routing models for emergency services with stochastic travel times. A. Ghosh, G. Rushton (eds.). *Spatial analysis and location-allocation models*, Van Nostrand Reinhold, New York, 224-265.
G. Laporte (1988). Location-routing problems. B.L. Golden, A.A. Assad (eds.). *Vehicle routing: methods and studies*, North Holland, Amsterdam, 163-198.

G. Laporte, Y. Nobert, D. Arpin (1986). An exact algorithm for solving a capacitated location. Routing problems. *Ann. Oper. Res.* 6, 293-310.
This paper considers the problem of finding optimal location of one or more depots as well as optimal delivery routes from the selected depots to the various clients. A branch and bound algorithm, using an initial relaxation of the subtour elimination

constraints, vehicle capacity constraints and chain barring constraints, is presented. Computational results are reported.

G. Laporte, Y. Nobert, S. Taillefer (1988). Solving a family of multi-depot vehicle routing and location-routing problems. *Transportation Sci.* 22, 161-172.
Three types of multiple depot asymmetrical location-routing problems are considered (capacity constrained and cost constrained VRPs, cost constrained LRPs). The authors show how, after an appropriate graph tranformation, they can be formulated as constrained assignment problems, and solved using an adaptation of Carpaneto and Toth branch and bound procedure for the asymetrical TSP. Problems involving up to 80 nodes are solved.

G. Laporte, F. Louveaux, H. Mercure (1989). Models and exact solutions for a class of stochastic location-routing problems. *European J. Oper. Res.* 39, 71-78.
In a stochastic environment, decisions regarding depot location, fleet size and planned routes have to be made without knowing the actual supplies. The total supply of a route may turn out to exceed the vehicle capacity, i.e. failures may occur. Two variants are considered, one that minimizes first-stage cost so that the probability of route failure does not exceed a preset treshold, one that minimizes first-stage cost, so that expected penalty of a route does not exceed a fraction of its planned cost.

I.M. Branco, J.D. Coelho (1990). The Hamiltonian p-median problem. *European J. Oper. Res.* 47, 86-95.
The Hamiltonian p-median problem consists of finding an optimal location of p depots and Hamiltonian circuits starting and ending at these depots and visiting all customers. Several heuristics are proposed. Results of computational experiments are reported.

P.H. Hansen, B. Hegedahl, S. Hjortkjaer, B. Obel (1994). A heuristic solution to the warehouse location-routing problem. *European J. Oper. Res.* 76, 111-127.
The paper presents a modified version of the heuristic methods by Perl and Daskin for the multiple depot location-routing problem.

5.3 Hub location problems

M.E. O'Kelly (1987). A quadratic integer program for the location of interacting hub facilities. *European J. Oper. Res.* 32, 393-404.
T. Aykin (1990). On "A quadratic integer program for the location of interacting hub facilities". *European J. Oper. Res.* 46, 409-411.
The first paper presents a quadratic 0/1 programming formulation of the p-hub location problems in which each node must be connected to exactly one hub and the total fixed costs plus transportation costs must be minimized. It also proposes two heuristics which perform a complete enumeration of the locational configurations and allocate nodes to the closest hub or to the first or second closest hubs. The second paper comments on these heuristics.

A.V. Iyer and H.D. Ratliff (1990). Accumulation point location on tree networks for guaranteed time distribution. *Management Sci.* 36, 958-969.

Two location problems of p hubs in order to minimize the largest travel time and with centralized or decentralized sorting system for the distribution are considered. Polynomial time algorithms are provided if the transportation configuration is tree structured.

J.G. Klincewicz (1991). Heuristics for the p-hub location problem. *European J. Oper. Res.* 53, 25-37.

Various heuristics for the p-hub location problem in which each node is assigned to exactly one hub and the total transportation cost is minimized are examined. Computational experiments compare these heuristics on problems with up to 52 nodes and $2 \leq p \leq 10$.

J.F. Campbell (1994). Integer programming formulations of discrete hub location problems. *European J. Oper. Res.* 72, 387-405.

Integer programming formulations are presented for four types of hub location problems: the p-hub median, the uncapacitated hub location problem, p-hub center problems and hub covering problems.

D. Skorin-Kapov, J. Skorin-Kapov (1994). On tabu search for the location of interacting hub facilities. *European J. Oper. Res.* 73, 502-509.

A tabu search heuristic is proposed for the p-hub location problem in which each node is connected to exactly one hub and the total transportation cost is minimized. Computational experiments are reported for instances for $p = 2, 3, 4$ and up to 25 nodes.

M. O'Kelly, D. Skorin-Kapov, J. Skorin-Kapov (1995). Lower bounds for the hub location problem. *Management Sci.* 41, 713-721.

The p hub location problem with single node connection and where the total transportation cost is minimized and distances satisfy the triangle inequality is considered. A lower bound based on a linearization of the objective function is proposed as well as its modification obtained by using information from a know feasible solution. Computational experiments are reported for problems with between 10 and 25 nodes and $p = 2, 3$ and 4.

17 Flows and Paths

Ravindra K. Ahuja
Indian Institute of Technology, Kanpur

Contents

Highways, telephone lines, electrical power systems, computer chips, water delivery systems, and rail lines: these physical networks, and many others, are familiar to all of us. In each of these problem settings, we often wish to send some goods (vehicles, messages, electricity, or water) from one point to another, typically as efficiently as possible—that is, along a shortest path or via some minimum cost flow pattern. Although these problems trace their roots to the work of Kirchhoff and other great scientists of the last century, the topic of network flows as we know it today has its origins in the 1940's survey paper, we present a selected bibliography of some important problems in the area of network flows.

We use the following notation in this chapter. We denote by $G = (N, A)$ a directed

Annotated Bibliographies in Combinatorial Optimization, edited by M. Dell'Amico, F. Maffioli and S. Martello

network, which is defined by a set N of n nodes and a set A of m directed arcs. Each arc (i,j) in A has an associated cost (or, length) c_{ij} that denotes the cost per unit flow on the arc. We also associate with each arc (i,j) a capacity u_{ij} that denotes the maximum amount that can flow on the arc. We also associate with each node i in N an integer $b(i)$ representing its supply/demand.

In this chapter, we shall present the literature on the following network flow problems:

Shortest path problem. In this problem, we wish to find a path of minimum cost (or length) from a specified source node s to every other node in the network.

Maximum flow problem. This problem seeks a feasible solution that sends the maximum amount of flow from a specified source node s to another specified sink node t.

Minimum cost flow problem. We wish to determine a least cost shipment of a commodity through a network in order to satisfy demands at certain nodes from available supplies at other nodes.

Nonlinear flow problem. The minimum cost flow problem assumes that the cost of flow on each arc is a linear function of the amount of flow. In case the cost of flow is a nonlinear function of the amount of flow, we get the nonlinear flow problem. The convex cost flow problem assumes that arc flow costs are convex, and the concave cost flow problem assumes that the arc flow costs are concave.

Generalized flow problem. In the minimum cost flow problem, arcs conserve flow (that is, the flow entering an arc equals the flow leaving the arc). In the generalized flow problems, arcs might "consume" or "generate" flow.

Multicommodity flow problem. The multicommodity flow problem arises when several commodities use the same underlying network. Different commodities have different origins and destinations, but the sharing of the common arc capacities binds the different commodities together.

This chapter focuses on the literature published on the above network flow problems after 1985. The references given in Section 1.2 provide reference notes on the literature published prior to 1985. Even after 1985, a vast literature has been published and our bibliography includes only a subset of the literature. In deciding what to include, we have used our taste and preferences. We offer our humble apologies to the authors whose papers could not be included due to the space limitation.

1 Books and Surveys

In this section, we list some books and survey papers that serve as general introduction to the area of network flows. In Section 1.1 we give a survey of books. The references for two additional books, by Evans and Minieka [1992] and by Johnson and McGeoch [1993], are given in Section 8. In Section 1.2, we give references of only those survey papers whose scope encompasses more than one network flow problem. Surveys on specific flow problems are dealt with in the section devoted to that problem.

1.1 Books

L.R. Ford, D.R. Fulkerson (1962). *Flows in Networks*, Princeton University Press, Princeton, NJ.

This is the first book in network flows. Widely considered a classic book, it presents a thorough discussion of the early developments in the area.

E.L. Lawler (1976). *Combinatorial Optimization: Networks and Matroids*, Holt, Rinehart and Winston, New York.

It is another classic book and summarizes the developments in the area up to 1976. The book presents an excellent introduction of matroids.

J.L. Kennington, R.V. Helgason (1980). *Algorithms for Network Programming*, Wiley-Interscience, New York.

This book takes a linear programming perspective, and presents a comprehensive discussion of simplex based algorithms for network flow problems.

C.H. Papadimitriou, K. Steiglitz (1982). *Combinatorial Optimization: Algorithms and Complexity*, Prentice Hall, Englewood Cliff, NJ.

This widely used book develops a primal-dual perspective and shows that many well-known algorithms can be conceived of as primal-dual algorithms.

R.E. Tarjan (1983). *Data Structures and Network Algorithms*, SIAM, Philadelphia, PA.

The book describes some important data structures and their applications in improving algorithms for several network flow problems, including the minimum spanning tree problem, shortest path problem, maximum flow problem, minimum cost flow problem, and matching problems.

R.T. Rockafellar (1984). *Network Flows and Monotropic Optimization*, John Wiley & Sons, New York.

Monotropic optimization problems are those problems where a separable convex function is minimized subject to linear constraints. The book focuses on monotropic optimization problems arising in the area of network flows.

M. Gondran, M. Minoux (1984). *Graphs and Algorithms*, Wiley-Interscience, New York.

A very useful reference book of algorithms for network and graph problems. It also contains a rich collection of applications.

U. Derigs (1988). *Programming in Networks and Graphs*, Springer Verlag, Berlin.

The book focuses on matching problems. It gives theory and algorithms for bipartite matching, nonbipartite matching, and general b-matching problems. The book emphasises successive shortest path algorithms for the matching problems.

T.H. Cormen, C.L. Leiserson, R. L. Rivest (1990). *Introduction to Algorithms*, MIT Press and McGraw-Hill, New York.

This is an excellent text and reference book on algorithms and data structures. The book is very comprehensive; it covers both the classical material and modern developments such as amortized analysis and parallel algorithms.

M.S. Bazaraa, J.J. Jarvis, H.D. Sherali (1990). *Linear Programming and Network Flows*, Second Edition, Wiley, New York.
This is the second edition of a popular book addressing linear programming and network flows.

G. Ruhe (1991). *Algorithmic Aspects of Flows in Networks*, Kluwer Academic Publishers, Dordrecht.
The book gives a thorough treatment of multicriteria flows, parametric flows, detecting embedded network structures, and networks with additional constraints.

D.P. Bertsekas (1991). *Linear Network Optimization Algorithms and Codes*, MIT Press, Cambridge, MA.
The book focuses on relaxation and auction algorithms for network flow problems. The book also provides a diskette containing the computer codes of algorithms presented in the book.

K.G. Murty (1992). *Network Programming*, Prentice Hall, Englewood Cliffs, NJ.
This book uses a linear programming perspective in presenting the subject material.

R.K. Ahuja, T.L. Magnanti, J.B. Orlin (1993). *Network Flows: Theory, Algorithms, and Applications*, Prentice Hall, Englewood Cliffs, NJ.
This book presents an integrative view of the classical work and the latest advances in the area of network flow algorithms, with special emphasis on the worst-case complexity. The book presents a comprehensive discussion of scaling algorithms, contains a rich collection of applications, and detailed reference notes.

1.2 Surveys

J.E. Aronson (1989). A survey of dynamic network flows. *Ann. Oper. Res.* 20, 1-66.
Dynamic network flow models describe network-structured, decision making problems over time. The paper presents a state-of-the-art survey of the results, applications, algorithms and implementations for dynamic network flows.

R.K. Ahuja, T.L. Magnanti, J.B. Orlin (1989). Network flows. G.L. Nemhauser, A.H.G. Rinooy Kan, M.J. Todd (eds.), *Handbooks in Operations Research and Management Science. Vol. 1: Optimization*, North-Holland, Amsterdam, 211-369.
This paper presents a survey of shortest path, maximum flow, minimum cost flow, and assignment problem with particular emphasis on recent developments. The special features of the paper include radix heap algorithm for shortest paths, preflow-push algorithms for maximum flows, and scaling algorithms for minimum cost flows. The paper also presents a detailed set of reference notes on these problems.

A.V. Goldberg, E. Tardos, R.E. Tarjan (1990). Network flow algorithms, B. Korte,

L. Lovasz, H.J. Proemel, and A. Schrijver (eds.), *Paths, Flows and VLSI Design*, Springer-Verlag, 101-164.

The paper presents a survey of network flow algorithms with a special focus on preflow-push algorithms for the maximum flow problem and the cost scaling algorithms for the minimum cost flow problem.

2 Shortest Paths

Researchers have studied several different types of shortest path problems. Shortest path problems for networks with nonnegative arc lengths have been generally solved using label-setting algorithms. Shortest path problems for arbitrary arc lengths have been typically solved using label-correcting algorithms. The problem of finding shortest paths between all pairs of nodes is known as the all-pairs shortest path problem. We give pointers to selected literature on these three problem classes. We also give selected references on various generalizations of the shortest path problem.

We refer the reader to the following annontated bibliography on shortest path problem of the literature prior to 1980.

N. Deo, C. Pang (1984). Shortest path algorithms: taxonomy and annotation. *Networks* 14, 275-323.

2.1 Label setting algorithms

H.N. Gabow (1985). Scaling algorithms for network problems. *J. Comput. Syst. Sci.* 31, 148-168.

The cost scaling technique is used to obtain an $O(m \log C)$ algorithm for solving the shortest path problem with nonnegative integral arc lengths, where C denotes the largest arc cost.

M.L. Fredman, R.E. Tarjan (1987). Fibonacci heaps and their uses in improved network optimization algorithms. *J. ACM* 34, 596-615.

The authors propose a new data structure for implementing priority queues, called Fibonacci heaps data structure, which supports arbitrary deletion from an n-item heap in $O(\log n)$ amortized time and all standard heap operations in $O(1)$ amortized time. Using Fibonacci heaps, the authors obtain an $O(m + n \log n)$ algorithm for the shortest path problem with nonnegative arc lengths.

R.K. Ahuja, K. Mehlhorn, J.B. Orlin, R.E. Tarjan (1990). Faster algorithms for the shortest path problem. *J. ACM* 37, 213-223.

The paper describes the implementation of Dijkstra's algorithm using the radix heap data structure running in $O(m + n \log C)$ time. A variant running in time $O(m + n(\log C)^{1/2})$ is also proposed.

2.2 Label correcting algorithms

F. Glover, D.D. Klingman, N.V. Philips, R.F. Schneider (1985). New polynomial shortest path algorithms and their computational attributes. *Management Sci.*

31, 1106-1128.

The paper presents six new variants of the partitioning shortest path algorithm (certain type of label correcting algorithm) for the shortest path problem. Extensive computational results are also presented.

G. Gallo, S. Pallottino (1986). Shortest path methods: a unifying approach. *Math. Program. Study* 26, 38-64.

All the shortest path algorithms are shown to derive from one single prototype method. Reoptimization methods are also discussed.

M.S. Hung, J.J. Divoky (1988). A computational study of efficient shortest path algorithms. *Computers Oper. Res.* 15, 567-576.

Five efficient shortest path algorithms are implemented and compared.

D. Goldfarb, J. Hao, S.R. Kai (1990). Efficient shortest path simplex algorithms. *Oper. Res.* 38, 624-628.

Two efficient variants of the primal simplex algorithm for the shortest path problem are analysed and shown to require $O(n^2)$ pivots and $O(n^3)$ time.

H.D. Sherali (1991). On the equivalence between some shortest path algorithms. *Oper. Res. Lett.* 10, 61-65.

The paper shows the equivalence between a particular implementation of the partitioning shortest path algorithm, Moore's algorithm, and a dynamic programming algorithm.

D. Goldfarb, J. Hao, S.R. Kai (1991). Shortest path algorithms using dynamic breadth-first search. *Networks* 21, 29-50.

A new $O(nm)$ label-correcting algorithm is presented for determining shortest paths in a directed network or identifying the presence of a negative cycle. Variants of this algorithm are presented and some of these are computationally very effective.

J.F. Mondou, G.T. Crainic, S. Nguyen (1991). Shortest path algorithms: a computational study with the C programming language. *Computers Oper. Res.* 18, 767-786.

The study evaluates the computational efficiency of eight shortest path algorithms when coded in C programming languages. It is shown that one may expect savings of 20% to 30% by using the C language.

D.P. Bertsekas (1991). An auction algorithm for shortest paths. *SIAM J. Optim.* 1, 425-447.

In this paper, the author adapts his auction algorithm for the assignment problem to the shortest path problem.

S. Pallottino and M.G. Scutellà (1991). Strongly polynomial algorithms for shortest paths. *Ricerca Operativa* 60, 33-53.

The paper describes strongly polynomial variants of shortest path auction algorithms due to Bertsekas (1991).

J.B. Orlin, R.K. Ahuja (1992). New scaling algorithms for the assignment and minimum mean cycle problems. *Math. Program. (A)* 54, 41-56.

The paper describes $O(n^{1/2}m \log nC))$ algorithms to solve the shortest path problem with arbitary arc costs (but no negative cycles) and the minimum mean cycle problem. These algorithms use scaling of arc costs.

D.P. Bertsekas (1993). A simple and fast label correcting algorithm for shortest paths. *Networks* 23, 703-709.

An implementation of label correcting method is described which tries to scan nodes with small labels as early as possible.

D.P. Bertsekas, S. Pallottino, M.G. Scutellà (1995). Polynomial auction algorithms for shortest paths. *Computational Opt. Appl.* 4, 99-125.

The paper combines the auction algorithms due to Bertsekas (1991) and Pallottino and Scutellà (1991) and obtains algorithms with improved worst-case complexity and, at the same time, with attractive empirical behaviour .

S. Pallottino, M.G. Scutellà (1996). Dual algorithms for the shortest path tree problem. To appear in *Networks.*

The authors describe a new family of dual ascent algorithms for the shortest path problem which are quite suitable for reoptimization approaches.

V. Cherkassky, A.V. Goldberg, T. Radzik (1996). Shortest path algorithms: theory and experimental evaluation. *Math. Prog.* 73, 129-174.

The paper describes the results of an extensive computational study of shortest path algorithms, including some very recent algorithms. The study also identifies strengths and weaknesses of various algorithms.

2.3 All-pairs shortest path problem

M. Jun, T. Takaoka (1989). An $O(n(n^2/p + \log p))$ parallel algorithm to compute the all pairs shortest path and the transitive closure. *J. Info. Process.* 12, 119-124.

Multiprocessors with shared memory structured as a complete binary tree are used to obtain an $O(n(n^2/p+\log p))$ all-pairs shortest path algorithm using $p < n^2$ processors.

G.E. Pantziou, G.P. Sprakis, C.D. Zaroliagis (1992). Efficient parallel algorithms for shortest paths in planar digraphs. *BIT* 32, 215-236.

A parallel algorithm is developed to solve the all-pairs shortest path problem in directed planar networks.

M.B. Habbal, H.N. Koutsopoulos, S.R. Lerman (1994). A decomposition algorithm for the all-pairs shortest path problem on massively parallel computer architectures. *Transportation Sci.* 28, 292-308.

The paper examines the solution of the shortest path problem on massively parallel architectures using a network decomposition strategy. Computational results on the connection machine are presented.

2.4 Additional topics

R.K. Ahuja (1988). Minimum cost-reliability ratio path problem. *Computers Oper. Res.* 15, 83-89.

It is shown that the optimal solution of the minimum cost-reliability path problem is an efficient point of a bicriterion shortest path problem. This result is used to develop a parametric algorithm for the minimum cost-reliability ratio path problem.

C.C. Skiscim, B.L. Golden (1989). Solving k-shortest and constrained shortest path problem efficiently. *Ann. Oper. Res.* 20, 249-282.

The authors examine the problem of finding the k-shortest paths, the k-shortest paths without repeated nodes, and the shortest path problem with a single side constraint. Two versions of algorithms for all these problems are compared.

N.E. Young, R.E. Tarjan, J.B. Orlin (1991). Faster parametric shortest path and minimum-balance algorithms. *Networks* 21, 205-221.

Fibonacci heaps are used to improve a parametric shortest path problem and a minimum balance problem. Algorithms for both the problems run in $O(nm + n^2/\log n)$ time.

A.P. Punnen (1991). A linear time algorithm for the maximum capacity path problem. *European J. Oper. Res.* 53, 402-404.

The paper gives an $O(m)$ algorithm to solve the maximum capacity path problem in undirected networks.

D. Burton, Ph.L. Toint (1992). On an instance of the inverse shortest paths problem. *Math. Program. (A)* 53, 45-61.

Given a solution x^* and a cost vector c, the inverse optimization problem is to find another cost vector d so that x^* is optimal with respect to d and "distance" of d from c is minimum. The paper formulates and solves a class of inverse shortest path problems with Euclidean distance measure.

C.T. Tung and K.L. Chew (1992). A multicriteria Pareto-optimal path algorithm. *European J. Oper. Res.* 62, 203-209.

The paper presents an algorithm to determine a Pareto-optimal path for each efficient objective vector for the multicriteria shortest path problem, and also gives pointers to the related research.

3 Maximum Flows and Minimum Cuts

In Section 3.1, we give recent references of papers that focus on developing algorithms for the maximum flow problem with improved worst-case complexity. Papers on the empirical testing of maximum flow algorithms are covered in Section 3.2. Determining maximum flow values (or, minimum cut capacities) between all pairs of nodes can be done more efficiently than applying the maximum flow (or, minimum cut) algorithm $n(n-1)/2$ times. This problem is known as the *multi-terminal flow problem*, and we study it in Section 3.3. The maximum flow problems on planar networks can be solved more efficiently; we provide selected references in this area in Section 3.4. In Section 3.5, we give references on further special cases and generalizations of the maximum flow problem.

3.1 Maximum flows: Theoretical developments

H.N. Gabow (1985). Scaling algorithms for network problems. *J. Comput. Syst. Sci.* 31, 148-168.

This landmark paper describes an application of the scaling technique for the maximum flow problem solving it in $O(nm \log < U >)$ time.

A.V. Goldberg, R.E. Tarjan (1988). A new approach to the maximum flow problem. *J. ACM* 35 (1988), 921-940.

The paper introduces the concept of distance labels, which later became the backbone of new maximum flow algorithms and obtains an $O(nm \log(n^2/m))$ preflow-push algorithm for the maximum flow problem.

J. Cheriyan, S.N. Maheshwari (1989). Analysis of preflow push algorithms for maximum network flow. *SIAM J. Comput.* 18, 1057-1086.

This paper shows that in the preflow-push algorithm, if the flow is pushed from nodes with highest distance labels, then it runs in $O(n^2\sqrt{m})$ time. Further, by constructing parametrized classes of networks it shows that the worst-case time bounds of several preflow-push algorithms are tight.

R.K. Ahuja, J.B. Orlin (1989). A fast and simple algorithm for the maximum flow problem. *Oper. Res.* 37, 748-759.

The paper develops an improvement of Goldberg and Tarjan's algorithm using the excess-scaling technique without using any sophisticated data structures. This algorithm runs in $O(nm + n^2 \log U)$ time.

R.K. Ahuja, J.B. Orlin, R.E. Tarjan (1989). Improved time bounds for the maximum flow problem. *SIAM J. Comput.* 18, 939-954.

The paper extends the ideas of Ahuja-Orlin excess-scaling algorithm to develop two faster algorithms running in time $O(nm+n^2 \log U/ \log\log U)$ and $O(nm+n^2\sqrt{\log U})$. Further improvements are obtained using dynamic trees in these algorithms.

J. Cheriyan, T. Hagerup (1989). A randomized maximum-flow algorithm. *Proc. 30th IEEE Symp. Found. Comput. Sci.*, 118-123.

The paper proposes a randomized algorithm for the maximum flow problem that has an expected running time of $O(nm)$ for all $m \geq n \log^2 n$.

N. Alon (1990). Generating pseudo-random permutations and maximum flow algorithms. *Inform. Process. Lett.* 35, 201-204.

The paper describes a derandomized version of [Cheriyan and Hagerup 1989] algorithm and obtains a deterministic maximum flow algorithm what runs in $O(nm)$ time for all $m = \Omega(n^{5/3} \log n)$ and $O(nm \log n)$ time for all other values of n and m.

D. Goldfarb, J. Hao (1990). A primal simplex algorithm that solves the maximum flow problem in at most nm pivots and $O(n^2m)$ time. *Math. Program.(A)* 47, 353-365.

The paper describes the first polynomial-time implementation of the primal simplex algorithm for the maximum flow problem. The algorithm performs $O(nm)$ pivots and can be implemented to run in $O(n^2m)$ time.

J. Cheriyan, T. Hagerup, K. Mehlhorn (1990). Can a maximum flow be computed in O(nm) time? *Proc. 17th Int. Coll. Automata, Lang. and Program.*, 235-248.

The paper obtains an $O(n^3/\log n)$ algorithm for the maximum flow problem.

R.K. Ahuja, J.B. Orlin (1991). Distance-directed augmenting path algorithms for maximum flow and parametric maximum flow problems. *Naval Res. Log. Quart.* 38, 413-430.

Distance label based augmenting path algorithms are developed and applied to unit capacity networks, bipartite networks, and networks with parametric capacities.

G. Mazzoni, S. Pallottino, M.G. Scutellà (1991). The maximum flow problem: a max-preflow approach. *European J. Oper. Res.* 53, 257-278.

The first part of the paper presents a unified framework from which the preflow-push algorithms as well as the augmenting path algorithms can be derived as special cases. The second part of the paper presents results from a wide-ranging computational experimentation of maximum flow algorithms.

V. King, S. Rao, R.E. Tarjan (1992). A faster deterministic maximum flow algorithm. *Proc. of the 3rd Annual ACM-SIAM Symp. on Discr. Algorithms*, 157-164.

The paper gives an $O(nm + n^{2+\epsilon})$ algorithm for the maximum flow problem. For $m > n^{1+\epsilon}$, the algorithm runs in $O(nm)$ time.

G.R. Wassi (1993). A new polynomial-time algorithm for maximum value flow with an efficient parallel implementation. *Naval Res. Log. Quart.* 40, 393-414.

The paper proposes a variant of Dinic's algorithm suitable for parallel implementation.

R.D. Armstrong, Z. Jin, D. Goldfarb, W. Chen (1994). A strongly polynomial dual simplex algorithm for the maximum flow problem. To appear in *Math. Program.*

The paper gives efficient dual simplex algorithms for the maximum flow problem.

J. Hao and J.B. Orlin (1994). A faster algorithm for finding the minimum cut in a directed graph. *J. Algorithms*17, 424-446.

The paper shows how to find a minimum cut (not a minimum s-t cut) in a directed or an undirected graph by solving $(2n-2)$ maximum flow problems in a total time of $O(nm\log(n^2/m))$.

R.K. Ahuja, J.B. Orlin (1996). Equivalence of primal and dual simplex algorithms for the maximum flow problem. To appear in *Oper. Res. Lett.*.

The paper shows that any primal simplex algorithm for the maximum flow problem can be converted into a dual simplex algorithm that performs the same number of pivots and runs in the same time. The converse result is also true though in a somewhat weaker form.

3.2 Maximum flows: Empirical developments

H. Imai (1983). On the practical efficiency of various maximum flow algorithms. *J. Oper. Res. Soc. Japan* 26, 61-82.

The paper reports the result of an extensive study of maximum flow algorithms developed upto the 1970 and reported that Dinic's and Karzanov's algorithms are the fastest maximum flow algorithms in practice.

U. Derigs, W. Meier (1989). Implementing Goldberg's max-flow algorithm—a computational investigation. *Z. Oper. Res.* 33, 383-403.

It is shown that implementations of Goldberg's algorithm outperform Dinic's and Karzanov's methods by a substantial margin.

D.S. Johnson, C.C. McGeoch (1993). *Network Flows and Matching: First DIMACS Implementation Challenge*, AMS, New York.

The book contains five papers of computational studies on maximum flow algorithms performed by different researchers. The algorithms investigated include the preflow-push algorithms and its several variants, excess scaling algorithms, as well as dynamic tree implementations of preflow push algorithms. Preflow push algorithms (without the dynamic tree data srtucture) are found to be consistently superior to other algorithms.

D.P. Bertsekas (1995). An auction algorithm for the max-flow problem. *J. Optim. Theory Appl.* 87, 69-101

The paper develops an auction algorithm for the maximum flow problem and finds that it is empirically faster than the highest-label preflow-push algorithm for many classes of networks.

R.K. Ahuja, M. Kodialam, A.K. Mishra, J.B. Orlin (1996). Computational testing of maximum flow algorithms. To appear in *European J. of Oper. Res.*

The paper presents the results of an extensive empirical study of maximum flow algorithms using representative operation counts and gives new insights into the behaviour of maximum flow algorithms.

3.3 All-pairs maximum flow and minimum cut problems

D. Gusfield (1990). Very simple methods for all pairs network flow analysis. *SIAM J. Comput.* 19, 143-155.

The paper describes a new method for solving the multi-terminal flow problem that is simpler than the existing methods.

C.K. Cheng, T.C. Hu (1990). Ancestor tree for arbitrary multi-terminal cut functions. R. Kannan and W.R. Pulleyblank (eds.). *Proc. 1st HPS Conference IPCO*, University of Waterloo Press, Waterloo.

This paper proposes another method to solve the all-pairs maximum flow problem.

M. Padberg, G. Rinaldi (1990). An efficient algorithm for the minimum capacity cut problem. *Math. Program. (A)* 47, 19-36.

The paper describes yet another method for the multiterminal cut problem with the same worst-case bound as several other existing methods (e.g., $O(n^4)$) but which is easier to implement and very efficient in practice.

D. Hartvigsen, R. Mardon (1994). The all-pairs min-cut problem and the minimum cycle basis problem on planar graphs. *SIAM J. Discr. Math.* 7, 403-418.

The all-pairs minimum cut problem in planar graphs is considered and is shown to be equivalent the minimum cycle basis problem on the dual graph.

3.4 Planar networks

R. Hassin (1981). Maximum flow in (s,t)-planar networks. *Inform. Process. Lett.* 13, 107-108.

The paper shows that solving a maximum flow problem in an undirected (s,t)-planar network can be transformed to a shortest path problem. This transformation yields an $O(n \log n)$ algorithm for the maximum flow problem.

R. Hassin, D.B. Johnson (1985). An $O(n \log^2 n)$ algorithm for maximum flow in undirected planar networks. *SIAM J. Comput.* 14, 612-624.

The paper generalizes the ideas in Hassin's (1981) paper and gives an $O(n \log^2 n)$ algorithm for determining maximum flow in undirected planar networks.

D.B. Johnson (1987). Parallel algorithms for minimum cuts and maximum flows in planar networks. *J. ACM* 34, 950-967.

Algorithms are given that compute maximum flow in planar directed networks either in $O((\log n)^3)$ parallel time using $O(n^4)$ processors, or in $O((\log n)^2)$ parallel time using $O(n^6)$ processors.

V. Adlakha, B. Gladysz, J. Kamburowski (1991). Minimum flows in (s,t) planar networks. *Networks* 21, 767-773.

The minimum flow problem is to find a flow of minimum value honoring arc's lower bounds. The paper presents a linear time algorithm for solving this problem in (s,t)-planar directed networks.

3.5 Special cases and additional topics

D.W. Matula (1987). Determining edge connectivity in O(nm). *Proc. 28th Annual IEEE Symp. Found. Comput. Sci.* 249-251.

The paper presents an $O(nm)$ algorithm to determine edge connectivity of an undirected network.

D. Gusfield, C. Martel, D. Fernandez-Baca (1987). Fast algorithms for bipartite network flow. *SIAM J. Comput.* 16, 237-251.

In a bipartite network $G = (N_1 \cup N_2, A)$ where $N_1 \ll N_2$, network flow algorithms can often be substantially speeded up. The paper describes several interesting applications of maximum flow problems on these types of bipartite networks, and also describes speedups in several augmenting path and preflow-push algorithms.

Y. Mansour, B. Schieber (1988). *Finding the edge connectivity of directed graphs*, Research Report RC 13556, IBM Thomas J. Watson Research Centre, Yorktown Heights, NJ.

The paper presents an $O(nm)$ algorithm to determine the edge connectivity of a directed network.

G. Gallo, M.D. Grigoriadis, R.E. Tarjan (1989). A fast parametric maximum flow algorithm and applications. *SIAM J. Comput.* 18, 30-55.

The paper considers a special case of the parametric maximum flow problem, where the arcs emanating from the source node are increasing functions of a parameter and arcs entering the sink node are decreasing function of the parameter. An $O(nm \log(n^2/m))$ algorithm is developed to solve this problem.

D. Fernandez-Baca, C.U. Martel (1989). On the efficiency of maximum flow algorithms on networks with small integer capacities. *Algorithmica* 4, 173-189.

Maximum flow algorithms with "small" integer capacities are presented and analyzed.

D.K. Wagner, H. Wan (1993). A polynomial-time simplex method for the maximum k-flow problem. *Math. Program. (A)* 60, 115-123.

The paper presents a network simplex algorithm for a generalization of the maximum flow problem.

H. Nagamochi, T. Ono, T. Ibaraki (1994). Implementing an efficient minimum capacity cut algorithm. *Math. Program.* 67, 325-341.

The paper algorithms for computing the minimum capacity cut of an undirected network. Computational results are also given.

D. Gusfield, E. Tardos (1994). A faster paramatric minimum-cut algorithm. *Algorithmica* 11, 278-290.

The paper improves and generalizes the algorithm of [Gallo, Grigoriadis, Tarjan 1989] for the parametric maximum flow problem.

R.K. Ahuja, J.B. Orlin, C. Stein, R.E. Tarjan (1994). Improved algorithms for bipartite network flow problems. *SIAM J. Comput.* 23, 906-933.

The paper improves the results of [Gusfield et al. 1987] and gives better speedups for maximum flow, parametric maximum flow and minimum cost flow problems. The paper considers dynamic tree implementations as well as parallel algorithms.

4 Minimum Cost Flows

The minimum cost flow problem is the central object of study in network flows and has been extensively studied in the past decade. We present here only a selected list of references focusing on (i) simplex based algorithms for minimum cost flow problem; (ii) combinatorial algorithms for minimum cost flow problem; and (iii) empirical investigations of minimum cost flow algorithms.

4.1 Simplex algorithms

J.B. Orlin (1984). *Genuinely polynomial simplex and non-simplex algorithms for the minimum cost flow problem*, Technical Report 1615-84. Sloan School of Management, MIT, Cambridge, MA.

The paper describes the first strongly-polynomial time simplex algorithm for the minimum cost flow problem. This is a dual simplex algorithm and performs $O(n^3 \log n)$ pivots for the uncapacitated minimum cost flow problem.

J.B. Orlin (1985). On the simplex algorithm for networks and generalized networks. *Math. Program. Study* 24, 166-178.

The paper investigates the behavior of the network simplex algorithm for the minimum cost flow problem with Dantzig's pivot rule (that is, selecting a nonbasic arc with the maximum violation as the entering arc). It shows that for the shortest path and assignment problems, the pivot rule yields polynomial time algorithms.

D. Goldfarb, J. Hao (1988). *Polynomial-time primal simplex algorithms for the minimum cost network flow problem*, Technical Report, Dept. of Industrial Engineering and Operations Research, Columbia University, New York.

The paper presents two variants of the primal network simplex algorithm which solve the minimum cost flow problem in at most $O(n^2 m^2 \log n)$ pivots. The algorithms are not genuine primal simplex algorithms in the sense that some pivots can increase the objective function value of the solution maintained.

A.I. Ali, R. Padman, H. Thiagarajan (1989). Dual algorithms for pure network problems. *Oper. Res.* 37, 159-171.

Computational results presented in the paper indicate that the dual reoptimization is faster than primal reoptimization.

D. Goldfarb, J. Hao, S. Kai (1990). Anti-stalling pivot rules for the network simplex algorithm. *Networks* 20, 79-91.

Stalling in the simplex algorithm is defined as an exponentially long sequence of consecutive degenerate pivots without cycling. The paper describes pivot rules that prevent both cycling and stalling.

R.E. Tarjan (1991). Efficiency of the primal network simplex algorithm for the minimum-cost circulation problem. *Math. Oper. Res.* 16, 272-291.

The paper describes a pivot rule for the network simplex algorithm for the minimum cost flow problem which performs at most $n^{(\log n)/2+O(1)}$ pivots. When the cost-increasing pivots are allowed, then a rule is proposed with a bound of $O(nm \min\{\log(nC), m \log n\})$ on the number of pivots.

R.K. Ahuja, J.B. Orlin (1992). The scaling network simplex algorithm. *Oper. Res.* 40, S5-S13.

The paper incorporates capacity scaling in the network simplex algorithm for the minimum cost flow problem and uses a pivot rule that always selects an arc with "sufficiently large" violation as the entering arc. Specializing these results for the assignment and shortest path problems, the authors obtain $O(nm \log C)$ algorithms for these problems.

J.B. Orlin, S.A. Plotkin, E. Tardos (1993). Polynomial dual network simplex algorithms. *Math. Program.* 60, 255-276.

The paper uses polynomial and strongly-polynomial capacity scaling algorithms for the transshipment problems to design a polynomial dual network simplex pivot rule with $O(m^2 \log n)$ pivots.

R.S. Barr, B.L. Hickman (1994). Parallel simplex for large pure network problems: Computational testing and sources of speedup. *Oper. Res.* 42, 65-80.

The paper reports a parallel implementation of the network simplex algorithm for the minimum cost flow problem that decomposes both the pivoting and pricing operations. The parallel algorithm is capable of substantial speedups as parallel computing units are added.

R.D. Armstrong, Z. Jin (1994). A new strongly polynomial dual network simplex algorithm. To appear in *Math. Program. (B)*.

The paper describes a new dual simplex algorithm for the minimum cost flow problem.

P.T. Sokkalingam, P. Sharma, R.K. Ahuja (1996). A new primal simplex pivot rule for the minimum cost flow problem. To appear in *Math. Program. (B)*.

The paper studies the minimum ratio pivot rule for the minimum cost flow problems and obtains a pseudo-polynomial time algorithm. Specializing this approach for the shortest path and assignment problems yields $O(nm + n^2 \log n)$ time algorithms for these problems.

D. Goldfarb and Z. Jin (1996). *Strongly polynomial excess scaling and dual simplex algorithms for the minimum cost flow problem.* Technical Report, Dept. of IE & OR, Columbia University, New York.

The paper describes two dual simplex algorithms for the minimum cost flow problem that solve it in $O(m^2(m + n \log n))$ time and perform $O(m^2 n)$ pivots.

J.B. Orlin (1996). A polynomial time primal network simplex algorithm for minimum cost flows. To appear in *Math. Program. (B)*.

This recent paper settled a long standing open question by developing the first polynomial-time primal simplex algorithm for the minimum cost flow problem.

4.2 Combinatorial algorithms

E. Tardos (1985). A strongly polynomial minimum cost circulation algorithm. *Combinatorica* 5, 247-255.

The paper settled a long-standing open problem by giving the first strongly polynomial-time algorithm for the minimum cost flow problem. The running time of the algorithm is $O(m^4)$.

S. Fujishige (1986). A capacity-rounding algorithm for the minimum cost circulation problem: a dual framework of the Tardos algorithm. *Math Program.* 35, 298-308.

The paper takes an approach that is a dual of [Tardos 1985] and obtains a strongly-polynomial time algorithm with better worst-case complexity. The suggested algorithm runs in $O(m^2(m + n \log n) \log n)$ time.

D.P. Bertsekas, J. Eckstein (1988). Dual coordinate step methods for linear network flow problems. *Math. Program. (B)* 42, 203-243.

The authors review a class of network flow methods which are amenable to distributed implementations, present some specific methods, and also discuss implementation issues.

D.P. Bertsekas, P. Tseng (1988). Relaxation methods for minimum cost ordinary and generalized network flow problems. *Oper. Res.* 36, 93-114.

Using ideas from Gauss-Seidel relaxation methods, the authors propose a new class of algorithms for the minimum cost flow problem. These methods are based on the iterative improvement of a dual cost, and their computational performance is excellent for many classes of network flow problems.

Z. Galil, E. Tardos (1988). An $O(n^2(m + n \log n) \log n)$ min-cost flow algorithm. *J. ACM* 35, 374-386.

The paper improves [Fujishige 1986] algorithm and obtains an $O(n^2(n + m \log n) \log n)$ algorithm to solve the minimum cost flow problem.

D.P. Bertsekas, D.A. Castanon (1989). The auction algorithm for the transportation problem. *Ann. Oper. Res.* 20, 67-96.

The paper generalizes the auction algorithm for the assignment problem and gives computational results.

A.V. Goldberg, R.E. Tarjan (1989). Finding minimum-cost circulations by canceling negative cycles. *J. ACM* 36, 873-886.

The paper describes several polynomial-time implementations of the well-known cycle-canceling algorithm for the minimum cost flow problem. It shows that if the algorithm always cancels negative cycles with minimum mean cost then it performs $O(\min\{nm \log(nC), nm^2 \log n\})$ iterations.

F. Barahona, E. Tardos (1989). Note on Weintraub's minimum cost circulation algorithm. *SIAM J. Comput.* 18, 579-583.

The paper describes another polynomial-time implementation of the cycle-canceling algorithm, which proceeds by augmenting flows along arc disjoint cycles with maximum improvement and can be implemented in $O(m^2(m + n \log n) \log((mCU))$ time.

H.N. Gabow, R.E. Tarjan (1989). Faster scaling algorithms for network problems. *SIAM J. Comput.* 18, 1013-1036.

The paper presents cost scaling based algorithms for the assignment problem, the transportation problem, and the minimum cost flow problem, and obtains the best available time bounds for several classes of problems.

A.V. Goldberg, R.E. Tarjan (1990). Solving minimum cost flow problem by successive approximation. *Math. Oper. Res.* 15, 430-466.

The paper describes several cost scaling algorithms for the minimum cost flow problem. The fastest of these runs in $O(nm \log(n^2/m) \log(nC))$ and uses both the finger tree and dynamic tree data structure.

K. Masuzava, S. Mizuno, M. Mori (1990). A polynomial-time interior point algorithm for minimum cost flow problems. *J. Oper. Res. Soc. Japan* 33, 157-167.

The paper describes an interior point algorithm for the minimum cost flow problem with polynomial running time.

A.B. Gamble, A.R. Conn, W.R. Pulleyblank (1991). A network penalty method. *Math. Program.* 50, 53-73.

The paper describes an adaptation of the nonlinear penalty function method for the minimum cost flow problem and presents computational results.

C. Wallacher, U. Zimmermann (1992). A combinatorial interior point method for network flow problems. *Math. Program. (A)* 56, 321-335.

Based on a variant of Karmarkar's algorithm, the authors develop a combinatorial interior point algorithm for the minimum cost flow problem running in time $O(mn^2L)$, where L denotes the total length of the input data.

R.K. Ahuja, A.V. Goldberg, J.B. Orlin, R.E. Tarjan (1992). Finding minimum-cost flows by double scaling. *Math. Program.* 53, 243-266.

The paper combines the capacity scaling technique due to Edmonds-Karp and cost scaling technique due to Goldberg-Tarjan to develop a double scaling algorithm for the minimum cost flow problem with worst-case complexity of $O(nm(\log\log U) \log(nC))$.

T.R. Ervolina, S.T. McCormick (1993). Canceling most helpful total cuts for minimum cost network flow. *Networks* 23, 41-52.

The authors describe a dual algorithm which proceeds by canceling "most helpful total cuts" (which are the cuts that lead to the maximum possible increase in the dual objective function). The suggested algorithm runs in $O(n \log(nC) S(n, m))$ time, where $S(n, m)$ is the time needed to solve a shortest path problem with nonnegative arc lengths.

J.B. Orlin (1993). A faster strongly polynomial minimum cost flow algorithm. *Oper. Res.* 41, 338-350.

The paper develops a clever variant of the well-known Edmonds-Karp scaling algorithm with a running time of $O(m(m + n \log n) \log n)$. This algorithm achieves currently the best time bound to solve the minimum cost flow problem in the worst case.

4.3 Empirical investigations

M.D. Grigoriadis (1986). An efficient implementation of the network simplex method. *Math. Program. Study* 26, 83-111.

This paper describes an efficient implementation of the network simplex algorithm for the minimum cost flow problem.

J.J. Divoky, M.S. Hung (1990). Performance of shortest path algorithms in network flow problems. *Management Sci.* 36, 661-673.

The minimum cost flow problems can be solved by successive augmentations along shortest paths. The paper investigates the performance of shortest path algorithms in this context.

R.G. Bland, D.L. Jensen (1992). On the computational behavior of a polynomial-time network flow algorithm. *Math. Program.* 54, 1-39.

The experiments reported in the paper indicate that the computational behavior of scaling algorithms may be much better than had been presumed earlier.

R.K. Ahuja, J.B. Orlin (1996). Use of representative counts in computational testings of algorithms. *ORSA J. Comput.* 8, No. 3.

Using *representative operations counts*, the paper describes a computational study of the network simplex algorithm for the minimum cost flow problem and finds that updating node potentials is the bottleneck step in the algorithm.

D.S. Johnson, C.C. McGeoch (eds.) (1993). *Network Flows and Matching: First DIMACS Implementation Challenge.* AMS, New York.

The book contains six papers of computational studies on minimum cost flow algorithms performed by different researchers. The algorithms investigated include the network simplex algorithms, relaxation algorithm, push-relabel algorithms, cycle-canceling algorithms, as well as interior point algorithms.

5 Nonlinear Cost Network Flows

The minimum convex cost flow problem is, in general, much easier than the minimum concave cost flow problem, or the problems with arbitrary cost structure. We present the literature on these two problem subtypes in two different subsections.

5.1 Convex cost flow problems

R.K. Ahuja, J.L. Batra, S.K. Gupta (1984). A parametric algorithm for the convex cost network flow and related problems. *European J. Oper. Res.* 16, 222-235.

The paper describes a parametric simplex algorithm for solving the convex cost network flow problems with separable piecewise linear costs. This suggested approach can also solve (i) certain classes of generalized flow problems; (ii) a time-cost tradeoff problem in CPM-networks; and (iii) a capacity expansion problem.

M. Minoux (1984). A polynomial algorithm for minimum quadratic cost flow problems. *European J. Oper. Res.* 18, 377-387.

The paper applies Edmonds-Karp capacity scaling algorithm to solve the convex cost flow problems with quadratic cost structure in polynomial time.

M. Minoux (1986). Solving integer minimum cost flows with separable convex cost objectivie polynomially. *Math. Program. Study* 26, 237-239.

The paper adapts the capacity scaling algorithm of Minoux (1984) to determine optimal integer flows in a convex cost network flow problem in polynomial time.

S.A. Zenios, J.M. Mulvey (1986). Relaxation techniques for strictly convex network problems. *Ann. Oper. Res.* 5, 517-538.

The paper is an extension of Bertsekas-Tseng approach for solving the minimum cost flow problem and uses Gauss-Seidel type relaxation techniques. Extension to generalized network is also discussed.

A. Tamir (1993). A strongly polynomial algorithm for minimum convex separable quadratic cost flow problems on two-terminal series-parallel networks. *Math. Program.* 59, 117-132.

The paper presents strongly-polynomial algorithms to find rational and integer flow vectors that minimize a convex separable quadratic cost function on two-terminal series-parallel graphs.

5.2 Concave and general cost flow problems

R.E. Erickson, C.L. Monma, A.F. Veinott, Jr. (1987). Send-and-split method for minimum-concave-cost network flows. *Math. Oper. Res.* 12, 634-664.

The authors give a dynamic programming method for the concave cost network flow problem which consists of repeatedly solving set-splitting and shortest path problems.

S.A. Zenios, R.A. Lasken (1988). Nonlinear network optimization on a massively parallel connection machine. *Ann. Oper. Res.* 14, 147-165.

The paper reviews applications of network optimization problems with nonlinear objectives with possible generalized arcs and with particular emphasis on large-scale implementations.

R. Katsura, M. Fukushima, T. Ibaraki (1989). Interior methods for nonlinear minimum cost network flow problems. *J. Oper. Res. Soc. Japan* 32, 174-199.

The paper describes feasible descent methods for nonlinear network flow problems which successively generate search directions based on the idea of the Newton method.

R.S. Dembo, J.M. Mulvey, S.A. Zenios (1989). Large-scale nonlinear network models

and their applications. *Oper. Res.* 37, 353-372.
The paper reviews applications of network optimization problems with nonlinear objectives and, possibly generalized arcs, and with special-emphasis on large-scale implementations.

Ph.L. Toint, D. Tuyttens (1990). On large scale nonlinear network optimization. *Math. Program. (B)* 48, 125-159.
The paper describes a quasi-Newton methods for solving large scale nonlinear network flow problems.

G.M. Guisewite, P.M. Pardalos (1990). Minimum concave-cost network flow problems: applications, complexity, and algorithms. *Ann. Oper. Res.* 25, 75-100.
The authors discuss a wide range of results for minimum concave-cost network flow problems including (i) related applications in production, inventory planning, and communication network design, (ii) new complexity results, and (iii) an overview of solution techniques.

S. Ibaraki, F. Masao, T. Ibaraki (1991). Dual-based Newton methods for nonlinear minimum cost network flow problems. *J. Oper. Res. Soc. Japan* 34, 263-286.
The paper describes several descent methods for solving nonlinear network flow problems which successively generate search directions and perform line searches.

G.M. Guisewite, P.M. Pardalos (1991). Single-source uncapacitated minimum concave-cost network flow problems. *Oper. Res.* 90, 703-713.
The authors investigate complexity issues and present computational results for the single-source uncapacitated version of the minimum concave-cost network flow problem.

P.T. Thach (1992). A decomposition method using a pricing mechanism for min concave cost flow problems with hierarchical structure. *Math. Program. (A)* 53, 339-359.
The paper develops a decomposition method for solving the concave network flow problem for directed and undirected networks with a hierarchical structure.

P.M. Pardalos, G.M. Guisewite (1993). Parallel computing in nonconvex programming (1993). *Ann. Oper. Res.* 43, 87-107.
The paper describes parallel neighborhood search algorithms for finding local optimal solutions of the minimum concave cost flow problems.

6 Generalized Flows

We will give reference notes for the literature on the generalized maximum flow problem (where we want to send desired flow without considering arc costs), and the generalized minimum cost flow problem (where we want to send the desired flow at minimum cost). Several combinatorial algorithms have been suggested for the generalized maximum flow problem, but the generalized minimum cost flow problem has mostly been solved by specializing linear programming techniques.

6.1 Generalized maximum flow problem

A.V. Goldberg, S.A. Plotkin, E. Tardos (1991). Combinatorial algorithms for the generalized circulation problem. *Math. Oper. Res.* 16, 351-381.

The paper studies the generalized maximum flow problem, and suggests the first polynomial-time combinatorial algorithms for solving it. The paper describes two algorithms: the first algorithm, called *Algorithm MCF*, is based on the repeated application of the a minimum cost flow subroutine, and the second algorithm, called the *Fat-Path algorithm*, proceeds by augmenting along fat (big improvement) paths.

T. Radzik (1993). Faster algorithms for the generalized network flow problem. *Proc. 34-th Annual IEEE Symp. Found. Comput. Sci.* 438-448.

The paper presents an improvement of the Fat-Path algorithm, due to Goldberg, Plotkin and Tardos (1991).

E. Cohen, N. Megiddo (1994). New algorithms for the generalized network flow problem. *Math. Program.* 64, 325-336.

The paper describes a new algorithm for the generalized network flow problem. This algorithm is then modified to obtain a strongly-polynomial approximation scheme for generalized flows.

D. Goldfarb, Z. Jin, J.B. Orlin (1996). Polynomial-time highest-gain augmenting path algorithms for the generalized circulation problem. Research Report, Sloan School of Management, MIT, Cambridge, MA.

The paper develops an $O(m^2(m + n \log n) \log B)$ algorithm for the generalized maximum flow problem, which is currently the fastest algorithm for solving it.

6.2 Generalized minimum cost flow problems

M.D. Chang, M. Engquiswt, R. Finkel, R.R. Meyer (1988). A parallel algorithm for generalized networks. *Ann. Oper. Res.*14, 125-145.

The paper describes three parallel primal simplex variants for solving the generalized network flow problem. Computational results indicate that speedups are almost linear in the number of processors.

P.M. Vaidya (1989). Speeding up linear programming using fast matrix multiplication. *Proc. 30th Annual IEEE Symp. Found. Comput. Sci.* 332-337.

The best bound until date for the generalized circulation problem is achieved in this paper by specializing the linear programming algorithm reported in this paper.

F. Glover, D. Klingman, N. Phillips (1990). Netform modeling and applications. *Interfaces* 20, 7-27.

This survey paper describes several applications of the generalized flow problems.

R.H. Clark, J.L. Kennington, R.R. Meyer, M. Ramamurti (1992). Generalized networks: parallel algorithms and an empirical analysis. *ORSA J. Comput.* 4, 132-145.

The paper develops simplex-based parallel algorithms for the generalized network flow problem. The largest test problem solved contained 30,000 nodes and 1.2 million arcs and displayed a speedup of 13 on 15 processors.
N.D. Curet (1994). An incremental primal-dual method for generalized flows. *Computers Oper. Res.* 21, 1051-1059.
The paper develops a primal-dual simplex method and finds it to be faster than the generalized network simplex algorithm.

6.3 Directed hypergraphs

Very recently, a new research area has been developed, mainly by G. Gallo and his associates at University of Pisa, Italy, on flow problems on directed hypergraphs. Whereas in a directed graph, each arc has one tail node and one head node, but in a directed hypergaph an arc may have multiple tail nodes and multiple head nodes. Given below are some sample papers in this area.
G. Gallo, G. Longo, S. Nguyen, S. Pallottino (1993). Directed hypergraphs and applications. *Discr. Appl. Math.*42, 177-201.
R. Cambini, G. Gallo, M.G. Scutellà (1996). Flows on hypergraphs. To appear in *Math. Program. (B).*
R.G. Jeroslow, R.K. Martin, R.R. Rardin, J. Wang (1992). Gainfree Leontief substitution flow problems. *Math. Program.* 57, 375-414.

7 Multicommodity Flows

The multicommodity flow problems have been studied quite extensively in the recent past. We classify the literature in two categories: where cost is relevant (that is, the multicommodity minimum cost flow problem) and where it is not (that is, the multicommodity maximum flow problem).

7.1 Multicommodity minimum cost flow problem

A.A. Assad (1978). Multicommodity network flows: a Survey. *Networks* 8, 37-91.
The paper discusses solution techniques for both linear and nonlinear multicommodity flow problems. For the former problem, it describes decomposition, partitioning, compact inverse methods, and primal-dual algorithms. For the latter problem, it describes a variety of feasible direction methods.

J.L. Kennington (1978). A survey of linear cost multicommodity network flows. *Oper. Res.* 26, 209-236.
This survey paper presents the best-known linear multicommodity models, properties exhibited by special cases of these models, and specializations of the linear programming methods that use its special structure.

I. Ali, D. Barnett, K. Farhangian, J. Kennington, B. Patty, B. Shetty, B. McCarl, P. Wong (1984). Multicommodity network problems: applications and computations. *IIE Trans.* 16, 127-134.

This study presents three real-world applications of multicommodity flow problems and also the results of a computational study.

G. Saviozzi (1986). Advanced start for the multicommodity network flow problem. *Math. Program. Study* 26, 221-224.
Advanced starting passes to solve the multicommodity network flow problems are proposed. The underlying idea is to use the Lagrangian relaxation on coupling (bundle) constraints and use subgradient optimization.

M. Minoux (1987). Network Synthesis and dynamic network optimization. *Ann. Discr. Math.* 31, 283-324.
This paper surveys the past work concerning two important and quite general problems arising in the area of distributed telecommunications (telephone and/or data processing) networks: (a) network synthesis under non-simultaneous single-commodity or multicommodity flow requirements; and (b) determining an optimal investment policy for meeting increasing multicommodity flow requirements over a given time period.

I.C. Choi, D. Goldfarb (1989). Solving multicommodity network flow problems by an interior point method. T. Coleman and Q. Li (eds.). *Proc. SIAM Workshop on Large-Scale Numerical Optimization.*
The paper describes an interior point algorithm for the muticommodity flow problem.

R.R. Meyer, G.L. Shultz (1991). An interior point method for block-angular optimization. *SIAM J. Optim.* 1, 121-152.
The paper describes a three-stage primal feasible log-barrier method for the solution of multicommodity flow problems.

Y.L. Chen, Y.H. Chin (1992). Multicommodity network flows with safety considerations. *Oper. Res.* 40, S48-S55.
Most of the previous research on the multicommodity flow problem has assumed that each arc has two associated quantitites: cost and capacity.
This paper adds the third attribute, the degree of difficulty, and shows that this generalized problem is polynomially equivalent to the original problem.

M.C. Pinar, S.A. Zenios (1992). Parallel decomposition of multicommodity network flows using a linear-quadratic penalty algorithm. *ORSA J. Comput.* 4, 235-249.
The paper describes a parallel algorithm for solving the multicommodity network flow problem.

C. Barnhart (1993). Dual-ascent methods for large-scale multicommodity flow problems. *Naval Res. Log. Quart.* 40, 305-324.
The paper describes a heuristic algorithm for the multicommodity flow problem which uses a dual-ascent heuristic to obtain an advanced starting solution and then uses a primal solution generator to improve the solution.

C. Barnhart, Y. Sheffi (1993). A network based primal-dual heuristic for the solution of multicommodity network flow problems. *Transportation Sci.* 27, 102-117.

A primal-dual type heuristic solution procedure is presented to solve large-scale multicommodity flow problems and is used to solve a large scale freight assignment problem encountered in the trucking industry.

J.M. Farvolden, W.B. Powell, I.J. Lustig (1993). A primal partitining solution for the arc-chain formulation of a multicommodity network flow problem. *Oper. Res.*, 41, 669-693.

The paper presents a new approach for the multicommodity network flow problem based upon primal partitioning and decomposition techniques which simplifies the computations required by the simplex method.

K.L. Jones, I.J. Lustig, J.M. Farvolden, W.B. Powell (1993). Multicommodity network flows: The impact of formulation on decomposition. *Math. Program.* 62, 95-117.

This paper investigates the impact of problem formulation of Dantzig-Wolfe decomposition for the multicommodity network flow problem. It is shown that solving the path-based problem formulations results in lower CPU times.

S.A. Zenios (1993). Data-parallel computing for network-structured optimization problems. *Computational Opt. Appl.* 3, 199-242.

This survey paper reviews several approaches suitable for implementation on data-parallel computers, for both single-commodity and multicommodity flow problems.

M.C. Pinar, S.A. Zenios (1994). A data-level parallel linear-quadratic penalty algorithm for multicommodity network flows. *ACM Trans. Math. Software* 20, 531-532.

The paper describes a software implementation of the parallel decomposition algorithm and applies it to a set of test problems drawn from a military airlift command application.

A.V. Karzanov (1994). Minimum cost multiflows in undirected networks. *Math. Program.* 66, 313-325.

The paper studies a special type of multicommodity flow problem which have half-integer optimal solution and gives a pseudo-polynomial time algorithm.

R. De Leone, S. Kontogiorgis, R.R. Meyer, A. Zakarian, G. Zakeri (1994). Coordination in coarse-grained decomposition. *SIAM J. Optim.* 4, 777-793.

The paper reviews three different approaches suitable for the parallel solution of multicommodity flow problems and presents computational results.

M. D. Grigoriadis, L.G. Kahchiyan (1994). Fast approximate schemes for convex programs with many blocks and coupling constraints. *SIAM J. Optim.* 4, 86-107.

The paper describes an approximation algorithm, based on an exponential-function barrier approach, suitable for the solution of block-angular convex programs such as multicommodity flow problems.

C.A. Hane, C. Barnhart, E.L. Johnson, R.E. Marsten, G.L. Nemhauser, G. Sigismondi (1995). The fleet assignment problem: solving a large-scale integer program.

Math. Program. 70, 211-232.
The paper shows how a careful utilization of standard optimization instruments can lead to the efficient solution of large-scale integer multicommodity flow problems.

D. Bienstock, O. Gunluk (1995). Computational experience with a difficult mixed-integer multicommodity flow problem. *Math. Program.* 68, 213-237.
The paper considers a specific type of integer multicommodity flow problem arising in the study of lightwave networks and describes a cutting plane algorithm.

R.R. Schneur, J.B. Orlin (1995). A scaling algorithm for multicommodity flow problems. To appear in *Oper. Res.*
The paper relaxes the capacity constraints in the formulation of the multicommodity flow problems and penalizes the violations of the capacities in the objective function. It uses a scaling techniques to increase the penalties.

7.2 Multicommodity maximum flow problem

K. Matsumoto, T. Nishizeki, N. Saito (1985). An efficient algorithm for finding multicommodity flows in planar networks. *SIAM J. Comput.* 14, 289-302.
The paper presents an efficient algorithm for finding multicommodity flows in planar networks. The algorithm runs in $O(kn + n^2(\log n)^{1/2})$ time and requires $O(kn)$ space, where k denotes the number of source-sink pairs.

K. Matsumoto, T. Nishizeki, N. Saito (1986). Planar multicommodity flows, maximum matchings and negative cycles. *SIAM J. Comput.* 15, 495-510.
The paper shows that the feasibility of a multicommodity flow problem on a class of planar undirected networks can be reduced to solving weighted matching problems.

H. Nagamochi and T. Ibaraki (1989). On max-flow min-cut and integral flow properties for multicommodity flows in directed networks. *Inform. Process. Lett.* 31, 279-285.
Integer multicommodity flow problem, in general, is a NP-complete problem. The paper studies special cases of the multicommodity flow problem which are polynomially solvable.

P. Klein, A. Agrawal, R. Ravi, S. Rao (1990). Approximation through multicommodity flows. *Proc. 31th Annual IEEE Symp. Found. Comput. Sci.* 43, 726-727.
The paper describes the use of multicommodity max-flow problems as a basic tool for the solution several NP-complete graph problems.

F. Shahrokhi and D.W. Matula (1990). The maximum concurrent flow problem. *J. ACM* 37, 318-334.
The maximum concurrent flow problem is a multicommodity flow problem in which every pair of nodes can send and receive flow concurrently, and the objective is to maximize the throughput. The paper presents a fully polynomial-time approximation scheme for solving this problem in the case of uniform capacities.

T. Leighton, F. Makedon, S. Plotkin, C. Stein, E. Tardos, S. Tragoudas (1991). Fast approximation algorithms for multicommodity flow problems. *Proc. 23rd Annual ACM Symp. Theory of Comput.* 101-111.

This paper develops approximation algorithms for the concurrent multicommodity flow problem for non-uniform capacities. The paper gives randomized as well as deterministic algorithms.

F. Granot, M. Penn (1992). On the integral plane two-commodity flow problem. *Oper. Res. Lett.* 11, 135-139.

The authors consider the maximum integral two-commodity flow problems in augmented planar networks and suggest an $O(n \log n)$ time algorithm.

P. Klein, S. Plotkin, C. Stein, E. Tardos (1994). Faster approximation algorithms for the unit capacity concurrent flow problem with applications to routing and finding sparse cuts. *SIAM J. Comput.* 23, 466-487.

This paper describes algorithms for approximately solving the concurrent multicommodity flow problem with uniform capacities which are faster than those suggested by Shahrokhi and Matula (1990).

S. Rajagopalan (1994). Two commodity flows. *Oper. Res. Lett.* 15, 151-156.

The paper studies the two-commodity flows in undirected networks and gives a sufficient condition for the existence of an optimal flow that is integral.

8 Software

M.M. Syslo, N. Deo, J.S. Kowalik (1983). *Discrete Optimization Algorithms*, Prentice Hall, Englewood Cliffs, NJ.

This book gives computer programs for several network flow and discrete optimization problems, and presents their computational results.

M.M. Syslo, N. Deo, J.S. Kowalik (1983). *Discrete Optimization Algorithms*, Prentice Hall, Englewood Cliffs, NJ.

This book gives computer programs for several network flow and discrete optimization problems, and presents their computational results.

B. Simeone, P. Toth, G. Gallo, F. Maffioli, S. Pallottino (eds.). *Fortran Codes for Network Optimization*, Ann. Oper. Res. 13, North-Holland, Amsterdam.

The book gives Fortran codes of several network flow problems and also gives their computational results.

D.P. Bertsekas (1991). *Linear Network Optimization Algorithms and Codes*, MIT Press, Cambridge, MA.

This book comes with a diskette containing the computer codes for assignment, transportation, and minimum cost flow problems.

J.R. Evans, E. Minieka (1992). *Optimization Algorithms for Networks and Graphs: Second Edition*, Marcel Dekkar, Inc., New York.

The software package NETSOLVE comes with this book and can solve a large variety of network flow problems.

D.S. Johnson, C.C. McGeoch (eds.) (1993). *Network Flows and Matching: First DIMACS Implementation Challenge*, AMS, New York.

The book summarizes the results of an Implementation Challenge at DIMACS held ar Rutgers University from November 1990 through August 1991. Several participants at the DIMACS challenge contributed bibliographies, problem instances, instance generators, problem-solving codes, and programming tools for the network flow and matching problems. These files are available in the public ftp directory *pub/netflow* on the Internet site *dimacs.rutgers.edu*. Inquiries regarding files in this directory can be sent to *netflow@dimacs.rutgers.edu*.

Acknowledgements

I thank Stefano Pallottino for his perceptive comments on an earlier draft of this bibliography and bringing to my notice many important references that I missed.

18 Network Design

Anantaram Balakrishnan
The Pennsylvania State University

Thomas L. Magnanti
Massachusetts Institute of Technology

Prakash Mirchandani
University of Pittsburgh

CONTENTS

Network design is a central problem in combinatorial optimization. The problem is deceptively easy to describe, but very difficult to solve. Given a set of available (perhaps capacitated) edges, the problem seeks a least cost network, composed of both variable flow costs and edge fixed costs, that meets prescribed flow requirements. The core network design problem, the minimum spanning tree problem, is well understood. Researchers know how to solve this problem efficiently and know how to describe its underlying mathematical (polyhedral) structure. Perhaps surprisingly, imposing minor and apparently innocuous modifications on this problem challenges the research community's state of knowledge and its computational abilities. The classical Steiner network problem is an example. This problem is theoretically difficult to solve (it is $\mathcal{NP}$-hard) and its polyhedral structure appears to be quite complicated. Imposing flow costs can also complicate the problem. In a network with unit flow requirements between every pair of nodes, finding a spanning tree that minimizes the sum of flow costs on its edges (with no fixed edge costs) also appears to be difficult to solve.

Annotated Bibliographies in Combinatorial Optimization, edited by M. Dell'Amico, F. Maffioli and S. Martello

Adding capacities on the flows adds even more complexity—complexity that is not well understood.

In 1985, when the research community last prepared a collection of annotated bibliographies on combinatorial optimization, the situation in network design was remarkably bleak. To its credit, the optimization community had at that time made considerable progress in solving certain core facility location problems (the uncapacitated warehouse location problem and the p-center problem) and it did understand and had solved the minimum spanning tree problem and some of its variants (for example, problems with a degree constraint imposed upon a single "root" node). But the community's ability to solve more general network design problems was embarassingly limited to problems with only a few nodes and edges. Since then, the community has made significant progress. It is now able to effectively solve (to near optimality with performance guarantees) large-scale Steiner tree problems with hundreds of nodes and certain network connectivity problems, uncapacitated network design problems, and hierarchical network design problems of a similar size. And, it has made progress in solving and understanding the polyhedral structure other more complex models. This paper summarizes some of the contributions to this literature, focusing on contributions made since 1985. As seen in our discussion, more general situations continue to pose significant challenges to the optimization community.

To set notation and define the class of problems we will be examining, we consider a network $G = (N, E)$ with node set N and (directed) edge set E on which we will route K commodities. Let $n = |N|$ and $m = |E|$. For each commodity $k = 1, \ldots, K$, $b^k \in R^n$ is a vector of node supplies and demands, $f^k \in R^m$ is a vector of edge flows, and $c^k \in R^m$ is a vector of edge costs per unit flow. x and F are m-dimensional vectors of edge design variables and edge installation costs (each component x_{ij} of x might be one or zero indicating the presence or absence of edge (i, j) in the network design). u^k and U are vectors of commodity specific and total edge capacities. If $\mathcal{N}$ denotes the node-edge incidence matrix of G, we can describe the network design problem as follows.

$$\text{minimize} \sum_{k=1}^{K} c^k f^k + Fx \tag{1}$$

subject to :

$$\mathcal{N} f^k = b^k \quad k = 1, \ldots, K \tag{2}$$

$$f^k \leq u^k x \quad k = 1, \ldots, K \tag{3}$$

$$\sum_{k=1}^{K} f^k \leq U x \tag{4}$$

$$f^k \geq 0 \quad k = 1, \ldots, K \tag{5}$$

$$x \in X. \tag{6}$$

We refer to this problem as the *multicommodity network design problem, MCND.*

Equations (2) are the familiar flow balance constraints. The "forcing" constraints (3) impose capacity limits $u^k x$ on the individual commodity flows, and the forcing constraints (4) impose capacity limits $U x$ on the total flows. In each case, if $x_{ij} = 0$, then the flow capacity of edge (i,j) is zero.

The constraints $x \in X$ impose topological restrictions on the design variables. Typically, they will state that $x \in \{0,1\}^m$ or that x are nonnegative integers (permitting $x_{ij} > 1$ corresponds to edge duplication or edge replication). They might also impose other restrictions on the design such as degree constraints on certain nodes, multiple choice constraints, and required precedence relationships.

The MCND model imposes directed capacity constraints (3) and (4). If the edges are undirected, we would replace each occurrence of f_{ij}^k in these constraints with $f_{ij}^k + f_{ji}^k$, thereby limiting the total flow in both directions on each undirected edge $\{i,j\}$.

When each commodity has a single source s_k and sink t_k and a demand of d_k, the max-flow min-cut theorem shows that the flow constraints are equivalent to an exponential family of cut constraints

$$\sum_{ij \in C} x_{ij} \geq d_k \text{ for all cuts } C \text{ separating } s_k \text{ and } t_k. \quad (7)$$

For problems with no flow costs, the so-called cutset models impose these constraints in place of the flow constraints (2).

If we add parallel edges with different capacities (U and U'), the formulation (1)-(6) is capable of modeling situations with multiple types of edge facilities (for example, different types of trucks in a transportation application or different type of transmission lines in a telecommunication application).

There are many ways to classify network design models. In some models, the underlying commodities flow at the same time (*simultaneous flows*) and in other models they do not (*non-simultaneous flows*). Some models have only fixed costs (each $c^k = 0$) and some have both fixed and variables costs. Some models are uncapacitated, some impose capacities on individual commodities, and some impose joint or bundle capacities shared by the commodities. Some models require a single route for each commodity while others permit bifurcated routes. In addition, we can distinguish models by any topological restrictions that they impose upon the network design. In particular, do they permit duplication of edges or not (x integer or binary), or do they impose connectivity restrictions or other topological restrictions?

In this chapter we focus on four types of models that have attracted much attention in the literature: uncapacitated network design (UND), capacitated network design (CND), network loading (NL), and network restoration (NR). In addition, we also discuss some related network design models and applications of network design. A companion chapter by Raghavan and Magnanti in this volume discusses the Steiner tree problem (ST), network survivability (NS), and multi-level network design (MLND). We will define each of the models we examine more formally as we discuss them. Table I specifies some of their characteristics.

Most of our discussion focuses on linear models. However, in a section on related models (Section 6), we consider problems with nonlinear costs (for example, to reflect congestion effects), though we make no attempt to provide a comprehensive account of these models.

Several solution methods, especially heuristic methods, linear programming-based polyhedral methods, problem decomposition, Lagrangian relaxation, and dual ascent procedures have become central to the network design literature. Our discussion focuses on the use of these solution strategies.

Table I. Network Design Taxonomy

	Uncapacitated Problems			Capacitated Problems			
	UND	**ST**	**MLND**	**CND**	**NL**	**NS**	**NR**
Cost (Fixed: F, Variable: V)	F,V	F	F,V	F,V	F,V	F	F
Capacity (Bundle: B, Commodity based: CB)	None	None	None	B	B	CB	CB
Topology (Tree: T, Connectivity: C, Multi-level: M, Duplication: D)	—	T	M	—	D	C	—
Flow (Simultaneous: S, Nonsimultaneous: N)	S	S	S	S	S	N	N

1 Surveys

The following surveys provide broad-based discussions of modeling and algorithmic issues for network design problems.

T.L. Magnanti, R.T. Wong (1984). Network design and transportation planning: Models and algorithms. *Transportation Sci.* 18, 1–55.

This lengthy survey examines the state of the art of optimal network design until 1984. It considers multicommodity fixed-charge models, concave cost models, and computational experience to that time. It shows that many classical problems in combinatorial optimization, for example, the traveling salesman problem, are special cases of formulation (1)-(6).

R.T. Wong (1985). Location and network design. M. O'hEigeartaigh, J. Lenstra, A. Rinnooy Kan (eds.). *Combinatorial Optimization: Annotated Bibliography*, John Wiley & Sons, New York, 129–147.

This annotated bibliography reviews about 100 papers in network design and facility location. It examines model formulations, approximation algorithms, bounding procedures, valid inequalities and facets, and optimization algorithms.

M. Minoux (1989). Network synthesis and optimum network design problems: models, solution methods and applications. *Networks* 19, 313–360.

Beginning with a general concave cost model, this survey describes a number of network design models: a fixed and variable cost model, capacitated minimum spanning trees, two level concentrator locations, and the network synthesis problem. It also discusses heuristic and bounding methods, relaxation approaches, and the generalized linear programming approach.

T. Magnanti, L. Wolsey (1995). Optimal trees. T. Magnanti, M. Ball, C. Monma, G. Nemhauser (eds.). *Network Models*, Handbooks in Operations Research and Management Science, 6, North Holland, Amsterdam, 503–615.

Network design problems defined on trees and whose solutions must be trees are typically easier to solve than those whose solutions can have more general network topologies. This survey summarizes known results about the polyhedral structure and alternative formulations for these special classes of problems. It illustrates the use of many methods from combinatorial optimization that are currently used in solving network design problems.

2 Uncapacitated Network Design

This section examines problems without flow capacities or equivalently MCND problems in which U_{ij} is at least the largest possible flow on edge (i, j) and u_{ij}^k is at least the largest possible flow of commodity k on edge (i, j). Note that, without edge capacities, we can normalize the demand for each commodity to one unit (and set $u_{ij}^k = 1$ and $U_{ij} = K$). For these problems, which we refer to as UND problems, it is important to distinguish two types of models: the disaggregate model (2)-(6), and aggregate models that either eliminate the commodity-based forcing constraints (3) and/or define commodities by their points of origin (or destination) instead of by origin-destination pair. The aggregate models have far fewer constraints and variables, but their linear programming relaxations provide a poorer approximation to the underlying mixed integer program.

Researchers have made the greatest algorithmic progress in solving these uncapacitated problems. As shown by the next paper, the problems can be very large and so pose significant algorithmic challenges.

A. Balakrishnan, T.L. Magnanti, R.T. Wong (1989). A dual-based procedure for large scale uncapacitated network design. *Oper. Res.* 37, 716–740.

The authors develop a family of dual-ascent algorithms for the UND problem. On problem instances with up to 45 nodes and 595 edges (1.98 million continuous variables and constraints), the method generates solutions that in almost all cases are guaranteed to be within 1 to 4 percent of optimality.

T.L. Magnanti, P. Mireault, R.T. Wong (1986). Tailoring Benders decomposition for uncapacitated network design. *Math. Program. Study* 26, 112–154.

Exploiting the fact that network flow problems are typically degenerate, the authors develop procedures for generating and selecting good (Pareto optimal) cuts to use in a Benders decomposition procedure for the UND problem. Limited computational experience demonstrates that Pareto optimal cuts can reduce the number of required iterations by up to two orders of magnitude.

R.K. Ahuja, V.V.S. Murty (1987). Exact and heuristic algorithms for the optimum communication spanning tree. *Transportation Sci.* 21, 163–170.

This paper uses maximum spanning tree subproblems to construct lower bounds and presents a two-phase heuristic algorithm, a greedy tree generation phase followed by a 1-opt tree improvement phase. Computational experience shows that while the lower bounds are not tight, the heuristic provides near-optimal solutions.

The following papers prove theoretical results about the UND polytope.

R.L. Rardin, U. Choe (1979). *Tighter relaxations of fixed charge network flow problems*,

Technical Report J-79-18, Industrial and Systems Engineering, Georgia Institute of Technology, Atlanta, GA.

Comparing the linear programming relaxations of the path and the edge-based formulations, the authors show that the edge-based formulation provides a stronger relaxation for the UND problem but that neither formulation dominates the other for capacitated models.

A. Balakrishnan (1987). LP extreme points and cuts for the fixed-charge network design problem. *Math. Program.* 39, 263–284.

For a path-based formulation of the UND problem, this paper characterizes fractional LP extreme points and develops two families of valid inequalities that exclude some fractional solutions.

J. Hellstrand, T. Larson, A. Migdalas (1992). A characterization of the uncapacitated network design polytope. *Oper. Res. Lett.* 12, 159–163.

The authors show that the linear programming relaxation polytope of the UND problem (with unit commodity demands) is quasi-integral. That is, any edge of the network design (integer) polytope is an edge of the LP relaxation polytope.

R.L. Rardin, L.A. Wolsey (1993). Valid inequalities and projecting the multicommodity extended formulation for uncapacitated fixed charge network flow problems. *European J. Oper. Res.* 71, 95–109.

For the single source UND problem, the authors introduce a rich family of *dicut* inequalities that completely describe the projection of the disaggregate formulation onto the space of the aggregate variables. These inequalities generalize facets for the uncapacitated lot-sizing problem (a special case of the single source UND problem).

D.S. Hochbaum, A. Segev (1989). Analysis of a flow problem with fixed charges. *Networks* 19, 291–312.

An UND problem with a single origin node is $\mathcal{NP}$-hard even when the fixed costs are a constant multiple of the flow costs. Establishing that the optimal solution to this UND version must be a branching from the origin node, the authors develop a Lagrangian relaxation approach and test it on problems with up to 50 nodes, generating integrality gaps of less than 3

B.W. Lamar, Y. Sheffi, W.B. Powell (1990). A capacity improvement lower bound for fixed charge network design problems. *Oper. Res.* 38, 704–710.

For the aggregate formulation of the uncapacitated network design problem, this paper describes an iterative procedure, using shortest path and knapsack solutions, to compute valid upper bounds on arc flows that provide successively tighter LP lower bounds on the optimal value.

A. Balakrishnan, T.L. Magnanti, P. Mirchandani (1996). Heuristics, LPs, and trees on trees: network design analyses. *Oper. Res.* 44, 478-496.

Building upon their previous work on "overlay" optimization problems, the authors develop the first known worst case performance bounds on a heuristic method and on the linear programming relaxation for the UND problem. Both bounds are proportional to the square root of the number of commodities in the problem.

An important generalization of the UND model, the *multi-level network design* (MLND) problem, considers multiple classes or levels of nodes and commodities with associated nested facility types (i.e., higher level facilities can carry lower level traffic but not vice versa). The Steiner network problem is a special case of this model. The chapter on network connectivity by Raghavan and Magnanti reviews the recent literature on the MLND problem.

3 Capacitated Network Design

We refer to the MCND model with genuine edge capacities U_{ij} and u_{ij}^k as *CND* to emphasize the imposition of capacities. This model occurs frequently in practice (Section 7), but is from a computational perspective one of the most difficult problems encountered in the field of combinatorial optimization. Despite the problem's importance, few optimization-based algorithms exist even for the simplest versions of this problem.

Researchers have noted several reasons for the problem's computational difficulty. First, the conventional flow-based formulation MCND leads to large, degenerate linear programs even for moderately sized networks. Second, the integrality gap for these problems is generally very high. Third, small changes in the problem structure can significantly alter the solution. Finally, for some versions of the problem, even finding a feasible solution is $\mathcal{NP}$-hard.

For a special (tree) version of the CND model, however, the research community has developed several heuristic, Lagrangian relaxation, and (recently) cutting plane solution methods. In this version of the problem, called the *capacitated minimum spanning tree (CMST)* problem, a root node (say, a centralized computer) must be linked by a *tree* to user nodes with specified demands. Each edge of the root node can carry only a finite amount of flow (usually equal for all root-incident edges). When all user demands equal one, we refer to the problem as the *C1MST* problem.

The capacitated vehicle routing problem is related to the CMST problem since after removing the last edge on each vehicle route, we can view any solution to this problem as a tree with a root node (the central distribution depot) and several subtrees, the total (customer) demands of which do not exceed the capacity of the edge (vehicle) connecting each of the subtrees to the root node. Another chapter of this volume provides annotated bibliographies for the vast vehicle routing literature.

In this section, we first review CMST papers and then discuss the general CND problem. The next section focuses on the specialized "Network Loading" version of the CND problem.

3.1 Capacitated minimum spanning trees

The following papers develop algorithmic approaches and theoretical results for the CMST and the C1MST models.

J.R. Araque, L. Hall, T.L. Magnanti (1990). *Capacitated trees, capacitated routing, and associated polyhedra*, Technical Report SOR-90-12, Program in Statistics and Operations Research, Princeton University, Princeton, New Jersey.

In an investigation of the polyhedral structure of capacitated minimum spanning trees and capacitated paths, the authors delineate a broad class of facets based upon several underlying network structures: *multi-stars*, *cliques*, and graphs called *ladybugs*. These results demonstrate the inherent complexity of the CMST polyhedral structure.

L. Hall, T.L. Magnanti (1992). A polyhedral intersection theorem for capacitated trees. *Math. Oper. Res.* 17, 398–410.

When the capacity of each edge incident to the root node is between three and $n/2$, the C1MST problem is $\mathcal{NP}$-hard. This paper provides a complete characterization of the underlying polytope when the capacity equals two, and shows that this (integral) polytope is the intersection of two well-known integral polytopes: a spanning tree polytope and a matching polytope.

J.R. Araque (1989). *Solution of a 48-city vehicle routing problem by branch-and-cut*, Technical report, Department of Applied Mathematics and Statistics, SUNY at Stonybrook, NY.

K. Malik, G. Yu (1993). A branch and bound algorithm for the capacitated minimum spanning tree problem. *Networks* 23, 525–532.

These papers develop related sets of cutset inequalities to strengthen the problem formulation. For the CMST problem, the second paper proposes a branch-and-bound procedure that uses Lagrangian reduced costs based upon the strong formulation. For C1MST test problems with up to 50 nodes, the authors generate solutions within 3 percent of optimality.

B. Gavish (1983). Formulations and algorithms for the capacitated minimal directed tree problem. *J. ACM* 30, 118–132.

L. Gouveia (1995). A $2n$ constraint formulation for the capacitated minimal spanning tree problem. *Oper. Res.* 43, 130–141.

The first paper proposes a flow-based formulation of the CMST. In a study of the C1MST problem, the second paper projects this formulation into a higher-dimensional space of edge variables indexed by the two incident nodes and the flow value, and tests a successful Lagrangian relaxation approach for problems with up to 80 nodes.

L. Hall (1996). Experience with a cutting plane algorithm for the capacitated spanning tree problem. *INFORMS J. Comput.* 8, 219–234.

Using the inequalities introduced by [Araque, Hall and Magnanti 1990] (see above), this paper describes an implementation of a cutting plane approach for the CMST and the C1MST problems. Besides comparing this method with other solution approaches for the problems, the paper provides an extensive list of references.

C. Bousba, L.A. Wolsey (1991). Finding minimum cost directed trees with demands and capacities. *Ann. Oper. Res.* 33, 285–303.

This paper considers a more general (directed) version of the CMST in which all edges are capacitated (not just the root-incident ones) and have flow costs as well. Strengthening two classes of formulations—a flow-based model and a subtree partition model—but indicating the advantages of the subtree partition formulation for their particular application, the authors develop and test a cutting plane algorithm on problems with up to 105 nodes.

K. Altinkemer, B. Gavish (1988). Heuristics with constant error guarantees for the design of tree networks. *Management Sci.* 34, 331–341.

This paper analyzes a heuristic based on the optimal partitioning of a traveling salesman tour and shows that the heuristic has a worst-case bound of 3 for the C1MST and a bound of 4 for the CMST problems.

3.2 General networks

The following papers consider the general CND problem. The last two papers study a multi-period version of the problem.

B. Gendron, T.G. Crainic (1996). *Relaxations for multicommodity network design problems*, Technical Report CRT-96-05, Centre de recherche sur les transport, Université de Montréal, Montréal.
B. Gendron, T.G. Crainic (1994). *Parallel implementations of bounding procedures for multicommodity capacitated network design problems*, Technical Report CRT-94-45, Centre de recherche sur les transport, Université de Montréal, Montréal.

Studying a general capacitated network design problem that allows multiple origin-destination pairs for each commodity, the authors propose three different formulations and several Lagrangian relaxations. Experiments on problems with up to 30 nodes and 100 commodities conclude that relaxing the *linking* constraints provides the best lower bound. Resource-based decomposition provides the upper bounds. The second paper describes a parallel implementation of these bounding procedures.

B. Gendron, T.G. Crainic, A. Frangioni (1996). *Multicommodity capacitated network design*, Technical report, Centre de recherche sur les transport, Université de Montréal, Montréal.

In the context of a general model for the CND problem, this paper provides an insightful survey of recent modeling and algorithmic results. The computational investigations conclude that an efficient solution procedure for the difficult capacitated design problems should judiciously combine cutting plane, Lagrangian, and heuristic approaches.

A. Dutta, J.-I. Lim (1992). A multiperiod capacity planning model for backbone communication networks. *Oper. Res.* 40, 689–705.

Using a Lagrangian approach for a multi-period model that adds new nodes and edges over time and incorporates capacity using a congestion constraint, the authors solve randomly generated instances of the problem having up to 30 nodes and a 6-year planning horizon.

S.-G. Chang, B. Gavish (1995). Lower bounding procedures for multiperiod telecommunications network expansion problems. *Oper. Res.* 43, 43–57.

This paper extends the dual-ascent algorithm for uncapacitated network design of [Balakrishnan, Magnanti and Wong 1989] (see §2) to multi-period network expansion problems. The Lagrangian bounding procedure uses newly proposed valid inequalities, and improves upon the bounds previously determined by the same authors.

4 Network Loading

Like the CND problem, the *network loading (NL)* problem seeks a minimum cost solution that allows prescribed point-to-point commodity demand to flow simultaneously on capacitated facilities that we install (load) on edges. However, the NL model has two distinct characteristics. First, we can load an *integral* number of p different types of available capacitated facilities on any edge. Second, the facility capacities are *modular*, that is, if $C_1 < C_2 < \ldots < C_p$ are the facility capacities, then C_{i+1} is a multiple of C_i. The number of facility types, p, depends on the particular application and is typically less than five, often one or two; when $p = 1$, all facilities have equal capacity. We incur two types of edge costs (that vary by edge): (i) a facility dependent fixed cost for loading each facility, and (ii) a commodity dependent flow cost for routing each commodity. The problem solution loads the least cost configuration of facilities on each edge to simultaneously satisfy all commodity demands. We can view the NL problem as the general MCND model with parallel edges (representing the different possible facilities) between nodes.

NL applications arise in many industrial settings, most notably in telecommunications and transportation. In telecommunications, the facility types are different transmission facilities (e.g., T1 and T3 lines) and in transportation, they might represent different types of vehicles (24 foot or 48 foot trailers) or services. A T3 line has 28 times the capacity of a T1 line, but costs less than 28 times as much. Therefore, the resulting step-wise cost function exhibits economies of scale.

The vast majority of solution methods for this problem are based on tightening the linear programming relaxation by adding valid inequalities and facets, and using this strengthened formulation in an optimization framework. This section focuses on papers that deal with NL problems without delay costs or constraints. Section 6 on network design variants reviews solution methods (primarily heuristic) for NL models incorporating nonlinear delay costs or constraints to represent congestion effects.

4.1 Polyhedral methods

The following papers use polyhedral methods to study the underlying structure and to develop solution methods for the NL problem.

T.L. Magnanti, P. Mirchandani, R. Vachani (1993). The convex hull of two core capacitated network design problems. *Math. Program.* 60, 233–250.

Studying the single facility type case, this paper develops facets and completely characterizes the convex hulls of the feasible solutions for two subproblems of the NL problem: (i) a single edge subproblem, and (ii) a three node subproblem. The use of these facets for network substructures, including those obtained by aggregating nodes, strenghtens the NL formulation.

T.L. Magnanti, P. Mirchandani, R. Vachani (1995). Modeling and solving the two-facility capacitated network loading problem. *Oper. Res.* 43, 142–157.

After strengthening the natural NL model to one whose LP relaxation solution value is at least as good as the Lagrangian relaxation bound, this paper describes computational experiments on problem instances with up to 15 nodes and a variety of demand patterns. The paper demonstrates the effectiveness of a cutting-plane based solution procedure for the NL problem and also computationally compares the linear programming and Lagrangian relaxation approaches.

D. Bienstock, O. Günlük (1996). Capacitated network design-polyhedral structure and computation. *INFORMS J. Comput.* 8, 243–259.

Generalizing and extending the inequalities in [Magnanti, Mirchandani and Vachani 1995] (see above) to situations with directed demand, flow costs, and existing capacities, this paper presents polyhedral results using *cutset*, *flow cutset*, and *three partition* inequalities. Tests on randomly generated test problems and two real-life problems with up to 16 nodes show the robustness of these inequalities to small perturbations in the demand values.

P. Mirchandani (1992). *Projections of the capacitated network loading problem*, Technical Report WP-706, Katz Graduate School of Business, University of Pittsburgh, Pittsburgh, PA 15260.

Projecting out the flow variables from the NL formulation and describing a formulation in the space of only the edge design variables, this paper presents several classes of *multi-partition* facets.

D. Bienstock, S. Chopra, O. Günlük, C.Y. Tsai (1995). *Minimum cost capacity installation for multicommodity flows*, Working paper, IEOR Department, Columbia University, New York, NY. To appear in *Math. Program.*

For the directed network case with a single facility type having a normalized capacity of one, the authors compare two formulations: a model with only design variables and an aggregated multicommodity flow formulation. They provide a complete description of the 3-node problem. ([Magnanti, Mirchandani and Vachani, 1993] (see above) had given the complete description for the 3-node undirected network case.)

F. Barahona (1996). Network design using cut inequalities. *SIAM J. Optim.* 6, 823–837.

The author avoids solving the difficult underlying multicommodity flow problem by proposing an NL relaxation based on aggregate demand and capacity constraints across cuts. Incorporating this relaxation in a heuristic setting, the paper reports computational results on problems with up to 64 nodes with and without bifurcated flows.

M. Stoer, G. Dahl (1994). A polyhedral approach for multicommodity survivable network design. *Numer. Math.* 68, 149–167.

By considering incremental capacities as 0-1 decision variables, the authors propose an NL formulation that allows a more general cost structure and that has a combinatorial flavor. Combining network loading with network survivability (to single node or single edge failures), they draw upon literature from the knapsack and survivability domains to strengthen the basic model by adding three types of facet-defining inequalities.

G. Dahl, M. Stoer (1996). *A Cutting Plane Algorithm for Multicommodity Survivable Network Design Problems*, Technical report, University of Oslo.

The authors incorporate survivability in the NL model by limiting the fraction of the total commodity demand that can flow through any node or edge while also ensuring that at least a prespecified proportion of the total demand is satisfied when an edge or node fails. They propose a cutting plane heuristic and perform computational tests on problems with up to 118 nodes and 134 edges.

4.2 Other solution methods

The following papers present other optimal and approximate approaches for the NL problem.

L.W. Clarke, P. Gong (1995). *Capacitated network design with column generation*, Technical Report LEC 96-06, The Logistics Institute Research Division, School of Industrial and Systems Engineering, Georgia Institute of Technology, Atlanta, Georgia 30332-0205.

The authors compare the edge-based modeling and solution approach with a path-based column generation approach for the non-bifurcated flow case. Using problem instances with up to 15 nodes, they conclude that the path-based approach is computationally superior because it allows for quicker solutions of the underlying linear programming relaxations.

B. Brockmüller, O. Günlük, L.A. Wolsey (1996). *Designing private line networks–Polyhedral analysis and computation*, Technical report, Cornell University, Ithaca, NY.

The authors consider the non-bifurcated flow version of the problem with the additional requirement that a commodity may flow only through a specified subset of nodes. Presenting a hierarchy of linear approximations to decrease the number of integer variables, they identify valid inequalities for a single edge (see also [Magnanti, Mirchandani and Vachani 1993], §§4.1), and optimally solve four real-life problems with up to 54 nodes and eight intermediate hub nodes.

Y. Mansour, D. Peleg (1994). *An approximation algorithm for minimum cost network design*, Technical report, Weizmann Institute of Science, Rehovot.

Using the concept of light-weight distance preserving spanners, the authors develop an approximation method for the single facility NL problem with worse-case bound that is logarithmic in the number of nodes for the general case and that is a constant for Euclidean problems. The authors assume that the design must connect or span all the nodes in the network.

F. Salman, J. Cheriyan, R. Ravi, S. Subramanian (1997). Buy-at-bulk network design: approximating the single-sink edge installation problem. *Proceedings of the Symposium on Discrete Algorithms*, 619–628.

Assuming a single facility type and a fixed cost on each edge as well as the facility loading costs, the authors develop approximation algorithms with worse case bounds for a single-sink problem that combines network loading and fixed cost network design. A planar division heuristic for a Euclidean version of the problem has a worse-case bound that is logarithmic in the total demand; in the general cost case, whenever the design must connect each source to the sink by a path with at most two edges, a set covering heuristic has a worse-case bound that is logarithmic in the number of nodes.

D. Bienstock (1997). *Experiments with a network design algorithm using ϵ-approximate linear programs*, Technical report, Columbia University.

The author develops and tests a new cutting plane algorithm for solving the single facility network loading problem. The algorithm uses an exponential penalty procedure to find ϵ-approximate solutions for the underlying linear programs and a careful rounding up procedure for generating feasible solutions from the approximate linear programming solutions. Computational experience on several problems, including one with as many as 200 nodes and 1828 edges, was quite successful.

4.3 Special cases and related models

The NL special case with one commodity is also $\mathcal{NP}$-hard, even in situations with mild assumptions. The following papers examine the computational complexity and polyhedral properties of the single commodity NL model.

T.L. Magnanti, P. Mirchandani (1993). Shortest paths, single origin-destination network design, and associated polyhedra. *Networks* 23, 103–121.

One of the first papers to study the NL problem with facilities at three different capacity levels, this paper characterizes the optimal solution properties for the single commodity case and, under reasonable cost assumptions, develops a formulation for which some optimal LP solution is integral. The authors show that the single commodity problem with two facility types is strongly $\mathcal{NP}$-hard if edges have existing capacities, or have upper bounds on the flow values.

S. Chopra, I. Gilboa, T. Sastry (1996). Extended formulations for one and two facility network design. W. Cunningham, S.T. McCormick, M. Queyranne (eds.). *Integer Programming and Combinatorial Optimization*, Lecture Notes in Computer Science, *5th International IPCO conference, Vancouver, Canada.* Springer, Berlin, 44–57.

The single commodity NL problem is $\mathcal{NP}$-hard for problems with (i) a single facility type and flow costs, and (ii) two facility types without flow costs. After proposing a pseudo-polynomial algorithm for these single commodity NL versions, the authors develop an extended formulation that is at least as strong as the formulation in [Magnanti and Mirchandani 1993] (see above).

Many other models, especially in the facility location and production literature, have modular parameters similar to the NL model. The following papers study the polyhedral structure of such problems.

J. Leung, T.L. Magnanti (1989). Valid inequalities and facets of the capacitated plant location problem. *Math. Program.* 44, 271–291.

J. Leung, T.L. Magnanti, R. Vachani (1989). Facets and algorithms for capacitated lot sizing. *Math. Program.* 45, 331–359.

T.L. Magnanti, R. Vachani (1990). A strong cutting plane algorithm for production scheduling with changeover costs. *Oper. Res.* 38, 456–473.

Y. Pochet, L.A. Wolsey (1993). Lot-sizing with constant batches: formulation and valid inequalities. *Math. Oper. Res.* 18, 767–785.

These papers study the polyhedral structure of capacitated problems in three different domains when all facility capacities are equal. The reported computational studies provide strong evidence of the efficacy of cutting plane methods in a variety of settings. The literature review in the second paper provides a comprehensive list of lot sizing papers and polyhedral methods papers published until about 1988.

Y. Pochet, L.A. Wolsey (1995). Integer knapsack and flow covers with divisible coefficients: polyhedra, optimization, and separation. *Discr. Appl. Math.* 59, 57–74.

This paper studies the polyhedral structure of three subproblems, including a knapsack covering problem with divisible items weights and an aggregated capacitated flow problem. It presents a complete characterization of the feasible set, and discusses a polynomial reformulation and efficient separation algorithm.

L.A. Wolsey (1989). Submodularity and valid inequalities in capacitated fixed charge networks. *Oper. Res. Lett.* 8, 119–124.
K. Aardal, Y. Pochet, L.A. Wolsey (1995). Capacitated facility location: valid inequalities and facets. *Math. Oper. Res.* 20, 562–582.
K. Aardal, Y. Pochet, L.A. Wolsey (1996). Erratum: capacitated facility location: valid inequalities and facets. *Math. Oper. Res.* 21, 253–356.

The second paper shows how knapsack cover and flow cover inequalities also define facets of the capacitated facility location problem with general facility capacities. It also shows that the flow cover inequalities generalize to the *effective capacity* inequalities which in turn have a *submodular* structure (see also the first paper). These papers also discuss the case with equal facility capacities.

5 Network Restoration

In network restoration, given a current flow, we wish to install sufficient spare capacity to be able to reroute the current flow on any single undirected edge, should that edge fail. The flows will be nonsimultaneous since we assume that only one edge fails at any time. In this case, we define a commodity k for each edge e of the network, set its demand d_e to be the current total flow on edge e, and impose the constraint that if $e = \{i, j\}$, then $f^e_{ij} = f^e_{ji} = 0$. That is, we cannot use edge e to reroute its own traffic.

In the *link* version of this problem, if an edge $\{i, j\}$ fails, we reroute the traffic on that edge so that it uses one or more paths from node i to node j (independent of where that flow originates or terminates). In the *path* version of this problem, if an edge $\{i, j\}$ fails and flow from origin s to destination t flows on that edge, then we find an alternative path from node s to node t to carry that flow. In both cases, the restored traffic cannot flow on edge $\{i, j\}$.

H. Sakauchi, Y. Nichimura, S. Hasegawa (1990). A self-healing network with an economical spare channel assignment. *1990 IEEE Global Telecommunications Conference*, 438–443.
M. Herzberg (1993). A decomposition approach to assign spare channels in self-healing networks. *1993 IEEE Global Telecommunications Conference*, 1601–1605.

The first paper proposes an Iterative Cutset Heuristic (ICH) which is an LP-based algorithm that iteratively generates cutset inequalities. The second paper exploits two simple graph structures, the triangle and the triangle pyramid, to improve the linear programming formulation.

W.D. Grover, T.D. Bilodeau, B.D. Venables (1991). Near optimal spare capacity planning in a mesh restorable network. *1991 IEEE Global Telecommunications Conference*, 2007–2012.
B.D. Venables, W.D. Grover, M.H. MacGregor (1993). Two strategies for spare capacity placement in mesh restorable networks. *1993 IEEE International Conference on Communications*, 267–271.

The two stage Spare Link Placement (SLP) heuristic procedure proposed in the first paper creates an initial feasible solution by adding paths of spare capacities and then attempts to reduce the overall amount of spare capacity by reallocating capacity from one edge to another. The second paper reports on computational testing of the ICH heuristic and three versions of the SLP heuristic.

T. Chujo, H. Komine, K. Miyazaki, T. Ogura, T. Soejima (1991). Distributed self-healing network and its optimum spare-capacity assignment algorithm. *Electronics and Communications in Japan, Part 1* 74, 1–9.
M. Herzberg, S.J. Bye, A. Utano (1995). The hop-limit approach for spare capacity assignment in survivable networks. *IEEE/ACM Transactions on Networking* 3, 775–784.

In contrast to the previous paper, these two papers restrict the choice of restoration paths by imposing hop constraints on paths used in the first phase of the SLP heuristic.

D.A. Dunn, W.D. Grover, M.H. MacGregor (1994). Comparison of k-shortest paths and maximum flow routing for network facility restoration. *1994 IEEE JSAC Special Issue on Integrity of Public Telecommunications Networks* 12, 88–99.

An incremental heuristic for link restoration solves a series of maximum flow problems to sequentially reroute traffic on the edges. In computational experiments, the authors show that using k-shortest paths instead of maximum flow computations leads to essentially the same solutions.

L. Lee, H.W. Chun (1992). An ANN approach to spare capacity planning. *1992 IEEE Region 10 International Conference*, 891–895.

This paper applies a neural network approach and obtains good results on a four node network.

J. Veerasamy, S. Venkatesan, J.C. Shah (1994). Effect of traffic splitting on link and path restoration planning. *1994 IEEE Global Telecommunications Conference*, 1867–71.

Most approaches to network restoration assume that rerouted traffic can be split on alternative paths. The authors show that the possibility of traffic splitting reduces the required capacity of the restoration network required by approximately 35 per cent, with larger savings for larger, denser networks.

J. Veerasamy, S. Venkatesan, J.C. Shah (1995). Spare capacity assignment in telecommunication networks using path restoration. *MASCOTS 1995*, 370–375

The authors investigate the computational performance of an incremental shortest path heuristics for the path restoration problem. They also show that for some test problems with 20 to 60 nodes, path restoration requires from 50 to 80 per cent less spare capacity than link restoration.

A. Lisser, R. Sarkissian, J.P. Vial (1995). *Survivability in telecommunication networks*, Technical report, France Telecom, CNET, Rue du General Leclerc, Issy les Moulineaux Cedex, France.

Assuming linear capacity costs and path restoration, the authors solve a linear program using the analytic center cutting plane method to decide the reserve capacity in a telecommunication network. The paper reports computational results for randomly generated problems and a problem instance from the literature.

A. Balakrishnan, T.L. Magnanti, J. Sokol, Y. Wang (1996). *Algorithms for Link*

Restoration in Telecommunication Networks, Operations Research Center, MIT, Cambridge, MA.

The authors propose models, heuristics with error bounds, and bad-case examples for their algorithms. An LP-based heuristic was able to solve (to within 0.5 percent of optimality) real world and randomly generated problems ranging from 10 nodes and 15 edges to 50 nodes and 150 edges. The paper also examines situations when the restoration network can use two types of facilities as in the network loading problem.

The literature on network restoration include papers on such topics as the impacts of failed networks, the nature of network failures (hardware and software), standards, performance evaluations of restoration methods, and restoration for various technologies (e.g., ring networks). The January 1994 issue of the *IEEE Journal on Selected Areas in Communications* contains 22 papers that cover many of these issues and serves as a valuable introduction to this topic.

6 Network Design Model Variants

Often, practical applications require various enhancements of the basic network design model. These variants typically include additional restrictions (e.g., hop constraints that limit the number of edges a commodity can traverse), enhanced cost structures (e.g., including delay costs), congestion effects (e.g., an additional constraint that models facility capacity utilization), multi-period extensions, and models with special topological structure (e.g., rings). The specialized algorithms and experience gained with the basic network design model be very valuable for solving these enhanced models. Our discussion of model variations in this section is only representative, not exhaustive, and we do not cover some of the variants already discussed in previous sections. Section 7 (Selected Applications) discusses additional model variants that are based upon actual applications.

R.E. Erickson, C.L. Monma, A.F. Veinott (1987). Send and split method for minimum concave-cost network flows. *Math. Oper. Res.* 12, 634–664.

This paper presents an efficient dynamic programming algorithm, called the send-and-split method, for finding the minimum-cost (extreme) flows for a single-commodity UND problem with concave edge costs (more general than the fixed plus variable costs in the MCND model). The running time of the method is polynomial in the number of nodes and edges, but exponential in the number of demand nodes and capacitated edges. Algorithmic refinements improve the running time for specialized network structures.

A. Balakrishnan, S.C. Graves (1989). A composite algorithm for a concave-cost network flow problem. *Networks* 19, 175–202.

Motivated by the less-than-truckload (LTL) consolidation problem, this paper develops a composite algorithm, combining problem reduction, Lagrangian lower bounds, and heuristics, to solve a network design problem with piecewise-linear concave costs. Computational tests on general as well as layered (LTL) networks establish the effectiveness of this algorithm.

A. Balakrishnan, K. Altinkemer (1992). Using a hop-constrained model to generate alternative communication network designs. *INFORMS J. Comput.* 4, 192–205.

The authors develop a Lagrangian-based algorithm for solving the network design problem with additional hop constraints that limit the maximum number of edges each commodity can traverse. They demonstrate the use of this model to study the tradeoff between connectivity and cost.

I. Althofer, G. Das, D. Dobkin, D. Joseph, J. Soares (1993). On sparse spanners of weighted graphs. *Discrete and Computational Geometry* 9, 81–90.
B. Chandra, G. Das, G. Narasimhan, J. Soares (1992). New sparseness results on graph spanners. *8th Symposium on Computational Geometry*, 192-201.

A light-weight distance preserving spanner (t-spanner) of a weighted graph is a subgraph satisfying the property that the weighted distance between any two nodes in it is at most t times the weighted distance in the original graph. A spanner is light if its total edge weight is a "small" multiple of weight of a minimum spanning tree on the original graph. In these papers, the authors present polynomial-time algorithms for finding sparse spanners for any given value of t. When the graphs are Euclidean, they obtain improved results.

S. Khuller, B. Raghavachari, N.E. Young (1995). Balancing minimum spanning and shortest path trees. *Algorithmica* 14, 305–322.

A light, approximate shortest-path tree (LAST) is a rooted tree that simultaneously approximates a shortest path tree and a minimum spanning tree. Given a positive constant C, a shortest path tree, and a minimum spanning tree, the authors develop a linear-time algorithm that finds a LAST satisfying the property that (i) the length of the path to each node on LAST is no more than $1 + \sqrt{2}C$ times the shortest path length, and (ii) the tree costs no more than $1 + \sqrt{2}/C$ times the cost of a minimum spanning tree.

The following network design variants stem from applications in computer and communication networks.

F. Glover, M. Lee, J. Ryan (1991). Least-cost network topology design for a new service: An application of tabu search. *Ann. Oper. Res.* 33, 351–362.

The authors study the problem of selecting a *platform* type and location, and routing traffic from end nodes to a central location on the chosen topology. The facilities—platforms and links—are available in different types, each with different capacities and fixed/variable costs. The authors test a two-phase tabu search method to find cost-effective network designs.

P.C. Fetterolf, G. Anandalingam (1992). A Lagrangian relaxation technique for optimizing interconnection of local area networks. *Oper. Res.* 40, 678–688.

Given a set of LANs (local area networks) and inter-LAN traffic requirements, the LAN interconnection problem requires selecting the locations and types of bridges to connect these LANs (to form a spanning tree of LANs) subject to LAN and bridge capacity constraints. The authors propose and test a Lagrangian relaxation algorithm, with an associated heuristic, to solve this capacitated network design problem.

H. Pirkul, V. Nagarajan (1992). Locating concentrators in centralized computer networks. *Ann. Oper. Res.* 36, 247–262.

Employing a two-phase heuristic that uses a sweep method and Lagrangian relaxation, the authors solve the cost minimization problem of locating concentrators, assigning user nodes to the concentrators, and interconnecting the concentrators to a central computer. The solution is a two-level tree network with "backbone" paths connecting all the concentrators to the central computer and direct links from each user node to its assigned concentrator.

M. Laguna (1994). Clustering for the design of SONET rings in interoffice telecommunications. *Management Sci.* 40, 1533–1541.

Given the projected traffic demands between various central offices and the available ADM (add-drop multiplexer) capacities and fixed costs, the logical ring design model assigns offices to rings to minimize the total fixed cost of ADMs plus variable costs of inter-ring traffic. The author uses a tabu search heuristic to solve this problem.

The following papers consider the network loading problem (Section 4), but assume that either the objective function or the constraints are nonlinear because of congestion effects.

A. Kershenbaum (1993). *Telecommunications network design algorithms*, McGraw-Hill Computer Science Series, McGraw-Hill, New York, NY.
A. Kershenbaum, P. Kermani, G.A. Grover (1991). Mentor: An algorithm for mesh network topological optimization and routing. *IEEE Trans. Comput.* 39, 503–513.

One chapter, "Mesh Topology Optimization", of Kershenbaum's book considers several heuristic approaches for solving a version of the network loading problem (and its relaxations) with a delay constraint. The approach also applies to situations that minimize total delay with a prespecified budget imposed upon the cost. Part of this chapter describes the fast *Mentor* heuristic that appeared in the second paper.

B. Gavish, K. Altinkemer (1990). Backbone network design with economic tradeoffs. *INFORMS J. Comput.* 2, 236–252.

This paper adds a delay cost (in addition to the fixed and flow costs) to the objective function of the NL problem and develops a Lagrangian relaxation approach for the non-bifurcated flows case.

L.J. LeBlanc, R.V. Simmons (1989). Continuous models for capacity design of large packet-switched telecommunication networks. *INFORMS J. Comput.* 1, 271–286.

Arguing that leased capacity can be augmented using public telephone lines (available at continuous levels) with high-speed modems, the authors develop a Lagrangian relaxation approach for the continuous relaxation of the NL model with a delay constraint.

I. Neuman (1992). A system for priority routing and capacity assignment in packet switched networks. *Ann. Oper. Res.* 36, 225–246.

The author proposes and tests a Lagrangian relaxation algorithm for the problem of selecting link capacities from a discrete set of options and routing flows with two priorities (each priority level has an associated delay cost) on selected paths from a prespecified set to minimize the total fixed, variable, and delay costs.

D. Bienstock, O. Günlük (1995). Computational experience with a difficult mixed-integer multicommodity flow problem. *Math. Program.* 68, 213–237.

The mixed integer program for a generalization of the uncapacitated network design problem, with node degree restrictions and a congestion constraint that minimizes the maximum edge flow, has large integrality gaps. Computational experiments on problems with up to 20 nodes use new families of facets described in the paper.

The following NL papers focus on multi-period expansion and time-varying demand.

S.H. Parrish, T. Cox, W. Kuehner, Y. Qiu (1992). Planning for optimal expansion of leased line communication networks. *Ann. Oper. Res.* 36, 347–364.

The authors consider a multiple period version of the two-facility NL problem with the added stipulation that the solution have a tree structure. For four test problems containing 7 to 13 nodes, their heuristic produced solutions within 13 percent of the LP value.

K. Chari, A. Dutta (1993). Design of private backbone networks—I: time varying traffic. *European J. Oper. Res.* 67, 428–442.
K. Chari, A. Dutta (1993). Design of private backbone networks—II: time varying grouped traffic. *European J. Oper. Res.* 67, 443–452.

These papers consider single-facility NL problems with time-dependent demand, and develop a Benders decomposition-based heuristic for bifurcated and non-bifurcated flows. For the non-bifurcated case, the second paper uses bin-packing heuristics (each T1 line corresponds to a bin). Test problem sizes range up to eight nodes and three busy-periods per day.

E. Rosenberg (1987). A nonlinear programming heuristic for computing optimal link capacities in a multi-hour alternate routing communications network. *Oper. Res.* 35, 354–364.

This paper also considers dynamic demand, routing a commodity either directly from its origin to its destination or via at most one intermediate node. The paper develops a nonlinear programming heuristic for solving a continuous relaxation of the model.

The chapter by Raghavan and Magnanti in this volume examines other models related to network design, incorporating survivability constraints and special topological restrictions (e.g., a tree-star configuration for digital data networks) .

7 Selected Applications

The core features of network design, namely, the joint optimization of facility choices and routing decisions with economies of scale in the facility costs (i.e., fixed and variable or, more generally, concave costs), arise in many different application contexts, particularly for infrastructure planning. Over the past decade, two contexts—telecommunications planning and supply chain optimization—have been especially active domains for applying network design. The research in these contexts includes modeling, analysis, algorithmic development, and actual implementation.

Traditionally, the telecommunications industry views the overall transmission system hierarchically as local area networks and their interconnection, local access networks connecting customer locations and distribution points to the central office, inter-office networks, and (long-distance) backbone networks. The emergence of the Synchronous Optical Network (SONET) standard and development of SONET-compatible equipment has blurred traditional hierarchical distinctions between the distribution, local access, and inter-office networks and also made it possible for telecommunication companies to deploy fault-tolerant (e.g., ring) fiber-optic network architectures that are cost-effective. These architectures impose new constraints on the design problem, and require new models and solution strategies.

Similarly, in manufacturing and logistics, companies are increasingly emphasizing the re-design of their networks of suppliers and facilities, considering the system-wide and global impacts—from procurement of raw materials to delivery of finished goods—of supply chain decisions. With economies of scale in production, inventories, and transportation (e.g., truck-load *TL* versus less-than-truckload *LTL*), the (capacitated) network design model is an appropriate platform for integrated supply chain planning models. For production-distribution networks with assembly (or disassembly) operations, the flow equations differ from the model we have been considering since we no longer maintain flow balances (to account for the fact that flows of components and subassemblies to finished goods are governed by bill of materials).

This section reviews selected application papers, primarily in telecommunications and supply chain optimization, that were motivated by real managerial problems (and tested using data from practice) or that describe an actual system implementation. We also include some overview and survey papers that provide a good perspective of modeling and optimization opportunities in these two application domains.

7.1 Telecommunications

B. Gavish (1991). Topological design of telecommunication networks—local access design methods. *Ann. Oper. Res.* 33, 17–71.

B. Gavish (1992). Topological design of computer communication networks—the overall design problem. *European J. Oper. Res.* 58, 149–172.

These two papers provide a good introduction to the evolution of network topologies and the hierarchy of network design decisions. The first paper reviews worst-case and empirical results for several local access models—the capacitated minimal spanning tree problem, the multicenter tree problem, the minimal cost loop problem, and the Telpak problem. The second paper presents an integrated model to select a minimum cost (including delay costs) configuration of network control processors, links, and routes. The author proposes a Lagrangian-relaxation scheme to solve this model and reports computational results.

A. Balakrishnan, T.L. Magnanti, A. Shulman, R.T. Wong (1991). Models for planning capacity expansion in local access telecommunication networks. *Ann. Oper. Res.* 33, 239–284.

This paper describes the underlying structure and technology of current local access telephone networks (connecting distribution points to the central office) and presents a modeling framework for addressing these problems

A. Balakrishnan, T.L. Magnanti, R.T. Wong (1995). A decomposition algorithm for local access telecommunications network expansion planning. *Oper. Res.* 43, 58–76.

Expanding existing local access tree networks to accommodate increasing telecommunication traffic entails either installing concentrators at nodes or adding more cables. The authors develop and test (on real problem instances) a Lagrangian relaxation-based approach, incorporating valid inequalities to strengthen the problem formulation and an efficient dynamic programming algorithm to solve uncapacitated local access network design subproblems.

G. Cho, D. Shaw (1994). *Limited column generation for local access network design problems: Formulation, algorithm, and implementation*, Technical report, School of Industrial Engineering, Purdue University, West Lafayette, Indiana.
O.E. Flippo, A.W.J. Kolen, A.M.C.A. Koster, R.L.M.J. van de Leensel (1996). *A dynamic programming algorithm for the local access network expansion*, Technical Report RM/96/027, University of Limburg, Maastricht.

For the local access network expansion problem introduced by [Balakrishnan, Magnanti and Wong 1995] (see above), the first paper discusses the application of a limited column generation technique with an embedded dynamic programming algorithm. Assuming a broader class of class of "decomposable" cable expansion and concentrator installation costs, the second paper presents and tests an effective pseudo-polynomial dynamic programming algorithm for the same problem.

C. Jack, S.-R. Kai, A. Shulman (1992). Design and implementation of an interactive optimization system for telephone network planning. *Oper. Res.* 40, 14–25.
C. Jack, S.-R. Kai, A. Shulman (1992). NETCAP-An interactive optimization system for GTE telephone network planning. *Interfaces* 22, 72–89.

These papers describe the development process and structure of NETCAP, an interactive planning tool developed by GTE Laboratories to plan multi-period expansion of the local access network. The system contains embedded optimization procedures—heuristics and a dynamic programming algorithm—to obtain cost-effective expansion plans quickly. The second paper describes some of the implementation difficulties and successes of NETCAP.

L.A. Cox, W.E. Kuehner, S.H. Parrish, Y. Qiu (1993). Optimal expansion of fiber-optic telecommunications networks in metropolitan areas. *Interfaces* 23, 35–48.

This paper describes a mixed integer programming model to plan least-cost expansion of metropolitan fiber-optic networks, considering multi-year demand scenarios, different types of capacitated facilities (cables and terminating equipment), equipment bay capacity, traffic protection requirements, and the option of introducing new links. Using a combination of heuristics and optimal subproblem solutions (using CPLEX), the tool was projected to provide USWest savings of over 100 million dollars.

S. Cosares, D.N. Deutsch, I. Saniee, O.J. Wasem (1995). SONET Toolkit: A decision support system for designing robust and cost-effective fiber-optic networks. *Interfaces* 25, 20–40.

The SONET Toolkit is a suite of optimization tools developed by Bellcore to determine a cost-effective augmentation or redesign of an existing network (including topological design, equipment sizing, and flow routing) in order meet projected demand growth and incorporate selective traffic protection requirements. This article describes the toolkit development process, an overview of its embedded models (some of which are network design-related), and an assessment of the toolkit's impact (10 to 30 percent cost savings, potentially amounting to hundreds of millions of dollars for all new SONET networks, and a drastic reduction in planning time).

T. Barnea, D. Benanav, K. Dutta, I. Eisenberg, J. Euchner, E. Gilbert, A. Goodarzi, E. Lee, Y.L. Lin, J. Martin, J. Peterson, R. Pope, R. Salgame, S. Sardana, G. Sevitsky

(1996). Arachne: planning the interoffice facilities network at NYNEX. *Interfaces* 26, 85–101.

Arachne, a tool for planning the interoffice network at NYNEX, uses problem decomposition, expert systems, and optimization subroutines to decide optimal facility expansion and routing strategies at all levels of the signal hierarchy for an asynchronous network. It has reduced planning time significantly, permitting additional sensitivity analyses, and saved NYNEX at least 33 million dollars in capital investment costs.

R.H. Cardwell, C.L. Monma, T.H. Wu (1989). Computer-aided design procedures for survivable fiber optic networks. *IEEE J. Selected Areas in Communications* 7, 1188–1197.

This early paper on survivable networks motivates survivability concerns in the context of new telecommunications technologies and discusses practical design strategies to address these concerns.

The chapter on Network Connectivity in this volume contains additional references on applications and systems for designing survivable telecommunication networks.

7.2 Supply chains

A. Geoffrion, G. Graves (1974). Multicommodity distribution system design by Benders decomposition. *Management Sci.* 29, 822–844.

This classic paper on distribution system design for Hunt-Wesson Foods using a fixed-charge multicommodity model represents a successful application of Benders' decomposition and also served as a precursor to several subsequent supply chain optimization models.

A. Geoffrion, R. Powers (1995). 20 years of strategic distribution systems design: an evolutionary perspective. *Interfaces* 25(5), 105–127.

This paper provides an excellent summary, based on the authors' considerable practical experience, of the trends in modeling and algorithmic development, software capabilities, computing and communications technology, and the management of the logistics function that have led to the successful applications of optimization models for distribution systems design.

G.G. Brown, G.W. Graves, M.D. Honczarenko (1987). Design and operation of a multicommodity production/distribution system using primal goal decomposition. *Management Sci.* 33, 1469–1480.

To address strategic and tactical production/distribution planning problems at NABISCO, the authors developed a fixed-charge network design model that decides what facilities to assign to each plant, where to produce each product, and how to distribute products. A specialized algorithm, incorporating a new class of goal decompositions, solved large problem instances effectively.

R.L. Breitman, J.M. Lucas (1987). PLANETS: a modeling system for business planning. *Interfaces* 17, 94–106.

PLANETS, a model-building/decision support system incorporating mathematical programming capabilities, helps managers at General Motors decide production, marketing, logistics, and resource allocation decisions.

M.A. Cohen, H.L. Lee (1989). Resource deployment analysis of global manufacturing and distribution networks. *J. Manufacturing and Operations Management* 2, 81–104.

The authors describe a fixed-charge network flow model with (production and supplier) capacity constraints, material balance, and offset requirements to maximize the total after-tax profits in the value-added chain. They solve the model by manually identifying various plausible manufacturing, supplier, and distribution scenarios, and solving LP subproblems. The authors use this model to analyze global resource deployment for a personal computer manufacturer.

T.J. Van Roy (1989). Multi-level production and distribution planning with transportation fleet optimization. *Management Sci.* 35, 1443–1453.

This paper describes an integrated optimization model—combining facility location, distribution, and fleet optimization—for liquified petroleum gas (LPG) products, developed for a petrochemical company and solved using MPSARX, a general-purpose mathematical programming software incorporating automatic reformulation and cut generation features.

W.B. Powell, Y. Sheffi (1989). Design and implementation of an interactive optimization system for the network design in the motor carrier industry. *Oper. Res.* 37, 12–29.

APOLLO, an interactive decision support system to route freight over a less-than-truckload motor carrier network, combines decomposition, heuristics, and user interaction to construct practical solutions to a large-scale mixed integer program containing embedded network design structure. This paper describes the planning problem, elements of the APOLLO system, implementation issues, and results.

J. Roy, T.G. Crainic (1992). Improving intercity freight routing with a tactical planning model. *Interfaces* 22, 31–44.

The authors apply a (nonlinear) network flow optimization methodology called NETPLAN to minimize transportation (TL and LTL), freight consolidation, and capacity/service penalty costs. The tactical planning model determines an optimal distribution pattern and service frequency, given the projected demand, cost parameters for different transportation modes, and the capacities of various terminals.

E.P. Robinson, L.L. Gao, S.D. Muggenborg (1993). Designing an integrated distribution system at DowBrands, Inc. *Interfaces* 23, 107–117.

The authors develop and implement a dual-based solution procedure for an uncapacitated network design model to design a two-echelon (central and regional distribution centers), multi-product distribution system incorporating TL shipments to central distribution centers, and TL or LTL shipments to customers. Using this model to design DowBrands' distribution system led to savings of over 1.5 million dollars per year.

J.F. Shapiro, V.M. Singhal, S.N. Wagner (1993). Optimizing the value chain. *Interfaces* 23, 102–117.

This paper describes the application (including an actual case study) of SLIM (Strategic Logistics Integrative Modeling System), a decision support system incorporating mixed integer programming solution capabilities to model, instantiate, and solve production and distribution planning problems.

B.C. Arntzen, G.G. Brown, T.P. Harrison, L.L. Trafton (1995). Global supply chain management at Digital Equipment Corporation. *Interfaces* 25, 69–93.

GSCM, a large scale optimization model developed to rationalize Digital Equipment Corporation's supply chain, selects a global manufacturing and distribution plan to minimize a weighted combination of the total flow time and cost. Embedded in this model is a fixed-charge network flow structure with additional restrictions on local content, offset trade, and facility capacity constraints as well as the product assembly structure. The model reportedly led to savings over 100 million dollars.

7.3 Other applications

J.M. Bloemhof-Ruwaard, M. Salomon, L.N.V. Wassenhove (1996). The capacitated distribution and waste disposal problem. *European J. Oper. Res.* 88, 490–503.

This paper discusses model formulations, lower and upper bounding procedures, and computational results for a two-level distribution and waste disposal problem. The objective is to minimize the fixed costs of plants and waste disposal units plus the variable costs of product and waste flows subject to capacity constraints on the facilities.

J.J. Jarvis, R.L. Rardin, V.E. Unger, R.W. Moore, C.C. Schimpeler (1978). Optimal design of regional wastewater systems: A fixed-charge network flow model. *Oper. Res.* 26, 538–550.

Using a network design model with piecewise linear concave costs, the authors design a cost-effective wastewater system for connecting collector nodes and treatment plant sites. After describing some problem reduction and variable elimination procedures, the authors discuss the application of this model to wastewater treatment system design in Jefferson County, Kentucky.

T.A. Feo, J.F. Bard (1989). Flight scheduling and maintenance base planning. *Management Sci.* 35, 1415–1432.

The authors model a combined flight scheduling and maintenance base planning problem as a network design problem with side constraints. They propose a two-phase heuristic to minimize the total cost of installing bases and maintenance activities, and illustrate its application using actual data from an airline.

D. De Wolf, Y. Smeers (1996). Optimal dimensioning of pipe networks with application to gas transmission networks. *Oper. Res.* 44, 596–608.

This paper describes a model to select optimal dimensions for pipes in a fluid transmission network, applied to the Belgian gas network. The model, incorporating both investment and operating costs, contains both flow conservation equations and constraints linking the flow variables to the pressure at each node (which are decision variables) and the diameters of the pipes.

19 Network Connectivity

S. Raghavan
US WEST Advanced Technologies

Thomas L. Magnanti
Massachusetts Institute of Technology

CONTENTS

Network design problems arise in a wide range of application domains. In many settings, one of the principle objectives is to connect designated network components (modules in VLSI design, component suppliers, assembly plants, warehouses, and customers in supply chain applications). In telecommunications, cable systems, and other contemporary applications, networks often provide time sensitive information and are, with modern technology (e.g., fiber optic cables), capable of carrying enormous amounts of information. Consequently, the failure of any link can have significant, perhaps even catastrophic, consequences and so we are naturally drawn to examine *Network Design Problems with Connectivity Requirements (NDC)*.

Informally, the NDC problem seeks a minimum cost network that satisfies requirements on the number of edge- or node-disjoint paths between pairs of nodes. Several examples of the NDC problem are classical problems in combinatorial optimization: (i) the minimum spanning tree problem, where we wish to design a minimum cost network that contains a single path from every node to every other node (the network contains all nodes and is connected), and (ii) the *Steiner tree*

Annotated Bibliographies in Combinatorial Optimization, edited by M. Dell'Amico, F. Maffioli and S. Martello

problem, where we seek a minimum cost network that connects a set of designated terminal nodes T. The resulting minimum cost network can use nodes, called *Steiner nodes*, other than the terminal nodes.

NDC problems appear both as stand alone problems and as subproblems in more complex network design applications. Because of the problem's importance in practice and because the underlying optimization problems are technically alluring, the combinatorial optimization research community has devoted considerable attention to solving these problems. When the prior version of this book appeared in 1985, knowledge about network connectivity was quite limited, mostly restricted to the Steiner tree problem. In the past decade, the research community has made noteworthy progress, both theoretically and practically. From the perspective of theory, researchers have added considerably to their understanding of several special cases of the NDC. For example, they have a good assessment of the complexity of these problems (i.e., which are polynomially solvable, which are approximable, and which are non-approximable). Researchers have also developed approximation algorithms for several classes of NDC problems, and have made significant progress in characterizing the polyhedral structure of these problems. From the practical standpoint, drawing upon the theoretical progress, the community has been able to successfully devise techniques to solve several large-scale NDC problems (for example, Steiner tree problems with several hundred nodes and several thousand edges). This paper summarizes some of these contributions.

In order to set notation and define the class of problems we consider, we formally state the *Network Design Problem with Connectivity Requirements* as follows. As data, we are given an *undirected* graph $G = (N, E)$ with node set N and edge set E, a cost vector $\mathbf{c} \in \mathcal{R}_+^{|E|}$ on the edges E, and a symmetric $|N| \times |N|$ matrix $\mathbf{R} = [r_{ij}]$ of connectivity requirements prescribing the minimum number of edge- or node-disjoint paths required between each pair i and j of nodes (we use a superscript $[n]$ to designate when the paths must be node disjoint): we wish to design a network that satisfies these requirements at minimum cost. We use the notation *edge connectivity* if the paths must be edge disjoint and *node connectivity* if the paths must be node disjoint.

The NDC problem is quite general and models a wide variety of problems. One important specialization is the *Survivable Network Design Problem (SND)* in which each node v in the graph has a connectivity requirement r_v and the network connectivity requirements are given by $r_{st} = \min\{r_s, r_t\}$. Several special cases of the SND problem are noteworthy.

Steiner Tree: Each node v has connectivity requirement $r_v = 0, 1$.

Minimum Cost k-Edge-Connected Spanning Subgraph Problem: Each node v has connectivity requirement $r_v = k$ and so the network remains connected after we delete any k or fewer edges.

Minimum Cost k-Node-Connected Spanning Subgraph Problem: Each node has requirement $k^{[n]}$ and so the network remains connected after we delete any k or fewer nodes.

Network Design with Low Connectivity Requirements (NDLC): For this special case of the SND problem, which is of particular interest to local telephone companies, the connectivity requirements r_v are restricted to either $\{0, 1, 2\}$ or $\{0, 1, 2^{[n]}\}$ (corresponding to edge- and node-connectivity versions of the NDLC problem).

One classification of NDC problems is useful. If the connectivity requirements imply that all nodes with a (positive) connectivity requirement must be connected, we say the problem is a *unitary NDC problem*. Otherwise, it is a nonunitary NDC problem. For example, the SND problem is a unitary NDC problem.

When the problem is defined on a directed graph (digraph), we obtain the *Directed Network Design Problem with Connectivity Requirements (DNDC)*. In this setting, the connectivity requirements are for arc- and node-disjoint paths.

Throughout this discussion, we assume the reader is familiar with standard graph-theory terminology. In addition, unless otherwise specified, we assume the NDC model does not permit edge replication. Typically, algorithms for the model without edge replication also apply to models with edge replication. To apply these algorithms, replace each edge $\{i, j\}$ by $\min(b_{ij}, r_{\max})$ copies of the edge (b_{ij} represents the number of parallel edges allowed between nodes i and j, and $r_{\max}$ represents the maximum connectivity requirement). In contrast, algorithms designed for situations that allow (unlimited) edge replication usually do not apply to situations that do not permit edge replication.

The rest of this discussion is organized as follows. After briefly reviewing books and surveys, in §2 we examine the Steiner tree problem. Next, in §3 we consider the *multi-level network design (MLND)*, a recent generalization of the NDC problem. In this setting, we have L facility types (i.e., edges) with edge costs satisfying $0 \leq c_{ij}^1 \leq c_{ij}^2 \cdots \leq c_{ij}^L$. The connectivity requirements between a pair of nodes have two attributes, one prescribing the number of edge- or node-disjoint paths between the nodes, and the other the lowest grade facility that can be used on these paths. Until now, most of the relevant literature on this problem has been on the two-level generalization of the Steiner tree problem, and for this reason we discuss the problem immediately after considering the Steiner tree problem. Then we resume our discussion of the NDC problem, summarizing research on connectivity augmentation problems in §4, structural properties and heuristics in §5, approximation algorithms in §6, the polyhedral approach in §7, and the directed NDC problem in §8. Finally, in §9 we discuss some variants (and applications) of the NDC problem.

In general, network design problems address issues other than connectivity. A companion chapter in this volume by Balakrishnan, Magnanti, and Mirchandani, complements this discussion by examining several more general issues.

1 Surveys

1.1 Steiner tree

The following papers and books contain surveys of Steiner tree problems.

P. Winter (1987). Steiner problems in networks: a survey. *Networks* 17, 129–167.
This paper summarizes exact algorithms and heuristics for the Steiner tree problem.

N. Maculan (1987). The Steiner problem in graphs. *Ann. Discr. Math.* 31, 185–212.

The author reviews several different formulations, and solution procedures based upon these formulations, for both undirected and directed Steiner tree problem.

F.K. Hwang, D.S. Richards, P. Winter (1992). *The Steiner Tree Problem*, Ann. Discr. Math. 53, North-Holland, Amsterdam.

This well written book contains a comprehensive review of the Euclidean Steiner tree problem, rectilinear Steiner tree problem, and the Steiner problem in networks. The section (chapters) on the Steiner problems in networks contains a lucid description of reduction tests, exact algorithms, heuristics, and special polynomially solvable cases.

T.L. Magnanti, L.A. Wolsey (1995). Optimal trees. M. Ball, T.L. Magnanti, C.L. Monma, G.L. Nemhauser (eds.). *Network Models*, Handbooks in Operations Research and Management Science, 7, North-Holland, Amsterdam, 503–615.

This survey examines two classes of network design problems: (i) those defined on trees, and (ii) those whose desired solution must be a tree (which includes the Steiner tree problem). The discussion emphasizes polyhedral results and the strength of alternative formulations, and demonstrates the use of many methods from combinatorial optimization for this problem class. The paper contains citations to about 100 papers in the literature.

1.2 Connectivity requirements

The following papers contain surveys of network design with connectivity requirements.

M. Grötschel, C.L. Monma, M. Stoer (1995). Design of survivable networks. M. Ball, T.L. Magnanti, C.L. Monma, G.L. Nemhauser (eds.). *Network Models*, Handbooks in Operations Research and Management Science, 7, North-Holland, Amsterdam, 617–672.

This paper examines the polyhedral structure of the survivable network design problem as well as computational experience in solving these problems using cutting plane methods. The first third of the paper contains a valuable survey of earlier results, including heuristic methods, and polynomially solvable cases.

A. Frank (1995). Connectivity and network flows. R.L. Graham, M. Grötschel, L. Lovász (eds.). *Handbook of Combinatorics*, 1, Elsevier Science and The MIT Press, Amsterdam and Cambridge, MA, 111–178.

This excellent survey provides a comprehensive overview of structural results in connectivity. The paper covers, among other topics, shortest paths, trees and arborescences, 2-, 3-, and higher level connectivity, and the disjoint path problem.

A. Frank (1994). Connectivity augmentation problems in network design. J.R. Birge, K.G. Murty (eds.). *Mathematical Programming: State of the Art 1994*, The University of Michigan, Ann Arbor, Michigan, 34–63.

The author examines those connectivity augmentation problems (see §4) that are solvable by exact algorithms running in strongly polynomial time.

M. Goemans, D. Williamson (1997). The primal-dual method for approximation and its application to network design problems. D. Hochbaum (ed.). *Approximation Algorithms for $\mathcal{NP}$-Hard Problems*, PWS Publishing Company, Boston, MA, 144–191.

This well written survey presents a generic primal-dual (dual-ascent) method for deriving approximation algorithms. The authors apply these algorithms, and derive approximation algorithms, for several network design problems, including the edge-connectivity version of the NDC problem. Most of the techniques and algorithms presented are very recent, having been obtained only in the last six years.

S. Khuller (1997). Approximation algorithms for finding highly connected subgraphs. D. Hochbaum (ed.). *Approximation Algorithms for $\mathcal{NP}$-Hard Problems*, PWS Publishing Company, Boston, MA, 236–265.

This paper contains a collection of approximation results on the following connectivity problems: (i) weighted and unweighted k-edge-connected spanning subgraph problems, (ii) weighted and unweighted k-node-connected spanning subgraph problems, (iii) connectivity augmentation problems, and (iv) the unweighted strongly connected spanning subgraph problem (i.e., the 1-arc-connected spanning digraph problem, a directed connectivity problem).

2 Steiner Trees

Before we begin to examine the Steiner tree problem, it will be useful to introduce two simple generalizations. Recall that the Steiner tree problem seeks a minimum cost tree that connects a set of designated terminal nodes T. The *node weighted Steiner tree problem* introduces node weights that represent the cost (or benefit) of including a Steiner node in the network. The *directed Steiner tree problem*, defined on a directed graph $D = (N, A)$ with associated arc costs, requires a minimum cost directed tree rooted at a specified root node r.

We can always transform the undirected Steiner tree problem (and the node weighted version) into a directed Steiner tree problem by

1. replacing every edge $\{i,j\} \in E$ by two directed arcs (i,j) and (j,i), each with cost $c_{(i,j)} = c_{(j,i)} = c_{\{i,j\}}$. (If the problem has node weights w_v for $v \in N$, then set $c_{(i,j)} = c_{\{i,j\}} + w_j$ and $c_{(i,j)} = c_{\{j,i\}} + w_i$.)

2. selecting one of the nodes in T as the root node and the remaining nodes as the terminal nodes in the directed Steiner tree problem.

Most of the research on Steiner tree problems has focused on the undirected version of the problem. However, solution methods for the directed problem have proven quite effective when applied to the undirected problem (after it has been transformed into a directed problem). Consequently, we will discuss both the undirected and directed problems together. For simplicity, we refer to the undirected Steiner tree problem (in networks) as the Steiner tree problem. We will explicitly qualify any other versions (e.g., directed, node weighted) as we discuss them.

2.1 Reductions

The research community has developed several procedures for reducing the size of the underlying network in which the Steiner tree problem must be solved. These procedures are useful because they can lead to substantial reductions in the problem size.

A. Balakrishnan, N.R. Patel (1987). Problem reduction methods and a tree generation algorithm for the Steiner network problem. *Networks* 17, 65–85.

By exploiting special properties of optimal Steiner tree solutions, the authors show how to reduce the problem. They also show that under some mild probabilistic assumptions, these reductions are asymptotically optimal (i.e., they obtain the optimal Steiner tree solution). They then describe an enumerative tree generation algorithm for the Steiner tree problem that is based on modeling the problem as a minimum cost spanning tree with additional degree constraints.

C.W. Duin, A. Volgenant (1989). Reduction tests for the Steiner problem in graphs. *Networks* 19, 549–568.

The authors briefly review reduction tests for the Steiner tree problem. Using a bottleneck approach, they then generalize these problem reduction methods and describe an edge inclusion test, an edge exclusion test, and a node exclusion test. In computational experiments, they solve a majority of Steiner tree test problems by using only these problem reduction methods.

2.2 Heuristics and approximation algorithms

One of the first heuristics for the Steiner tree problem, a minimum *spanning tree heuristic*, was also one of the first approximation algorithms in network design.

H. Takahashi, A. Matsuyama (1980). An approximate solution for the Steiner problem in graphs. *Math. Japonica* 6, 573–577.
L.T. Kou, G. Markowsky, L. Berman (1981). A fast algorithm for Steiner trees. *Acta Inform.* 15, 141–145.
J. Plesník (1981). A bound for the Steiner tree problem in graphs. *Math. Slovaca* 31, 155–163.

These three papers independently proposed slightly different versions of a minimum spanning tree heuristic, each with a worst-case ratio of $2 - (2/|T|)$.

Subsequently, researchers proposed several variations of these spanning tree heuristics with the same worst case ratio $2 - (2/|T|)$. These heuristics start with a forest containing the nodes in T (and none of the edges), and repetitively add edges and nodes to obtain a tree that spans T (the methods have the property that if $T = N$, they obtain the optimal spanning tree). The methods differ in how they add edges and nodes. [Hwang, Richards, Winter 1992] (see §§1.1) provide an excellent survey of these heuristics.

P. Winter, J.M. Smith (1992). Path-distance heuristics for the Steiner tree problem in undirected networks. *Algorithmica* 7, 309–327.

The authors experimentally compare the performance of the different variations of the tree heuristics.

M. Bern, P. Plassmann (1989). The Steiner tree problem with edge lengths 1 and 2. *Inform. Process. Lett.* 32, 171–176.

The authors show that when all the edge lengths are either 1 or 2, the problem is $\mathcal{NP}$-hard. They prove that the worst case of the average distance heuristic (a tree heuristic proposed by Rayward-Smith in 1983, whose worst case is also $2 - (2/|T|))$ for this problem is 4/3.

The worst case bound of $2 - 2/|T|$ stood for many years as the best known bound of any heuristic until Zelikovsky recently improved upon it.

A. Zelikovsky (1993). An 11/6-approximation algorithm for the network Steiner problem. *Algorithmica* 9, 463–470.

The paper describes a heuristic to iteratively determine the choice of Steiner nodes in the Steiner tree. The procedure starts with the minimum spanning tree on T, and considers connecting a Steiner node to a 3-node subset of T. The benefit of adding a Steiner node is determined by the decrease in the cost of the tree (i.e., cost of edges deleted minus cost of edges added). The procedure iteratively adds a Steiner node that provides maximum benefit (over all possibilities) until adding a Steiner node provides no benefit. This heuristic has a worst case ratio of 11/6.

P. Berman, V. Ramaiyer (1994). Improved approximation algorithms for the Steiner tree problem. *J. Algorithms* 17, 381–408.
A. Zelikovsky (1996). *Better approximation bounds for the network and Euclidean Steiner tree problems*, Technical Report CS-96-06, Department of Computer Science, University of Virginia, Charlottesville, Virginia.
M. Karpinski, A. Zelikovsky (1997). New approximation algorithms for the Steiner tree problems. *J. Comb. Opt.* 1, 1–19.

Berman and Ramaiyer improve the worst-case ratio of the algorithm of [Zelikovsky 1993] (see above) by (i) generalizing it to consider k-node subsets of T, and (ii) delaying the decision to add Steiner nodes until the end of the algorithm. They show that when $k = 4$, the algorithm's worst case ratio is 16/9, and its asymptotic worst-case ratio, i.e., when $k \rightarrow \infty$, is ≈ 1.746. Zelikovsky further improves the asymptotic worst case ratio of the algorithm to $1 + \log 2 \approx 1.693$, by determining the benefit of adding a Steiner node as the ratio of the cost of edges deleted to the cost of edges added. Karpinski and Zelikovsky add a preprocessing step that improves the asymptotic bound to ≈ 1.644.

2.3 Lagrangian relaxation and dual-ascent methods

Obtaining good lower bounds for the Steiner tree problem (and in general for NDC problems) is of considerable importance. Lower bounding procedures can be used to assess the performance of heuristics for large problems when it might not be possible to compute the exact solution due to time or space constraints of an exact solution procedure. Lower bounding procedures can often be used to develop enumerative branch and bound algorithms to solve the Steiner tree problem exactly.

R.T. Wong (1984). A dual ascent approach for Steiner tree problems on a directed graph. *Math. Program.* 28, 271–287.

The author describes a dual-ascent (primal-dual) solution procedure for the directed Steiner tree problem using a directed flow formulation for the problem. In his computational work, he solves undirected Steiner tree problems by converting them to a directed Steiner tree problem. The solution procedure, which finds both a lower bound and a feasible heuristic solution for the problem, was very successful in solving the undirected problems to near-optimality (generally within a 2 percent guarantee).

J.E. Beasley (1984). An algorithm for the Steiner problem in graphs. *Networks* 14, 147–159.
J.E. Beasley (1989). An SST-based algorithm for the Steiner problem in graphs. *Networks* 19, 1–16.

These two papers propose different Lagrangian relaxations, incorporated in a branch and bound framework. The first paper proposes two relaxations based upon an undirected flow-based formulation for the Steiner tree problem. The second paper develops Lagrangian bounds by relaxing degree constraints in a formulation that models the problem as a minimum cost spanning tree with additional degree constraints. The bounds provided by the relaxation proposed in the second paper are stronger; by using this relaxation, the author successfully solves problems with up to 2500 nodes and 62500 edges.

W. Liu (1990). A lower bound for the Steiner tree problem in directed graphs. *Networks* 20, 765–778.

The author describes a new integer programming formulation for the directed Steiner tree problem, based upon the observation that every directed Steiner tree contains a two-terminal directed Steiner tree for every pair of terminal nodes. He then describes a dual-ascent algorithm based on this formulation, and shows that for the directed Steiner tree problem this algorithm obtains significantly better lower bounds than Wong's dual-ascent algorithm.

2.4 Polyhedral approach

Y.P. Aneja (1980). An integer linear programming approach to the Steiner problem in graphs. *Networks* 10, 167–178.

This paper modifies a well-known cutting plane algorithm to solve a set covering formulation (i.e., the cutset formulation) of the Steiner tree problem.

M.X. Goemans, Y.-S. Myung (1993). A catalog of Steiner tree formulations. *Networks* 23, 19–28.

The authors establish connections between several different formulations of the Steiner tree problem and certain of its extensions. They also show that the choice of the root node does not affect the value of the LP-relaxation of the bidirected formulation of the Steiner tree problem.

S. Chopra, M.R. Rao (1994). The Steiner tree problem I: formulations, compositions and extension of facets. *Math. Program.* 64, 209–230.
S. Chopra, M.R. Rao (1994). The Steiner tree problem II: properties and classes of facets. *Math. Program.* 64, 231–246.

In these two companion papers, the authors investigate the facial structure of the polyhedron of the undirected Steiner tree problem. They describe several different classes of facet-defining inequalities—including partition and odd-hole inequalities. They also show that the bidirected cut formulation for the Steiner tree problem is stronger than the undirected cut formulation. In the second paper, they briefly describe some facets for the directed Steiner tree problem.

S. Chopra, E. Gorres, M.R. Rao (1992). Solving the Steiner tree problem on a graph using branch-and-cut. *ORSA J. Comput.* 4, 320–335.

This paper presents computational experience with a branch and cut algorithm, based upon the bidirected formulation, for the Steiner tree problem.

M.X. Goemans (1994). The Steiner tree polytope and related polyhedra. *Math. Program.* 63, 157–182.

Given an undirected graph and associated edge and node costs, in the r-tree problem we wish to find a minimum cost tree containing node r. This problem models both the node weighted Steiner tree problem and the Steiner tree problem. The author describes a formulation for the r-tree problem consisting of edge and node variables, and shows that this formulation describes the r-tree polytope on series-parallel graphs. By projection, he then derives a new class of facet defining inequalities for the Steiner tree polytope, called combinatorial design inequalities.

M. Fischetti (1991). Facets of two Steiner arborescence polyhedra. *Math. Program.* 51, 401–419.

The author describes new classes of facets for both the directed Steiner tree polytope and its dominant.

2.4.1 Polyhedra for special cases

A. Prodon (1992). Steiner trees with n terminals among $n+1$ nodes. *Oper. Res. Lett.* 11, 125–133.

For the special case when the number of terminal nodes is one less than the total number of nodes, the author presents an extended formulation that completely describes the Steiner tree polytope.

F. Margot, A. Prodon, T.M. Liebling (1994). Tree polytopes on 2-trees. *Math. Program.* 63, 183–192.

The paper describes an extended formulation for the Steiner tree problem using the natural edge variables and additional node variables, and shows that the linear programming relaxation of this formulation completely characterizes the Steiner tree problem on series-parallel graphs.

M.X. Goemans (1994). Arborescence polytopes for series-parallel graphs. *Discr. Appl. Math.* 51, 277–289.

The author characterizes the polytopes of the directed Steiner tree problem and some of its extensions on bidirected series-parallel graphs.

M. Ball, W. Liu, W. Pulleyblank (1989). Two terminal Steiner tree polyhedra. B. Cornet, H. Tulkens (eds.). *Contributions to Operations Research and Econometrics – The Twentieth Anniversary of CORE*, MIT Press, Cambridge, MA, 251–284.

The authors provide an extended formulation for the directed Steiner tree problem with two terminal nodes and show that when the arc costs are non-negative, the LP-relaxation of this formulation provides an optimal integer solution for the problem. By projecting onto the space of the natural variables, they obtain a complete description of the dominant of the two-terminal directed Steiner tree polytope.

3 Multi-level Network Design

MLND applications arise in telecommunications, transportation, and electric power distribution. In contemporary communication networks, certain nodes might represent business or government customers requiring high bandwidth (e.g., fiber optic) connections while other nodes correspond to customers with lower bandwidth needs that can be accommodated using copper cables. Similarly, a transportation system might need to connect large metropolitan areas via all-weather highways while smaller access roads are adequate to reach small towns.

To date the relevant literature on the MLND problem has focused on a particular multi-level generalization of the Steiner tree problem. In this setting with L facility types, each node i has a level $l_i \in \{1, \ldots, L\}$, and a multi-level network design must connect all nodes at each level l using level l or higher level (grade) facilities. For the remainder of this discussion, we use the term MLND to refer to this multi-level Steiner tree problem.

When the network only contains two facility types, the MLND problem is called the *two-level network design (TLND)*. In this setting, we refer to level 2 nodes as primary nodes, and level 1 nodes as secondary nodes. A further specialization of the TLND, the so-called *Hierarchical Network Design (HND)* problem, is defined over a two-level network containing only two primary nodes and $(|N| - 2)$ secondary nodes. Even with very special cost structures—for example, with unit primary costs and binary secondary costs or the proportional costs case in which the ratio of primary to secondary costs is the same for all edges—this simple HND problem is $\mathcal{NP}$-hard. Consequently, research on this problem class has focused on developing and testing heuristic and optimization-based methods, and characterizing the worst-case performance of (tree-based) heuristics for the MLND problem and its special cases.

J.R. Current, C.S. ReVelle, J.L. Cohon (1986). The hierarchical network design problem. *European J. Oper. Res.* 27, 57–66.

This paper, one of the first to motivate, define, and formulate the HND problem, presents a HND heuristic method that identifies the K shortest paths between the two primary nodes, augments each of these paths to cover all of the remaining (secondary) nodes, and chooses the best among these K solutions. The method found the optimal solution for two test problems, the larger containing 21 nodes and 39 edges.

H. Pirkul, J. Current, V. Nagarajan (1991). The hierarchical network design problem: A new formulation and solution procedures. *Transportation Sci.* 25, 175–182.

To facilitate using the Lagrangian relaxation method, the authors develop a new HND formulation and use it to solve problems with up to 85 nodes. Their approach solves 95 percent of their test problems to optimality.

C. Duin, A. Volgenant (1989). Reducing the hierarchical network design problem. *European J. Oper. Res.* 39, 332–344.

The authors show how to transform the TLND problem into an equivalent directed Steiner tree problem (with twice the number of nodes and edges), extend several Steiner tree problem preprocessing and reduction tests to this problem, propose a Lagrangian relaxation solution methodology, and report results of computational tests to assess the effectiveness of the reduction strategies on random problems containing up to 40 nodes and 780 edges.

C. Duin, A. Volgenant (1991). The multi-weighted Steiner tree problem. *Ann. Oper. Res.* 33, 451–469.

For the TLND problem (which the authors refer to as the multi-weighted Steiner tree problem), this paper describes two classes of greedy/local improvement heuristics that utilize shortest path and spanning tree subroutines, and extends some Steiner tree problem reduction tests to the TLND model.

A. Balakrishnan, T.L. Magnanti, P. Mirchandani (1994). A dual-based algorithm for multi-level network design. *Management Sci.* 40, 567–581.

This paper develops preprocessing rules and a dual-ascent algorithm to determine both lower bounds and heuristic solutions for the MLND problem, and reports computational results on two-level test problems containing up to 500 nodes and 5000 edges. The method produces solutions that are within 0.9 percent of optimality for problem instances with varying cost structures.

A. Balakrishnan, T.L. Magnanti, P. Mirchandani (1994). Modeling and heuristic worst-case performance analysis of the two-level network design problem. *Management Sci.* 40, 846–867.

After establishing the LP-equivalence of two flow-based problem formulations of the TLND problem—an enhanced undirected version and a directed formulation—the authors develop tight worst-case heuristic bounds for a composite method that selects, for any given problem instance, the best among alternative heuristic solutions obtained using spanning and Steiner tree subproblem solutions. For the proportional costs case with optimal subproblem solutions, the method produces a design whose cost is guaranteed to be no more than 4/3 the cost of the optimal solution.

P. Mirchandani (1996). The multi-tier tree problem. *INFORMS J. Comput.* 8, 202–218.

Building upon the analytical results and algorithmic experience for TLND problems, this paper derives new worst-case bounds for a recursive heuristic, describes problem reduction strategies, and develops a dual-based algorithm for MLND problems with more than two levels. Extensive computational testing for large-scale problems with up to 2000 nodes, 6000 edges, and 5 levels confirm the algorithm's effectiveness.

J. Singhal (1990). Two-level hierarchical transportation networks: a new set of problems and practical applications. *Decision Sci.* 21, 171–182.

The author introduces the following Euclidean version of the HND problem that arises in the context of rural network planning: given a set of points in the plane, select a primary link (straight line) in the plane and connect each point to this primary link via a secondary link (perpendicular to the primary link) to minimize the weighted sum of the primary and secondary links. The special structure of this problem permits it to be solved using a one-dimensional search procedure.

Before concluding this section, we note that generalizations of the classical network design problem lead to more general versions of the MLND problem. These variants and generalizations include models with flow costs, directed edges, routing restrictions, and connectivity requirements. Some of the solution methods discussed in this section apply to these more general situations. For example, the dual-ascent algorithms described in the paper by Balakrishnan, Magnanti, and Mirchandani and in the paper by Mirchandani apply to MLND problems with flow costs.

4 Connectivity Augmentation Problems

We now resume our discussion of the NDC problem. In connectivity augmentation problems, the underlying graph is complete and the edge costs are either 0 or 1. These problems have the following interpretation. Starting from the graph defined by the zero-cost edges, add the fewest possible edges (augment the graph) to satisfy the connectivity requirements. If all edge costs are 1 (i.e., we start with an empty graph), these problems are referred to as uniform cost problems.

K.P. Eswaran, R.E. Tarjan (1976). Augmentation problems. *SIAM J. Comput.* 5, 653–665.

This paper introduces augmentation problems and studies the problem of making a given graph 2-edge-connected or 2-node-connected as well as the problem of making a digraph strongly connected. The authors provide efficient algorithms for both these problems and show that weighted versions of these problems (i.e., when edge costs are arbitrary) are $\mathcal{NP}$-complete.

S. Sridhar, R. Chandrasekaran (1992). Integer solution to synthesis of communication networks. *Math. Oper. Res.* 17, 581–585.

The authors consider the edge-connectivity version of the uniform cost NDC problem, assuming that the graph is complete and edge replication is permitted (i.e., the integer version of the classical network synthesis problem introduced and solved by Gomory and Hu). The authors show how to modify the well-known Gomory-Hu algorithm so that it provides integer capacities for the edges.

A. Frank (1992). Augmenting graphs to meet edge-connectivity requirements. *SIAM J. Discr. Math.* 5, 25–53.

This excellent paper contains a plethora of results for situations with complete graphs (digraphs) with edge (arc) replication. It provides a polynomial-time algorithm for solving the edge-connectivity augmentation, thereby generalizing several earlier results (which we have not mentioned). It also solves the problem of augmenting a digraph to a k-arc-connected digraph (i.e., the network has k arc-disjoint paths from every node to every other node), again generalizing all known results in this area.

The node-connectivity augmentation problem is considerably more difficult than the edge-connectivity augmentation problem. As reviewed in the survey paper by [Frank 1994] (see §§1.2), there are very few polynomially solvable cases.

When the underlying graph is not complete, and edge replication is prohibited, the augmentation problem is $\mathcal{NP}$-Hard. As a result, researchers have focused on designing approximation algorithms for this problem.

S. Khuller, U. Vishkin (1994). Biconnectivity approximations and graph carvings. *J. ACM* 41, 214–235.

The authors consider the 2-edge- and 2-node-connected spanning subgraph problem when the underlying graph is arbitrary, edge replication is prohibited, and all edge costs are 1. These problems are $\mathcal{NP}$-hard. They describe a $\frac{3}{2}$-approximation algorithm for the 2-edge-connected spanning subgraph problem and a $\frac{5}{3}$-approximation algorithm for the 2-node-connected spanning subgraph problem. They also give a 2-approximation algorithm for the minimum cost k-edge-connected spanning subgraph problem.

N. Garg, V. Santosh, A. Singla (1993). Improved approximation algorithms for biconnected subgraphs via better lower bounding techniques. *Proc. 4th Annual ACM-SIAM Symp. Discr. Algorithms*, 103–111.

By adding an additional edge discarding step to Khuller and Vishkin's algorithm for the uniform cost 2-node-connected case, the authors improve its worst case approximation ratio to $\frac{3}{2}$. The paper also sketches a $\frac{5}{4}$-approximation algorithm (full details are not provided) for the 2-edge-connected spanning subgraph problem.

H. Nagamochi, T. Ibaraki (1992). A linear-time algorithm for finding a sparse k-connected spanning subgraph of a k-connected graph. *Algorithmica* 7, 583–596.

This paper describes a 2-approximation algorithm for the uniform cost case of both the k-edge- and k-node-connected spanning subgraph problems when the underlying graph is arbitrary and edge replication is prohibited.

J. Cheriyan, M.Y. Kao, R. Thurimella (1993). Scan-first search and sparse certificates: an improved parallel algorithm for k-vertex connectivity. *SIAM J. Comput.* 22, 157–174.

This paper describes a 2-approximation algorithm for the uniform cost k-node-connected problem when the underlying graph is arbitrary and edge replication is not permitted. The algorithm is novel since it has particularly efficient implementations in the parallel, distributed, and sequential models of computation.

5 Structural Properties and Heuristics

Structural properties of solutions are very useful in constructing heuristics. Often they can lead to improved formulations for a combinatorial optimization problem. For this reason, many researchers have investigated structural properties of the NDC problem.

G.N. Frederickson, J. JáJá (1982). On the relationship between the biconnectivity augmentation and traveling salesman problems. *Theoretical Comp. Sci.* 19, 189–201.

The authors show that when the cost function satisfies the triangle inequality, a 2-edge-connected graph can be made 2-node-connected without increasing (and possibly decreasing) its cost. (This result implies that when the cost function satisfies the triangle inequality, and the connectivity requirements $r_{ij} \in \{0, 1, 2\}$, edge-connectivity requirements and node-connectivity requirements are equivalent.) They also show that Christofides' heuristic for the TSP is a $\frac{3}{2}$-approximation algorithm for the 2-node-connected spanning subgraph problem.

C.L. Monma, B.L. Munson, W.R. Pulleyblank (1990). Minimum-weight two-connected spanning networks. *Math. Program.* 46, 153–171.

When designing a minimum cost 2-node-connected spanning subgraph for cost function satisfying the triangle inequality, the authors prove that for some optimal solution, all vertices have degree 2 or 3, and show that the cost of an optimal traveling salesman tour is at most $\frac{4}{3}$ times the cost of an optimal 2-node-connected solution.

D. Bienstock, E.F. Brickell, C.L. Monma (1990). On the structure of minimum-weight k-connected spanning networks. *SIAM J. Discr. Math.* 3, 320–329.

When the cost function satisfies the triangle inequality, the authors establish structural properties of the optimal solution for the problem of designing minimum cost k-edge-connected and k-node-connected spanning subgraphs,

M.X. Goemans, K.T. Talluri (1991). 2-change for k-connected networks. *Oper. Res. Lett.* 10, 113–117.

The authors derive a structural property that proves that 2-opt is a feasible heuristic for the minimum cost k-edge-connected spanning subgraph problem when k is even or 1.

C.L. Monma, D.F. Shallcross (1989). Methods for designing communication networks with certain two-connected survivability constraints. *Oper. Res.* 37, 531–541.
C.-W. Ko, C.L. Monma (1989). *Heuristic methods for designing highly survivable communication networks*, Technical report, Bellcore, Morristown, New Jersey.

The first paper describes local improvement heuristics, designed to work well on sparse graphs, for the network design problem with low connectivity requirements. Using similar ideas, the second paper proposes heuristics for designing k-edge-connected and k-node-connected graphs.

6 Approximation Algorithms

G.N. Frederickson, J. JáJá (1981). Approximation algorithms for several graph augmentation problems. *SIAM J. Comput.* 10, 270–283.

This paper considers three problems: the minimum cost 2-edge-connected spanning subgraph problem, the minimum cost 2-node-connected spanning subgraph problem, and the minimum cost strongly connected subgraph problem (a directed problem). For the minimum cost 2-node connected spanning subgraph problem, the authors provide a 3-approximation algorithm, and for the other two problems, they present 2-approximation algorithms.

M.X. Goemans, D.J. Bertsimas (1993). Survivable networks, linear programming relaxations and the parsimonious property. *Math. Program.* 60, 145–166.

The authors consider the edge-connectivity version of the SND problem for situations with edge-replication. They show that when the edge costs satisfy the triangle inequality, the value of the LP-relaxation of a simple cutset model is unaffected by the addition of degree constraints that state that the degree of a node is equal to its connectivity requirement. They describe and present worst-case analysis for two simple heuristics (i) a generalization of the tree heuristic for the Steiner tree problem, and (ii) an improved method combining Christofides heuristic and the tree heuristic. Their analysis establishes worst-case bounds on the ratio between the optimal IP/LP ratio of the cutset formulation.

M.X. Goemans, D.P. Williamson (1995). A general approximation technique for constrained forest problems. *SIAM J. Comput.* 24, 296–317.

The authors describe a primal-dual algorithm for a wide class of problems that includes the edge-connectivity version of the NDC problem with edge replication. They describe a primal-dual algorithm that has the same worst-case ratio as the tree heuristic described by Goemans and Bertsimas. Their analysis also establishes worst-case bounds on the ratio between the optimal IP/LP ratio of the cutset formulation.

D.P. Williamson, M.X. Goemans, M. Mihail, V. Vazirani (1995). A primal-dual approximation algorithm for generalized Steiner network problems. *Combinatorica* 15, 435–454.
H.N. Gabow, M.X. Goemans, D.P. Williamson (1993). An efficient approximation algorithm for the survivable network design problem. *Proc. 3rd MPS Conf. Integer Prog. and Comb. Opt.*, 57–74.
M.X. Goemans, A. Goldberg, S. Plotkin, D.B. Shmoys, E. Tardos, D.P. Williamson (1994). Improved approximation algorithms for network design problems. *Proc. 5th Annual ACM-SIAM Symp. Discr. Algorithms*, 223–232.

The first paper describes the *first polynomial time approximation algorithm* for a class of problems which includes the edge-connectivity version of the network design problem with connectivity requirements. The algorithm has a worst case ratio of $2r_{\max}$ (recall $r_{\max}$ is the maximum connectivity requirement). The analysis also shows that the worst-case ratio of the IP/LP ratio of the cut formulation is at most $2r_{\max}$. The second paper describes a different implementation of the algorithm that improves its worst case running time. The third paper changes the order in which the algorithm considers the connectivity requirements to improve the worst case ratio to $2\mathcal{H}(r_{\max})$ (where $\mathcal{H}(n) = 1 + \frac{1}{2} + \cdots + \frac{1}{n}$), which is tight for this algorithm. This worst-case ratio also applies to the IP/LP ratio of the cut formulation.

A. Balakrishnan, T.L. Magnanti, P. Mirchandani (1994). *Doubling or splitting: strategies for survivable network design*, Technical Report OR 297-94, Operations Research Center, MIT, Cambridge, MA.

The authors describe a $\frac{3}{2}$-approximation algorithm for the NDLC problem. They also describe a new extended formulation for the edge-connectivity version of the NDC problem, examine the tightness of its LP-relaxation, and use this analysis to establish an improved heuristic to LP worst-case ratio of the algorithms of [Goemans and Bertsimas' 1993] and [Goemans and Williamson 1995] (see above) when at least one node has unit connectivity requirement.

R. Ravi, D. Williamson (1995). An approximation algorithm for minimum-cost vertex connectivity problems. *Proc 6th Annual ACM-SIAM Symp. Discr. Algorithms*, 332–341.

The authors present a $2\mathcal{H}(k)$-approximation algorithm for the minimum cost k-node-connected spanning subgraph problem, and a 3-approximation algorithm for the NDC problem when $r_{ij} \in \{0, 1, 2^{[n]}\}$.

T.L. Magnanti, S. Raghavan (1997). *A dual-ascent algorithm for low-connectivity network design*, Technical report, Operations Research Center, MIT, Cambridge, MA.

This paper presents a dual-ascent algorithm for the edge-connectivity version of the NDLC problem, based upon a directed formulation for the problem. The solution procedure provides both a lower bound and a feasible heuristic solution to the problem. Computational experiments report on successful solutions to problems with 300 nodes and 3000 edges to within 4 percent of optimality.

S. Khuller, B. Raghavachari (1996). Improved approximation algorithms for uniform connectivity problems. *J. Algorithms*, 21, 434–450.

This paper contains three approximation results: (i) an (1.85)-approximation algorithm for the minimum cost k-edge-connected spanning subgraph problem when all edge costs are equal, and the underlying graph is arbitrary; (ii) a $(2 + \frac{1}{|N|})$-approximation algorithm for the minimum cost 2-node-connected spanning subgraph problem; and (iii) a 2-approximation algorithm for the minimum cost k-node-connected spanning subgraph problem when the edge costs satisfy the triangle inequality.

7 Polyhedral Approach

M. Grötschel, C.L. Monma, M. Stoer (1992). Facets for polyhedra arising in the design of communication networks with low-connectivity requirements. *SIAM J. Optim.* 2, 474–504.
M. Grötschel, C.L. Monma, M. Stoer (1992). Computational results with a cutting plane algorithm for designing communication networks with low-connectivity constraints. *Oper. Res.* 40, 309–330.

These two papers investigate the polyhedral structure of the low edge- and node-connectivity network design problems (i.e., the NDLC problem). The first paper focuses on the polyhedral structure of the problem. The second paper uses several valid inequalities developed in the first paper in a cutting plane algorithm to solve to optimality some real-world problems with about 100 nodes and 200 edges.

M. Grötschel, C.L. Monma, M. Stoer (1995). Polyhedral and computational investigations for designing communication networks with high survivability requirements. *Oper. Res.* 43, 1012–1024.

Continuing along the lines of their earlier work, the authors investigate the polyhedral structure of SND problems whose highest connectivity requirements are three or more. They generalize several valid inequalities for the low-connectivity case, and in some cases present conditions under which these inequalities define facets. They incorporate these valid inequalities in a cutting plane algorithm that easily solves minimum cost k-edge-connected spanning subgraph problems with random edge costs to optimality, and obtains lower bounds that are usually within 4 percent for real-world high-connectivity problems.

S. Boyd, T. Hao (1993). An integer polytope related to the design of survivable communication networks. *SIAM J. Discr. Math.* 6, 612–630.

The authors introduce a class of valid inequalities for the 2-edge-connected spanning subgraph polytope and describe necessary and sufficient conditions for these valid inequalities to define facets.

T.L. Magnanti, S. Raghavan (1997). *Strong formulations for network design problems with connectivity requirements*, Technical report, Operations Research Center, MIT, Cambridge, MA.

The authors describe a directing procedure for unitary NDC problems, and present a new directed formulation for the unitary NDC problem that is stronger than a natural undirected formulation. They then project out several classes of valid inequalities—partition inequalities, odd-hole inequalities, and combinatorial design inequalities—that generalize known classes of valid inequalities for the Steiner tree problem to the unitary NDC problem.

7.1 Polyhedral results for series-parallel graphs

A.R. Mahjoub (1994). Two-edge connected spanning subgraphs and polyhedra. *Math. Program.* 64, 199–208.
M. Baïou, A.R. Mahjoub (1993). *The 2-edge connected Steiner subgraph polytope of a series-parallel graph*, Technical report, Départment d'informatique, Université de Bretagne Occidentale, 6 Avenue Victor Le Gorgeu, B.P. 452, 29275 Brest Cedex, France.

The first paper shows that the cut inequalities completely describe the 2-edge-connected spanning subgraph polytope on series-parallel graphs. The second paper describes an additional class of cut inequalities whose addition describes the Steiner 2-edge-connected spanning subgraph polytope on series-parallel graphs.

C.R. Coullard, A. Rais, R.L. Rardin, D.K. Wagner (1990). *The 2-connected-spanning-subgraph polytope for series-parallel graphs*, Technical Report 90-12, Industrial Engineering Department, Purdue University, West Lafayette, Indiana.
C.R. Coullard, A. Rais, R.L. Rardin, D.K. Wagner (1991). *The 2-connected Steiner subgraph polytope for series-parallel graphs*, Technical Report 91-32, Industrial Engineering Department, Purdue University, West Lafayette, Indiana.

These two papers describe the polyhedra of the 2-node-connected problem (both the spanning and Steiner versions). The first paper shows that the well-known cut formulation (containing simple cut and node-cut inequalities) completely describes the 2-node-connected spanning subgraph polytope on series-parallel graphs. The second paper describes an additional class of inequalities, and shows that the addition of these inequalities provides a complete description of the Steiner 2-node-connected spanning subgraph polytope on series-parallel graphs.

S. Chopra (1994). The k-edge-connected spanning subgraph polyhedron. *SIAM J. Discr. Math.* 7, 245–259.

The author studies the minimum cost k-edge-connected spanning subgraph problem with edge-replication. He provides a complete linear inequality description of the polytope associated with the k-edge-connected spanning subgraph problem on an outerplanar graph (an outerplanar graph is a special type of series-parallel graph) when k is odd.

8 Directed Network Connectivity

In this section we review the few known results for the directed NDC problem. We describe polynomially solvable cases, approximation algorithms, and polyhedral results.

[Frank 1992] (see §4) generalized most results for digraph augmentation problems by describing an algorithm for solving the k-arc-connected spanning digraph augmentation problem when the underlying digraph is complete, and arc-replication is permitted.

A. Frank, É. Tardos (1989). An application of submodular flows. *Linear Alg. App.* 114/115, 329–348.

Using a tricky reduction to submodular flows, the authors describe an algorithm for the minimum cost node-disjoint k-branching problem, that is, the problem of finding

k node-disjoint directed paths between a specified root node and every other node in the graph.

The first approximation result for directed problems appears to be the 2-approximation algorithm by [Frederickson and JáJá 1981] (see §6) for the minimum cost strongly-connected spanning subgraph problem (also known as the 1-arc-connected spanning subgraph or the equivalent subgraph problem). We reviewed the paper containing this result in §6. When the arc costs are uniform, the following two papers describe approximation algorithms with better worst-case ratios.

S. Khuller, B. Raghavachari, N. Young (1995). Approximating the minimum equivalent digraph. *SIAM J. Comput.* 24, 859–872.
S. Khuller, B. Raghavachari, N. Young (1996). On strongly connected digraphs with bounded cycle length. *Discr. Appl. Math.* 69, 281–289.

Both these papers consider the uniform cost case of the minimum cost strongly-connected spanning subgraph problem. The first paper describes an approximation algorithm with worst-case ratio $\pi^2/6 \approx 1.645$. The second paper improves this worst-case ratio to $\pi^2/6 - 1/36 \approx 1.617$.

S. Chopra (1992). Polyhedra of the equivalent subgraph problem and some edge connectivity problems. *SIAM J. Discr. Math.* 5, 321–337.

This paper describes several facet-defining inequalities (including directed cut inequalities) for the minimum weight strongly-connected spanning subgraph problem. It then shows how to use a directed cut formulation for the edge-connectivity version of the NDLC problem with edge replication (i.e., the undirected problem), and reports that the lower bounds obtained by solving the LP relaxation of the directed cut formulation are consistently within 1 percent of the optimal objective value.

S. Chopra (1992). The equivalent subgraph polytope and directed cut polyhedra on series-parallel graphs. *SIAM J. Discr. Math.* 5, 475–490.

The author shows that directed cut inequalities completely describe the equivalent subgraph polyhedron on strongly-connected series-parallel graphs. He also shows that families of minimal directed cuts and minimal equivalent subgraphs form a pair of blocking clutters, thereby showing that a class of inequalities called minimal equivalent subgraph inequalities completely describes the directed cut polyhedron on series-parallel graphs.

G. Dahl (1993). Directed Steiner problems with connectivity requirements. *Discr. Appl. Math.* 47, 109–128.
G. Dahl (1994). The design of survivable directed networks. *Telecommunication Sys.* 2, 349–377.

The first paper studies valid inequalities and facets for a special case of the directed network design problem with connectivity requirements (DNDC) when all directed paths emanate from a specified root node. The second paper discusses two models for the more general DNDC problem, one a natural directed cut (dicut) model and the other a directed flow formulation. For the dicut model, the author presents several classes of strong valid inequalities and describes a cutting plane solution algorithm. He also gives a general method for obtaining facets for the DNDC polytope from facets for the special case described in the first paper.

9 Variants

Variants of NDC problems often arise is practice. In this section we briefly describe some variants (and applications) of NDC problems that are of considerable interest to the research community.

The first few papers describe applications of the NDC problem in the design of telecommunication networks.

R.H. Cardwell, C.L. Monma, T.H. Wu (1989). Computer-aided design procedures for survivable fiber optic networks. *IEEE J. Select. Areas Commun.* 7, 1188–1197.

This paper describes methods for designing low cost survivable telecommunications networks. It contains an insightful description of telecommunications technology, survivability issues, and how these problems are solved in practice.

M. Mihail, D. Shallcross, N. Dean, M. Mostrel (1996). A commercial application of survivable network design: ITP/INPLANS CCS network topology analyzer. *Proc. 7th Annual ACM-SIAM Symp. Discr. Algorithms*, 279–287.

This paper discusses an application of the network design problem with connectivity requirements that arises in the design of low cost common channel signaling networks.

Y. Lee, L. Lu, Y. Qiu, F. Glover (1994). Strong formulations and cutting planes for designing digital data networks. *Telecommunication Sys.* 2, 261–274.

Y. Lee, S. Y. Chiu, J. Ryan (1996). A branch and cut algorithm for a Steiner tree-star problem. *INFORMS J. Comput.* 8, 194–201.

These two papers study the digital data design network design problem, a node weighted Steiner tree problem in which terminal nodes must have unit degree, and can be connected only to Steiner nodes. The first paper compares several formulations for the problem, and the second paper presents a branch and cut algorithm that is based on a formulation similar to one described by [Goemans 1994] (see §§2.4). The authors solve problems with up to 100 nodes to optimality.

M.B. Rosenwein, R.T. Wong (1995). A constrained Steiner tree problem. *European J. Oper. Res.* 81, 430–439.

The authors compare and provide computational experience for two different Lagrangian relaxations applied to a Steiner tree problem with an additional budget constraint.

Analysts frequently use hop constraints, which limit the number of edges (nodes) on a path between nodes in a network, to model delay and reliability (fewer hops implies less delay and a lower route loss probability) in communication networks. In one real-world application of hop constraints, the design of long distance telephone networks, planners typically design networks so that the traffic they carry will not require more than two hops (to ensure that a solution will not overutilize scarce switching resources at the nodes). In the past few years, several researchers have studied hop constrained network design problems. The following paper is a representative example.

L. Gouveia (1996). Multicommodity flow models for spanning trees with hop constraints. *European J. Oper. Res.* 95, 178–190.

This paper considers the problem of designing a minimum cost spanning tree with an additional hop constraint on the path from each node to a designated root node.

After describing two equivalent multicommodity flow models—one undirected and one directed—for the problem, it adopts a Lagrangian solution approach for solving problems with up to 40 nodes.

M. Grötschel, A. Martin, R. Weismantel (1996). Packing Steiner trees: polyhedral investigations. *Math. Program.* 72, 101–123.
M. Grötschel, A. Martin, R. Weismantel (1996). Packing Steiner trees: a cutting plane algorithm and computational results. *Math. Program.* 72, 125–145.
These two papers examine the Steiner packing problem: Given a graph $G = (N, E)$ with edge costs c_e, edge capacities w_e, and node sets (nets) $T_1, T_2, \ldots, T_P$, find Steiner trees $S_1, S_2, \ldots, S_P$ so that S_k is a Steiner tree for net T_k, no more than w_e of the Steiner trees use edge $e \in E$, and the sum of the costs of the trees is minimized. This model applies to some routing problems in VLSI design. The first paper studies the polyhedral structure of the Steiner tree packing polyhedron, emphasizing the so-called joint-facets obtained by considering more than a single net. The second paper presents a branch-and-cut algorithm for the Steiner tree packing problem based upon the previous valid inequalities. This approach solves many switchbox routing problems (a special application of this problem) from the literature to optimality.

In many practical situations, due to the decentralization and distributed nature of modern communication networks, we need to design networks that satisfy certain connectivity requirements, and also satisfy the criterion of minimizing the maximum degree of the nodes in the network. The following survey paper summarizes approximation results in this area.

B. Raghavachari (1997). Algorithms for finding low degree structures. D. Hochbaum (ed.). *Approximation Algorithms for $\mathcal{NP}$-Hard Problems*, PWS Publishing Company, Boston, MA, 266–295.

The multicut problem is somewhat dual to the problem of augmenting a network to satisfy connectivity requirements. In this problem we wish to select a subset of edges or nodes to delete from the network to ensure that certain pairs of nodes are disconnected. A simple case, the problem of finding a minimum cost set of edges to disconnect a pair of nodes s and t is the minimum cut problem. Directed generalizations lead to the feedback arc set and feedback node set problem (where we wish to select a subset of edges or nodes respectively whose deletion creates an acyclic digraph). Researchers have proposed numerous other generalizations, e.g., multiway cut, Steiner multicut, etc., of these problems. The following paper provides an up to date review of approximation results for some of these problems.

D.B. Shmoys (1997). Cut problems and their application to divide and conquer. D. Hochbaum (ed.). *Approximation Algorithms for $\mathcal{NP}$-Hard Problems*, PWS Publishing Company, Boston, MA, 192–235.

20 Linear Assignment

Mauro Dell'Amico
Università di Modena

Silvano Martello
Università di Bologna

CONTENTS

The *Assignment Problem* (AP) is one of the most popular and intensively studied topics in combinatorial optimization. Given an $n \times n$ cost matrix $[c_{ij}]$, the problem is to match each row to a different column in such a way that the sum of the selected costs is a minimum. Formally,

$$z = \min \sum_{i=1}^{n} \sum_{j=1}^{n} c_{ij} x_{ij}, \tag{1}$$

$$\sum_{j=1}^{n} x_{ij} = 1 \quad (i = 1, \ldots, n), \tag{2}$$

$$\sum_{i=1}^{n} x_{ij} = 1 \quad (j = 1, \ldots, n), \tag{3}$$

$$x_{ij} \in \{0, 1\} \quad (i, j = 1, \ldots, n) \tag{4}$$

where $x_{ij} = 1$ iff row i is assigned to column j. The costs are assumed to be, in general, real numbers. However, there are a family of algorithms based on *cost scaling*, which apply only to instances with integer costs and have weakly polynomial time complexity (logarithmic function of the input data): for such cases we will denote by C the maximum $|c_{ij}|$ value.

It is well-known that the constraint matrix defined by (2), (3) is totally unimodular.

Annotated Bibliographies in Combinatorial Optimization, edited by M. Dell'Amico, F. Maffioli and S. Martello

Hence AP is equivalent to the linear program given by (1), (2), (3) and

$$x_{ij} \geq 0 \quad (i,j = 1,\ldots,n). \tag{5}$$

By associating dual variables u_i and v_j with constraints (2) and (3), respectively, the corresponding dual is

$$w = \max \textstyle\sum_{i=1}^{n} u_i + \sum_{j=1}^{n} v_j, \tag{6}$$

$$u_i + v_j \leq c_{ij} \quad (i,j = 1,\ldots,n). \tag{7}$$

A pair of solutions respectively feasible for the primal and the dual is optimal if and only if (*complementary slackness*)

$$x_{ij}(c_{ij} - u_i - v_j) = 0 \quad (i,j = 1,\ldots,n). \tag{8}$$

Most algorithms for AP include a preprocessing phase in which a dual feasible solution is determined as

$$u_i = \min_j \{c_{ij}\} \quad (i = 1,\ldots,n) \quad (\textit{row reduction}),$$

$$v_j = \min_i \{c_{ij} - u_i\} \quad (j = 1,\ldots,n) \quad (\textit{column reduction})$$

and a partial primal solution is defined by assigning, in some order, an unassigned row i to an unassigned column j if $c_{ij} = u_i + v_j$ (see (8)).

The assignment problem is also known as the *Weighted Bipartite Matching Problem*. Let $G = (U \cup V, E)$ be a bipartite graph with vertex sets $U = V = \{1, 2, \ldots, n\}$, edge set $E = \{(i,j) : i \in U, j \in V, c_{ij} > 0\}$ and $|E| = m$. The problem is then to find a perfect matching of minimum cost. It can be shown that a feasible basis of the linear program equivalent to AP induces a *spanning tree* on G.

After the section on books and surveys, in §2 we examine a selection of fundamental results obtained from the origins to 1981: in this period all basic approaches to the solution of the problem have been introduced. More recent results, classified according to the algorithmic approach adopted, are discussed in the following four sections. Starting from the Eighties, parallel algorithms for AP have also appeared: these are examined in §7. Information on available software is given in §8. In the Appendix (§9) assignment problems with different objective functions are considered.

1 Books and Surveys

E.L. Lawler (1976). *Combinatorial Optimization: Networks and Matroids*, Holt, Rinehart and Winston, New York.

This classic book presents an improved version of the Hungarian algorithm (see [Kuhn 1955, 1956], §2) giving an $O(n^3)$ time complexity.

R.E. Burkard (1979). Travelling salesman and assignment problems: a survey. P.L. Hammer, E.L. Johnson, B.H. Korte (eds.). *Discrete Optimization I*, Ann. Discr. Math. 4, North-Holland, Amsterdam, 193–215.

The first survey on the problem presents an interesting summary of results on the structure of the associated polytope.

R.E. Burkard, U. Derigs (1980). *Assignment and Matching Problems: Solution Methods with FORTRAN Programs*, Springer-Verlag, Berlin.

This book on various assignment-type problems includes, among others, a FORTRAN program implementing a variant of the shortest augmenting path algorithm presented in [Tomizawa 1971] (see §2).

U. Derigs (1985). The shortest augmenting path method for solving assignment problems –motivation and computational experience. C. L. Monma (ed.). *Algorithms and Software for Optimization – Part I*, Ann. Oper. Res. 4, Baltzer, Basel, 57–102.

All the classical algorithms are discussed and related to the shortest augmenting path technique. The results of an extensive computational experience, performed over fourteen FORTRAN codes, are analyzed.

S. Martello, P. Toth (1987). Linear assignment problems. S. Martello, G. Laporte, M. Minoux, C. Ribeiro (eds.). *Surveys in Combinatorial Optimization*, Ann. Discr. Math. 31, North-Holland, Amsterdam, 259–282.

AP and other linear assignment-type problems are studied. The performance of different algorithms is analyzed through computational experiments.

D.P. Bertsekas (1991). *Linear Network Optimization: Algorithms and Codes*, The MIT Press, Cambridge, MA.

A state-of-the art book on relaxation and auction techniques. It includes several implementations of algorithms for AP and the corresponding FORTRAN listings (also avilable on diskette).

M. Akgül (1992). The linear assignment problem. M. Akgül, H.W. Hamacher, S. Tüfekçi (eds.). *Combinatorial Optimization*, Springer-Verlag, Berlin, 85–122.

An excellent survey which analyzes all exact sequential solution approaches and discusses their relationships.

D.S. Johnson, C.C. McGeoch (eds.) (1993). *Network Flows and Matching: First DIMACS Implementation Challenge*, AMS, Providence, RI.

This volume includes several papers on implementations of algorithms for AP. The most interesting ones are commented in the following sections.

U. Faigle (1994). Some recent results in the analysis of greedy algorithms for assignment problems. *Z. Oper. Res.* 15, 181–188.

A specialized survey on the probabilistic analysis of simple heuristic algorithms for AP. Both online and offline algorithms are considered.

2 The First Fifty Years (1931–1981)

E. Egerváry (1931). Matrixok kombinatorius tulajdonságairol (in Hungarian). *Matematikai és Fizikai Lapok* 38, 16–28. (English translation by H.W. Kuhn (1955). On combinatorial properties of matrices. *Logistic Papers* 11, paper 4, 1-11, George Washington University).
D. König (1936). *Theorie der Endlichen und Unendlichen Graphen* (in German), Akademische Verlagsgesellschaft M.B.H., Leipzig.
Two fundamental theorems on square matrices are presented by these Hungarian mathematicians. In 1955 Kuhn derived the *Hungarian* algorithm from them.

T.E. Easterfield (1946). A combinatorial algorithm. *J. London Math. Soc.* 21, 219–226.
A non-polynomial algorithm is presented, based on iterated column reduction.

G. Birkhoff (1946). Tres observaciones sobre el algebra lineal. *Revista Facultad de Ciencias Exactas, Puras y Aplicadas Universidad Nacional de Tucuman, Serie A (Matematicas y Fisica Teorica)* 5, 147–151.
A fundamental result on doubly stochastic matrices is presented, from which one has directly the proof that the extreme points of the assignment polytope are integral.

D.F. Votaw, A. Orden (1952). The personnel assignment problem. *Symposium on Linear Inequalities and Programming*, SCOOP 10. U.S. Air Force, 155–163.
The name of the problem appears for the first time.

H.W. Kuhn (1955). The Hungarian method for the assignment problem. *Naval Res. Log. Quart.* 2, 83–97.
H.W. Kuhn (1956). Variants of the Hungarian method for the assignment problem. *Naval Res. Log. Quart.* 3, 253–258.
These papers present the first polynomial-time algorithm for the problem. A partial solution, in which $k \leq n$ rows are assigned to k columns, is initially obtained through row and column reduction of the cost matrix. The number of assignments is iteratively increased through zero cost augmenting paths and dual variable updatings. The time complexity of this *primal–dual* algorithm is $O(n^4)$. The genesis of the Hungarian method was later reported in
H.W. Kuhn (1991). On the origin of the Hungarian method. J.K. Lenstra, A.H.G. Rinnooy Kan, A. Schrijver (eds.). *History of Mathematical Programming*, North-Holland, Amsterdam, 77–81.

J. Munkres (1957). Algorithms for the assignment and transportation problems. *J. SIAM* 5, 32–38.
Improvements on the Hungarian algorithm are discussed.

R. Silver (1960). Algorithm 27: assignment. *Commun. ACM* 3, 603–604.
R. Silver (1960). An algorithm for the assignment problem. *Commun. ACM* 3, 605–606.
The first computer code for the problem: an Algol program implementing a variation of the Munkres algorithm.

J.M. Kurtzberg (1962). On approximation methods for the assignment problem. *J. ACM* 9, 419–439.

One of the earliest results on the *probabilistic analysis* of approximation algorithms. Bounds are obtained on the expected value of the optimal solution and the approximate solution of an $O(n^2)$ heuristic algorithm, when the costs are uniformly distributed in the interval [0,1]. It is proved that the relative error tends to 0 as $n \to \infty$. Further results on the expected value of the optimal solution are presented in
W.E. Donath (1969). Algorithm and average-value bounds for assignment problems. *IBM J. Res. Develop.* 13, 380–386.
D.W. Walkup (1979). On the expected value of a random assignment problem. *SIAM J. Comput.* 8, 440–442.

M.L. Balinski, R.E. Gomory (1964). A primal method for the assignment and transportation problems. *Management Sci.* 10, 578–593.

This algorithm maintains a feasible assignment and a dual (unfeasible) solution satisfying complementary slackness. The assignment is iteratively improved through alternating paths. This is the first polynomial-time *primal algorithm* for the problem. The time complexity is $O(n^4)$.

E.A. Dinic, M.A. Kronrod (1969). An algorithm for the solution of the assignemt (*sic.*) problem. *Sov. Math. Dokl.* 10, 1324–1326.

This paper was ignored for a long time, although it gives the first algorithm with $O(n^3)$ time complexity. This is also the first *dual* algorithm for the problem. A dual solution (u, v) with $u = 0$ determines a semi-assignment (each row is assigned to the column producing the minimum reduced cost): dual variables v are updated through alternating paths, until a feasible primal assignment is obtained.

N. Tomizawa (1971). On some techniques useful for solution of transportation network problems. *Networks* 1, 173–194.

In spite of its title, this paper presents a *shortest augmenting path* technique for the min-cost flow problem. When applied to the assignment problem it produces a primal-dual algorithm with $O(n^3)$ time complexity. The same result was independently obtained in
J. Edmonds, R.M. Karp (1972). Theoretical improvements in algorithmic efficiency for network flow problems. *J. ACM* 19, 248–264.

M.L. Balinski, A. Russakoff (1974). The assignment polytope. *SIAM Rev.* 16, 516–525.

A series of fundamental results is proved. It is shown in particular that the "Hirsch conjecture" holds, i.e., any two extreme points of the assignment polytope are joined by a path of at most two edges.

R.E. Machol, M. Wien (1976). A hard assignment problem. *Oper. Res.* 24, 190–192.
R.E. Machol, M. Wien (1977). Errata. *Oper. Res.* 24, 364.

A famous class of difficult instances ($c_{ij} = (i-1)(j-1) \forall i, j$) is introduced.

R.S. Barr, F. Glover, D. Klingman (1977). The alternating basis algorithm for assignment problems. *Math. Program.* 13, 1–13.

This paper presents a specialization of the primal simplex algorithm to AP. Theoretical results are given, allowing only a subset of bases (*alternating path bases*, independently developed by Cunningham for network flows) to be considered, although the worst-case time complexity remains non-polynomial.

M.S. Hung, W.O. Rom (1980). Solving the assignment problem by relaxation. *Oper. Res.* 28, 969–982.

This algorithm is closely related to the Dinic-Kronrod approach. Dual alternating path bases are updated through shortest paths until a semi-assignment becomes feasible.

R.M. Karp (1980). An algorithm to solve the $m \times n$ assignment problem in expected time $O(nm \log n)$. *Networks* 10, 143–152.

A variant of the Tomizawa-Edmonds-Karp algorithm is presented. It is shown that its expected time complexity is $O(n^2 \log n)$ if the costs in each row are drawn independently from a common distribution.

D.P. Bertsekas (1981). A new algorithm for the assignment problem. *Math. Program.* 21, 152–171.

This dual algorithm has a high average efficiency, although its worst-case time complexity is exponential (pseudo-polynomial). At each stage the dual values v_j corresponding to assigned columns are increased as much as possible without violating the complementary slackness conditions. The approach is known as the *auction algorithm*, as the unassigned rows can be imagined as competitors in an auction and the columns as items for sale.

3 Shortest Path (Primal–Dual) Algorithms

M. Engquist (1982). A successive shortest path algorithm for the assignment problem. *INFOR* 20, 370–384.
L.F. McGinnis (1983). Implementation and testing of a primal-dual algorithm for the assignment problem. *Oper. Res.* 31, 277–299.
G. Carpaneto, P. Toth (1983). Algorithm 50: algorithm for the solution of the assignment problem for sparse matrices. *Computing* 28, 83–94.

Three FORTRAN implementations (of the Dinic-Kronrod algorithm, of the Hungarian algorithm and of the Lawler version of the Hungarian algorithm, respectively) are presented and computationally compared with the algorithm in [Barr, Glover and Klingman 1977] (see §2) as well as with other algorithms.

R.E. Tarjan (1983). *Data Structures and Network Algorithms*, CBMS-NSF Regional Conference Series in Applied Mathematics 44, SIAM, Philadelphia, PA.
M.L. Fredman, R.E. Tarjan (1984). Fibonacci heaps and their uses in improved network optimization algorithms. *Proc. 25th Annual IEEE Symp. Found. Comput. Sci.*, 338–346. Also in *J. ACM* 34, 596–615 (1987).

Data structures for shortest path algorithms are introduced. Their use for AP produces algorithms having $O(nm \log_{(2+m/n)} n)$ and $O(n^2 \log n + nm)$ time complexity, respectively, particularly efficient for sparse matrices. The latter value is the best strongly polynomial time bound known for the problem.

J. Aráoz, J. Edmonds (1985). A case of non-convergent dual changes in assignment problems. *Discr. Appl. Math.* 11, 95–102.

The Hungarian algorithm converges in a finite number of steps, independently of the input data. If the data are rational numbers the same holds for any primal–dual algorithm, independently of the labelling technique used. This paper presents an instance including irrational numbers, and shows that a non-Hungarian primal–dual algorithm may run forever without finding the optimal solution.

H.N. Gabow (1985). Scaling algorithms for network problems. *J. Comput. Syst. Sci.* 31, 148–168.

This is the first application of a scaling approach to AP. The proposed algorithm, based on recursive costs halving, applies only to instances with integer costs. Its time complexity is weakly polynomial, namely $O(n^{3/4} m \log C)$.

P. Carraresi, C. Sodini (1986). An efficient algorithm for the bipartite matching problem. *European J. Oper. Res.* 23, 86–93.

F. Glover, R. Glover, D. Klingman (1986). Threshold assignment algorithm. *Math. Program. Study* 26, 12–37.

These two algorithms, independently obtained, are very close to each other. An original shortest path labelling technique is used to improve the average performance of the algorithm.

U. Derigs, A. Metz (1986). An efficient labeling technique for solving sparse assignment problems. *Computing* 36, 301–311.

An effective implementation of the shortest augmenting path technique.

R. Jonker, A. Volgenant (1986). Improving the Hungarian assignment algorithm. *Oper. Res. Lett.* 5, 171–175.

Improvements for the Hungarian algorithm are presented and evaluated through computational experiments.

G. Carpaneto, P. Toth (1987). Primal-dual algorithms for the assignment problem. *Discr. Appl. Math.* 18, 137–153.

R. Jonker, A. Volgenant (1987). A shortest augmenting path algorithm for dense and sparse linear assignment problems. *Computing* 38, 325–340.

Efficient implementations, including effective initialization procedures, are presented. The latter paper also gives the corresponding Pascal code. These have been, for some years, the fastest algorithms available. Computational experiments concerning the latter algorithm can be found in [Goldberg and Kennedy 1995] (see §5) and in [Kennington and Wang 1991] and [Zaki 1995] (see §7.3).

H.N. Gabow, R.E. Tarjan (1989). Faster scaling algorithms for network problems. *SIAM J. Comput.* 18, 1013–1036.

A new scaling technique produces a weakly polynomial algorithm whose time complexity, $O(\sqrt{n}m\log(nC))$, improves that in [Gabow 1985] (see above). This is the best time bound currently known for a cost-scaling algorithm for AP.

R. Jonker, A. Volgenant (1989). Teaching linear assignment by Mack's algorithm. J. K. Lenstra, H. Tijns, A. Volgenant (eds.). *Twenty-Five Years of Operations Research in the Netherlands: Papers Dedicated to Gijs de Leve*, CWI Tract, 70, Centre for Mathematics and Computer Science, Amsterdam, 54–60.

This paper discusses and improves Mack's so-called Bradford method, which is equivalent to the Hungarian algorithm, but more comprehensible. See also
C. Mack (1969). The Bradford method for the assignment problem. *New J. Statis. and Oper. Res.* 1, 17–29.

W.M. Nawijn, B. Dorhout (1989). On the expected number of assignments in reduced matrices for the linear assignment problem. *Oper. Res. Lett.* 8, 329–335.

A simple initialization procedure is described and the expected number of assignments obtained is determined under the assumption that the costs are a family of independent identically distributed random variables with continuous distribution functions.

Y. Lee, J.B. Orlin (1994). On very large scale assignment problems. W.W. Hager, D.W. Hearn, P.M. Pardalos (eds.). *Large Scale Optimization: State of the Art*, Kluwer Academic Publishers, Dordrecht, 206–244.

A modified Dijkstra algorithm is used to obtain a shortest augmenting path algorithm for AP, which finds an optimal assignment on a restricted graph. Under certain conditions this solution is also optimal for the original graph.

A. Volgenant (1996). Linear and semi-assignment problems: a core oriented approach. *Computers Oper. Res.* 23, 917–932.

A sparsification technique is applied to the algorithm developed in [Jonker and Volgenant 1987] (see above) to obtain a code which does not need to store the entire cost matrix. Computational comparisons with the codes of [Carpaneto and Toth 1987] and [Jonker and Volgenant 1987] are given.

4 Primal Algorithms

M.S. Hung (1983). A polynomial simplex method for the assignment problem. *Oper. Res.* 31, 595–600.

This is a polynomial-time specialization of the primal simplex algorithm for instances of AP with integer costs. The number of pivots is bounded by $O(n^3 \log \Delta)$, where Δ is the difference between the initial and final solution value.

J.B. Orlin (1985). On the simplex algorithm for networks and generalized networks. *Math. Program. Study* 24, 166–178.

The [Hung 1983] algorithm above is weakly polynomial. This paper presents a strongly polynomial primal simplex algorithm requiring at most $O(n^2 m \log n)$ pivots. However, Roohy-Laleh, in an unpublished Ph.D. thesis (Carleton University, Ottawa,

1980), had presented a primal pivot rule giving, for AP, a bound of $O(n^3)$ on the number of pivots.

R.K. Ahuja, J.B. Orlin (1992). The scaling network simplex algorithm. *Oper. Res.* Suppl. 1, S5–S13.

A scaling version of Dantzig's pivot rule is presented. The resulting simplex algorithm solves AP in $O(n^2 \log C)$ pivots and $O(nm \log C)$ time.

M. Akgül (1993). A genuinely polynomial primal simplex algorithm for the assignment problem. *Discr. Appl. Math.* 45, 93–115.

This is the best primal simplex algorithm currently known. It requires at most $O(n^2)$ pivots, and can be implemented to run in $O(n^3)$ time for dense matrices or in $O(n^2 logn + nm)$ time for sparse matrices.

5 Dual Algorithms

M.L. Balinski (1984). The Hirsch conjecture for dual transportation polyhedra. *Math. Oper. Res.* 9, 629–633.

This paper shows that any two extreme points of the transportation polytope are joined by a path of at most $(m-1)(n-1)$ extreme edges, i.e., the Hirsh conjecture holds. This is proved through an algorithm which finds the required path using a signature to uniquely identify each extreme point. (The *signature* of a tree is the vector of the degrees of the vertices in U.)

M.L. Balinski (1985). Signature methods for the assignment problem. *Oper. Res.* 33, 527–536.

Signatures are used to obtain a dual algorithm for AP which requires $(n-1)(n-2)/2$ pivots. An $O(n^3)$ time implementation is then obtained through techniques pointed out to the author by W.H. Cunningham and D. Goldfarb in private communications. A variant of the algorithm and an $O(nm + n^2 \log n)$ implementation for the case of sparse graphs are presented in
D. Goldfarb (1985). Efficient dual simplex algorithms for the assignment problem. *Math. Program.* 33, 187–203.

G. Kindervater, A. Volgenant, G. de Leve, V. van Gijlswijk (1985). On dual solutions of the linear assignment problem. *European J. Oper. Res.* 19, 76–81.

A simple approach to generate alternative dual solutions is presented.

M.L. Balinski (1986). A competitive (dual) simplex method for the assignment problem. *Math. Program.* 34, 125–141.

The [Balinski 1985] and [Goldfarb 1985] algorithms above do not implement a pure dual simplex method since they may pivot on an edge with zero or positive flow. Here a genuinely simplex algorithm is obtained by introducing *strong dual feasible trees* and combining them with the signature approach. The time complexity is $O(n^3)$.

P. Kleinschmidt, C.W. Lee, H. Schannat (1987). Transportation problems which can be solved by the use of Hirsch-paths for the dual problems. *Math. Program.* 37, 153–168.

It is proved that the [Balinski 1985] algorithm above is equivalent, cycle by cycle, to the [Hung and Rom 1980] algorithm (see §2).

M. Akgül (1988). A sequential dual simplex algorithm for the linear assignment problem. *Oper. Res. Lett.* 7, 155–158.
M. Akgül (1989). Erratum. A sequential dual simplex algorithm for the linear assignment problem. *Oper. Res. Lett.* 8, 117.

This algorithm generates a series of problems defined over subgraphs of the original graph. Each problem is solved through the technique used in [Balinski 1986] (see above). The time complexity is $O(n^3)$ for dense graphs and $O(nm+n^2 \log n)$ for sparse graphs.

D.P. Bertsekas, J. Eckstein (1988). Dual coordinate step methods for linear network flow problems. *Math. Program.* 42, 203–243.

The idea of *ε-relaxation* of the complementary slackness conditions is introduced. For AP, this results in an algorithm in which a condition of type (8) is considered satisfied if $c_{ij} - u_i - v_j \leq \varepsilon$. An improvement of the [Bertsekas 1981] auction algorithm (see §2) is then presented: applying ε-scaling, a time complexity of $O(mn \log(nC))$ is obtained.

K. Paparrizos (1988). A non-dual signature method for the assignment problem and a generalization of the dual simplex method for the transportation problem. *RAIRO Rech. Oper.* 22, 269–289.

At each stage of this dual algorithm two phases are executed: a decomposition of the current dual-feasible tree and a linking phase which builds a new (improved) tree. During the linking phase the dual-feasibility may be violated. The overall time complexity is $O(n^4)$. An improved $O(n^3)$ version is presented in
M. Akgül, O. Ekin (1991). A dual feasible forest algorithm for the linear assignment problem. *RAIRO Rech. Oper.* 25, 403–411.

K. Paparrizos (1991). An infeasible (exterior point) simplex algorithm for assignment problems. *Math. Program. (A)* 51, 45–54.
K. Paparrizos (1991). A relaxation column signature method for assignment problems. *European J. Oper. Res.* 50, 211–219.

Variants of a dual-type simplex algorithm are presented. The approach is related to that presented in [Akgül 1988] (see above). The time complexity is $O(n^3)$.

J.B. Orlin, R.K. Ahuja (1992). New scaling algorithms for the assignment and minimum cycle mean problems. *Math. Program.* 54, 41–56.

The approach presented in this paper applies a scaling technique to a hybrid of the auction algorithm and the shortest path algorithm. It has the same time complexity as the [Gabow and Tarjan 1989] algorithm (see §3), while using much simpler data structures.

D.A. Castañon (1993). Reverse auction algorithms for assignment problems. D.S. Johnson, C.C. McGeoch (eds.). *Network Flows and Matching: First DIMACS*

Implementation Challenge, AMS, Providence, RI, 407–430.

An implementation of the auction algorithm is presented. An analysis of the sensitivity to the choice of the scale factor is given. For the computational performance of the algorithm see also below, [Goldberg and Kennedy 1995].

K.G. Ramakrishnan, N.K. Karmarkar, A.P. Kamath (1993). An approximate dual projective algorithm for solving assignment problems. D.S. Johnson, C.C. McGeoch (eds.). *Network Flows and Matching: First DIMACS Implementation Challenge*, AMS, Providence, RI, 431–452.

A specialization of the interior-point method to AP is presented. Extensive computational experiments are performed on large-size instances. (See also below, [Goldberg and Kennedy 1995].)

B.L. Schwartz (1994). A computational analysis of the auction algorithm. *European J. Oper. Res.* 42, 161–169.

An analysis of the expected performance of the auction algorithm is presented. The analysis is based on computational experiments performed for a problem of military interest, the so-called Weapon-Target Assignment. (The actual experimental results, however, are not given, as this was forbidden by the US Department of Defense.)

A.V. Goldberg, R. Kennedy (1995). An efficient cost scaling algorithm for the assignment problem. *Math. Program.* 71, 153–177.

A recent technique for the maximum flow problem, the *push-relabel* method, is used within a cost scaling algorithm for AP which also includes several heuristics. The resulting code is compared, through extensive computational experiments, with the algorithms by [Jonker and Volgenant 1987] (see §3), [Castañon 1993] and [Ramakrishnan, Karmarkar and Kamath 1993] (see above).

A.V. Goldberg, R. Kennedy (1997). Global price updates help. *SIAM J. Discr. Math.* (to appear).

The authors present an algorithm that is faster than the one in the previous paper by a factor of $n^{1/2}$ if it includes periodic global updates of the dual variables, but that has the standard time bound $O(mn \log(nC))$ without global updates. Unfortunately the theoretically faster algorithm does not perform well in practice.

6 Approximation Algorithms and Probabilistic Analysis

Most results of a probabilistic nature have been obtained under the assumption that the costs are independent uniformly distributed random variables on the interval [0,1]: this is denoted in the following as the [0,1]-*assumption*.

D. Avis (1983). A survey of heuristics for the weighted matching problem. *Networks* 13, 475–493.

In the sections devoted to the assignment problem, the author analyzes the algorithms by [Kurtzberg 1962] (see §2) and a few other results obtained before 1981.

Worst-case results and average-case results (under the [0,1]-assumption) are presented.

D. Avis, L. Devroye (1985). An analysis of decomposition heuristic for the assignment problem. *Oper. Res. Lett.* 3, 279–283.

The decomposition algorithm presented in [Kurtzberg 1962] (see §2) is analyzed. A worst-case bound is obtained for the maximization version of AP. A probabilistic analysis of the algorithm is provided for the case in which the [0,1]-assumption holds.

D. Avis, C.W. Lai (1988). The probabilistic analysis of a heuristic for the assignment problem. *SIAM J. Comput.* 17, 732–741.

A new heuristic with $O(n^2)$ time complexity is presented, and the expected value of its solution is determined under the [0,1]-assumption.

B. Kalyanasundaram, K. Pruhs (1993). Online weighted matching. *J. Algorithms* 14, 478–488.

In the *online assignment problem* we have n time intervals. During the i-th interval the weights of a row of the cost matrix are divulged and one unmatched column must be assigned to the row. The worst-case performance of different algorithms is studied under the restriction that the costs are positive and satisfy the triangle inequality.

R.M. Karp, A.H.G. Rinnooy Kan, R. V. Vohra (1994). Average analysis of a heuristic for the assignment problem. *Math. Oper. Res.* 19, 513–522.

This paper presents an $O(n \log n)$ time heuristic for perfect matching in a special class of random bipartite graphs, and derives an $O(n^2)$ algorithm for AP. A bound on the expected solution value is determined under the [0,1]-assumption.

7 Parallel Algorithms

7.1 Theoretical results

J.R. Driscoll, H.N. Gabow, R. Shrairman, R.E. Tarjan (1988). Relaxed heaps: an alternative to Fibonacci heaps with applications to parallel computation. *Commun. ACM* 31, 1343–1354.

New priority queue data structures are introduced which can be used to obtain an efficient parallel implementation of the Dijkstra algorithm. Use of this technique in a shortest path approach to AP gives an $O(nm/p)$ time complexity on an EREW PRAM with $p \leq m/(n \log n)$ processors, under the assumption that the costs are non-negative.

H.N. Gabow, R.E. Tarjan (1988). Almost-optimum speed-ups of algorithms for bipartite matching and related problems. *Proc. 20th Annual ACM Symp. Theory of Comput.*, 514–527.

The assignment problem is solved through a cost scaling algorithm that runs on an EREW PRAM with p processors. The algorithm requires $O(\sqrt{n}m \log(nC) \log(2p)/p)$ time and $O(m)$ space, for $p \leq m/(\sqrt{n} \log^2 n)$. For $p = 1$ this gives the best time bound known for a cost scaling sequential algorithm (see [Gabow and Tarjan 1989], §3).

A.V. Goldberg, S.A. Plotkin, P.M. Vaidya (1988). Sublinear-time parallel algorithms for matching and related problems. *Proc. 29th Annual IEEE Symp. Found. Comput. Sci.*, 174–184.

This is the first deterministic sublinear time algorithm for instances of AP with integer costs. It runs on a CRCW PRAM with no more than n^3 processors, and has time complexity $O(n^{2/3} \log^3 n \log(nC))$. A sequential version has the same time complexity as the algorithm by [Gabow and Tarjan 1989] (see §3).

C.N.K. Osiakwan, S. Akl (1990). A perfect speedup parallel algorithm for the assignment problem on complete weighted bipartite graphs. *Proc. of IEEE Parbase 90*, 293–304.

A parallelization of the Hungarian algorithm is presented that runs in $O(n^3/p+n^2p)$ time on an EREW PRAM computer with $p \leq \sqrt{n}$ processors.

A.V. Goldberg, S.A. Plotkin, D.B. Shmoys, E. Tardos (1992). Using interior point methods for fast parallel algorithms for bipartite matching and related problems. *SIAM J. Comput.* 21, 140–150.

Interior point techniques are applied in the context of parallel computation. The resulting algorithm solves instances of AP with integer costs on a CRCW PRAM with m^3 processors in $O(\sqrt{m} \log^2 n \log(nC))$ time. This compares favourably with the algorithm in [Goldberg, Plotkin and Vaidya 1988] (see above) for sparse graphs.

J.B. Orlin, C. Stein (1993). Parallel algorithms for the assignment and minimum-cost flow problems. *Oper. Res. Lett.* 14, 181–186.

Two well-known randomized parallel algorithms for the minimum-cost matching problem on general graphs need a number of processors that is proportional to C. This paper presents a scaling method which, combined with such algorithms, produces algorithms for AP in which the number of processors needed is independent of C.

7.2 Implementations of the auction algorithm

D.P. Bertsekas (1988). The auction algorithm: a distributed relaxation method for the assignment problem. R.R. Meyer, S.A. Zenios (eds.). *Parallel Optimization on Novel Computer Architectures*, Ann. Oper. Res., 14, Baltzer, Basel, 105–123.

D.P. Bertsekas, D.A. Castañon (1991). Parallel synchronous and asynchronous implementations of the auction algorithm. *Parallel Comput.* 17, 707–732.

The bidding phase of the [Bertsekas 1981] auction algorithm (see §2) can be implemented in two ways. One possibility is to consider a single unassigned row at a time, resulting in the updating of a single dual variable. In the second approach all unassigned rows are considered at a time, so several dual variables are updated. (The two techniques are known as the *Gauss-Seidel* and the *Jacobi* version, respectively, because of their similarity with analogous methods for solving systems of nonlinear equations.) Parallel implementations of such approaches are discussed in these two works. Computational experiments on a MIMD computer (the Encore Multimax) are presented in the second paper. Further details on parallel implementations can be found in

D.P. Bertsekas, J.N. Tsitsiklis (1989). *Parallel and Distributed Computation: Numerical Methods*, Prentice-Hall, Englewood Cliffs, NJ.

C. Philips, S.A. Zenios (1989). Experiences with large scale network optimization on the connection machine. R. Sharada, B.L. Golden, E. Wasil, W. Stuart, O. Balci (eds.). *Impact of Recent Computer Advances on Operations Research*, North-Holland, Amsterdam, 169–180.
J.M. Wein, S.A. Zenios (1991). On the massively parallel solution of the assignment problem. *J. Parallel Distrib. Comput.* 13, 228–236.

Different implementations of the auction algorithm (Gauss-Seidel, Jacobi and hybrid versions, see above) are computationally experimented on an SIMD computer (the Connection Machine CM-2) with up to 32K processors. A table in the second paper gives comparisons with the codes of [Kempka, Kennington and Zaki 1991], [Kennington and Wang 1991], [Balas, Miller, Pekny and Toth 1991] and [Zaki 1995] (see below).

D. Kempka, J.L. Kennington, H. Zaki (1991). Performance characteristics of the Jacobi and Gauss-Seidel versions of the auction algorithm on the Alliant FX/8. *ORSA J. Comput.* 3, 92–106.

The two versions of the auction algorithm are computationally tested on an Alliant FX/8 computer with eight processors and vector-concurrent capabilities, both with and without ε-scaling (see [Bertsekas and Eckstein 1988], §5).

7.3 Implementations of shortest path algorithms

J.L. Kennington, Z. Wang (1991). An empirical analysis of the dense assignment problem: sequential and parallel implementations. *ORSA J. Comput.* 3, 299–306.

A simple parallel version of the code presented in [Jonker and Volgenant 1987] (see §3) is obtained by executing the vectorial operations in parallel. The code is tested on a Symmetry S81 with twenty processors.

E. Balas, D. Miller, J. Pekny, P. Toth (1991). A parallel augmenting shortest path algorithm for the assignment problem. *J. ACM* 38, 985–1004.

A nontrivial parallelization of the shortest path approach is presented, in which all processors construct simultaneously trees rooted at different nodes. Computational experiments are performed on a Butterfly GP1000 computer with fourteen processors.

D.P. Bertsekas, D.A. Castañon (1993). Parallel asynchronous Hungarian methods for the assignment problem. *ORSA J. Comput.* 3, 261–274.

A parallelization of the shortest augmenting path approach is presented, in which each processor finds a path rooted at a different node but the processors execute the iterations asynchronously. It is shown that the synchronous version of the algorithm is equivalent to the algorithm by [Balas, Miller, Pekny and Toth 1991] (see above). Computational results, obtained on an Encore Multimax computer with eight processors, for three versions of the algorithm (two synchronous) are given.

H. Zaki (1995). A comparison of two algorithms for the assignment problem. *Computational Opt. Appl.* 41, 23–45.

The Gauss-Seidel implementation of the auction algorithm without ε-scaling (presented in [Kempka, Kennington and Zaki 1991], see §7.2) is computationally compared on the Alliant FX/8 with a parallel implementation of the [Jonker and

Volgenant 1987] algorithm (see §3) in which: (a) the vector capabilities of the machine are used to speed-up the initialization phase; (b) in the augmentation phase a single path is determined using all processors jointly.

The following result was obtained outside the combinatorial optimization "community":

G.M. Megson, D.J. Evans (1990). A systolic array solution for the assignment problem. *The Computer J.* 33, 562–569.

The design of a systolic architecture for solving AP through the Hungarian algorithm is presented. The resulting time complexity is $O(n^2)$. It is shown that the design is suitable for Wafer Scale integration.

8 Software

Listings of FORTRAN programs for AP can be found in [Burkard and Derigs 1980] and [Bertsekas 1991] (see §1); a Pascal code is given in [Jonker and Volgenant 1987], [Volgenant 1996] (see §3). Other codes are included in the following references.

G. Carpaneto, S. Martello, P. Toth (1988). Algorithms and codes for the assignment problem. B. Simeone, P. Toth, G. Gallo, F. Maffioli, S. Pallottino (eds.). *FORTRAN Codes for Network Optimization*, Ann. Oper. Res. 13, Baltzer, Basel, 193–223.

The FORTRAN source code is also provided in the diskette included in the book.

V. Lotfi (1989). A labeling algorithm to solve the assignment problem. *Computers Oper. Res.* 16, 397–408.

A QuickBasic listing of a simple implementation of the Hungarian algorithm is given.

AP instances are available through anonymous ftp at the following addresses:

`graph.ms.ic.ac.uk`, directory `pub`, file `assigninfo.txt`
(this is the OR-LIBRARY, also available at `http://mscmga.ms.ic.ac.uk`);

`ftp.zib-berlin.de`, directory `pub/mp-testdata/assign`
(this is the ELIB library, also available at `ftp://ftp.zib-berlin.de`).

The source codes of the algorithms by [Goldberg and Kennedy 1995] (see §5) can be obtained as a `tar uuencoded` file by sending an empty E-mail message, with subject `send csas.tar`, to `ftp-request@theory.stanford.edu`

9 Appendix: Bottleneck Assignment (and Other Variations)

In this section we discuss relevant results on assignment problems with different objective functions, still solvable in polynomial time.

The most famous variation of AP is its *bottleneck* version (BAP), in which the objective function is

$$z = \min \max_{i,j=1,\ldots,n} \{c_{ij}x_{ij}\} \tag{9}$$

The first algorithm for the problem is the primal approach in
O. Gross (1959). *The Bottleneck Assignment Problem*, Tech. Rep. P-1630, The Rand Corporation, Santa Monica, CA.

R. Garfinkel (1971). An improved algorithm for the bottleneck assignment problem. *Oper. Res.* 19, 1747–1751.
G. Carpaneto, P. Toth (1981). Algorithm for the solution of the bottleneck assignment problem. *Computing* 27, 179–187.
Threshold algorithms are presented, in which the Hungarian approach is applied to the cost matrix entries not exceeding a given value, which is iteratively increased until a feasible solution is found.

U. Derigs, U. Zimmermann (1978). An augmenting path method for solving linear bottleneck assignment problems. *Computing* 19, 285–295.
U. Derigs (1984). Alternate strategies for solving bottleneck assignment problems – analysis and computational results. *Computing* 33, 95–106.
These papers give algorithms based on the determination of bottleneck augmenting paths. A FORTRAN code is provided in [Burkard and Derigs 1980] (see §1).

R.D. Armstrong, Z. Jin (1992). Solving linear bottleneck assignment problems via strong spanning trees. *Oper. Res. Lett.* 12, 179–180.
A dual simplex algorithm is introduced.

H.N. Gabow, R.E. Tarjan (1988). Algorithms for two bottleneck optimization problems. *J. Algorithms* 9, 411–417.
A.P. Punnen, K.P.K. Nair (1994). Improved complexity bound for the maximum cardinality bottleneck bipartite matching problem. *Discr. Appl. Math.* 55, 91–93.
Efficient algorithms are obtained by combining the threshold and bottleneck augmenting paths approaches. The algorithm in the second paper has time complexity $O(n\sqrt{nm})$, the best time bound known for BAP.

U. Pferschy (1995). The random linear bottleneck assignment problem. E. Balas, J. Clausen (eds.). *Integer Programming and Combinatorial Optimization*, Lecture Notes in Computer Science, 920, Springer-Verlag, Berlin, 145–156.
The asymptotic behaviour of BAP is studied and upper and lower bounds on the expected optimal solution value are given.

R.E. Burkard, F. Rendl (1991). Lexicographic bottleneck problems. *Oper. Res. Lett.* 10, 303–307.
A generalized version of BAP is introduced, in which the goal is the *lexicographic* minimization of the objective function. Two general approaches to the solution of the problem are presented.

R.E. Burkard, W. Hahn, U. Zimmermann (1977). An algebraic approach to assignment problems. *Math. Program.* 12, 318–327.
R.E. Burkard, U. Zimmermann (1980). Weakly admissible transformations for solving algebraic assignment and transportation problems. *Math. Program. Study* 12, 1–18.

The first paper introduces the *algebraic assignment problem* (AAP), arising when the entries of the cost matrix are replaced by elements from some totally ordered commutative semigroup. AAP includes as special cases AP, BAP and the lexicographic bottleneck assignment problem. The second paper describes an $O(n^3)$ shorthest path algorithm for AAP.

S. Martello, W.R. Pulleyblank, P. Toth, D. de Werra (1984). Balanced optimization problems. *Oper. Res. Lett.* 3, 275–278.

The *balanced assignment problem* is introduced. It calls for an assignment minimizing the objective function $z = \max_{i,j=1,\ldots,n} \{c_{ij}x_{ij}\} - \min_{i,j=1,\ldots,n} \{c_{ij}x_{ij}\}$.

Acknowledgements

This work was supported by Ministero dell'Università e della Ricerca Scientifica e Tecnologica, and by Consiglio Nazionale delle Ricerche, Italy. We would like to thank an anonymous referee for his insightful comments.

21 Quadratic and Three-Dimensional Assignments

Rainer E. Burkard
Institute of Mathematics, University of Technology, Graz

Eranda Çela
Institute of Mathematics, University of Technology, Graz

CONTENTS

Given two $n \times n$ matrices A and B the quadratic assignment problem (QAP) of size n can be stated as follows

$$\min_{\phi \in \mathcal{S}_n} \sum_{i=1}^{n} \sum_{j=1}^{n} a_{\phi(i)\phi(j)} b_{ij} ,$$

where $\mathcal{S}_n$ is the set of permutations ϕ of $\{1, 2, \dots, n\}$. Initially the QAP arose as a mathematical model of a location problem concerning economic activities. In the context of location problems which still remain a major application of the QAP, n facilities and n locations are given. Matrix A is the flow matrix, i.e. a_{ij} is the flow of materials moving from facility i to facility j, and matrix B is the distance matrix, i.e. b_{kl} is the distance between facilities k and l. The cost of simultaneously assigning facility i to location k and facility j to location l is $a_{ij}b_{kl}$. The objective of the QAP consists of finding an assignment of the facilities to the locations with

Annotated Bibliographies in Combinatorial Optimization, edited by M. Dell'Amico, F. Maffioli and S. Martello

the minimum overall cost. Nowadays, a large variety of other *practical applications* of the QAP are known, including such areas as scheduling, manufacturing, parallel and distributed computing, statistical data analysis and chemistry. From the *theoretical point of view*, other combinatorial optimization and graph theoretical problems can be formulated as QAPs. Just to mention some well known examples consider the traveling salesman problem, the turbine problem, the linear ordering problem, graph partitioning problems, subgraph isomorphism and maximum clique problems. Due to its theoretical and practical relevance, but also due to its *complexity*, the QAP has been subject of extensive research since its first occurrence in 1957. In the last decade we have seen a dramatic increase of the size of NP-hard combinatorial optimization problems which can be efficiently solved in practice. Unfortunately, the QAP *is not* one of them; QAP instances of size larger than 20 are still considered intractable. Thus, the QAP still remains a challenging problem from both theoretical and practical point of view.

The research done on the QAP covers more or less all of its aspects. With the intention to identify new structural combinatorial properties a number of alternative formulations for the QAP have been given. Ranging from equivalent Boolean linear and mixed integer linear programming (MILP) formulations to the trace formulation, they have led to diverse lower bounding procedures and exact solution methods for this problem. It is probably remarkable that quite different approaches have been applied to this end: combinatorial methods, eigenvalue computation and subgradient and nonsmooth optimization techniques. The resulting lower bounds have been incorporated in cutting planes and branch and bound algorithms for the QAP, the latter being considered the most efficient. Recently, parallel implementation of branch and bound methods have enabled the solution of test instances of size 22. However, even the most sophisticated implementations of exact algorithms fail in solving real size QAPs and heuristics still remain the unique mean to solve medium to large size instances of the problem. Among the large variety of heuristics proposed for the QAP the so called metaheuristics, tabu search, simulated annealing and genetic algorithms, seem to be the most efficient. As these methods are based on neighborhood search, they are also appropriate for parallel implementations. This enables in turn the heuristic solution of real life problems. Unfortunately, there is no guarantee on the quality of the solutions produced by these methods and no tight bounds for large sized QAPs are known. This is not surprising when considering that even the approximation problem for QAPs is in general NP-hard. On the other side, under certain probabilistic conditions the random QAP becomes in some sense trivial as the size of the problem increases.

Another research direction on QAPs concerns restricted versions of the problem. Clearly, most of the efforts focus on identifying polynomially solvable cases of the QAP. However, the identification of provably NP-hard cases helps on understanding structural properties of the problem. Recently, QAPs whose coefficient matrices have a special combinatorial structure have been investigated leading to some new polynomially solvable cases. However, only a few results of this type are know and a lot remains to do in this direction.

Another object of research work related to QAPs concerns its generalizations. Given the large area of applications of this problem, its numerous generalizations and related problems should not be surprising. The generalizations may be related to the structure of the problem coefficients or to the set of feasible solutions. Two well known examples are probably the *biquadratic assignment problem* (BiQAP) and the *semi-quadratic assignment problem* (SQAP).

Another widely known and well studied assignment problem is the *multidimensional assignment problem* (MAP), in particular the *three-dimensional assignment problem* (3-DAP). There are two well distinguished versions of the 3-DAP: the *axial* 3-DAP and the *planar* 3-DAP. In the 3-DAP of size n we are given three disjoint sets I, J, K of cardinality n each and a weight c_{ijk} associated with each ordered triplet $(i, j, k) \in I \times J \times K$. In the axial 3-DAP we want to find a minimum (maximum) weight collection of n pairwise disjoint triplets as above, whereas in the planar 3-DAP the goal is to find n^2 triplets forming n disjoint sets of n disjoint triplets each. The multidimensional assignment problem arises as a generalization of the axial 3-DAP when n-tuples are considered instead of triplets. Both the axial and the planar 3-DAP are known to be NP-hard and have several applications in scheduling and time-tabling problems. A recent application of the MAP concerns data association problems in multitarget tracking and multisensor data fusion. The axial (planar) 3-DAP is a close relative to the (solid) transportation problem. This relationship has been helpful on studying the facial structure of these problems. Some classes of facet defining inequalities and corresponding separation algorithms have been derived.

Among algorithms known for 3-DAPs some branch and bound methods involving Lagrangean relaxation and subgradient optimization can be mentioned, the axial problem being the mostly studied. Recently, a tabu search algorithm for the planar 3-DAP has been proposed.

Finally, some investigations have been done on special cases of the axial 3-DAP. These investigations concern problems whose coefficients have a special structure. e.g. are decomposable or fulfill the triangle inequality, or possess Monge-like properties. It turns out that in most of the cases the problems remain NP-hard, unless their coefficients fulfill additional, more restrictive conditions. Such additional conditions lead then to polynomially solvable and polynomially approximable cases, respectively.

There exists an abundant literature on the QAP and its generalizations. In drawing up this bibliography, we have concentrated on publications that appeared in 1985 or later, focusing on the most recent contributions. However, seminal work related to the roots of the QAP or review articles which contain a large number of pointers to relevant previous work have also been mentioned. When reviewing papers related to algorithmic aspects of the problem, we have only reported on those which present the best computational results, unless relevant theoretical contribution is provided. We hope to have not overlooked any important contribution on the considered problems. However, we would be pleased to hear about additional relevant work in this area and we would highly appreciate any related pointers.

1 Books and Surveys

The following survey papers can serve as a general introduction to quadratic assignment problems. Covering all aspects of research on QAPs, these papers provide also a large number of pointers to the roots of the QAP and to earlier surveys which are not listed in this section.

G. Finke, R.E. Burkard, F. Rendl (1987). Quadratic assignment problems. S. Martello, G. Laporte, M. Minoux, C. Ribeiro (eds.). *Surveys in Combinatorial Optimization*, Ann. Discr. Math. 31, North-Holland, Amsterdam, 61–82.

This survey focuses on the "trace formulation" of the QAP. The eigenvalue approach for the lower bound computation in the case of symmetric QAPs is introduced together with a reduction scheme for improving the resulting bounds.

S.W. Hadley, F. Rendl, H. Wolkowicz (1990). Bounds for the quadratic assignment problem using continuous optimization techniques. *Integer Programming and Combinatorial Optimization*, University of Waterloo Press, 237–248.

This article reviews lower bounding procedures for the QAP based on continuous optimization techniques. Eigenvalue techniques, reduced gradient methods, trust region methods, sequential quadratic programming and subdifferential calculus are applied to approximations and relaxations of the QAP.

R.E. Burkard (1991). Locations with spatial interactions: the quadratic assignment problem. P.B. Mirchandani, R.L. Francis (eds.). *Discrete Location Theory*, John Wiley & Sons, New York, 387–437.

This survey resumes known results related to (mixed) integer programming formulations, bounding procedures, exact algorithms and heuristics for QAPs. Some typical applications of the QAP are described and a number of papers describing less typical applications are referenced. Moreover, the asymptotic behavior of the QAP is described.

P.M. Pardalos, F. Rendl, H. Wolkowicz (1994). The quadratic assignment problem: a survey and recent developments. *DIMACS Series Discr. Math. Theor. Comp. Sci.* 16, 1–42.

This is the most recent survey on the QAP. It appeared as an introductory article in the proceedings book of the DIMACS workshop on quadratic assignment and related problems. It focuses on recent results concerning computation of lower bounds, computational complexity, heuristic approaches for the QAP and its generalizations.

Quadratic Assignment and Related Problems, Proc. DIMACS Workshop on Quadratic Assignment Problems, P.M. Pardalos, H. Wolkowicz (eds.). DIMACS Series Discr. Math. Theor. Comp. Sci. 16.

This book offers a collection of up-to-date contributions on computational approaches to the QAP and its applications.

M.W. Padberg, M.P. Rijal (1996). *Location, Scheduling, Design and Integer Programming*, Kluwer Academic Publishers, Boston.

This book analyzes various classes of $0-1$ quadratic problems with special ordered set constraints, in general, and the QAP, in particular. The goal is to investigate the polyhedra related to these problems in order to develop practical solution methods, in the vein of branch and cut approaches. The authors propose new MILP formulations for the QAP and for the symmetric QAP. Further, the dimension of the QAP polytope is computed, simple facet defining inequalities and other valid inequalities, the so-called *clique* and *cut* inequalities, are identified.

2 Roots of the Quadratic Assignment Problem (QAP): Basic Facts and Complexity

T.C. Koopmans, M.J. Beckmann (1957). Assignment problems and the location of economic activities. *Econometrica* 25, 53–76.

This is the first occurrence of the standard QAP formulation

$$\min_{\phi} \sum_{i,j=1}^{n} a_{\phi(i)\phi(j)} b_{ij} ,$$

where ϕ ranges over all permutations of $\{1, 2, \ldots, n\}$. The QAP is derived as a mathematical formulation of a problem arising along with the location of economic activities.

P.C. Gilmore (1962). Optimal and suboptimal algorithms for the quadratic assignment problem. *SIAM J. Appl. Math.* 10, 305–313.

This paper introduces the so called *Gilmore-Lawler bounds* which still remain one of the most important and frequently used bounds for the QAP. Based on these bounds, two heuristic approaches are proposed.

E.L. Lawler (1963). The quadratic assignment problem. *Management Sci.* 9, 586–599.

A more general QAP is introduced, where the objective is the minimization of a double sum of the form $\sum_{i,j=1}^{n} d_{\pi(i)\pi(j)ij}$ over all permutations ϕ of $\{1, 2, \ldots, n\}$. The problem coefficients d_{ijkl} form an array with n^4 elements. Moreover, the author derives an equivalent integer programming formulation for this problem and describes the computation of lower bounds.

C.E. Nugent, T.E. Vollmann, J. Ruml (1968). An experimental comparison of techniques for the assignment of facilities to locations, *Oper. Res.* 16, 150–173.

An improvement method combined with random elements is proposed. This method is compared with deterministic improvement algorithms on a set of QAP test instances. Nowadays these instances are known as *Nugent's problems* and are frequently used for experimental purposes.

G.W. Graves, A.B. Whinston (1970). An algorithm for the quadratic assignment problem. *Management Sci.* 17, 453–471.

The authors derive formulas for the mean and the variance of the objective function value of the QAP. Moreover, enumerative algorithms are proposed which exploit this statistical information.

R.E. Burkard (1974). Quadratische Bottleneckprobleme. *Oper. Res. Verfahren* 18, 26–41.

The QAP with bottleneck objective function is introduced. The goal is to minimize $\max_{1\leq i,j\leq n} a_{\phi(i)\phi(j)}b_{ij}$ over all permutations ϕ of $\{1,2,\ldots,n\}$. Moreover, the author proposes lower bounds for the *bottleneck QAP* to be incorporated in branch and bound algorithms.

S. Sahni, T. Gonzalez (1976). P-complete approximation problems. *J. ACM* 23, 555–565.

The computational complexity of the QAP is investigated showing that this problem is strongly NP-hard. Moreover, it is shown that the existence of a polynomial ϵ-approximate algorithm for QAPs implies $\mathcal{P} = \mathcal{NP}$.

M. Queyranne (1986). Performance ratio of polynomial heuristics for triangle inequality quadratic assignment problems. *Oper. Res. Lett.* 4, 231–234.

The author considers QAPs with coefficient matrices fulfilling the triangle inequality. It is shown that for such QAPs no polynomial heuristic algorithm with bounded asymptotic performance ratio exists unless $\mathcal{P} = \mathcal{NP}$.

K.A. Murthy, P. Pardalos and Y. Li (1992). A local search algorithm for the quadratic assignment problem. *Informatica* 3, 594-538.

A new neighborhood for QAPs is proposed which is similar to the Kernighan-Lin neighborhood for the graph partitioning problem. It is shown that the corresponding local search problem is PLS-complete.

3 Linearizations of the QAP

The QAP can be equivalently formulated as a Boolean, an integer or a mixed integer linear program. There exists a large number of such equivalent formulations for the QAP. This approach is particularly fruitful concerning the computation of lower bounds.

L. Kaufman, F. Broeckx (1978). An algorithm for the quadratic assignment problem using Benders' decomposition. *European J. Oper. Res.* 2, 204–211.

The authors propose an equivalent formulation for the QAP as a mixed integer linear program with n^2 real variables, n^2 integer variables and $O(n^2)$ constraints. This is one of the "smallest" linearizations of the QAP with respect to the number of variables and constraints.

M.S. Bazaraa, H.D. Sherali (1980). Benders' partitioning scheme applied to a new formulation of the quadratic assignment problem. *Naval Res. Log. Quart.* 27, 29–41.

An equivalent formulation of the QAP as a mixed integer linear program with a highly specialized structure is proposed. This formulation which involves n^2 Boolean variables, $n^2(n-1)/2$ real variables and $2n^2$ linear constraints, permits the effective use of the partitioning scheme of Benders.

W.P. Adams, H.D.Sherali (1986). A tight linearization and an algorithm for zero-one quadratic programming problems. *Management Sci.* 32, 1274–1290.

A linearization for a class of linearly constrained 0–1 quadratic programming problems containing the QAP is proposed. It is shown that this linearization is tighter than other ones existing in the literature. Moreover, an implicit enumeration algorithm which makes use of the strength of this linearization is derived.

W.P. Adams, T.A. Johnson (1994). Improved linear programming-based lower bounds for the quadratic assignment problem. *DIMACS Series Discr. Math. Theor. Comp. Sci.* 16, 43–76.

A new mixed 0-1 linear formulation for the QAP is proposed. By appropriately surrogating selected constraints and combining variables, most of the known linear formulations for the QAP can be obtained. Moreover, most of the resulting bounding techniques can be described in terms of the Lagrangean dual of this new formulation of the QAP. A dual-ascent procedure is proposed for suboptimally solving a relaxation of the dual of the new QAP formulation deriving also new lower bounds.

M. Jünger, V. Kaibel (1996). *A basic study of the QAP polytope*, Technical Report 96.215, Angewandte Mathematik und Informatik, Universität zu Köln.

Basic structural properties of the QAP polytope, partially and independently considered also by Padberg and Rijal (1996), are investigated. The QAP polytope is transformed isomorphically into a space which is different from the one where this polytope is naturally defined. This transformation enables a simpler computation of the dimension of the QAP polytope and leads to interesting investigations concerning its combinatorial and facial structure.

M. Jünger, V. Kaibel (1996). *On the SQAP polytope*, Technical Report 96.241, Angewandte Mathematik und Informatik, Universität zu Köln.

The authors investigate the symmetric QAP (SQAP) polytope and apply a transformation similar to the one applied for the general QAP polytope. This transformation enables the computation of the dimension of the SQAP polytope, and thereby, the proof of a recent conjecture of Padberg and Rijal (1996). Moreover, the trivial faces of the SQAP polytope are investigated, and the first class of non-trivial facet defining inequalities, the so called *curtain inequalities*, is identified.

4 Lower Bounds for the QAP

N. Christofides, M. Gerrard (1981). A graph theoretic analysis of bounds for the quadratic assignment problem. P. Hansen (ed.). *Studies on Graphs and Discrete Programming*, North-Holland, 61–68.

The authors consider a special version of the QAP, where the feasible solutions correspond to isomorphisms of graphs. This version of the QAP is polynomially solvable in the case that the coefficient matrices are weighted adjacency matrices of isomorphic trees or other simple graphs, eg. wheels or cycles. The latter solvable cases can be used to generate lower bounds for the general QAP.

The following three papers deal with Lagrangean techniques for computing lower bounds.

A.M. Frieze, J. Yadegar (1983). On the quadratic assignment problem. *Discr. Appl. Math.* 5, 89–98.

The relationship between the Gilmore-Lawler bounds for the QAP on reduced matrices and a Lagrangean relaxation of a particular mixed $0-1$ linear formulation for the QAP is investigated. The Gilmore-Lawler bounds obtained by involving an "optimal" reduction are dominated by the continuous relaxation of the proposed linear formulation for the QAP.

A.A. Assad, W. Xu (1985). On lower bounds for a class of quadratic $0,1$ programs. *Oper. Res. Lett.* 4, 175–180.

P. Carraresi, F. Malucelli (1992). A new lower bound for the quadratic assignment problem. *Oper. Res.* 40, Suppl. No. **1**, S22–S27.

In these two papers iterative methods are used for generating non-decreasing sequences of lower bounds for the QAP. In each iteration the problem is reformulated and a lower bound for the new formulation is computed. The reformulation is based on a Lagrangean dual-ascent procedure and on the information given by the dual variables arising along with the lower bound computation, respectively.

The following four papers use the trace formulation of the QAP for generating lower bounds based on eigenvalue computations.

F. Rendl (1985). Ranking scalar products to improve bounds for the quadratic assignment problem. *European J. Oper. Res.* 20, 363–372.

The author reconsiders the eigenvalue bounds for QAPs proposed by Finke, Burkard and Rendl (1987) as described in §1. In the case when the linear term resulting after the reduction mostly influences the objective function, the eigenvalue bound can be improved by ranking the k-best solutions of the linear term.

S.W. Hadley, F. Rendl, H. Wolkowicz (1992). Symmetrization of nonsymmetric quadratic assignment problems and the Hoffman-Wielandt inequality, *Linear Alg. Appl.* 167, 53–64.

A technique is proposed to transform a nonsymmetric QAP into an equivalent QAP on Hermitian matrices. The eigenvalue bound for symmetric QAPs is extended to the general problem and Hoffman-Wielandt-type eigenvalue inequalities for general matrices are derived.

F. Rendl, H. Wolkowicz (1992). Applications of parametric programming and eigenvalue maximization to the quadratic assignment problem. *Math. Program.* 53, 63-78.

The classical eigenvalue bounds for QAPs on symmetric matrices can be improved by applying special reduction schemes. The authors derive an "optimal" reduction taking simultaneously into account the quadratic term and the linear term of the objective function. This involves a steepest-ascent algorithm based on subdifferential calculus.

S.W. Hadley, F. Rendl, H. Wolkowicz (1993). A new lower bound via projection for the quadratic assignment problem. *Math. Oper. Res.* 17, 727–739.

The standard eigenvalue bounds for the QAP are improved. The new bounds make use of a tighter relaxation on orthogonal matrices with constant row and column sums. The additional constraints of the new relaxation are projected into the space of orthogonal matrices of size $n-1$, where n is the size of the given QAP. For bounding the quadratic part of the projected program standard eigenvalue approaches are used.

S.E. Karisch, F. Rendl (1995). Lower bounds for the quadratic assignment problem via triangle decompositions. *Math. Program.* 71, 137-151.

QAPs where one of the coefficient matrices is the distance matrix of a grid graph are considered. The problem is decomposed into a trivially solvable QAP and a so called "residual QAP". A lower bound for the residual problem is computed via projection and nonsmooth optimization techniques are used to derive an appropriate decomposition.

M.G.C. Resende, K.G. Ramakrishnan, Z. Drezner (1995). Computing lower bounds for the quadratic assignment problem with an interior point algorithm for linear programming. *Oper. Res.* 43, 781–791.

The linear relaxation of a common MILP formulation for the QAP is solved by an interior point method. This approach produces the best existing bounds for most of the test instances from QAPLIB (see §9). However, these bounds are extremely expensive in terms of computation time.

Q. Zhao, S.E. Karisch, F. Rendl, H. Wolkowicz (1996). *Semidefinite relaxations for the quadratic assignment problem*, Technical Report DIKU-TR-96/32, Department of Computer Science, University of Copenhagen.

The authors propose semidefinite programming relaxation for the QAP and solve them by using interior point methods and cutting plane approaches. The resulting bounds are competitive with the best existing bounds for almost all test instances from QAPLIB (see §9 below), in terms of quality. However, these bounds are very expensive in terms of computation time, which makes them inappropriate for use within branch and bound algorithms.

P. Hahn, T. Grant (1996). Lower bounds for the quadratic assignment problem based upon a dual formulation, to appear in *Oper. Res.*

The lower bounding technique proposed by the authors combines Gilmore-Lawler bound computations and reduction ideas in an iterative approach. The novelty consists of a clever redistribution of the cost coefficients after each iteration. Basically, the algorithm solves approximately the dual of a linear relaxation of a MILP formulation for the QAP in an iterative, steepest-ascent-like, approach. The produced bounds are competitive with those of Resende et al. (1995) in terms of quality, and clearly beat the latter in terms of computation time.

5 Exact Algorithms for the QAP

A large variety of exact algorithms has been proposed for the QAP among which the branch and bound methods generally yield the better results. Many of these algorithms are mentioned and/or described in the surveys cited in §1. A newer development is derived by Edwards in the following paper.

C.S. Edwards (1980). A branch and bound algorithm for the Koopmans-Beckmann quadratic assignment problem. *Math. Program. Study* 13, 35–52.

The branch and bound approach is based on the trace formulation of the QAP which allows to effectively use a binary branching rule.

The performance of branch and bound algorithms depends significantly on the efficience and on the quality of the involved lower bounds. The performance of such algorithms can be also improved by a smart use of the available hardware in parallel implementations. The following three papers describe some parallel branch and bound algorithms for the QAP.

C. Roucairol (1987). A parallel branch and bound algorithm for the quadratic assignment problem. *Discr. Appl. Math.* 18, 221–225.
P. Pardalos, J. Crouse (1989). A parallel algorithm for the quadratic assignment problem. *Proc. Supercomputing Conf.*, 351–360.
J. Clausen, M. Perregaard (1994). Solving large quadratic assignment problems in parallel. Technical Report DIKU TR-94-22, Department of Computer Science, University of Copenhagen.

T. Mautor, C. Roucairol (1994). A new exact algorithm for the solution of quadratic assignment problems, *Discr. Appl. Math.* 55, 281–293.
It is shown how to exploit the symmetries on the cost matrix in order to reduce the branch and bound tree. The proposed branch and bound algorithm uses polytomic branching rule and outperforms most of the other branch and bound schemes existing at that time.

N. Christofides, E. Benavent (1989). An exact algorithm for the quadratic assignment problem on a tree. *Oper. Res.* 37, 760–768.
A special case of the QAP is considered, where the flow matrix is the weighted adjacency matrix of a tree. A branch and bound method for this NP-hard special case is derived. An integer programming formulation for this problem is given and its Lagrangean relaxation is solved by using a dynamic programming scheme. This approach produces tight lower bounds.

M.E. Dyer, A.M. Frieze, C.J.H. McDiarmid (1986). On linear programs with random costs. *Math. Program.* 35, 3–16.
The authors derive an interesting result on the size of branch and bound trees for random QAPs. The considered lower bounds arise as solutions of an LP relaxation of the Boolean linear programming formulation of the QAP given by Frieze and Yadegar (1983) and described in §4. It is shown that in case of binary branching the number of the explored branching nodes grows super-exponentially with probability tending to 1 as the site of the problem approaches infinity.

6 Heuristics for the QAP

There is a large variety of heuristics for the QAP ranging from construction and deterministic improvement methods to tabu search and simulation based algorithms. In the following we will only mention some of the most recent approaches.

Y. Li, P.M. Pardalos, M. Resende (1994). A greedy randomized adaptive search procedure for the quadratic assignment problem. *DIMACS Series Discr. Math. Theor. Comp. Sci.* 16, 237–261.
P.M. Pardalos, L.S. Pitsoulis, M.G.C. Resende (1995). A parallel GRASP implementation for the quadratic assignment problem. A. Ferreira, J.D.P. Rolim (eds.). *Parallel Algorithms for Irregular Problems: State of the Art*, Kluwer Academic Publishers, Dordrecht.

In the first paper the authors propose an improvement method which combines greedy elements with probabilistic aspects. The so called GRASP shows a good computational behavior in many QAP instances from QAPLIB (see §9 below). In the second paper a parallel implementation of GRASP is presented. In particular, efficient techniques for solving large-scale sparse quadratic assignment problems on a parallel computer are discussed. This parallel implementation of GRASP shows a good behavior on test instances from QAPLIB and achieves an average speedup that is almost linear in the number of processors.

6.1 Simulated annealing approaches

R.E. Burkard, F. Rendl (1983). A thermodynamically motivated simulation procedure for combinatorial optimization problems. *European J. Oper. Res.* 17, 169–174.

This is one of the first applications of *simulated annealing* (SA) to the QAP. It is shown that SA outperforms most of the existing heuristics for the QAP at that time.

M.R. Wilhelm, T.L. Ward (1987). Solving quadratic assignment problems by "simulated annealing". *IIE Trans.* 19, 107-119.

SA is improved by introducing "equilibria" components which comply with the statistic mechanics background of the underlying Metropolis algorithm.

D.T. Connolly (1990). An improved annealing scheme for the QAP. *European J. Oper. Res.* 46, 93–100.

A new element of the annealing scheme, the so called *optimal temperature*, is introduced. The corresponding algorithm yields a promising improvement of the trade-off between computation time and solution quality.

6.2 Tabu search approaches

One of the first applications of tabu search to the QAP and a parallel implementation of tabu search for QAPs can be found in the following two papers, respectively.

J. Skorin-Kapov (1990). Tabu search applied to the quadratic assignment problem. *ORSA J. Comput.* 2, 33–45.
J. Chakrapani, J. Skorin-Kapov (1993). Massively parallel tabu search for the quadratic assignment problem. *Ann. Oper. Res.* 41, 327–342.

The performance of tabu search algorithms depends very much on the size of the tabu list and on the way this list is handled. Two of the most effective strategies leading to a good trade-off between the diversification and the intensification of the search are presented in the following two papers.

E. Taillard (1990). Robust taboo search for the quadratic assignment problem. *Parallel Comput.* 17, 443–455.
R. Battiti, G. Tecchiolli (1994). The reactive tabu search. *ORSA J. Comput.* 6, 126–140.

6.3 Genetic algorithms

We believe that among genetic approaches for QAPs, the following is a remarkable contribution.

R.K. Ahuja, J.B. Orlin, A. Tivari (1995). A greedy genetic algorithm for the quadratic assignment problem. *Working paper*, Sloan School of Management, MIT, Cambridge, MA.
This genetic algorithm attempts to strike a balance between diversity and a bias towards fitter individuals. Appropriate greedy elements are combined to this end with genetic ingredients like new crossover schemes, tournamenting, periodic local optimization and an immigration rule that promotes diversity.

The following paper presents a comparison of the performance of 8 different heuristic approaches on QAP instances from QAPLIB (see §9 below). Among others, simulated annealing, tabu search, genetic algorithms, and greedy approaches are tested. In most of the cases, tabu search produces the best trade-off between solution quality and computation time.

V. Maniezzo, M. Dorigo, A. Colorni (1993). ALGODESK: an experimental comparison of eight evolutionary heuristics applied to the quadratic assignment problem. *European J. Oper. Res.* 81, 188–204.

7 Asymptotic Behavior of the QAP

Under natural probabilistic constraints on the input data, the QAP show an interesting asymptotic behavior. Namely, the ratio between the "best" and the "worst" value of the objective function approaches 1 with probability tending to 1 as the size of the problem approaches infinity. This behavior was first shown by Burkard and Fincke for sum and bottleneck objectives:

R.E. Burkard, U. Fincke (1983). The asymptotic probabilistic behavior of quadratic sum assignment problems. *Z. Oper. Res.* 27, 73–81.
R.E. Burkard, U. Fincke (1982). On random quadratic bottleneck assignment problems. *Math. Program.* 23, 227–232.

J.B.G. Frenk, M. van Houweninge, A.H.G. Rinnooy Kan (1985). Asymptotic properties of the quadratic assignment problem. *Math. Oper. Res.* 10, 100-116.
The range of the convergence in the above mentioned behavior is improved from "with probability" to "almost sure".

W.T. Rhee (1988). A note on asymptotic properties of the quadratic assignment problem. *Oper. Res. Lett.* 7, 197–200.

The results of Frenk, Houweninge and Rinnooy Kan (1988) are improved by providing a simpler proof and sharper estimations for the almost sure convergence in the asymptotic behavior of the QAP.

W.T. Rhee (1988). Stochastic analysis of the quadratic assignment problem. *Math. Oper. Res.* 16, 223–239.

The maximization version of the QAP is considered. A simple greedy approach is described which produces a good approximation of the optimal solution with overhelming probability. This complies with the previous results on the asymptotic behavior of the QAP.

E. Bonomi, J.-L. Lutton (1986). The asymptotic behavior of quadratic sum assignment problems: a statistical mechanics approach. *European J. Oper. Res.* 26, 295–300.

A statistical mechanics approach, based on the Boltzmann distribution and the Metropolis algorithm, is applied to study the asymptotic behavior of the QAP. The authors derive in this way the same result as Burkard and Fincke (1983) and perform numerical experiments which confirm this theoretical result.

In the following two papers a combinatorial condition which guarantees an analogous asymptotic behavior for general combinatorial optimization problems is singled out. The first paper shows convergence with probability whereas in the second one an almost sure convergence is proven.

R.E. Burkard, U. Fincke (1985). Probabilistic asymptotic properties of some combinatorial optimization problems. *Discr. Appl. Math.* 12, 21–29.

W. Szpankowski (1995). Combinatorial optimization problems for which almost every algorithm is asymptotically optimal! *Optimization* 33, 359–367.

8 Polynomially Solvable Cases of the QAP

The first polynomially solvable cases of the QAP were identified by Christofides and Gerrard (1976) as mentioned in §4. These results are extended and generalized to minimal vertex series-parallel (MVSP) digraphs as described in the following paper.

F. Rendl (1986). Quadratic assignment problems on series parallel digraphs. *Z. Oper. Res.* 30, 161–173.

It is shown that the general version of the QAP on isomorphic MVSPs is NP-hard. However, MVSP digraphs which do not contain the bipartite digraph $K_{2,2}$ as vertex induced subgraph lead to polynomially solvable cases. A polynomial time algorithm is proposed for solving the latter.

Recent investigation on polynomially solvable cases of the QAP rely on special combinatorial properties of the involved coefficient matrices, as shown in the following two papers.

R.E. Burkard, E. Çela, G. Rote, G.J. Woeginger (1995). The quadratic assignment problem with an anti-Monge and a Toeplitz matrix: easy and hard cases. SFB Report 341, Institute of Mathematics.
R.E. Burkard, E. Çela, V.M. Demidenko N.N. Metelski, G.J. Woeginger (1997). Easy and hard cases of the quadratic assignment problem: a survey. SFB Report 104, Institute of Mathematics, Graz University of Technology.

S. Arora, A. Frieze, H. Kaplan (1996). A new rounding procedure for the assignment problem with applications to dense graph arrangement problems. *Proc. 37-th IEEE Symp. Found. Comput Sci.*, 21–30.

A new rounding procedure is used to approximately optimize a polynomial over a set of variables fulfilling the assignment constraints. This yields, among others, an ϵ-approximation scheme for *dense* quadratic assignment problems, where "dense" means that the optimal objective function value of the QAP of size n is in $\Omega(n^2)$ and the entries of the coefficient matrices are in $O(1)$. For the general dense QAP this approximation scheme runs in $O(n^{O(\log n/\epsilon^2)})$ time. In the special case of a linear arrangement problem a polynomial time approximation scheme (PTAS) is derived.

9 Codes and Data for the QAP

R.E. Burkard, U. Derigs (1980). Assignment and Matching Problems: solution Methods with FORTRAN-Programs. *Lecture Notes Econ. Math. Sys.* 184, Springer-Verlag, Berlin.

This book contains FORTRAN codes for exact and heuristic algorithms for the QAP. A pointer to the source files can be found at
`http://fmatbhp1.tu-graz.ac.at/~karisch/qaplib/`

R.E. Burkard, S.E. Karisch, F. Rendl (1991). QAPLIB—a quadratic assignment problem library, *European J. Oper. Res.* 55, 115–119. Updated version March 1996.

This paper describes a library of test instances for the QAP. For each instance the best known solutions and the corresponding objective function values are given. The updated version provides also the best known lower bounds for each test instance. This library can be found at `http://fmatbhp1.tu-graz.ac.at/~karisch/qaplib/` and is also available per anonymous ftp at `ftp.tu-graz.ac.at/pub/papers/qaplib`.

M.G.C. Resende, P.M. Pardalos, Y. Li (1996). Algorithm 754: Fortran subroutines for approximate solution of dense quadratic assignment problems. *ACM Trans. Math. Software* 22, 104–118.

The paper describes a set of Fortran subroutines to be used within a GRASP algorithm for solving dense instances of the QAP, where at least one of the coefficient matrices is symmetric. The design and the implementation of the code are described in detail. Moreover, extensive computational experiments are reported, evaluating the quality of the solutions as a function of the running time.

The following two papers propose algorithms for generating QAP instances with known optimal solution.

G.S. Palubeckis (1988). Generation of quadratic assignment test problems with known optimal solutions. *U.S.S.R. Comput. Maths. Math. Phys.* 28, 97–98 (in Russian).
Y. Li, P.M. Pardalos (1992). Generating quadratic assignment test problems with known optimal permutations. *Computational Opt. Appl.* 1, 163–184.

The authors show that the test instances generated by the Palubeckis algorithm are "easy" in the sense that the corresponding optimal value of the objective function can be computed in polynomial time. The proof relies on the fact that the involved coefficient matrices are Euclidean. The Palubeckis' idea is then generalized to generate test instance with known optimal solution whose coefficient matrices are not Euclidean.

10 Generalizations

A natural generalization of the QAP, the so called *biquadratic assignment problem* (BiQAP), arises in the VLSI design. The BiQAP coefficients are organized in four dimensional arrays and an instance of size n looks as follows:

$$\min_{\pi \in \mathcal{S}_n} \sum_{i,j,k,l=1}^{n} a_{\pi(i)\pi(j)\pi(k)\pi(l)} b_{ijkl}$$

where $\mathcal{S}_n$ is the set of permutations of $\{1, 2, \ldots, n\}$. The following two papers generalize previous work on QAPs to derive linearizations, lower bounds and heuristic approaches for the BiQAP. Moreover, the asymptotic behavior of the BiQAP is investigated and it is shown that it is analogous to that of the QAP.

R.E. Burkard, E. Çela, B. Klinz (1994). On the biquadratic assignment problem. *DIMACS Series Discr. Math. Theor. Comp. Sci.* 16, 117–144.
R.E. Burkard, E. Çela (1995). Heuristics for biquadratic assignment problems and their computational comparison. *European J. Oper. Res.* 83, 283–300.

A number of applications for the so called *semi-quadratic assignment problem* (SQAP) are described in the following three papers. The SQAP has the same objective function as the QAP, whereas the feasible solutions do not need to be permutations but simply injective functions mapping $\{1, 2, \ldots, n\}$ into itself. The following references provide also pointers to bounding procedures, heuristics and polynomially special cases of the SQAP.

R.J. Freeman, D.C. Gogerty, G.W. Graves, R.B.S. Brooks (1966). A mathematical model of supply support for space operations. *Oper. Res.* 14, 1–15.
V.F. Magirou, J.Z. Milis (1989). An algorithm for the multiprocessor assignment problem. *Oper. Res. Lett.* 8, 351–356.
F. Malucelli, D. Pretolani (1993). *Lower bounds for the quadratic semi-assignment problem*, Technical Report 955, Centre des Recherches sur les transports, Université de Montréal.

11 The 3-Dimensional Assignment Problem (3-DAP): Basic Facts and Complexity

The multidimensional assignment problem (MAP) and some of its applications were introduced in:

W.P. Pierskalla (1968). The multidimensional assignment problem. *Oper. Res.* 16, 422–431.

The simplest MAP is the 3-DAP. Moreover, most of the results obtained for the 3-DAP can be naturally extended to the MAP, too.

O. Leue (1972). Methoden zur Lösung 3-dimensionaler Zuordnugsprobleme. *Angewandte Mathematik* 14, 154-162.

Two versions of the 3-DAP are stated: the axial 3-DAP and the planar 3-DAP. The feasible solutions of the axial 3-DAP are pair of permutations of $\{1, 2 \ldots, n\}$, whereas there is a one-to-one relation between latin squares and the feasible solutions of the planar 3-DAP.

A.M Frieze (1974). A bilinear programming formulation of the 3-dimensional assignment problems. *Math. Program.* 7, 376–379.

The author considers a slightly generalized form of the axial 3-DAP and gives an equivalent formulation of this problem as a bilinear program. This formulation is exploited then to derive a necessary optimality condition for the axial 3-DAP.

While the axial 3-DAP is NP-hard in the strong sense by a standard observation, the NP-hardness of the planar 3-DAP is shown in the following paper:

A.M. Frieze (1983). Complexity of a 3-dimensional assignment problem. *European J. Oper. Res.* 13, 161–164.

V.A. Jemelichev, M.M. Kovaliev, M.K. Kravtsov (1984). *Polytopes, Graphs and Optimization*, Cambridge University Press.

Multidimensional assignment problems are closely related to the multi-index transportation problems. This book provides a detailed analysis of the multi-index transportation problem concerning in particular its polyhedral structure.

12 The Facial Structure of the 3-DAP

The facial structure of the MAP plays an important role in deriving efficient algorithms of branch and cut type. The facial structure of the axial 3-DAP has been investigated in the following three papers. The authors identify facet defining equalities for the corresponding polytope and derive also separation algorithms for these facets.

E. Balas, M.J. Saltzman (1989). Facets of the three-index assignment polytope. *Discr. Appl. Math.* 23, 201–229.

E. Balas, L. Qi (1993). Linear-time separation algorithms for the three-index

assignment polytope. *Discr. Appl. Math.* 43 1–12.
L. Qi, E. Balas, G. Gwan (1994). A new facet class and a polyhedral method for the three-index assignment problem. D.-Z. Du, J. Sun (eds.). *Advances in Optimization and Approximation*, Kluwer Academic, 256–274.

The facial structure of planar 3-DAP has been investigated in the following two papers.

R.E. Burkard, R. Euler, R. Grommes (1986). On latin squares and the facial structure of related polytopes. *Discr. Math.* 62, 155-181.
R. Euler (1987). Odd cycles and a class of facets of the axial 3-index assignment polytope. *Applicationes Mathematicae (Zastosowania Matematyki)* XIX, 375–386.

13 Algorithms for the 3-DAP

P. Hansen, L. Kaufman (1973). A primal-dual algorithm for the three dimensional assignment problem. *Cahiers Centre Études Rech. Oper.* 15, 327–336.

The authors apply a modified version of the Hungarian method for the linear (2-dimensional) assignment problem to the axial 3-DAP.

R.E. Burkard, K. Fröhlich (1980). Some remarks on three dimensional assignment problems. *Methods Oper. Res.* 36, 31–36.

An exact solution method for the axial 3-DAP is derived. This method combines reduction steps with lower bounds computation by subgradient optimization within a branch and bound scheme.

A.M. Frieze, J. Yadegar (1981). An algorithm for solving 3-dimensional assignment problems with applications to scheduling a teaching practice. *Oper. Res.* 32, 989–995.

The authors propose a subgradient optimization method for solving a Lagrangean relaxation of a slightly generalized maximization version of the axial 3-DAP. This algorithm produces quite good solutions on test instances with real life and random input data.

E. Balas, M.J. Saltzman (1991). An algorithm for the three-index assignment problem. *Oper. Res.* 39, 150–161.

A branch and bound algorithm for the axial 3-DAP is derived. The computation of lower bound involves subgradient techniques for solving a Lagrangean relaxation of the problem which incorporates a class of facets defining inequalities. A novel branching strategy exploits the structure of the 3-DAP to reduce the size of the enumeration tree.

R.E. Burkard, R. Rudolf (1993). Computational investigations on 3-dimensional axial assignment problems. *Belgian J. Oper. Res. Stat. Comp. Sci.* 32, 85–98.

This computational study compares different branching rules and bounding procedures for the axial 3-DAP.

A. Poore (1994). Partitioning multiple data sets: multidimensional assignment and

Lagrangean relaxation. *DIMACS Series Discr. Math. Theor. Comp. Sci.* 16, 317–342.
A Lagrangean relaxation method for a class of MAPs is proposed. The relaxed problem is again a MAP and its maximization involves nonsmooth optimization techniques. The algorithm is illustrated and tested on instances arising as mathematical modes of real life data association problems.

A. Poore, A. Robertson (1995). A new Lagrangean relaxation based algorithm for a class of multidimensional assignment problems. *Computational Opt. Appl.* (to appear)
A new Lagrangean relaxation method for sparse MAPs is proposed. The relaxed problem is a linear (two-dimensional) assignment problem, whereas the computation of the Lagrangean multipliers involves non-smooth optimization methods.

An interesting application of the MAP arises along with data association in multi-target tracking as described in the following two papers.

A. Poore (1994). Multidimesional assignment formulation of data association problems arising from multitarget and multisensor tracking. *Computational Opt. Appl.* 3, 27–57.
A. Poore (1995). Multidimensional assignment and multitarget tracking. *DIMACS Series Discr. Math. Theor. Comp. Sci.* 19, 169–196.

Compared to the axial 3-DAP less work has been done on the planar 3-DAP. The three following papers describe branch and bound and heuristic methods for this problem.

M. Vlach (1967). A branch and bound method for the three index assignment problem. *Ekonomicko-Matematicky Obzor* 3, 181–191.
A straightforward branch and bound method for the planar 3-DAP is described.

D. Magos, P. Miliotis (1994). An algorithm for the planar three-index assignment problem. *European J. Oper. Res.* 77, 141–153.
A branch and bound algorithm for the planar 3-DAP is proposed and tested. It involves polytomic branching and an improvement method for the computation of the upper bounds. The computation of lower bound is based on Lagrangean relaxations solved by subgradient methods.

D. Magos (1996). Tabu search for the planar three-index assignment problem. *J. Global Opt.* 8, 35–48.
A tabu search algorithm for the planar 3-DAP is proposed and tested on problems of size 5 to 14. The algorithm combines standard tabu search elements such as fixed (variable) tabu list size and frequency-based memory with a new neighborhood structure in the set of latin squares.

14 Polynomially Solvable Cases of the 3-DAP

Polynomially solvable special cases of the axial 3-DAP were singled out in the following two papers.

R.E. Burkard, R. Rudolf, G.J. Woeginger (1996). Three-dimensional axial assignment problems with decomposable cost-coefficients. *Discr. Appl. Math.* 65, 123-139.

This paper investigates 3-DAPs of the form

$$\min_{\phi,\psi\in\mathcal{S}_n} \sum_{i=1}^{n} a_i b_{\phi(i)} c_{\psi(i)}$$

where $\mathcal{S}_n$ is the set of permutations of $\{1, 2, \ldots, n\}$, and shows that in general this problem is NP-hard. Additional conditions on the problem coefficients (a_i), (b_i) and (c_i), $1 \leq i \leq n$, lead, however, to polynomially solvable special cases. Finally, it is shown that the maximization version of the 3-DAP is also polynomially solvable provided that all coefficients are non-negative.

D. Fortin, R. Rudolf (1995). *Weak algebraic Monge arrays*. SFB Report 33, Institute of Mathematics, Graz University of Technology.

The authors generalize Monge properties to multidimensional arrays and give an explicit optimal solution for MAPs on arrays having such properties.

Other types of special cost coefficients for the MAP and the axial 3-DAP are considered in the following two papers, respectively. Though for the considered special cost coefficients the problems remain NP-hard, polynomial approximation schemes can be given.

H.-J. Bandelt, Y. Crama, F.C.R. Spieksma (1991) Approximation algorithms for multidimensional assignment problems with decomposable costs. *Discr. Appl. Math.* 49, 25–50.

Y. Crama, F.C.R. Spieksma (1992). Approximation algorithms for three-dimensional assignment problems with triangle inequalities. *European J. Oper. Res.* 60, 273–279.

Acknowledgment

We would like to thank Rüdiger Rudolf for helpful suggestions and remarks concerning references on three dimensional assignment problems.

22 Cutting and Packing

Harald Dyckhoff
Aachen University of Technology (RWTH)

Guntram Scheithauer
Dresden University of Technology

Johannes Terno
Dresden University of Technology

Contents

The field of *Cutting and Packing* (C&P) encompasses a variety of multiple connected, geometrical as well as combinatorial problems, models and algorithms on theory and practice. The main objective is the efficient arrangement of geometrically described pieces in larger domains. C&P are hereby two equivalent ways to characterize a given or desired arrangement. The following list of problems strongly related to C&P illustrates this topic: *cutting stock* and *trim loss problem*, *bin packing*, *strip packing*, *knapsack problem*; *vehicle loading*, *pallet loading*, *container loading*, *design*, *layout*, *nesting* and *partitioning problem*; *capital budgeting*, *change making*, *line balancing*, *memory allocation* and *multiprocessor scheduling problem*.

C&P problems are stated and treated in different disciplines such as Management Science, Engineering Science, Information and Computer Sciences, Mathematics as well as Operational Research. They appear in many practical branches such as the

Annotated Bibliographies in Combinatorial Optimization, edited by M. Dell'Amico, F. Maffioli and S. Martello

glass, steel, wood and paper industry, clothing and leather industry, cargo loading and logistics.

Many heuristic and exact algorithms and solution methods have been developed since the mid fifties of this century. Such approaches concern the application of branch-and-bound, dynamic programming, simulation, dialog techniques, revised simplex method, priority rules, enumeration and greedy techniques as well as various heuristics.

In drawing up this bibliography, we have concentrated on publications that appeared in 1985 or later (with some exceptions). We have exercised some judgement in determining which publications to include. If any reader feels we have overlooked an important contribution, we would be pleased to hear from her or him. For the literature not included we refer to the books and surveys listed in §1, especially to the book of [Dyckhoff and Finke 1992].

In this paper we have included more than 150 annotated references. A BiBTeX data base containing more than 1200 entries related to C&P can be obtained from the authors.

1 History and Development of C&P

In the following some milestones in the development of C&P are considered.

L.V. Kantorovich (1939 Russian, 1960 English). Mathematical methods of organising and planning production. *Management Sci.* 6, 366–422.
R.L. Brooks, C.A.B. Smith, A.H. Stone, W.T. Tutte (1940). The dissection of rectangles into squares. *Duke Mathematical Journal* 7, 312–340.

These are the first mathematical articles in the field of cutting and packing.

P.C. Gilmore, R.E. Gomory (1961/63). A linear programming approach to the cutting stock problem (Part I/II). *Oper. Res.* 9, 849–859, and *Oper. Res.* 11, 863–888.
P.C. Gilmore, R.E. Gomory (1965). Multistage cutting stock problems of two and more dimensions. *Oper. Res.* 13, 94–120.
P.C. Gilmore, R.E. Gomory (1966). The theory and computation of knapsack functions. *Oper. Res.* 14, 1045–1075.

For the first time, the articles present techniques which can be used in practice for problems of medium size. The authors describe the meanwhile classical *linear programming approach* to one-dimensional cutting stock problems (successive use of new cutting patterns via *column generation* in the revised simplex method). Furthermore more-dimensional guillotine patterns are introduced and optimal patterns are determined using dynamic programming. The papers also show the importance of the *knapsack problem* as a subproblem occurring in many cutting stock problems.

J.C. Herz (1972). Recursive computational procedure for two-dimensional stock cutting. *IBM J. Res. Develop.* 16, 462–469.

For the first time, Herz uses integral combinations of piece lengths and widths (so-called *raster points*) to *normalize* packing patterns which leads to an essential reduction of computational amount. The *symmetrization* and the use of *bounds* are also considered.

N. Christofides, C. Whitlock (1977). An algorithm for two-dimensional cutting problems. *Oper. Res.* 25, 31–44.

The authors give a branch-and-bound approach for the so-called *constrained guillotine cutting problem*. Constrained means that the number of times a piece is contained in a cutting pattern is restricted. They consider an enumeration of the possible cuts that can be made on the stock rectangle. Upper bounds are calculated by means of transportation routine and dynamic programming routine.

P.Y. Wang (1983). Two algorithms for constrained two-dimensional cutting stock problems. *Oper. Res.* 31, 573–586.

Wang discusses the generation of guillotine patterns by successively combining of pieces together.

Y.G. Stoyan, N.I. Gil (1976). *Methods and Algorithms for the Arrangement of Planar Geometric Objects*, Naukova Dumka, Kiev (in Russian).
Y.G. Stoyan, V.Z. Sokolovski (1980). *Solution of Some Multi-Extremal Problems by Method of Decreasing Neighborhoods*, Naukova Dumka, Kiev (in Russian).
E.A. Mukhachova (1984). *Efficient Cutting of Industrial Materials*, Moskva: Izd. mašinostroenie (in Russian).

These books contain some of the work done in the former USSR published in Russian. Problems of arranging rectangular and non-rectangular pieces on strips and other domains are considered theoretically as well as algorithmically and for applications.

J. Terno, R. Lindemann, G. Scheithauer (1987). *Cutting Problems and Their Practical Solution*, Verlag Harri Deutsch, Thun and Frankfurt/Main (in German).

The authors collect various cutting problems and algorithms for practical solution. For solving rectangular cutting stock problems efficient dynamic programming recursions are given, introducing the so-called *reduced raster point sets.*

An important step to support the international research on C&P was the creation of the informal *Special Interest Group on Cutting and Packing (SICUP)* by H. Dyckhoff and G. Wäscher in 1988. SICUP has an own bulletin published approximately twice a year. Special streams of papers on C&P have been initiated by SICUP at international conferences.

H. Dyckhoff (1990). A typology of cutting and packing problems. *European J. Oper. Res.* 44, 145–160.

Dyckhoff presents a typology for C&P problems which is based on the investigation of characteristics of their geometrical and logical structure and of their appearance in reality:

- Dimensionality: One-, two-, three- or n-dimensional problems.
- Type of assignment: An essential characterization results from whether a complete stock of large objects or small items must be assigned or only a suitable selection.
- Characteristics of large objects and small items: e. g. the type of quantity measurement, figure (shape), assortment, availability.
- Pattern restrictions: Technological conditions of the C&P process like guillotine cutting, stage number, frequency of patterns or the number of items, etc.

– Objectives: Objectives are the minimization of material or space, trim loss minimization, value maximization, etc.

H. Dyckhoff, U. Finke (1992). *Cutting and Packing in Production and Distribution*, Physica-Verlag, Heidelberg.

The book contains approximately 700 entries in the field of C&P. The authors develop a characterization scheme similar to the above mentioned and give a systematic survey of the existing literature.

A.I. Hinxman (1980). The trim-loss and assortment problems: a survey, *European J. Oper. Res.* 5, 8–18.
R.W. Haessler, P.E. Sweeney (1991). Cutting stock problems and solution procedures. *European J. Oper. Res.* 64, 141–150.
K.A. Dowsland, W.B. Dowsland (1992) Packing problems. *European J. Oper. Res.* 56, 2–14.
C.H. Cheng, B.R. Feiring, T.C.E. Cheng (1994). The cutting stock problem – a survey. *Int. J. Prod. Econ.* 36, 291–305.

The research on C&P is focused in these surveys.

The rapid expansion of literature related to C&P leads to the edition of special issues of some journals:
H. Dyckhoff, G. Wäscher (eds.) (1990). Cutting and packing, *European J. Oper. Res.* 44/2.
H. Dyckhoff, K.P. Schuster (eds.) (1991). Verpackungslogistik/Packaging Logistics, *Oper. Res. Spekt.* 13/4.
Y. Lirow (ed.) (1992). Cutting stock: geometric resource allocation, *Math. Comput. Mod.* 16/1.
S. Martello (ed.) (1994). Knapsack, packing and cutting, *INFOR* 32/3, 32/4.
E.E. Bischoff, G. Wäscher (eds.) (1995). Cutting and packing, *European J. Oper. Res.* 84/3.

2 Knapsack Packing

In *knapsack packing* a selection from a large supply of items of heterogeneous figures must be assigned to only a limited stock of objects. Usually one object is considered. Modifications arise if the number of pieces of a certain type can be bounded, the objects may contain some forbidden regions for some of the items, or the sizes of the pieces may vary in given tolerances, etc.

2.1 Knapsack problems

K. Dudzinski, S. Walukiewicz (1987). Exact methods for the knapsack problem and its generalizations. *European J. Oper. Res.* 28, 3–21.

A unified approach and a summary of the most important results concerning the exact methods for solving the (binary) knapsack problem and its generalizations are given.

M. Ozden (1988). A solution procedure for general knapsack problems with a few constraints. *Computers Oper. Res.* 15/2, 145–155.

The procedure developed uses a forward state generation scheme of the dynamic programming along with a facility to eliminate the nonpromising states by calculating an upper bound on the best return that can be achieved with the remaining variables.

S. Martello, P. Toth (1990). *Knapsack Problems – Algorithms and Computer Implementations*, John Wiley & Sons, Chichester.

The book gives a comprehensive compilation of problems related to the *Knapsack Problem* and solution methods. It contains chapters on *Knapsack*, *Bounded knapsack*, *Subset-sum*, *Change-making*, *Multiple knapsack*, *Generalized assignment* and *Bin packing problem*. 13 codes are contained on an assigned floppy disc.

E.A. Boyd (1992). A pseudo-polynomial network flow formulation for exact knapsack separation. *Networks* 22, 503–514.

The NP-complete separation problem for the knapsack polyhedron is formulated as a side-constrained network flow problem with a pseudo-polynomial number of vertices and edges.

M.M. Kovalev, V.M. Kotov (1992). Evaluation of gradient algorithms for knapsack and travelling salesman problems. Report 9203-dO, Belarussian St. Univ. Minsk.

A survey of worst case performance of *greedy algorithms* for the two model problems *Knapsack* and *Maximum travelling salesman* is given in the paper.

L. Amado, P. Barcia (1993). Matroidal relaxations for 0–1 knapsack problems. *Oper. Res. Lett.* 14, 147–152.

Conditions for a knapsack to be greedy solvable are studied, and necessary and sufficient conditions, verifiable in polynomial time, are presented. Furthermore, a family of matroidal relaxations for the 0–1 knapsack problem is investigated.

S.N.N. Pandit, M. Ravi Kumar (1993). A lexicographic search for strongly correlated 0–1 knapsack problems. *Opsearch* 30, 97–116.

The algorithm consists of a lexicographic search in which, for different possible values of an integer k, the best solution which inserts exactly k items in the knapsack is determined. The algorithm efficiently solves strongly correlated instances.

A. Freville, G. Plateau (1994). An efficient preprocessing procedure for the multidimensional 0-1 knapsack problem *Discr. Appl. Math.* 49, 189–212.

The algorithm provides sharp lower and upper bounds on the optimal value, and also a tighter equivalent representation by reducing the continuous feasible set and by eliminating constraints and variables.

T.E. Gerasch, P.Y. Wang (1994). A survey of parallel algorithms for one-dimensional integer knapsack problems. *INFOR* 32, 163–186.

The article surveys several methods that can be used to solve integer knapsack problems on a variety of parallel computing architectures.

E. Chajakis, M. Guignard (1994). The setup knapsack problem: exact algorithms, upper bounds and heuristics. *INFOR* 32, 124-142.
D. Krass, S. Sethi, G. Sorger (1994). Some complexity issues in a class of knapsack problems: What makes a knapsack problem hard?. *INFOR* 32, 149–162.
Structural properties of the knapsack problem are investigated in these papers.

G. Gens, E. Levner (1994). A fast approximation algorithm for the subset-sum problem. *INFOR* 32, 143–148.
A new fully polynomial approximation scheme is presented.

R.E. Burkard, U. Pferschy (1995). The inverse-parametric knapsack problem. *European J. Oper. Res.* 83, 376–93.
This paper deals with parametric knapsack problems where the costs resp. weights are replaced by linear functions depending on a parameter. The aim is to find the smallest parameter such that the optimal solution value of the knapsack problem is equal to a prespecified solution value.

S. Martello, P. Toth (1996). Upper bounds and algorithms for hard 0–1 knapsack problems. *Oper. Res.* (to appear).
Upper bounds are presented, obtained from the mathematical model of the problem by adding valid inequalities on the cardinality of an optimal solution, and relaxing it in a Lagrangian fashion.

2.2 Related one-dimensional C&P problems

B.R. Sarker (1988). An optimum solution for one-dimensional slitting problems: a dynamic programming approach. *J. Oper. Res. Soc.* 39, 749–755.
K. Richter (1992). Solving sequential interval cutting problems via dynamic programming. *European J. Oper. Res.* 57, 332–338.
G. Scheithauer (1995). The solution of packing problems with pieces of variable length and additional allocation constraints. *Optimization* 34, 81–96.
These papers deal with practical applications in which additional constraints and/or generalized objectives arise when defects of the stock material or other quality conditions have to be regarded.

S.U. Foronda, H.F. Carino (1991). A heuristic approach to the lumber allocation problem in hardwood dimension and furniture manufacturing. *European J. Oper. Res.* 54, 151–162.
C. Carnieri, G.A. Mendoza, L.G. Gavinho (1994). Solution procedures for cutting lumber into furniture parts. *European J. Oper. Res.* 73, 495–501.
Here case studies are given.

2.3 Two-dimensional rectangular C&P problems

In these problems, small rectangles (pieces) of various sizes must be assigned to a choice of large rectangles (objects). Mostly, one object is considered. In case of orthogonal cutting or packing the pieces have to be arranged parallel to the edges of the objects. A guillotine cut means a straight line from an edge of the object to the opposite one, parallel to the others. A problem is said to be *constrained* if the number of pieces of a certain type is restricted additionally in the pattern looked for.

P. Ghandforoush, J.J. Daniels (1992). A heuristic algorithm for the guillotine constrained cutting stock problem. *ORSA J. Comput.* 4, 351–356.
B. MacLeod, R. Moll, M. Girkar, N. Hanifi (1993). An algorithm for the 2D guillotine cutting stock problem. *European J. Oper. Res.* 68, 400–412.
New heuristics are given for the constraint guillotine cutting of a rectangle. In the second paper an $O(n^3)$ heuristic is presented.

J.F. Oliveira, J.S. Ferreira (1990). An improved version of Wang's algorithm for two-dimensional cutting problems. *European J. Oper. Res.* 44, 256–266.
K.V. Viswanathan, A. Bagchi (1993). Best-first search methods for constrained two-dimensional cutting stock problems. *Oper. Res.* 41, 768–776.
N. Christofides, E. Hadjiconstantinou (1995). An exact algorithm for orthogonal 2-D cutting problems using guillotine cuts. *European J. Oper. Res.* 83, 21–38.
V.P. Daza, A.G. de Alvarenga, J. de Diego (1995). Exact solutions for constrained two-dimensional cutting problems. *European J. Oper. Res.* 84, 633–644.
These papers give exact approaches for solving the constraint guillotine cutting problem.

P.K. Agarwal, M.T. Shing (1992). Oriented aligned rectangle packing problem. *European J. Oper. Res.* 62, 210–220.
Only guillotine cuts with respect to the stock rectangle are allowed in order to get feasible patterns.

G. Scheithauer, J. Terno (1988). Guillotine cutting of defective boards. *Optimization* 19, 111–121.
A dynamic programming approach for three-stage guillotine cutting of rectangles is developed regarding non-usable areas on the stock material.

J.E. Beasley (1985). An exact two-dimensional non-guillotine cutting tree search procedure. *Oper. Res.* 33, 49–64.
A Lagrangian relaxation of a zero-one integer programming formulation of the problem is developed and used as a bound in a tree search procedure. Subgradient optimization is used to optimize this bound. Problem reduction tests derived from both the original problem and the Lagrangian relaxation are given.

M.N. Arenales, R.N. Morabito (1995). An AND/OR graph approach to the solution of two-dimensional non-guillotine cutting problems. *European J. Oper. Res.* 84, 599–617.
A new graph-theoretic approach is considered.

A.H.G. Rinnooy Kan, J.R. De Witt, R.T. Wijmenga (1987). Non-orthogonal two-dimensional cutting patterns. *Management Sci.* 33, 670–684.

The packing of small identical rectangles within a strip is considered allowing a non-orthogonal position of the items.

D. Fayard, V. Zissimopoulos (1995). An approximation algorithm for solving unconstrained two-dimensional knapsack problems. *European J. Oper. Res.* 84, 618–632.

An effective heuristic for solving two-dimensional cutting problems is presented. The algorithm selects a subset of generated strips by solving a sequence of one-dimensional knapsack problems.

3 Pallet Loading

The *pallet loading problem* (PLP) looks for an optimal arrangement of a selection of identical small items to one large object. In general the optimal patterns have typical block or parquet structure. In some cases additional criteria are applied to determine a suitable pattern.

H. Carpenter, W.B. Dowsland (1985). Practical considerations of the pallet loading problem. *J. Oper. Res. Soc.* 36/6, 489–497.

Consideration is given to the development of loading patterns into pallet stacks, and criteria which might be applied to determine the sensibility of these stacks for storage and transportation. A technique is developed which allows the stability and clampability to be tested, and this is applied to layouts produced by an algorithm.

H. Isermann (1987). A planning system for optimizing the pallet loading with congruent items (German). *Oper. Res. Spekt.* 9, 235–249.

A new LP-based upper bound is proposed for the PLP which consists of an optimal packing of identical small rectangles on a pallet.

K.A. Dowsland (1987). An exact algorithm for the pallet loading problem, *European J. Oper. Res.* 31, 78–84.

A branch-and-bound algorithm is developed which is based on a transformation of the PLP to the *independent set problem.*

W.B. Dowsland (1991). Sensitivity analysis for pallet loading. *Oper. Res. Spekt.* 13, 198–203.

Based on a *pallet chart* the paper includes sensitivity analysis with respect to changes of box and pallet dimensions.

H. Exeler (1991). Upper bounds for the homogeneous case of a two-dimensional packing problem. *Z. Oper. Res.* 35, 45–58.

The upper bound for an instance E of the PLP is based on the minimization of the area bound with respect to all instances equivalent to E.

B. Ram (1992). The pallet loading problem: a survey. *Int. J. Prod. Econ.* 28, 217–25.
This paper provides a survey of recent research relating to considerations in pallet packing, the models and solution procedures used in pallet packing, and implementation of these approaches in palletization stations. (47 refs.)

P.K. Agrawal (1993). Minimizing trim loss in cutting rectangular blanks of a single size from a rectangular sheet using orthogonal guillotine cuts. *European J. Oper. Res.* 64, 410–422.
A branch-and-bound algorithm is proposed.

A.G. Tarnowski, J. Terno, G. Scheithauer (1994). A polynomial time algorithm for the guillotine pallet loading problem. *INFOR* 32, 275–287.
A polynomial time algorithm is presented having complexity $O(\Theta \log_2^2 w)$ where $\Theta = \log_2 \max\{L, W, l, w\}$ and $L \times W$ and $l \times w$ are the sizes of the pallet and boxes, respectively.

J. Nelissen (1994). New Approaches for Solving Pallet Loading Problems. Verlag Shaker, Aachen (in German).
G. Naujoks (1995). Optimal Utilization of Storage. Deutscher Universitätsverlag, Wiesbaden (in German).
These Ph.D. theses contain surveys of optimization algorithms and extensive computational results.

J. Nelißen (1995). How to use structural constraints to compute an upper bound for the pallet loading problem. *European J. Oper. Res.* 84, 662–680.
A new upper bound method is discussed. It is based on a set of structural constraints for a solution which realizes the assumed upper bound.

W.B. Dowsland (1995). Improved palletization efficiency – the theoretical basis and practical application. *Int. J. Prod. Res.* 33, 2213-2222.
This covers the application of a pallet chart to practical problems where product volume might be variable.

G. Scheithauer, J. Terno (1996). The G4-heuristic for the pallet loading problem. *J. Oper. Res. Soc.* 47, 511–522.
The new heuristic is based on structure investigations and it dominates all heuristics known so far. Moreover, the G4-heuristic produces in any case optimal patterns if not more than 50 items can be packed.

4 Bin Packing

In the *bin packing problem* numerous small, heterogeneously-shaped (one-, two- or three-dimensional) items have to be assigned to a selection of objects. In many cases a successive assignment of the items to the objects is an essential restriction of the problem.

J.J. Bartholdi, H.V. Vate, J. Zhang (1989). Expected performance of the shelf heuristic for 2-dimensional packing. *Oper. Res. Lett.* 8, 11–16.

The presented shelf heuristic is an on-line algorithm. The expected performance is analyzed and the choice of the shelf sizes is discussed.

E.G. Coffman Jr., G.S. Lueker (1991). *Probabilistic Analysis of Packing and Partitioning Algorithms*, John Wiley & Sons, New York.

The book examines techniques useful in the probabilistic analysis of algorithms, focusing on applications to bin packing and partitioning. Since the book concentrates as much on techniques as on results it is a useful introduction to probabilistic analysis even for those readers with interests in other problem areas.

S. Anily, J. Bramel, D. Simchi-Levi (1994). Worst-case analysis of heuristics for the bin packing problem with general cost structures. *Oper. Res.* 42, 287–298.

The cost of a bin is a concave function in the number of items in the bin. The objective is to store the items in such a way that total cost is minimized.

M. Girkar, B. MacLeod, R. Moll (1992). Performance bound for bottom-left guillotine packing of rectangles. *J. Oper. Res. Soc.* 43, 169–175.
D. Kröger (1995) Guillotine-able bin packing: a genetic approach. *European J. Oper. Res.* 84, 645–661.

Heuristics for the guillotine bin packing problem are proposed.

G. Galambos, A. van Vliet(1994). Lower bounds for 1-, 2- and 3-dimensional on-line bin packing algorithms. *Computing* 52, 281–97.

Lower bounds for the asymptotic worst case ratio of on-line algorithms for different kind of bin packing problems are discussed.

I. Schiermeyer (1994). Reverse-fit: a 2-optimal algorithm for packing rectangles. Working Paper. To appear in Lecture Notes in Computer Science, Proc. of the 2nd European Symposium on Algorithms.

A *level-oriented* algorithm is described and analyzed, called *Reverse-Fit*, for packing rectangles into a unit-width, infinite-height bin so as to minimize the total height of the packing.

D. Simchi-Levi (1994). New worst-case results for the bin-packing problem. *Naval Res. Log. Quart.* 41, 579–585.

New worst-case results are presented for a number of classical heuristics. It is shown that the first-fit and best-fit heuristics have an absolute performance ratio of no more than 1.75, and first-fit decreasing and best-fit decreasing heuristics have an absolute performance ratio of 1.5.

S. Martello, D. Vigo (1997). Exact solution of the two-dimensional finite bin packing problem. *Management Science* (to appear).

A known lower bound is analyzed and its worst case performance is determined. Moreover, new lower bounds are proposed which are used within a branch-and-bound algorithm. Results of numerical tests with instances involving up to 120 pieces are presented.

5 Stock Cutting

Stock cutting denotes those C&P problems in which many items of only several distinct shapes must be assigned to a selection of objects. Contrary to bin packing problems these groups of identical items allow in general the multiple use of certain patterns in the solution.

5.1 One-dimensional cutting stock problems

R.E. Johnston (1986). Rounding algorithms for cutting stock problems. *Asia–Pacific J. Oper. Res.* 3, 166–171.

The "rounding" algorithm performs the task of producing an integer solution from the (fractional) LP solution while at the same time reducing the number of distinct cutting patterns and removing, where possible, any pattern with small quantity.

H. Stadtler (1988). A comparison of two optimization procedures for 1- and 1,5-dimensional cutting stock problems. *Oper. Res. Spekt.* 10, 97–111.

The so-called *complete cut* and *one-cut* models are compared.

R.W. Haessler (1992). One-dimensional cutting stock problems and solution procedures. *Math. Comput. Mod.* 16, 1–8.

Problems and solution procedures are considered in which not only trim loss is regarded.

Z. Sinuany-Stern, I. Weiner (1994). The one-dimensional cutting stock problem using two objectives. *J. Oper. Res. Soc.* 45, 231–236.

The primary objective is to minimize the trim loss in a given piece of metal work requiring metal sections of various lengths. The secondary objective is to organize the cutting so that the maximum quantity of leftovers is accumulated in the last bar(s).

F.J. Vasko, M.L. Cregger, D.D. Newhart, K.L. Stott (1993). A real-time one-dimensional cutting stock algorithm for balanced cutting patterns. *Oper. Res. Lett.* 14, 275–282.

In the steel industry when a finished structural shape exits the mill upon which it was produced it is cut into customer order lengths. The actual length of the finished beam bar may not be known precisely until immediately before cutting.

In the following papers the difference between the optimal values of the integer and the corresponding LP-relaxation problem is investigated. If this difference is less than 2 then the so-called *Modified integer round-up property* holds.

O. Marcotte (1985). The cutting stock problem and integer rounding. *Math. Program.* 33/1, 82–92.

G. Scheithauer, J. Terno (1992). About the gap between the optimal values of the integer and continuous relaxation one-dimensional cutting stock problem. *Oper. Res. Proc. 1991*, Springer-Verlag, Berlin Heidelberg, 439–444.
G. Wäscher, T. Gau (1994). Generating almost optimal solutions for the integer one-dimensional cutting stock problem. Working paper no. 94/06, TU Braunschweig.
G. Scheithauer, J. Terno (1995). The modified integer round-up property of the one-dimensional cutting stock problem. *European J. Oper. Res.* 84, 562–571.
V. Nica (1994). General counterexample to the integer round-up property. Working Paper, Dep. Econ. Cyb., Academy of Econ. Studies, Bucharest.

T. Gau, G. Wäscher (1995). CUTGEN1: A problem generator for the standard one-dimensional cutting stock problem. *European J. Oper. Res.* 84, 572–579.
A problem generator for the standard one-dimensional cutting stock problem is developed.

G. Wäscher, T. Gau (1996). Heuristics for the one-dimensional cutting stock problem: a computational study. *Oper. Res. Spekt.* 18, 131–144.
The authors compare various heuristic approaches and show that two of them are clearly superior to the others as well as solve almost any of 4,000 randomly generated test problems to an optimum.

5.2 Two-dimensional cutting stock problems

A.A. Farley (1988). Practical adaptations of the Gilmore–Gomory approach to cutting stock problems. *Oper. Res. Spekt.* 10, 113–123.
The adaptations considered relate to initial pattern generation, multiple solutions per knapsack problem, explicit valuation of undersupply and oversupply, and waste having a value.

F.J. Vasko, F.E. Wolf, K.L. Stott (1989). A practical solution to a fuzzy two-dimensional cutting stock problem. *Fuzzy Sets and Systems* 29/3, 259-275.
A problem formulation in a fuzzy environment is presented which addresses a number of practical aspects also of interest besides the waste minimization.

A.A. Farley (1992). Limiting the number of each piece in two-dimensional cutting stock patterns *Math. Comput. Mod.* 16, 75–87.
F.J. Vasko, F.E. Wolf, K.L. Stott, O. Ehrsam (1992). Bethlehem Steel combines cutting stock and set covering to enhance customers service. *Math. Comput. Mod.* 16, 9–17.
Case studies are presented.

E.A. Mukhachova, V.A. Zalgaller (1993). Linear programming cutting problems. *Int. J. Software Knowledge Eng.* 3, 463–476.
Different cutting and related problems are considered like forming problems (closed packing problems) and problems of cutting planning. For solving these planning problems, linear or integer programming is used.

F. Chauny, R. Loulou (1994). LP-based method for the multi-sheet cutting stock problem. *INFOR* 32, 253–264.

An LP-solution is rounded off according to a special procedure and some further improvements are given via a branch-and-bound approach.

P.H. Vance, C. Barnhart, E.L. Johnson, G. Nemhauser (1994). Solving binary cutting stock problems by column generation and branch-and-bound. *Computational Opt. Appl.* 3, 111–130.

The authors present an algorithm to obtain optimal integer solutions.

G. Scheithauer (1994). On the MAXGAP problem for cutting stock problems. *J. Inf. Proc. Cyb. EIK* 30, 111–117.

The maximum difference between the optimal values of the integer and the corresponding LP-relaxation problem is investigated for two-dimensional cutting stock problems.

J.M.V. de Carvalho, A.J.G, Rodrigues (1995). An LP-based approach to a two-stage cutting stock problem. *European J. Oper. Res.* 84, 580–589.

An LP model for a two-stage cutting stock problem is described that arises in a make-to-order steel company – a case study.

J. Riehme, G. Scheithauer, J. Terno (1996). The solution of two-stage guillotine cutting stock problems having extremely varying order demands. *European J. Oper. Res.* 91, 543–552.

A new approach (called strip approach) is proposed in which the two-dimensional problem is considered as two one-dimensional problems.

5.3 Sequencing and assortment problems

Besides the waste minimization a suitable sequence of cutting the patterns may have remarkable effects in various practical situations. Furthermore, choosing appropriate sizes of the stock material may yield a waste reduction.

O.B.G. Madsen (1988). An application of travelling-salesman routines to solve pattern–allocation problems in the glass industry. *J. Oper. Res. Soc.* 39/3, 249–256.

A two-stage procedure to solve the two-dimensional pattern-allocation problem is suggested. The first stage consists of solving the cutting stock problem without the sequencing constraint. In the second stage a sequencing problem (formulated as a traveling salesman problem) is used for the ordering of the cutting patterns in an optimal or near-optimal way.

D.W. Pentico (1988). The discrete two-dimensional assortment problem. *Oper. Res.* 36/2, 324–332.

In this two-dimensional assortment problem, a unit of product that is larger or better on two measurements or quantities may be used to substitute, at some cost, for a unit of demand that is smaller or inferior on both measurements or quantities. The problem is to determine which dimensional combinations to stock in order to minimize the combined stocking and substitution costs. Three heuristics are compared.

B.J. Yuen (1991). Heuristics for sequencing cutting patterns. *European J. Oper. Res.* 55, 183–190.
A. Corominas, J. Bautista (1994). Procedures for solving a 1-dimensional cutting problem. *European J. Oper. Res.* 77, 154–168.
B.J. Yuen, K.V. Richardson (1995). Establishing the optimality of sequencing heuristics for cutting stock problems. *European J. Oper. Res.* 84, 590–598.
Various problems of a suitable *sequencing* of patterns are also investigated in the above papers.

D.D. Gemmill, J.L. Sanders (1991). A comparison of solution methods for the assortment problem. *Int. J. Prod. Res.* 29, 2521–2527.
In this article a comparison is made between three common heuristic methods for solution of the assortment problem.

C.S. Chen, S. Sarin, B. Ram (1993). A mixed-integer programming model for a class of assortment problems. *European J. Oper. Res.* 65, 362–367.
A mixed-integer non-linear programming model is formulated to compute a large rectangle of minimum area containing a given set of small rectangles.

P.K. Agrawal (1993). Determining stock-sheet-sizes to minimize trim loss. *European J. Oper. Res.* 64, 423–431.
A heuristic to find the best sizes of stock sheets to minimize the trim loss is given.

H.H. Yanasse (1994). A search strategy for the one-size assortment problem. *European J. Oper. Res.* 74, 135–142.
The problem of determining the optimal length of a board with fixed width from which panels have to be cut to satisfy some specified demand with minimum waste is analyzed. The proposed procedure generates a lower bound on the best solution that can be obtained for each of the board length values within the allowed range.

F.J. Vasko, F.E. Wolf (1994). A practical approach for determining rectangular stock sizes. *J. Oper. Res. Soc.* 45, 281–286.
The authors address the problem of determining what rectangular sizes should be stocked in order to satisfy a bill of materials composed of smaller rectangles.

6 General Two-Dimensional C&P

In contrast to the so-called *rectangular case* where only rectangular items and objects are considered, now items with other regular or irregular shapes have to be arranged on not-necessary rectangular-shaped objects. The packing of items within a strip of minimal length is included as well as the packing of a maximum number of circles, etc.

6.1 Strip packing and layout problems

J. Blazewicz, M. Drozdowski, B. Soniewicki, R. Walkowiak (1989). Two-dimensional cutting problem: basic complexity results and algorithms for irregular shapes. *Found. Cont. Eng.* 14, 137–160.
Some basic complexity results and solution approaches are summarized.

J. Blazewicz, A. Piechowiak, R. Walkowiak (1992). Defining two-dimensional irregular figures for automatic cutting system. *Found. Comp. Decis. Sc.* 17, 3–12.
An approach for interactive defining of two-dimensional irregular shapes is presented.

A. Elomri, P. Morel, L. Pun, G. Doumeingts (1994). Solving a cutting problem based on existing patterns. *European J. Oper. Res.* 77, 169–178.
Z. Li, V. Milenkovic (1995). Compaction and separation algorithms for non-convex polygons and their applications. *European J. Oper. Res.* 84, 539–561.
These papers describe heuristic approaches for problems occuring in the clothing industry.

6.2 Non-rectangular C&P problems

G. Scheithauer (1990). Optimal cutting of trapezoids. *Wiss. Z. TU Dresden* 39, 191–195 (in German).
The computation of optimal two-stage cutting patterns for trapezoids is considered.

H. Lutfiyya, B. McMillin, P. Poshyanonda, C. Dagli (1992). Composite stock cutting through simulated annealing. *Math. Comput. Mod.* 16, 57–74.
For placements of various shapes, it is shown how to determine a cost function, annealing parameters and performance.

R.E. Johnston (1992). Dimensional efficiency in cable manufacturing problems and solutions. *Math. Comput. Mod.* 16, 19–35.
This paper describes the principal dimensional efficiency and cutting stock problems associated with the manufacture of communication and power cables. Some unique mathematical programming problems are identified and practical solution approaches to some of the new problems are presented.

A. Arbel (1993). Large-scale optimization methods applied to the cutting stock problem of irregular shapes. *Int. J. Prod. Res.* 31, 483–500.
The paper describes a real feasibility study of applying large-scale optimization methods to the cutting stock problem of irregular shapes.

G. Scheithauer, J. Terno (1993). Modelling of packing problems. *Optimization* 28, 63–84.
Mixed-integer models are presented for two-dimensional packing problems of convex and non-convex polygons.

K.A. Dowsland (1991). Optimising the palletisation of cylinders in cases. *Oper. Res. Spekt.* 13, 204–212.
H. Isermann (1991). Heuristics for solving two-dimensional packing problems with round pieces. *Oper. Res. Spekt.* 13, 213–223 (in German).
Heuristics to pack identical cylinders on a pallet are proposed.

J.A. George, J.M. George, B.W. Lamar (1995) Packing different-sized circles into a rectangular container. *European J. Oper. Res.* 84, 693–712.
Heuristics for the *cylindrical bin packing problem* are presented.

R. Peikert (1994). Densest packing of identical circles within a square. *El. Math.* 49,16–31 (in German).
Optimal solutions and a method to prove their optimality are presented.

K.A. Dowsland and W.B. Dowsland (1995). Solution approaches to irregular nesting problems. *European J. Oper. Res.* 84, 506-521.
A survey is given of some of the literature concerning cutting and packing problems with irregular pieces.

7 Three-Dimensional C&P

Three-dimensional C&P problems cover the so-called *container loading* and the *multi-container loading problem.* In general, a selection of various rectangular-shaped boxes have to be packed within a container maximizing the volumetric utilization or minimizing the number of containers required to pack all boxes.

D. Sculli, C.F. Hui (1988). Three-dimensional stacking of containers. *Omega* 16/6, 585–594.
Results of a simulation study into the stacking and handling of containers with the same dimensions are reported. The measures of performance include volumetric utilization, wasteful handling ratios, storage ratio, and rejection ratio.

W. Schneider (1988). Trim-loss minimization in a crepe-rubber mill; optimal solution versus heuristic in the 2 (3)-dimensional case. *European J. Oper. Res.* 34/3, 273–281.
A three-dimensional cutting problem is investigated – a case study.

N. Ivancic, K. Mathur, B.B. Mohanty (1989). An integer programming based heuristic approach to the three-dimensional packing problem. *J. Manufacturing and Operations Management* 2/4, 268 298.
A three-dimensional packing problem is investigated where one tries to pack various types of boxes (with different sizes) in various types of containers. The total cost of containers used is minimized.

R.W. Haessler, F.B. Talbot (1990). Load planning for shipments of low density products. *European J. Oper. Res.* 44, 289–299.
The paper presents a complex computer-based heuristic procedure for sizing cus-

tomer orders and developing three-dimensional load diagrams for rail and truck shipment.

H. Gehring, K. Menschner, M. Meyer (1990). A computer-based heuristic for packing pooled shipment containers. *European J. Oper. Res.* 44, 277–288.
The rectangular boxes are packed in a layer-wise manner.

E.E. Bischoff (1991). Stability aspects of pallet loading. *Oper. Res. Spekt.* 13, 189–197.
Practical aspects are considered.

W.B. Dowsland (1991). Three-dimensional packing–solution approaches and heuristic development. *Int. J. Prod. Res.* 29, 1673-1685.
The application of methods to 3D problems is presented including that of a pallet chart.

K. Li, K.H. Cheng(1992). Heuristic algorithms for on-line packing in three dimensions. *J. Algorithms* 13, 589–605.
Several on-line packing algorithms, called level-strip algorithms, are proposed and analyzed. Each of these algorithms has a separate heuristic in each dimension.

G. Scheithauer (1992). Algorithms for the container loading problem. In *Oper. Res. Proc. 1991*, Springer-Verlag, Berlin Heidelberg, 445–452.
A heuristic based on the forward state strategy is proposed.

J.A. George (1992). A method for solving container packing for a single size of box. *J. Oper. Res. Soc.* 43, 307–312.
A heuristic and upper bounds are presented.

R.D. Tsai, E.L. Malstrom, W. Kuo (1993). Three-dimensional palletization of mixed box sizes. *IIE Transactions* 25, 64–75.
The modeling procedure presented converts the three-dimensional pallet loading problem into a standard mixed 0-1 integer programming model.

G. Abdou, M. Yang (1994). A systematic approach for the three-dimensional palletization problem. *Int. J. Prod. Res.* 32, 2381–2394.
A heuristic has been developed to cope with the three-dimensional palletization problem in which boxes have different base dimensions and can be grouped by the same height (layered-pallet-loading technique).

B.B. Mohanty, K. Marthur, N.J. Ivancic (1994). Value considerations in three-dimensional packing –a heuristic procedure using the fractional knapsack problem. *European J. Oper. Res.* 74, 143–151.
The heuristic presented is based on an integer programming formulation that uses the fractional knapsack problem as a column generation subproblem.

R. Morabito, M.N. Arenales (1994). An AND/OR graph approach to the container loading problem. *Int. Trans. Oper. Res.* 1, 59–73.

Some heuristics are presented together with some computational results.

B.K.A. Ngoi, M.L. Tay, E.S. Chua (1994). Applying spatial representation techniques to the container packing problem. *Int. J. Prod. Res.* 32, 111–123.
The paper describes an efficient method of packing boxes into a container using a unique spatial representation technique.

E.E. Bischoff, F. Janetz, M.S.W. Ratcliff (1995). Loading pallets with non-identical items. *European J. Oper. Res.* 84, 681–692.
The new heuristic aims to generate efficient loading arrangements which at the same time also provide a high degree of inherent stability.

E.E. Bischoff, M.S.W. Ratcliff (1995). Issues in the development of approaches to container loading. *Omega* 23, 377–390.
Two approaches are proposed to obtain stable, evenly distributed packing patterns and to cater for multi-drop loads.

C.S. Chen, S.M. Lee, Q.S. Shen (1995). An analytical model for the container loading problem. *European J. Oper. Res.* 80, 68–76.
The problem is formulated as a zero-one mixed integer programming model. It includes the consideration of multiple containers, multiple carton sizes, carton orientations, and the overlapping of cartons in a container.

8 Related Topics

In this section various references concerning special topics related to C&P are collected.

B.S. Baker, S.J. Fortune, S.R. Mahaney (1986). Polygon containment under translation. *J. Algorithms* 7, 532–548.
The problem of deciding whether a polygon P can be translated to fit inside another polygon P' is considered. Polynomial algorithms are presented for two cases: when both P and P' are rectilinearly convex, and when P is convex and P' is arbitrary.

D.M. Mount, R. Silverman (1990). Packing and covering the plane with translates of a convex polygon. *J. Algorithms* 11, 564–580.
It is shown that, given a convex polygon P with n vertices, the densest lattice packing of P in the plane can be found in $O(n)$ time.

H. Frank (1991). Covering polygons with rectangles via edge coverings of bipartite permutation graphs. *J. Inform. Process. Cyb. EIK* 27, 891–409.
Optimal algorithms are presented for solving the rectangle covering problem for monotone orthogonal polygons.

Y. Crama, J. Mazzola (1994). On the strength of relaxations of multidimensional knapsack problems. *INFOR* 32, 219–225.

J. Thiel, S. Voss (1994). Some experiences on solving multi-constraint zero-one knapsack problems with genetic algorithms. *INFOR* 32, 226–242.
Theoretical investigations and heuristic approaches are presented for the multidimensional knapsack problem.

E. Girlich, A.G. Tarnowski (1994). Efficiency of greedy algorithms for cutting stock problems. *Oper. Res. Spekt.* 16, 211–21.
An adaptation of the greedy algorithm for the one, two and three dimensional cutting stock problem is proposed. The efficiency of the algorithms is measured in terms of the worst case bound and the time complexity.

P. Harche, G.L. Thompson (1994). The column subtraction algorithm: an exact method for solving weighted set covering, packing and partitioning problems. *Computers Oper. Res.* 21, 689–705.
The authors present a new and conceptually simple approach for finding exact solutions to large set covering, packing and partitioning problems.

R.B. Grinde, T.M. Cavalier (1995). A new algorithm for the minimum-area convex enclosing problem. *European J. Oper. Res.* 84, 522–538.
Given two simple polygons, the algorithm finds the relative position of one with respect to the other such that the area of their convex enclosure is minimized.

G. Galambos, H. Kellerer, G. Woeginger (1993). A lower bound for on-line vector-packing algorithms. *Acta Cybernetica* 11, 23–34.
Lower bounds on the asymptotic worst case ratio of any on-line d-dimensional vector packing algorithm are given.

F.C.R. Spieksma (1994). A branch-and-bound algorithm for the two-dimensional vector packing problem. *Computers Oper. Res.* 21, 19–25.
A heuristic adapted from the first fit decreasing rule is proposed, and lower bounds for optimal solutions are investigated.

U. Betke, M. Henk, J.M. Wills (1994). Finite and infinite packings. *J. reine angew. Math.* 453, 165–191.
Finite and infinite packings of translates for centrally symmetric convex bodies in the d-dimensional Euclidean space are considered.

H.-G. Bigalke, H. Wippermann (1994). *Regular Parqueting*, Wissenschaftsverlag, Mannheim (in German).
The book contains the mathematical background of regular parqueting as well as various applications in crystallography and industry.

S. Dempe, E. Müller (1995). Stability analysis for a special interval cutting problem. *European J. Oper. Res.* 87, 188–199.
For a special cutting problem the regions of stability of an optimal solution are obtained by a slight modification of a dynamic programming procedure used for solving the unperturbed problem.

T.J. Richardson (1995). Optimal packing of similar triangles. *J. Combin. Theory, Series A* 69, 288–300.

The author proves that any set of similar triangles with total area equal to α can be packed, using only translations and reflections, inside a similar triangle of area 2α. The bound 2α cannot be improved in general.

9 Software

In this section various papers are contained concerning C&P software, concepts of program packages, comparisons of algorithms and numerical experience.

H. Dyckhoff, U. Finke, H.-J. Kruse (1988). Empirical investigation of cutting stock software. *Oper. Res. Spekt.* 10, 237–247 (in German).

The paper documents empirical investigation and presents the results for 28 programs of different firms. (12 refs.)

E.E. Bischoff, M.D. Marriott (1990). A comparative evaluation of heuristics for container loading. *European J. Oper. Res.* 44, 267-276.

Results are presented of a comparative analysis of fourteen heuristic packing rules.

J. Käschel, A. Mädler, K. Richter (1991). Software for packing logistics. *Oper. Res. Spekt.* 13, 229–238 (in German).

Several commercial software packages for pallet loading are analyzed and compared.

M. Dincbas, H. Simonis, P. Van Hentenryck (1992). Solving a cutting stock problem with the constraint logic programming language CHIP. *Math. Comput. Mod.* 16, 95–105.

The authors present two approaches to solve a two-dimensional cutting problem in CHIP and compare them with the standard ones. It turns out that, although CHIP greatly simplifies the problem statement, it is comparable in efficiency to specialized programs.

Y. Lirov (1992). Knowledge based approach to the cutting stock problem. *Math. Comput. Mod.* 16, 107–125.

In this survey, three major trends of engineering activities are identified and discussed in detail: integration of applications, shift from optimization to control, and construction of new related problems. Moreover a learning expert system CUT is proposed which combines Simulated Annealing and Case Based Reasoning.

J.M.V. De Carvalho, A.J.G. Rodrigues (1994). A computer based interactive approach to a two stage cutting stock problem. *INFOR* 32, 243–252.

An interactive system, developed for a Portuguese steel manufacturer, is presented.

H.J. Fraser, J.A. George (1994). Integrated container loading software for pulp and paper industry. *European J. Oper. Res.* 77, 466–474.

A container loading software package for a pulp and paper manufacturer is described.

G. Scheithauer, J. Terno, J. Riehme, U. Sommerweiß (1996). A new heuristic approach for solving the multi-pallet packing problem. Working paper MATH-NM-03-1996, TU Dresden.

Based on a branch-and-bound concept, boxes of various sizes are packed on a minimum number of pallets regarding several packing constraints of practical interest.

23 Set Covering Problem

Sebastián Ceria
Columbia University, New York

Paolo Nobili
CNR, Roma

Antonio Sassano
Università di Roma "La Sapienza"

CONTENTS

Set covering problems are, with little doubt, one of the most important classes of combinatorial optimization problems. The reason is that the underlying combinatorial structure is very general, and many combinatorial optimization problems can be modeled as special cases of covering problems. A partial list includes: set packing, min-cut, node (edge) packing (covering) on graphs, monotone traveling salesman, knapsack and acyclic sub-digraph problems.

In addition, set covering problems are extremely relevant because of the great variety of real life problems that can be efficiently solved once formulated as set covering problems.

In this survey, we have focused on what we consider the main aspects of the theoretical and practical developments which are specific of the set covering model. Although strongly related, we have chosen not to mention the results concerning set partitioning and set packing problems.

The interested reader can refer to the following papers that give a broad introduc-

Annotated Bibliographies in Combinatorial Optimization, edited by M. Dell'Amico, F. Maffioli and S. Martello

tion on the set covering problem and outline its relationship with other combinatorial problems. The first one is considered to be a landmark paper in the area.

E. Balas, M.W. Padberg (1972). On the set covering problem. *Oper. Res.* 20, 1152–1161.
E. Balas, M.W. Padberg (1975). On the set covering problem: II. An algorithm for set partitioning. *Oper. Res.* 23, 74–90.
E. Balas, M.W. Padberg (1976). Set partitioning: a survey. *SIAM Rev.* 18, 710–760.
M.W. Padberg (1979). Covering, packing and knapsack problems. P.L. Hammer, E.L. Johnson, B.H. Korte (eds.). *Discrete Optimization I*, Ann. Discr. Math. 4, North-Holland, Amsterdam, 265–287.

In what follows, we provide the definition of the basic objects that we will deal with in the rest of this paper.

Let $M = \{1, \ldots, m\}$ be a finite set and let $\mathcal{S} = \{S_j\}_{j \in N}$, for $N = \{1, \ldots, n\}$, be a given finite collection of subsets of M, $S_j \subseteq M, j \in N$. We say that $F \subseteq N$ *covers* M if $\cup_{i \in F} S_j = M$. In the *set covering problem* we are given a cost c_j for every set S_j and the goal is to find a minimum cost cover of M.

A set covering problem can be easily formulated as a 0–1 integer program. Namely, let A be a 0–1 $m \times n$ matrix, whose elements (a_{ij}), $i \in M, j \in N$, are defined by

$$a_{ij} = \begin{cases} 1 & \text{if } i \in S_j \\ 0 & \text{if } i \notin S_j. \end{cases}$$

For any $j \in N$, let

$$x_j = \begin{cases} 1 & \text{if } j \in F \\ 0 & \text{if } j \notin F \end{cases}$$

and c_j be the cost of set S_j. Then, the integer programming formulation of the set covering problem is

$$\begin{aligned} &\text{Min } cx \\ &\text{subject to} \\ &\qquad Ax \geq \mathbf{1} \\ &\qquad x_i \in \{0,1\}, \qquad i = 1, \ldots, n \end{aligned}$$

where $\mathbf{1}$ is an m-vector of all ones.

The linear programming relaxation is defined in the usual way, by relaxing the integrality restrictions and replacing them with simple upper and lower bounding constraints. The set covering polytope, that we denote by S^c, is the set given by the convex-hull of the incidence vectors of all covers of M, or equivalently,

$$S^c = \text{conv}(x : Ax \geq \mathbf{1}, x_j \in \{0,1\}, j \in N).$$

If we view the rows $r_1, \ldots, r_m$ of matrix A as incidence vectors of subsets $R_1, \ldots, R_m$ of N, and the costs c_j as weights of the elements of N, we get a different interpretation of the integer programming formulation given above. Namely, the set covering problem can be stated as the problem of finding a subset of N with minimum total weight which has non-empty intersection with every subset R_i. This variant is typically referred to as the *hitting set problem.* A family $\mathcal{E}$ of finite sets is said to be a *clutter* if no two sets $E_1, E_2 \in \mathcal{E}$ satisfy $E_1 \subset E_2$. Since a set having non-empty intersection with R_i also has non-empty intersection with any set containing R_i, we can safely remove from A a row r_j if it is *dominated* by r_i, i.e., $r_j \geq r_i$, componentwise. Hence, we can assume without loss of generality that the family $\mathcal{R} = \{R_1, \ldots, R_m\}$ is a clutter.

An *independence system* is a pair $(E, \mathcal{I})$, where E is a *finite ground set* and $\mathcal{I}$ is a family of subsets of E called *independent sets.* The family $\mathcal{I}$ is closed under inclusion, i.e., for any independent set $I \in \mathcal{I}$ every subset $I' \subseteq I$ is also independent ($I' \in \mathcal{I}$). A subset of E not belonging to $\mathcal{I}$ is called a *dependent set,* a minimal (with respect to set inclusion) dependent set is a *circuit.* It is apparent that a set is independent if and only if it contains no circuit, hence the independence system is characterized by the family of its circuits, which constitutes a clutter.

If every element of E is given a non-negative weight, the optimization problem associated with an independence system (*independence system problem*) asks for an independent set $I \in \mathcal{I}$ having maximum total weight. Equivalently, one can formulate the problem of finding the complement $C = E - I$ of an independent set, such that C has minimum total weight. Since I is independent if and only if $E - I$ has non-empty intersection with every circuit, the latter is a set covering formulation.

1 Optimization Methods

In the late 60's and early 70's specialized versions of implicit enumeration were used to solve set covering problems:

J.F. Pierce (1968). Application of combinatorial programming to a class of all zero-one integer programming problems. *Management Sci.* 15, 191–209.
J.F. Pierce and J. Lasky (1973). Improved combinatorial programming algorithms for a class of all zero–one integer programming problems. *Management Sci.* 19, 528–543.

R.S. Garfinkel and G. Nemhauser (1972). Optimal Set Covering: a Survey. A.M. Geoffrion, (ed.). *Perspectives on Optimization: A Collection of Expository Articles,* Addison-Wesley, Reading, MA, 164–183.

A survey of the early work in the optimal solution of this class of problems.

Several attempts were made to exploit the special structure of the set covering model to obtain effective algorithms based on branch and bound, implicit enumeration or dynamic programming. Interesting examples of this approach can be found in the following papers.

M. Bellmore, H.D. Ratliff (1971). Set covering and involutory bases. *Management Sci.* 18, 194–206.
C.E. Lemke, H.M. Salkin, K. Spielberg (1971). Set covering by single-branch enumeration with linear-programming subproblems. *Oper. Res.* 19, 998–1022.
F. Glover (1971). A note on extreme-point solutions and a paper by Lemke, Salkin, and Spielberg. *Oper. Res.* 19, 1023–1025.

N. Christofides, S. Korman (1975). A computational survey of methods for the set covering problem. *Management Sci.* 21, 591–599.

J. Etcheberry (1977). The set covering problem: a new implicit enumeration algorithm. *Oper. Res.* 25, 760–772.

An implicit enumeration algorithm for the set covering problems which exploits the structure of the problem to solve the linear programming relaxation more efficiently.

The effectiveness of set covering algorithms substantially improved in early 80's due to the seminal contributions of Egon Balas and his co-workers.

E. Balas (1980). Cutting planes from conditional bounds: a new approach to set covering. *Math. Program.* 12, 19–36.

E. Balas, A. Ho (1980). Set covering algorithms using cutting planes, heuristics, and subgradient optimization: a computational study. *Math. Program.* 12, 37–60.

In these papers theoretical foundations and algorithmic applications are discussed of a family of valid inequalities for the set covering polytope derived from a disjunctive programming approach. In particular, the algorithms that are described use several heuristics to find primal and dual feasible solutions, subgradient optimization to compute lower bounds and implicit enumeration to find optimal solutions.

1.1 Cutting planes

Other results on solution methods based on cutting planes are included in

M.L. Balinski and R.E. Quandt (1964). On an integer program for a delivery problem. *Oper. Res.* 12, 300-304.

H.M. Salkin, R.D. Koncal (1973). Set covering by an all integer algorithm: computational experience. *J. ACM* 20, 189–193.

Gomory cuts are used to solve some set covering problems to optimality.

1.2 Branch and bound

J.E. Beasley (1987). An algorithm for set covering problem. *European J. Oper. Res.* 31, 85–93.

J.E. Beasley, K. Jørnsten (1992). Enhancing an algorithm for set covering problems. *European J. Oper. Res.* 58, 293–300.

An algorithm for the optimal solution of set covering problems and subsequent enhancements which give a better computational performance are presented in these papers. Computational results on randomly generated large instances are also reported.

M.L. Fisher, P. Kedia (1990). Optimal solution of set covering/partitioning problems using dual heuristics. *Management Sci.* 36, 674–687.

This paper describes a branch and bound algorithm for the solution of mixed set covering/partitioning models (which include, as special cases, the set covering problems). The main feature of the algorithm is the use of continuous heuristics for the dual of the linear programming relaxation to provide lower bounds. The heuristics are adaptations of the greedy and 3-opt methods.

E. Balas, M. Carrera (1991, revised 1995). *A dynamic subgradient-based branch and bound procedure for set covering*, Management Sciences Research Report MSRR-568(R), Graduate School of Industrial Administration, Carnegie Mellon University, Pittsburgh. To appear in *Oper. Res.*

A branch and bound algorithm which solves Lagrangian relaxations at the nodes of the search tree is presented. It uses an upper/lower bounding procedure combining the standard subgradient method with primal and dual heuristics to dynamically modify the Lagrange multipliers, tighten the bounds and fix variables. An extensive computational testing is reported which shows very good performance of the method.

E. El-Darzi, G. Mitra (1992). Solution of set-covering and set-partitioning problems using assignment relaxations. *J. Oper. Res. Soc.* 43, 483–493.
E. El-Darzi, G. Mitra (1995). Graph theoretic relaxations of set covering and set partitioning problems. *European J. Oper. Res.* 87, 109–121.

In the first of these two papers, a tree search method which makes use of assignment relaxations is devised and computational experience on a set of test problems is reported. In the second paper, several relaxations of the set covering and partitioning problems to be used as bounding procedures in tree search methods are reviewed: network, matching, graph covering, assignment, shortest route and minimum spanning tree relaxations.

R.A. Rushmeier, G.L. Nemhauser (1993). Experiments with parallel branch-and-bound algorithms for the set covering problem. *Oper. Res. Lett.* 13, 277–286.

Eight branch-and-bound algorithms are tested on randomly generated set covering problems. The algorithms are based on a parallel multi-task model for integer optimization and use two types of tasks, linear programming and subgradient optimization with Lagrangian relaxation.

F. Harche, G.L. Thompson (1994). The column subtraction algorithm: an exact method for solving weighted set covering, packing and partitioning problems. *Computers Oper. Res.* 21, 689–705.

The method described in this paper optimally solves set covering, packing and partitioning problems, working as follows: first the linear programming relaxation is solved using the *pivot and probe* algorithm by Sethi and Thompson (*Math. Program.* 29, 219–223); then a branch and bound method, which proceeds by subtracting non-basic columns of the final simplex tableau from the right hand side column, is used to produce optimal integer solutions. An efficient greedy heuristic is also described.

C. Mannino, A. Sassano (1995). Solving hard set covering problems. *Oper. Res. Lett.* 18, 1–5.

A method is described for solving difficult minimum cardinality set covering problems associated with Steiner triple systems, which were proposed by Fulkerson, Nemhauser and Trotter in 1974. Reference to such problems can be found in §4.

1.3 Branch and cut

P. Nobili, A. Sassano (1992). A separation routine for the set covering polytope. *Proc. 2nd IPCO Conf.*, 201–219.

The *projection* operation is described as a tool for generating valid cuts and a practical projection separator is devised. Some computational results obtained with an experimental branch and cut procedure incorporating such a separator are reported.

E. Balas, S. Ceria, G. Cornuéjols (1995). Mixed 0-1 programming by lift-and-project in a branch-and-cut framework. (to appear in *Management Sci.*).

Very good results on difficult test problems are also obtained by the "Lift-and-Project" procedure, which is a branch and cut algorithm using general cutting planes.

1.4 Polynomially solvable cases

It is possible to define several classes of set covering problems which are solvable in polynomial time by imposing that the constraint matrices have special structure.

In this survey we will not be exhaustive on such cases, since many special 0,1 matrices will be described in other chapters of this book. In particular, we will not consider set covering problems arising from balanced matrices or from matrices whose rows are the incidence vectors of matroid circuits.

However, the special case of *regular set covering problems* is worth mentioning, since it has been the subject of extensive research and has important applications. A set covering problem is regular if there exists an ordering of the variables such that for every feasible solution x with $x_i = 1$, $x_j = 0$ and $i < j$, the vector $x + u_j - u_i$ is also a feasible solution, where u_i is the i-th unit vector. An example of regular set covering problem is defined by a matrix whose rows are the incidence vectors of the minimal covers of a knapsack.

Regular set covering problems were first studied in

P.L. Hammer, U.N. Peled, M.A. Pollatschek (1979). An algorithm to dualize a regular switching function. *IEEE Trans. Comput.* C-28, 238–243,

where a polynomial time algorithm for such problems was proposed. Subsequently, a more effective algorithm ($O(mn^3)$) was proposed in

U.N. Peled, B. Simeone (1985). Polynomial time algorithms for regular set-covering and threshold synthesis. *Discr. Appl. Math.* 12, 57–69.

Independently, the following two papers improved the complexity bound to $O(mn)$.

U.N. Peled, B. Simeone (1986). *Computing the dual of a regular boolean function in* $O(mn)$, Research Report in Computer Science No. 21, Department of Mathematics, Computer Science and Statistics, Univ. of Illinois, Chicago.

P. Bertolazzi, A. Sassano (1987). An $O(mn)$ algorithm for regular set-covering problems. *Theor. Comput. Sci.* 54, 237–247.

P. Bertolazzi, A. Sassano (1988). A class of polynomially solvable set-covering problems. *SIAM J. Discr. Math.* 1, 306–316.

A class of clutters is described for which the corresponding set covering problems are solvable in polynomial time and which provides a common generalization of regular set covering problems and problems with constraint matrices whose rows are the incidence vectors of matroid bases.

2 Heuristics

Heuristic methods for finding good feasible solutions for set covering problems have played a crucial role in the area. Many of the older techniques are of greedy type, where a solution is built by selecting one column at a time according to a specific merit function. More recently, very effective heuristic procedures based on Lagrangian relaxation, local search and other techniques have been proposed.

2.1 Greedy heuristics

V. Chvátal (1979). A greedy heuristic for the set-covering problem. *Math. Oper. Res.* 4, 233–235.

In this paper a greedy heuristic for the weighted set covering problem is presented and a bound is established on the ratio between the optimum and the solution obtained by the heuristic. Chvátal bound reduces, in the unweighted case, to the bound computed in the following papers:

D.S. Johnson (1974). Approximation algorithms for combinatorial problems. *J. Comput. System Sci.* 9, 256–278.

L. Lovász (1975). On the ratio of optimal integral and fractional covers. *Discr. Math.* 13, 383–390.

M.L. Fisher, L.A. Wolsey (1982). On the greedy heuristic for continuous covering and packing problems. *SIAM J. Alg. Disc. Meth.* 3, 584–591.

The results of Chvátal, Johnson and Lovász are generalized to the continuous case (i.e., the variables are not required to be 0 or 1).

It is also worth mentioning that [Balas and Ho 1980] (see §1) propose several merit functions for column selection in the greedy procedure.

F.J. Vasko, G.R. Wilson (1984). An efficient heuristic for large set covering problems. *Naval Res. Log. Quart.* 31, 163–171.

F.J. Vasko, F.E. Wolf (1988). Solving large set-covering problems on a personal computer. *Computers Oper. Res.* 15, 115–121.

Heuristic solution procedures for the weighted and unweighted set covering problems as well as statistical evaluation of the quality of the solutions are presented in the above papers. The procedures are essentially greedy algorithms augmented by a 1-opt local search.

T.A. Feo, M.G.C. Resende (1989). A probabilistic heuristic for a computationally difficult set covering problem. *Oper. Res. Lett.* 8, 67–71.

A non-deterministic variation of the Chvátal greedy heuristic is presented which is based on random selection steps.

O. Goldschmidt, D.S. Hochbaum, G. Yu (1993). A modified greedy heuristic for the set covering problem with improved worst case bound. *Inform. Process. Lett.* 48, 305–310.

An improved version of the greedy algorithm is described, whose worst case approximation ratio is bounded by $\sum_{i=1}^{d} 1/i - 1/6$, where d is the maximum number of sets covering an element.

2.2 Lagrangian heuristics

J.E. Beasley (1990). A Lagrangian heuristic for set-covering problems. *Naval Res. Log. Quart.* 37, 151–164.

A heuristic procedure based on Lagrangian relaxation is presented and computational results on randomly generated instances are reported.

L.A.N. Lorena, F. Belo Lopes (1994). A surrogate heuristic for set covering problems. *European J. Oper. Res.* 79, 138–150.

A dual heuristic for set covering which compares favorably with standard Lagrangian-based heuristics is presented and tested on several large-scale randomly-generated problems.

A. Caprara, M. Fischetti, P. Toth (1996). A heuristic algorithm for the set covering problem. *Proc. 5th IPCO Conf., Lecture Notes Comp. Sci.* 1084, 72–84.

The above paper describes the prize-winner algorithm in a competition issued by the Italian Railways for solving large-scale set covering problems arising from crew-scheduling. The algorithm is based on a Lagrangian heuristic complemented by an effective column-generation scheme.

S. Ceria, P. Nobili, A. Sassano (1995). *A Lagrangian-based heuristic for large-scale set covering problems*, Technical report R.406, Istituto di Analisi dei Sistemi ed Informatica, CNR, Roma. To appear in *Math. Progr.*

A Lagrangian-based heuristic which finds both a primal and a dual solution of the linear relaxation of a set covering problem is presented. The primal solution is also used to guide the search for a good quality integral solution. Computational results on large-scale set covering problems are reported.

2.3 Local search

Local (neighbourhood) search heuristics iteratively improve a starting feasible solution by a sequence of *moves*. Each move consists of replacing the current solution x by another one chosen in a suitable subset called the *neighborhood* of x. To this class of methods belong exchange heuristics, simulated annealing, genetic algorithms and tabu search techniques. These methods have been very successful in finding good feasible solutions for many combinatorial problems. For the set covering problem heuristics of this kind have been proposed only in the last few years.

J.E. Beasley, P.C. Chu (1995). *A genetic algorithm for the set covering problem.* Working paper, The Management School, Imperial College, London.

A heuristic based on genetic algorithms is described. Its main features are a new crossover-fusion operator and the idea of using variable mutation rate. The performance of the heuristic is evaluated on a large set of randomly generated problems.

L.W. Jacobs, M.J. Brusco (1995). Note: a local-search heuristic for large set covering problems. *Naval Res. Log. Quart.* 42, 1129–1140.

A heuristic based on the simulated annealing algorithm is described. Computational experiments show that the quality of the solutions obtained by such heuristic are competitive with other approaches.

K. Nonobe, T. Ibaraki (1996). *Tabu search approach to CSP (constraint satisfaction problem) as a general problem solver*, Technical report #96004, Department of Applied Mathematics and Physics, Graduate School of Engineering, Kyoto University, Kyoto.

This paper describes a general tabu search heuristic whose application to hard set covering problems is particularly effective.

2.4 Other techniques

D.S. Hochbaum (1982). Approximation algorithms for the set covering and vertex cover problems. *SIAM J. Comput.* 11, 555–556.

This paper describes a polynomial time ($O(n^3)$) heuristic, based on the linear programming relaxation, whose worst case approximation ratio is bounded by the maximum number of sets covering an element.

C. Vercellis (1984). A probabilistic analysis of the set covering problem. F. Archetti, F. Maffioli (eds.). *Stochastics and Optimization*, Ann. Oper. Res. 1, J.C. Baltzer AG, Scientific Publishing Company, Basel, 255–271.

The performances of two simple randomized algorithms are analyzed.

N. Karmarkar, M.G.C. Resende, K.G. Ramakrishnan (1991). An interior point algorithm to solve computationally difficult set covering problems. *Math. Program.* 52, 597–618.

This paper presents an interior point heuristic to find good feasible solutions to set covering problems. The procedure generates a sequence of feasible points, the direction used to determine the new iterate is computed by solving a nonconvex quadratic program on an ellipsoid. The approach is illustrated by application to difficult set covering instances associated with Steiner triple systems.

M. Afif, M. Hifi, V.T. Paschos, V. Zissimopoulos (1995). A new efficient heuristic for the minimum set covering problem. *J. Oper. Res. Soc.* 46, 1260–1268.

This paper describes a polynomial time heuristic algorithm for the unweighted set covering problem based on a transformation into a max-flow problem.

D. Wedelin (1995). An algorithm for large scale 0-1 integer programming with application to airline crew scheduling. R.E. Burkard, T. Ibaraki, M. Queyranne (eds.). *Mathematics of Industrial Systems I*, Ann. Oper. Res. 57, Baltzer Science Publishers, Amsterdam, 283–301.

An efficient heuristic with applications to crew-scheduling is discussed.

3 Polyhedral Results

This section concerns the structure of the polytope S^c. The polyhedral structure of a related problem, finding a maximum weight independent set in an *independence system* has been thoroughly investigated in the literature. The latter polytope is equivalent to the covering polytope, since there is an affine transformation which maps one into another.

Due to this equivalence, in this survey we have collected results relative to both problems.

3.1 The structure of the set covering polytope

D.R. Fulkerson (1971). Blocking and anti-blocking pairs of polyhedra. *Math. Program.* 1, 168–194.

This paper lays the foundation of the study of the polyhedral structure of the set covering problem. In particular, it points out the crucial role of min-max relationships in combinatorial optimization and their connection with the algebraic and geometric properties of blocking pairs of clutters.

G.L. Nemhauser, L.E. Trotter (1974). Properties of vertex packing and independence system polyhedra. *Math. Program.* 6, 48–61.

In this paper a lifting theorem and a class of rank facet-defining inequalities (called elsewhere *generalized cliques*) are given for the independence system problem.

In the late 80's, the remarkable success of the polyhedral approach in the solution of many combinatorial problems rejuvenated the interest in the study of the polyhedral structure of the set covering (independence system) polytope.

R. Euler, M. Jünger, G. Reinelt (1987). Generalization of cliques, odd cycles and anticycles and their relation to independence system polyhedra. *Math. Oper. Res.* 12, 451–462.

The authors study many different facet-defining inequalities for the independence system polytope.

E. Balas, S.M. Ng (1989). On the set covering polytope: I. All the facets with coefficients in $\{0, 1, 2\}$. *Math. Program.* 43, 57–69.

E. Balas, S.M. Ng (1989). On the set covering polytope: II. Lifting the facets with coefficients in $\{0, 1, 2\}$. *Math. Program.* 45, 1–20.

In these papers a characterization of those facets of the set covering polytope which are defined by inequalities with coefficients equal to 0, 1 or 2 is given. It is shown that such facets are obtained as liftings of facet-defining inequalities for lower dimensional polytopes that have only three nonzero coefficients. The lifting coefficients are given by closed form expressions.

A. Sassano (1989). On the facial structure of the set covering polytope. *Math. Program.* 44, 181–202.

General properties of the set covering polytope and several classes of rank facet-defining inequalities are discussed. Moreover, two lifting procedures for valid inequalities are presented.

G. Cornuéjols, A. Sassano (1989). On the 0, 1 facets of the set covering polytope. *Math. Program.* 43, 45–55.

This paper describes necessary and sufficient conditions for rank inequalities to define facets of the set covering polytope. Moreover, a polynomial characterization of a class of rank facets is also given.

P. Nobili, A. Sassano (1989). Facets and lifting procedures for the set covering polytope. *Math. Program.* 45, 111–137.

Some new classes of facet-defining inequality and procedures for lifting inequalities are described and the strong connections between several combinatorial problems and the set covering problem are pointed out.

M. Laurent (1989). A generalization of antiwebs to independence systems and their canonical facets. *Math. Program.* 45, 97–108.

A class of facet-defining inequalities for the independence system polytope is described (*antiwebs*) which generalize inequalities for the vertex packing polytope. Moreover, a sufficient condition for a rank inequality to define a facet is given.

R. Euler, A.R. Mahjoub (1991). On a composition of independence systems by circuit identification. *J. Combin. Theory (B)* 53, 235–259.

This paper describes a composition operation for general independence systems which generalizes operations for bipartite subgraph and acyclic subdigraph independence systems. As a consequence it is shown that the K_3-cover problem is polynomially solvable in graphs noncontractible to $K_5 - e$.

P. Nobili, A. Sassano (1993). The anti-join composition and polyhedra. *Discr. Math.* 119, 141–166.

A binary composition operation for clutters and the associated polyhedra is described. As a consequence a characterization of the facet-defining inequalities for the cycle and cocycle polyhedra associated with graphs noncontractible to the four-wheel W_4 is given.

P. Nobili, A. Sassano (1993). Polyhedral properties of clutter amalgam, *SIAM J. Discr.*

Math. 6, 139–151.
Another composition operation for clutters and polyhedra is described.

3.2 Ideality

Let $\mathcal{F}$ be a clutter, $b(\mathcal{F})$ its blocker and let A $(m \times n)$ and B $(p \times n)$ be the matrices whose rows are the incidence vectors of the members of $\mathcal{F}$ and $b(\mathcal{F})$, respectively. We say that $\mathcal{F}$ is *Ideal* (has the *Max-Flow-Min-Cut property*) if and only if

$$\min\{wx : Ax \geq \mathbf{1}\} = \min\{wx : Ax \geq \mathbf{1}, \quad x \in \{0,1\}^n\} \tag{1}$$

for each vector $w \in \Re^n$. In a remarkable paper,

A. Lehman (1979). On the width-length inequality. *Math. Program.* 17, 403–417,

Alfred Lehman proved that a matrix is ideal if and only if

$$\sum_{e \in E} l_e w_e \geq 1 \tag{2}$$

where $l \in \Re^{|E|}$ and $w \in \Re^{|E|}$ are solutions of the systems $Ax \geq \mathbf{1}$ and $Bx \geq \mathbf{1}$, respectively. In the same paper, Lehman described strong properties of minimally non-ideal matrices.

In 1990 Seymour and Padberg, independently, gave alternative proofs of Lehman's result:
P.D. Seymour (1990). On Lehman's width-length characterization. W. Cook, P.D. Seymour (eds.). *Polyhedral Combinatorics*, DIMACS Series in Discr. Math. and Th. Comp. Sc. 1, AMS, New York, 107–117.
M.W. Padberg (1993). Lehman's forbidden minor characterization of ideal 0–1 matrices. *Discr. Math.* 111, 409–420.

G. Cornuéjols, B. Novick (1994). Ideal 0, 1 matrices. *J. Combin. Theory (B)* 60, 145–157.
Many classes of minimally non-ideal matrices are introduced and a conjecture is formulated on the structure of minimally non-ideal matrices. In particular, it is shown that the conjecture holds for the class of *circulant* matrices.

M. Conforti, G. Cornuéjols (1995). A class of logic problems solvable by linear programming. *J. ACM* 42, 1107–1113.
The concept of ideality is generalized to the case in which the matrix A has $0, \pm 1$ entries.

The properties of ideal $0, \pm 1$ matrices have been studied in the following papers:
J. Hooker (1992). *Resolution and the integrality of satisfiability polytopes*, Working paper, Graduate School of Industrial Administration, Carnegie Mellon University, Pittsburgh.
B. Guenin (1994). *Perfect and ideal* $0, \pm 1$ *matrices*, Working paper, Graduate School of Industrial Administration , Carnegie Mellon University, Pittsburgh.

P. Nobili, A. Sassano (1995). (0, ± 1) ideal matrices. *Proc. 4th IPCO Conf.*, *Lecture Notes Comp. Sci.* 920, 344–359. To appear in *Math. Program.*

A complete review of ideal matrices can be found in Chapter 11.

4 Test Problems

D.R. Fulkerson, G.L. Nemhauser, L.E. Trotter (1974). Two computationally difficult set covering problems that arise in computing the 1-width of incidence matrices of Steiner triple systems. *Math. Program. Study* 2, 72–81.

Two minimum cardinality set covering problems are proposed which are very difficult to solve. The smallest one has 117 constraints and 27 variables, the larger one has 330 constraints and 45 variables. The constraint matrices of such problems are incidence matrices of Steiner triple systems.

J.E. Beasley (1990). OR-Library: distributing test problems by electronic mail. *J. Oper. Res. Soc.* 41, 1069–1072.
E. El-Darzi, G. Mitra (1990). Set covering and set partitioning: a collection of test problems. *Omega* 18, 195–201.

The authors describe two libraries of test problems and give details on how to get them.

5 Applications

In this section we list some real life problems solved by means of the set covering model.

C. Torregas, R. Swain, C. ReVelle, L. Bergman (1971). The location of emergency service facilities. *Oper. Res.* 19, 1363–1373.

The problem of locating emergency facilities is formulated as a set covering problem. A simple solution procedure based on linear programming is described.

M.S. Daskin, E.D. Stern (1981). A hierarchical objective set covering model for emergency medical service vehicle deployment. *Transportation Sci.* 15, 137–152.

This paper describes the application of the set covering model to the problem of locating ambulances and reports computational experience on data for the city of Austin, Texas.

F.J. Vasko, F.E. Wolf, K.L. Stott (1987). Optimal selection of ingot sizes via set covering. *Oper. Res.* 35, 346–353.
F.J. Vasko, F.E. Wolf, K.L. Stott, Jr. (1989). A set covering approach to metallurgical grade assignment. *European J. Oper. Res.* 38, 27–34.

The above two papers discuss successful applications of the set covering model to the solution of industrial problems at Bethlehem Steel Co. The first one deals with the selection of ingot sizes. It describe a two phase approach, in which several feasible ingot sizes are first generated (phase 1) and then a few of them are selected by solving a set covering problem (phase 2).

The second paper is a case study on the application of a heuristic algorithm (OPT-SOL) based on the set covering model to solve large instances of the metallurgical grade assignment problem.

J. Current, M. O'Kelly (1992). Locating emergency warning sirens. *Decision Sci.* 23, 221–234.

This paper describes an application of the set covering model to the design of a warning siren system for a city of 16,000 people encompassing approximately 17 square miles.

S. Nicoloso, P. Nobili (1992). A set covering formulation of the matrix equipartition problem. *Proc. 15th IFIP Conf. on Syst. Modelling Optim.*, *Lecture Notes Control Inform. Sci.* 180, 189–198.

Here a set covering model is used to solve a matrix equipartition problem arising in the design of VLSI devices and flexible manufacturing systems.

K.L. Hoffman, M.W. Padberg (1993). Solving airline crew scheduling problems by branch-and-cut. *Management Sci.* 39, 657–682.

Large set partitioning problems arising in the airline industry are solved to optimality by means of a branch-and-cut procedure which uses both cuts derived from set packing and set covering polytopes.

A. Paias, J. Paixão (1993). State space relaxation for set covering problems related to bus driver scheduling. *European J. Oper. Res.* 71, 303–316.

It describes a relaxation technique for a dynamic programming formulation of set covering problems. Such technique provides lower bounds and heuristic feasible solutions.

24 Combinatorial Topics in VLSI Design

Rolf H. Möhring
Technische Universität Berlin

Dorothea Wagner
Universität Konstanz

CONTENTS

The design of integrated circuits is one of the broadest and most varied areas of application for combinatorial optimization methods and discrete algorithms. Many problems coming up during the layout process of chips or boards admit, for example, abstract formulations as graph problems. Over a period of more than one decade, mathematicians and theoretical computer scientists have intensively studied graph-theoretic and combinatorial aspects of circuit layout. In many cases, their investigations have led to exact and elegant models which help to understand the highly complex problems coming up in practice. Moreover, theoretical results have revealed that many of these problems may be attacked using classical optimization methods. This article concentrates on some of the most prominent and attractive combinatorial problems in VLSI-design.

The first section about the *general layout problem* deals with an overall approach to find a layout for an integrated circuit or network. The physical components of

Annotated Bibliographies in Combinatorial Optimization, edited by M. Dell'Amico, F. Maffioli and S. Martello

the circuit or network are considered as given together with their interconnection structure, the so-called *net lists*. Then the general layout problem consists in placing the components and routing the nets between these components simultaneously. In its most general form, it can be modeled as a hypergraph or graph embedding problem. The main optimization goal is to minimize the layout area.

The starting point for the routing phase is a predetermined placement of the physical components on the chip. The *routing problem* consists in specifying the course of the wires connecting these under a prescribed routing model. From the mathematical point of view, the *knock-knee model* which corresponds to *edge-disjoint* Steiner tree problems in planar graphs is maybe the most attractive one. Under this model the *layer assignment* is a non-trivial task and leads to a very interesting combinatorial problem. For the more practical *Manhattan model* not that many theoretical results are known. On one hand, the layout problem under the Manhattan model is computationally hard. On the other hand, the layer assignment problem is trivial, since every Manhattan layout can be easily wired in two layers. An attractive combinatorial problem in connection to Manhattan routing is the *via minimization problem.* The most restrictive routing model is the *single-layer model*, where only one layer is available for the layout.

Section 4 deals with special layout technologies that usually occur in *logic synthesis*, i.e. in the construction of the physical components that are given for the layout and routing phases. This so-called *linear layout methods* reveal some rich and surprising connections to independent and currently very active areas of graph theory such as graph searching, Robertson-Seymour theory, embedding graphs into interval graphs, graph separation, and matching with side constraints.

1 Books and Surveys

1.1 Books

C. Mead, L. Conway (1979). *Introduction to VLSI Systems*, Wesley, Reading, MA.

The book of Mead and Conway may be viewed as the first book that presents essential features of the integrated circuit technology in a language appropriate for the computing theoretician.

J.D. Ullman (1984). *Computational Aspects of VLSI*, Computer Science Press, Rockville, Maryland.

Ullman covers many computational issues and algorithms underlying VLSI-design tools.

T. Lengauer (1990). *Combinatorial Algorithms for Integrated Circuit Layout*, Wiley-Teubner, Chichester.

Lengauer presents a unified account of the combinatorial knowledge on circuit layout obtained by practitioners in computer-aided circuit design as well as by computer scientists and mathematicians.

F.P. Preparata (ed.) (1984). *VLSI Theory*, vol. 2 of Advances in Computer Research, JAI Press Inc., Greenwich, London.

This book is a collection of fundamental research papers on algorithmic and combinatorial problems in VLSI-design.

B. Korte, L. Lovász, H.J. Prömel, A. Schrijver (1990). *Paths, Flows and VLSI-Layout*, Springer-Verlag, Berlin.

A collection of surveys is put together covering a wide range of topics that are of more or less interest in the context of VLSI-design. Some of the most attractive combinatorial and graph-theoretic problems underlying VLSI-layout problems, such as Steiner tree problems, homotopic routing, path and flow problems, or separator and cut problems are discussed.

1.2 Surveys

R.H. Möhring, D. Wagner, F. Wagner (1995). VLSI network design: a survey. M. Ball, T. Magnanti, C. Monma, G. Nemhauser (eds.). *Handbooks in Operations Research/Management Science, Volume on Networks*, North-Holland, Amsterdam, 625–712.

This article gives a survey on most of the topics mentioned in this annotated bibliography. It particularly describes new developments in routing and linear layout technologies in more detail.

H. Ripphausen-Lipa, D. Wagner, K. Weihe (1995). Survey on efficient algorithms for disjoint paths problems in planar graphs. W. Cook, L. Lovász, P. Seymour (eds.). *DIMACS-Series in Discrete Mathematics and Theoretical Computer Science, Volume 20 on the "Year of Combinatorial Optimization"*, AMS, 295–354.

Algorithms for various disjoint paths problems in planar graphs are reviewed, in particular for problems motivated by VLSI-design.

A. Martin, R. Weismantel (1993). Packing paths and Steiner trees: routing of electronic circuits. N. Temme, W. Hundsdorfer, B. Gerards, A. Siebes, R. Veltkamp (eds.). *CWI Quarterly, Volume 6, Number 3*, CWI, 185–204.

Steiner tree packing problems motivated from VLSI-design are discussed from a polyhedral point of view, and a short overview on the underlying technologies and design rules is given.

A.S. LaPaugh, R.Y. Pinter (1990). Channel routing for integrated circuits. *Annual Reviews Comp. Science* 4, 307–363.

A detailed overview of the major research directions in channel routing is given. Different relevant layout models are considered.

G. Di Battista, P. Eades, R. Tamassia, I.G. Tollis (1994). Algorithms for drawing graphs: an annotated bibliography. *Computational Geometry* 4, 235–282.

There are many algorithmic results on graph drawing under different embedding constraints and optimization criteria. This paper contains references on orthogonal graph embedding, especially for planar graphs. Although it is not addressed to readers interested in VLSI design, it covers problems that are closely related to layout problems arising during the design of integrated circuits.

2 The General Layout Problem

In its most general form, the layout of a circuit can be modeled as a *hypergraph* respectively *graph embedding* problem. A hypergraph or graph is embedded into a rectangular grid such that vertices are mapped on grid points and edges are realized by paths that are in some sense disjoint.

2.1 Layout of graphs

M.R. Kramer, J. van Leeuwen (1984). The complexity of wire-routing and finding minimum area layouts for arbitrary VLSI-circuits. F.P. Preparata (ed.). *Advances in Computer Research, vol. 2: VLSI Theory*, JAI Press Inc., Greenwich, London, 129–146.
M. Formann, F. Wagner (1990). The VLSI layout problem on various embedding models. R. H. Möhring (ed.). *16th Int. Workshop Graph-Theoretic Concepts in Comput. Sci., WG'90*, Lecture Notes in Computer Science 484. Springer-Verlag, Berlin, 130–139.

Minimizing the *layout area* is $\mathcal{NP}$-hard in grid graphs. It remains $\mathcal{NP}$-hard if the graph to be laid out is planar.

L.G. Valiant (1981). Universality considerations in VLSI circuits. *IEEE Trans. Comput.* C-30, 135–140.

$\Omega(n^2)$ is a lower bound for the layout area of a graph with n vertices.

2.2 Orthogonal graph embedding

In the context of *graph drawing* the problem of embedding a graph in a rectangular grid is also referred to as orthogonal graph embedding.

G. Kant (1993). *Algorithms for drawing planar graphs*, PhD Thesis, University of Uetrecht.
G. Kant (1992). Drawing planar graphs using the *lmc*-odering. *33th Annual IEEE Symp. on Found. of Comput. Sci., FOCS'92*, 101–110.
T. Biedl, G. Kant (1994). A better heuristic for orthogonal graph drawing. J. van Leeuwen (ed.). *2nd Annual European Symp. Algorithms, ESA'94*, Lecture Notes in Computer Science 855. Springer-Verlag, Berlin, 24–33.
G. Di Battista, A. Garg, G. Liotta, R. Tamassia, E. Tassinari, F. Vargiu (1995). An experimental comparison of three graph drawing algorithms. *11th Annual ACM Symp. Computational Geometry, SCG'95*, 306–315.

Kant covers many interesting algorithmic aspects of drawing graphs in grids, especially of drawing planar graphs. There is a simple algorithm with running time $O(n)$ to embed a graph in an $n \times n$ grid with at most $2n$ bends. In the last paper, several orthogonal graph drawing algorithms are compared on a large set of test data by an extensive experimental study. However, the properties of the drawings measured in this study are also "aesthetic" properties being of no interest in VLSI applications.

2.3 Divide-and-conquer approach

L.G. Valiant (1981). Universality considerations in VLSI circuits. *IEEE Trans. Comput.* C-30, 135–140.
C.E. Leiserson (1983). *Area-Efficient VLSI-Computation*, MIT Press, Cambridge, MA.
A general divide-and-conquer layout algorithm based on *minimum graph bisection*, i.e. partitioning the graph into two equal-sized parts by a removal of a minimum number of edges, is suggested independently by Valiant and Leiserson.

F.T. Leighton (1982). A layout strategy for VLSI which is provably good. *14th Annual ACM Symp. Theory of Comput., STOC'82*, 85–98.
S.N. Bhatt, F.T. Leighton (1984). A framework for solving VLSI graph layout problems. *J. Comput. Syst. Sci.* 28, 300–343.
A divide-and-conquer approach based on partition trees, so-called *bifurcators* is introduced. Guarantees for the layout area required for graphs with bifurcators of restricted size are proved.

2.4 Cuts and separators

M.J. Garey, D.S. Johnson, L. Stockmeyer (1976). Some simplified $\mathcal{NP}$-complete graph problems. *Theor. Comp. Sci.* 1, 237–267.
The minimum bisection problem is $\mathcal{NP}$-hard.

N. Garg, V.V. Vazirani, M. Yannakakis (1993). Approximate max-flow min-(multi)cut theorems and their applications. *25th Annual ACM Symp. Theory of Comput., STOC'93*, 698–707.
P. Klein, S. Plotkin, S. Rao (1993). Planar graphs, multicommodity flow, and network decomposition. *25th Annual ACM Symp. Theory of Comput., STOC'93*, 682–690.
P. Klein, S. Plotkin, C. Stein, E. Tardos (1994). Faster approximation algorithms for the unit capacity concurrent flow problem with applications to routing and finding sparse cuts. *SIAM J. Comput.* 23, 466–487.
One possibility to attack the bisection problem is the use of *uniform multicommodity flow techniques*. Approximation algorithms for a relaxation of the minimum bisection problem are developed that fulfill polylogarithmic quality guarantees.

D. Wagner, F. Wagner (1993). Between min cut and graph bisection. A.M. Borzyszkowski, S. Sokolowski (eds.). *18th Annual Symp. Math. Found. of Comput. Sci., MFCS'93*, Lecture Notes in Computer Science 711. Springer-Verlag, Berlin, 744–750.
Another approach to overcome this problem is to relax the condition on the size of the subgraphs the graph is partitioned in.

H. Nagamochi, T. Ibaraki (1992). Computing edge-connectivity in multigraphs and capacitated graphs. *SIAM J. Discr. Math.* 5, 54–66.
M. Stoer, F. Wagner (1994). A simple min cut algorithm. J. van Leeuwen (ed.). *2nd Annual European Symp. Algorithms, ESA'94*, Lecture Notes in Computer Science 855. Springer-Verlag, Berlin, 141–147.

Neglecting the balance condition leads to the classical *minimum cut* problem that can be solved using flow algorithms. Nagamoshi and Ibaraki present the first algorithm not based on flow techniques with an improved running time of $O(nm+n^2 \log n)$. This algorithm is simplified substantially by Stoer and Wagner.

R.J. Lipton, R.E. Tarjan (1979). A separator theorem for planar graphs. *SIAM J. Appl. Math.* 36, 177–189.
S.B. Rao (1987). Finding near optimal separators in planar graphs. *28th Annual IEEE Symp. on Found. of Comput. Sci., FOCS'87*, 225–237.
S.B. Rao (1992). Faster algorithms for finding small edge cuts in planar graphs. *24th Annual ACM Symp. Theory of Comput., STOC'92*, 229–240.

The *separator theorem for planar graphs* induces a layout of area $O(n\log^2 n)$ for planar graphs of maximum degree four. On the other hand, there are planar graphs that require $\Omega(n\log^2 n)$ layout area [Ullman 1984] (see §1.1). The approach presented by Rao leads far beyond the capabilities of the planar separator theorem as it gives quite good optimality guarantees.

B.W. Kernighan, S. Lin (1970). An efficient heuristic procedure for partitioning graphs. *The Bell Syst. Tech. J.* 49, 291–307.
C.M. Fiduccia, R.M. Mattheyses (1982). A linear time heuristic for improving network partitions. *19th Annual Design Automation Conference, DAC'82*, 175–181.
D.S. Johnson, C. Aragon, L.A. McGeoch, C. Schevon (1989). Optimization by simulated annealing: an experimental evatuation; Part I, Graph Partitioning. *Oper. Res.* 37, 865–892.

The by far most often used and very flexible partitioning algorithm for general graphs is the *heuristic* of Kernighan and Lin. It runs through a small number of "rounds", where each round costs $O(n^3)$ elementary operations. In the second paper, this number is brought down to a linear number of operations. The third paper compares this heuristic with the graph partitioning version of *simulated annealing*.

2.5 Layout of hypergraphs

In general, the layout problem consists in embedding a hypergraph in a grid. An elegant and general way to apply graph partitioning algorithms to hypergraphs would be to model hypergraphs by graphs with the same cut properties.

E. Ihler, D. Wagner, F. Wagner (1993). Modeling hypergraphs by graphs with the same mincut properties. *Inform. Process. Lett.* 45, 171–175.

For an edge-weighted hypergraph there is no edge-weighted graph such that the weight of the edges cut by any bipartition of the graph is the same as the weight of the hyperedges cut by the same bipartition in the hypergraph.

R. Klimmek, F. Wagner (1996). *A Simple Hypergraph Min Cut Algorithm*, Preprint, Freie Universität Berlin.

The minimum cut algorithm of Stoer and Wagner for graphs can be generalized to determine a minimum cut of a hypergraph in time $O(nm + n^2 \log n)$, where m is here the sum of the cardinalities of the hyperedges.

3 Routing

The routing problem consists in realizing k nets by *Steiner trees* (in case of multiterminal nets) respectively *paths* (for two-terminal nets) connecting the terminals of the nets. The underlying routing graph is mostly planar or even a grid graph. Special cases of routing graphs are *switchboxes*, *rectangles* and *channels*.

3.1 Edge-disjoint Steiner trees in planar graphs

The edge-disjoint routing problem is $\mathcal{NP}$-complete in planar graphs, even if the nets are all two-terminal nets [Kramer and van Leeuwen 1984] (see §2.1).

N. Robertson, P. Seymour (1990). Graph minors VIII, a Kuratowski theorem for general surfaces. *J. Combin. Theory* B 48, 255–288.
N. Robertson, P. Seymour (1990). An outline of a disjoint paths algorithm. B. Korte, L. Lovász, H. J. Prömel, A. Schrijver (eds.). *Paths, Flows and VLSI-Layout*, Springer-Verlag, Berlin, 267–292.

If the nets are all two-terminal nets and k is *fixed*, respectively the total *number of terminals* is *fixed*, the problem is polynomially solvable.

P. Seymour (1981). On odd cuts and plane multicommodity flows. *Proc. London Math. Society* 42, 178–192.
K. Matsumoto, T. Nishizeki, N. Saito (1986). Planar multicommodity, maximum matchings and negative cycles. *SIAM J. Comput.* 15, 495–510.

The two-terminal case is also polynomially solvable if the graph together with the *demand* edges induced by the nets is *planar* and *even*, i.e. all vertices have even degree.

M. Middendorf, F. Pfeiffer (1993). On the complexity of the disjoint path problem. *Combinatorica* 13, 97–107.

The latter version becomes $\mathcal{NP}$-complete if either planarity or the evenness condition is dropped.

H. Okamura, P. Seymour (1981). Multicommodity flows in planar graphs. *J. Combin. Theory* B 31, 75–81.
K. Matsumoto, T. Nishizeki, N. Saito (1985). An efficient algorithm for finding multicommodity flows in planar networks. *SIAM J. Comput.* 14, 289–302.
M. Becker, K. Mehlhorn (1986). Algorithms for routing in planar graphs. *Acta Inform.* 23, 163–176.
M. Kaufmann, G. Klär (1991). A faster algorithm for edge-disjoint paths in planar graphs. W.L. Hsu, R.C.T. Lee (eds.). *2nd Annual Int. Symp. Algorithms, ISA'91*, Lecture Notes in Computer Science 557. Springer-Verlag, Berlin, 336–348.

Based on the *theorem of Okamura and Seymour*, algorithms of time complexity $O(n^2)$ solving the *even one-face paths problem* are presented. In the last paper the time complexity is improved to $O(n^{\frac{5}{3}}(\log\log n)^{\frac{1}{3}})$.

D. Wagner, K. Weihe (1995). A linear time algorithm for edge-disjoint paths in planar graphs. *Combinatorica* 15, 135–150.

A simple linear-time algorithm solving this problem is given that is based on right-first search and yields a new proof for the theorem of Okamura and Seymour.

K. Okamura (1983). Multicommodity flows in graphs. *Discr. Appl. Math.* 6, 55–62.
H. Suzuki, T. Nishizeki, N. Saito (1989). Algorithms for multicommodity flows in planar graphs. *Algorithmica* 4, 471–501.
A. Schrijver (1989). The Klein bottle and multicommodity flow. *Combinatorica* 9, 375–384.

Necessary and sufficient solvability conditions respectively polynomial algorithms are given for more general instances where terminals are allowed to lie on at most *two face boundaries*.

M. Kaufmann, F. Maley (1993). Parity conditions in homotopic knock-knee routing. *Algorithmica* 9, 47–63.

Prescribed *homotopies* do not simplify the problem of finding edge-disjoint paths. In fact, the edge-disjoint paths problem with prescribed homotopy remains $\mathcal{NP}$-hard even for grid graphs.

3.2 Knock-knee routing in grids

M. Sarrafzadeh (1987). Channel routing problem in the knock-knee mode is $\mathcal{NP}$-complete. *IEEE Trans. Comput. Aided Design* CAD-6, 503–506.

The *multiterminal routing* problem is $\mathcal{NP}$-complete for instances containing five-terminal nets, even if the instance is a *channel.*

M. Grötschel, A. Martin, R. Weismantel (1996). Packing Steiner trees: polyhedral investigations. *Math. Program.* 72, 101–123.
M. Grötschel, A. Martin, R. Weismantel (1996). Packing Steiner trees: a cutting plane algorithm and computational results. *Math. Program.* 72, 125–145.

A *polyhedral approach* for packing Steiner trees in grids is introduced and applied for solving routing problems in switchboxes.

K. Mehlhorn, F. Preparata, M. Sarrafzadeh (1986). Channel routing in knock-knee mode: simplified algorithms and proofs. *Algorithmica* 1, 213–221.
S. Gao, M. Kaufmann (1994). Channel routing of multiterminal nets. *J. ACM* 41, 791–818.

Approximation algorithms for *multiterminal net channels* are presented. Upper bounds for the channel width required are proved.

For *even two-terminal net* instances in *general grid graphs*, the algorithm of Wagner and Weihe is an improvement on the algorithms known before with respect to running time [Wagner and Weihe 1995] (see §3.1).

T. Nishizeki, N. Saito, K. Suzuki (1985). A linear time routing algorithm for convex grids. *IEEE Trans. Comput. Aided Design* CAD-4, 68–76.
F. Wagner, B. Wolfers (1991). Short wire routing in convex grids. W. Hsu, R. Lee (eds.). *2nd Annual Int. Symp. Algorithms, ISA '91*, Lecture Notes in Computer Science

557. Springer-Verlag, Berlin, 72–83.

Weakly even two-terminal net instances in *convex grids* can be solved in time $O(n)$ as well. A modification of the algorithm by Nishizeki, Saito and Suzuki constructs layouts with *short wires* and *small number of bends*.

A. Frank (1982). Disjoint paths in a rectilinear grid. *Combinatorica* 2, 361–371.
K. Mehlhorn, F. Preparata (1986). Routing through a rectangle. *J. ACM* 33, 60–85.

Necessary and sufficient conditions for the solvability of two-terminal net *rectangle* problems are given. Mehlhorn and Preparata present an algorithm with sublinear running time, i. e. $O(k \log k)$ (k number of nets). It guarantees a total of only $O(k)$ bends.

M. Sarrafzadeh (1987). Channel routing with provably short wires. *IEEE Trans. Circuits and Systems* CAS-34, 1133–1135.
M. Formann, D. Wagner, F. Wagner (1993). Routing through a dense channel with minimum total wire length. *J. Algorithms* 15, 267–283.
D. Wagner (1991). A new approach to knock-knee channel routing. W. Hsu, R. Lee (eds.). *2nd Annual Int. Symp. Algorithms, ISA'91*, Lecture Notes in Computer Science 557. Springer-Verlag, Berlin, 83–93.

In these papers efficient algorithms for *minimizing wire length* in channels containing only two-terminal nets are presented.

3.3 Layer assignment

W. Lipski, Jr., F.P. Preparata (1987). A unified approach to layout wirability. *Math. Systems Theory* 19, 189–203.
W. Lipski, Jr. (1984). On the structure of three-layer wirable layouts. F.P. Preparata (ed.). *Advances in Computer Research, VOL 2: VLSI Theory*, JAI Press Inc., Greenwich, London, 231–243.
M.L. Brady, D.J. Brown (1984). VLSI routing: four layers suffice. F.P. Preparata (ed.). *Advances in Computer Research, VOL 2: VLSI Theory*, JAI Press Inc., Greenwich, London, 245–257.

Most results on the layer assignment problem for knock-knee layouts are based on the combinatorial framework introduced by Lipski and Preparata. It is $\mathcal{NP}$-complete to decide if a knock-knee layout is three-layer wirable. Every knock-knee layout is wirable in four layers.

F. Preparata, W. Lipski, Jr. (1984). Optimal three-layer channel routing. *IEEE Trans. Comput.* C-33, 427–437.
R. Kuchem, D. Wagner, F. Wagner (1996). Optimizing area for three-layer channel routing. *Algorithmica*, 15, 495–519.
D. Wagner (1993). Optimal routing through dense channels. *Int. J. on Comp. Geom. and Appl.* 3, 269–289.

Preparata and Lipski prove that every two-terminal net channel routing problem admits a three-layer wirable layout. But these layouts are not area-optimal. Area-optimal three-layer wirable layouts can be constructed in time $O(k \log k)$. For dense channels, Wagner presents an algorithm with running time $O(k)$ that constructs layouts with nearly minimum total wire length and number of bends that are wirable in only three layers.

M.L. Brady, M. Sarrafzadeh (1990). Stretching a knock-knee layout for multilayer wiring. *IEEE Trans. Comput.* C-39, 148–152.
M. Kaufmann, P. Molitor (1991). Minimal stretching of a layout to ensure 2-layer wirability. *INTEGRATION, the VLSI journal* 12, 339–352.
M. Sarrafzadeh, D. Wagner, F. Wagner, K. Weihe (1994). Wiring knock-knee layouts: A global approach. *IEEE Trans. Comput.* C-43, 581–589.

Stretching techniques for three-layer respectively two-layer wirability are considered. The last paper presents a global approach to attack the three-layer wiring problem that performs very well in practice.

3.4 Manhattan channel routing

T. Szymanski (1985). Dogleg channel routing is $\mathcal{NP}$-complete. *IEEE Trans. Comp. Aided Design* CAD-4, 31–41.
M. Middendorf (1996). On Manhattan channel routing with single sided nets. *Chicago J. of Theoretical Comput. Sci.* 6, 1–19.

The channel *width minimization* problem for multiterminal net channel routing problems is $\mathcal{NP}$-hard. It remains $\mathcal{NP}$-hard for the two-terminal case.

B.S. Baker, S.N. Bhatt, F.T. Leighton (1984). An approximation algorithm for Manhattan routing. F. P. Preparata (ed.). *Advances in Computer Research, VOL 2: VLSI Theory*, JAI Press Inc., Greenwich, London, 205–229.
C. Wieners-Lummer (1992). Manhattan routing with good theoretical and practical performance. *Discr. Appl. Math.* 40.

The notion of *flux* as a lower bound for the channel width is introduced and based on this concept an *approximation* algorithm respectively an improved version thereof is presented.

3.5 Via minimization

N.J. Naclerio, S. Masuda, K. Nakajima (1989). The via minimization problem is $\mathcal{NP}$-complete. *IEEE Trans. Comput.* C-38, 1604–1608.
H. Choi, K. Nakajima, C.S. Rim (1989). Graph bipartization and via minimization. *SIAM J. Discr. Math.* 2, 38–47.

In general, for two-layer wirable layouts the via minimization problem is $\mathcal{NP}$-hard, even if the layout is grid based or vias are restricted to lie at junctions, or if the junction degree is limited to four.

A. Hashimoto, J. Stevens (1971). Wire routing by optimizing channel assignment within large apertures. *8th Annual Design Automation Conference, DAC'71*, 155–169.
K.R. Stevens, W.M. VanCleemput (1979). Global via elimination in generalized routing environment. *Proceedings IEEE Int. Conf. Circuits and Computers*, 689–692.

M.J. Cielski, E. Kinnen (1981). An optimum layer assignment for routing in ICs and PCBs. *18th Annual Design Automation Conference, DAC'81*, 733–737.
K.C. Chang, D.H.-C. Du (1987). Efficient algorithms for layer assignment problems. *IEEE Trans. Comput. Aided Design* CAD-6, 67–78.
M. Grötschel, M. Jünger, G. Reinelt (1988). Via minimization with pin preassignment and layer preference. *Angew. Math. Mech.* 69, 393–399.
F. Barahona, M. Grötschel, M. Jünger, G. Reinelt (1988). An application of combinatorial optimization to physics and circuit layout design. *Oper. Res.* 36, 493–513.

The via minimization problem can be formulated as a graph-theoretic *maximum cut* problem. Based on this formulation, several heuristics, approximate methods and exact methods using integer programming respectively a cutting plane algorithm are proposed.

Y. Kajitani (1980). On via hole minimization of routing on a 2-layer board. *Proc. IEEE Int. Conf. Circuits and Computers*, 295–298.
R.W. Chen, Y. Kajitani, S.P. Chan (1983). A graph-theoretic via minimization algorithm for two-layer printed circuit board. *IEEE Trans. Circuits and Systems* CAS-30, 284–299.
R.Y. Pinter (1984). Optimal layer assignment for interconnect. *Journal of VLSI and Computer Systems* 1, 123–137.
P. Molitor (1987). On the contact minimization problem. *4th Annual Symp. on Theoretical Aspects of Comput. Sci., STACS'87*, 159–165.
N.J. Naclerio, S. Masuda, K. Nakajima (1987). Via minimization for gridless layouts. *24th Annual Design Automation Conference, DAC'87*, 159–165.
Y.S. Kuo, T.C. Chen, W. Shih (1988). Fast algorithm for optimal layer assignment. *25th Annual Design Automation Conference, DAC'88*, 554–559.

It transforms to a *maximum cut* problem in *planar* graphs if the maximum junction degree is three, or if vias are not allowed to be placed on junctions. Several algorithms with running time $O(n^3)$ based on a maximum cut algorithm for planar graphs or related methods have been proposed. This running time is improved to $O(n^{\frac{3}{2}} \log n)$ in the last paper.

3.6 Single-layer routing

J. Lynch (1975). The equivalence of theorem proving and the interconnection problem. *ACM SIGDA Newsletter* 5, 31–65.
R. Rockafellar (1984). *Network flows and monotropic optimization*, Wiley, New York.

The single-layer routing problem is $\mathcal{NP}$-hard even for two-terminal nets, or if the underlying graph is planar with maximum degree three respectively a grid graph [Kramer and van Leeuwen 1984] (see §2.1). In case the graph together with the demand edges is planar, the problem is still $\mathcal{NP}$-hard [Middendorf and Pfeiffer 1993] (see §3.1).

B. Reed, N. Robertson, A. Schrijver, P. Seymour (1993). Finding disjoint trees in graphs on surfaces. N. Robertson (ed.). *Graph structure theory, Proc. AMS-IMS-SIAM Joint Summer Research Conference on Graph Minors*, 295–301.
A. Schrijver (1990). Homotopic routing methods. B. Korte, L. Lovász, H.J. Prömel,

A. Schrijver (eds.). *Paths, Flows and VLSI-Layout*, Springer-Verlag, Berlin, 329–371.

However, if the *number of nets* is *fixed* or all terminals can be covered by a *fixed number of faces* the problem is solvable in polynomial time. But for both algorithms the order constant of the running time depends heavily on the fixed parameters.

H. Suzuki, T. Akama, T. Nishizeki (1990). Finding Steiner forests in planar graphs. *1st Annual ACM-SIAM Symp. Discr. Algorithms, SODA'90*, 444–453.
H. Ripphausen-Lipa, D. Wagner, K. Weihe (1993). Linear time algorithms for disjoint two-face paths problems in planar graphs. K. Ng, P. Raghavan, N. Balasubramanian, F. Chin (eds.). *4th Annual Int. Symp. Algorithms, ISAAC'93*, Lecture Notes in Computer Science 762. Springer-Verlag, Berlin, 343–352.

If we assume that all terminals lie on at most *two face boundaries* of a planar graph, the vertex-disjoint Steiner tree problem can be solved in time $O(n \log n)$ using divide-and-conquer techniques, respectively in time $O(n)$ using right-first search.

R. Cole, A. Siegel (1983). River routing every which way, but loose. *24th Annual IEEE Symp. on Found. of Comput. Sci., FOCS'83*, 65–73.
F.M. Maley (1990). *Single-Layer Wire Routing and Compaction*, PhD Thesis, MIT, Cambridge, MA.
J.-M. Ho, A. Suzuki, M. Sarrafzadeh (1993). An exact algorithm for single-layer wire length minimization. *IEEE Trans. Circuits and Systems* CAS-12, 175 –180.
C.E. Leiserson, F. Maley (1985). Algorithms for routing and testing routability of planar VLSI layouts. *17th Annual ACM Symp. Theory of Comput., STOC'85*, 69–78.
C.E. Leiserson, R.Y. Pinter (1983). Optimal placement for river routing. *SIAM J. Comput.* 12, 447–462.

In these papers routability of two-terminal nets subject to prescribed *homotopies* is studied.

A. Mirzaian (1987). River routing in VLSI. *J. Comput. Syst. Sci.* 34, 43–54.
A. Siegel, D. Dolev (1988). Some geometry for general river routing. *SIAM J. Comput.* 17, 583–605.
A. Mirzaian (1989). A minimum separation algorithm for river routing with bounded number of jogs. *Proc. ICCAD'89*, 10–13.
R.I. Greenberg, F.M. Maley (1992). Minimum separation for single-layer channel routing. *Inform. Process. Lett.* 43, 201–205.
T.C. Tuan, K.H. Teo (1991). On river routing with minimum number of jogs. *IEEE Trans. Comput. Aided Design* CAD 10, 270–273.
S. Tragoudas, I.G. Tollis (1993). River routing and density minimization for channels with interchangeable terminals. *INTEGRATION, the VLSI journal* 15, 151–178.

Optimization criteria for single-layer channel routing problems, as the *optimum offset problem*, *minimum separation problem* or minimizing the *number of jogs* are considered in these papers.

4 Linear VLSI Layout Styles

Cell synthesis is the part of the layout process in which the physical building blocks for the placement and routing phase are constructed. It typically employs one-dimensional layout styles (hence the term *linear*) such as Weinberger arrays, gate matrix layout and PLA-folding.

4.1 The VLSI background

D.G. Gajski, Y.-L.S. Lin (1988). Module generation and silicon compilation. B. Preas, M. Lorenzetti (eds.). *Physical Design Automation of VLSI Systems*, Benjamin/Cummings, Menlo Park, CA, 283–345.
R.K. Brayton, C. McMullen, G.D. Hachtel, A. Sangiovanni-Vincentelli (1984). *Logic Minimization Algorithms for VLSI Synthesis*, Kluwer Acad. Publ., Dordrecht.

These papers survey the general technical background of cell synthesis and linear layout styles.

W. Weinberger (1967). Large scale integration of MOS complex logic: a layout method. *IEEE Journal of Solid-State Circuits* SC-2, 182–190.
A. Lopez, H. Law (1980). A dense gate matrix layout method for MOS VLSI. *IEEE Trans. on Electronic Devices* ED-27, 1671–1675.
H. Fleisher, L.I. Maissel (1975). An introduction to array logic. *IBM J. Res. Develop.* 19, 98–109.

The papers that introduced the layout styles Weinberger arrays, gate matrix layout and PLA folding.

R.H. Möhring (1990). Graph problems related to gate matrix layout and PLA folding. G. Tinhofer, E. Mayr, H. Noltemeier, M. Sysło (eds.). *Computational Graph Theory*, Springer-Verlag, Wien, 17–51.

A common combinatorial model for all these layout styles as a matrix permutation problem with side constraints is presented. This is then used to obtain unified proofs for the different graph-theoretic models discussed below.

4.2 Graph problems related to linear layout styles

There is a large variety of independently investigated graph problems that are equivalent or closely related to linear layout styles.

T. Ohtsuki, H. Mori, E. Kuh, T. Kashiwabara, T. Fujisawa (1979). One-dimensional logic gate assignment and interval graphs. *IEEE Trans. Circuits and Systems* CAS-26, 675–684.
O. Wing (1982). Automated gate matrix layout. *Proc. IEEE Int. Symp. Circuits and Systems*, 681–685.

The first papers to show the equivalence of minimizing the number of tracks in a Weinberger array or a gate matrix layout with the *interval graph augmentation problem*:

Embed a graph G (the *incompatibility graph* associated with the layout problem) as a subgraph into an interval graph H on the same vertex set such that the clique size $\omega(H)$ of H (the maximum size of a clique of H) is as small as possible.

M.R. Fellows, M.A. Langston (1988). Nonconstructive tools for proving polynomial-time decidability. *J. ACM* 35, 727–739.

This paper observes the equivalence of the gate matrix layout problem with determining the *pathwidth* of the incompatibility graph G. This relates gate matrix layout to the theory of Robertson and Seymour about graph minors.

H.L. Bodlaender (1993). A tourist guide through treewidth. *Acta Cybernetica* 11, 1–23.

This paper gives a survey covering the most important aspects about treewidth and pathwidth.

T. Kloks (1994). *Treewidth, Computations and Approximations*, Lecture Notes in Computer Science, Springer-Verlag, Berlin, vol. 842.

A research monograph about pathwidth and treewidth. It contains structural properties and algorithms for computing or approximating treewidth and pathwidth for special classes of graphs.

A. Parra, P. Scheffler (1995). How to use the minimal separators of a graph for its chordal triangulations. *Proc. 22nd Int. Coll. Automata, Lang. and Program. CALP'95*, 123–134.

Relates pathwidth and treewidth to the structure of separators of the underlying graph. Is the most elegant presentation after several earlier results most of which are covered by [Kloks 1994] (see above).

L.M. Kirousis, C.H. Papadimitriou (1985). Interval graphs and searching. *Discr. Math.* 55, 181–184.
N. Megiddo, S.L. Hakimi, M.R. Garey, D.S. Johnson, C.H. Papadimitriou (1988). The complexity of searching a graph. *J. ACM* 35, 18–44.
D. Bienstock, P. Seymour (1991). Monotonicity in graph searching. *J. Algorithms* 12, 239–245.

These papers reveal the interesting relationship between pathwidth (viewed as interval graph augmentation problem) and graph searching.

L.M. Kirousis, C.H. Papadimitriou (1986). Searching and pebbling. *Theor. Comp. Sci.* 47, 205–218.

Establishes the equivalence of node search and vertex separation number of a graph.

H. Kaplan, R. Shamir, R.E. Tarjan (1994). Tractability of parameterized completion problems on chordal and interval graphs: Minimum fill-in and physical mapping. *35th Annual IEEE Symp. Found. Comput. Sci. FOCS'94*, 780–791.

Determining the bandwidth of a graph corresponds to a special interval graph augmentation problem. The resulting interval graph must be proper.

G.D. Hachtel, A.R. Newton, A.L. Sangiovanni-Vincentelli (1982). An algorithm for optimal PLA folding. *IEEE Trans. Comput. Aided Design Integrated Circuits* CAD-1, 63–77.

PLA folding problems can be seen as more restrictive gate matrix layout problems. This is the first paper that models arbitrary PLA-folding problems as matching problems. Constructing a minimum area layout is equivalent to finding a maximum cardinality matching in the complement of a digraph D such that the matching edges together with the arcs of D do not contain an alternating cycle.

T.C. Hu, Y.S. Kuo (1987). Graph folding and programmable logic array. *Networks* 17, 19–37.
S.S. Ravi, E.L. Lloyd (1988). The complexity of near-optimal programmable logic array folding. *SIAM J. Comput.* 17, 696–710.
E. Mach, T. Wolf (1995). Graph models for PLA folding problems. *Int. J. Syst. Sci.* 26, 1439–1445.
These papers deal with variations of the matching formulation and related graph-theoretic formulations. For instance, for a restricted version of PLA-folding, called *bipartite block-folding*, a maximum folding corresponds to finding a complete bipartite subgraph $K_{m,m}$ of a bipartite graph with m as large as possible.

4.3 Complexity results

T. Kashiwabara, T. Fujisawa (1979). $\mathcal{NP}$-completeness of the problem of finding a minimum-clique-number interval graph containing a given graph as a subgraph. *Proc. 1979 Int. Symp. Circuits and Systems*, 657–660.
[Kirousis and Papadimitriou 1986] (see §4.2)
S. Arnborg, D.G. Corneil, A. Proskurowski (1987). Complexity of finding embeddings in a k-tree. *SIAM J. Algebraic Discr. Meth.* 8, 277–284.
M. Luby, U. Vazirani, V. Vazirani, A.L. Sangiovanni-Vincentelli (1982). Some theoretical results on the optimal PLA folding problem. *Proc. 1982 Int. Symp. Circuits and Systems*, 165–170.
These papers contain the first $\mathcal{NP}$-completeness results for interval graph augmentation, node searching, pathwidth (even for complements of bipartite graphs), and PLA-folding, respectively. They were obtained independently without the knowledge about the relationship between these problems (see §4.2).

R. Müller, D. Wagner (1991). α-vertex separator is $\mathcal{NP}$-hard even for 3-regular graphs. *Computing* 46, 343–353.
Block folding is $\mathcal{NP}$-complete for graphs with degree at most k for any fixed $k \geq 3$.

J. Gustedt (1993). On the pathwidth of chordal graphs. *Discr. Appl. Math.* 45, 233–248.
Pathwidth is already $\mathcal{NP}$-complete for special chordal graphs.

B. Monien, I.H. Sudborough (1988). Min cut is $\mathcal{NP}$-complete for edge weighted trees. *Theor. Comp. Sci.* 58, 209–229.
Vertex separation (and thus pathwidth and gate matrix layout) is already $\mathcal{NP}$-complete for planar graphs.

P. Damaschke (1994). PLA folding in special graph classes. *Discr. Appl. Math.* 51, 63–74.
Gives sharper $\mathcal{NP}$-completeness results for PLA-folding.

H.L. Bodlaender, R.H. Möhring (1993). The pathwidth and treewidth of cographs. *SIAM J. Discr. Math.* 6, 181–188.
H.L. Bodlaender, T. Kloks, D. Kratsch (1995). Treewidth and pathwidth of permutation graphs. *SIAM J. Discr. Math.* 8, 606–616.
[Parra and Scheffler 1995] (see §§4.2)
[Hu and Kuo 1987] (see §§4.2)
[Damaschke 1994] (see above)

A selection of articles with polynomial algorithms for special graph classes.

4.4 Algorithms

D.F. Wong, H.W. Leong, C.L. Liu (1988). *Simulated annealing for VLSI design*, Kluwer Acad. Publ., Dordrecht.
Y.-Y. Yang, C.-M. Kyung (1990). An efficient algorithm for optimal PLA folding. *INTEGRATION the VLSI journal* 9, 271–285.
A.G. Ferreira, S.W. Song (1992). Achieving optimality for gate matrix layout and PLA folding: a graph theoretic approach. *INTEGRATION the VLSI journal* 14, 173–195.
S.S. Ravi, E.L. Lloyd (1993). Graph theoretical analysis of PLA folding heuristics. *J. Comput. Syst. Sci.* 46, 326–348.

Due to the VLSI background, many algorithms have been proposed and studied in the design automation literature. The majority of them can be classified as (sometimes a combination of) heuristics, branch-and-bound algorithms, or dynamic programming algorithms. These are a few representative references.

N. Deo, M.S. Krishnamoorty, M.A. Langston (1987). Exact and approximate solutions for the gate matrix layout problem. *IEEE Trans. Comput. Aided Design Integrated Circuits* CAD-6, 79–84.

One of the few papers about approximation algorithms. Unless $\mathcal{P} \neq \mathcal{NP}$, there is no approximation algorithm for gate matrix layout with a constant absolute performance guarantee.

H.L. Bodlaender, J.R. Gilbert, H. Hafsteinsson, T. Kloks (1991). Approximating treewidth, pathwidth, and minimum elimination tree height. G. Schmidt, R. Berghammer (eds.). *Proc. 17th Int. Workshop on Graph-Theoretic Concepts in Computer Science WG'91*, Lecture Notes in Computer Science, vol. 570, 1–12, Springer-Verlag, Berlin.

Establishes a relative performance ratio of $O(\log^2 n)$ for the pathwidth of a graph with n vertices. The existence of approximation algorithms with a constant performance ratio is open.

[Ravi and Lloyd 1988] (see §§4.2)

Different PLA-folding models are equivalent with respect to approximability. A constant relative performance ratio for block folding implies already the (unlikely) existence of a polynomial time approximation scheme.

25 Computational Molecular Biology

Martin Vingron
Deutsches Krebsforschungszentrum, Heidelberg

Hans-Peter Lenhof
Max-Planck-Institut für Informatik, Saarbrücken

Petra Mutzel
Max-Planck-Institut für Informatik, Saarbrücken

Contents

Computational Biology is a fairly new subject that arose in response to the computational problems posed by the analysis and the processing of biomolecular sequence and structure data. The field was initiated in the late 60's and early 70's largely by pioneers working in the life sciences. Physicists and mathematicians entered the field in the 70's and 80's, while Computer Science became involved with the new biological problems in the late 1980's. Computational problems have gained further importance in molecular biology through the various genome projects which produce enormous amounts of data.

For this bibliography we focus on those areas of computational molecular biology

Annotated Bibliographies in Combinatorial Optimization, edited by M. Dell'Amico, F. Maffioli and S. Martello

that involve discrete algorithms or discrete optimization. Thus, there are areas of computational molecular biology we neglected, e.g., databases for molecular and genetic data and genetic linkage algorithms. Of the literature on the protein folding problem we included only papers which cover combinatorial aspects of the problem. Due to the availability of review papers and a bibliography from 1984 some older papers will not be explicitely mentioned in this bibliography.

1 Books and Surveys

In this section, we list some books and surveys that serve as a general introduction to the field.

M.S. Waterman (1995). *Introduction to Computational Biology*, Chapman & Hall, London.

This is a recent book on computational molecular biology containing a wealth of material on all the subjects dealt with in this bibliography.

J.R. Jungk, R.M. Friedman (1984). Mathematical tools for molecular genetics data: an annotated bibliography. *Bull. Math. Biol.* 46, 699–744.

This annotated bibliography summarizes the literature up to 1984.

R. Doolittle (ed.) (1990). Molecular Evolution: Computer Analysis of Protein and Nucleic Acid Sequences. *Meth. Enzymol.* 183.
R. Doolittle (ed.) (1996). Computer Methods for Macromolecular Sequence Analysis. *Meth. Enzymol.* 266.

These volumes are collections of articles and give a good overview of the approaches and the software that are in use in molecular biology.

P.A. Pevzner, M.S. Waterman (1995). Open combinatorial problems in computational molecular biology. In *Proc. 3rd Israel Symp. Theory of Comput. and Systems*, Tel Aviv, Israel. IEEE Computer Society Press, New York.

This collection of open problems is an excellent summary of the area.

E.S. Lander, R. Langridge, D.M. Saccocio (1991). A report on computing in molecular biology: Mapping and interpreting biological information. *Commun. ACM* 34, 33–39.

This brief overview gives a taste of the problems in computational molecular biology.

The journal *Algorithmica* devoted a special issue to computational molecular biology: volume 13, numbers 1/2, 1995 by E.W. Myers (ed.). Similarly, *Discrete Applied Mathematics* published a special issue on this topic: volume 71, issues 1-3, 1996, S. Istrail, P.A. Pevzner, R. Shamir (eds.).

2 Sequence Alignment and Evolution

The sequence of a DNA molecule can be modeled as a string over a 4-letter alphabet, each letter representing one of the four nucleotides that make up DNA. Proteins, the other class of biological macromoleculas are linear chains of amino acids and are represented as strings over a 20-letter alphabet. Sequence alignment deals with comparing different DNA or different protein sequences. This is done by writing one on top of the other padding them with spaces ("indels", for insertion or deletion) to achieve identical length. In DNA, a simple criterion to distinguish among the many possibilities of this arrangment counts the number of unequal letters ending up on top of each other minus the number of spaces that were introduced. For protein sequence comparison the pairs of matched letters are weighted and the adjacent spaces are summarized into blocks which receive a penalty. More formally, let A be a finite alphabet and let $S = \{s_1, s_2, \ldots, s_k\}$ be a set of finite strings over A. We define a new alphabet $\hat{A} = A \cup \{-\}$ by adding to A the symbol dash "–" to represent indels. A set $\hat{S} = \{\hat{s}_1, \hat{s}_2, \ldots, \hat{s}_k\}$ of strings over the alphabet $\hat{A}$ is called an alignment of the set S, if the following properties hold: (1) The strings in $\hat{S}$ have the same length. (2) Ignoring dashes, string $\hat{s}_i$ is identical with string s_i (for all $i \in \{1, 2, \ldots, k\}$). Hence an alignment can be interpreted as an array with k rows. The ith row contains string $\hat{s}_i$. Each column must contain at least one letter of a string in S (columns filled only with dashes are forbidden). The score of an alignment is based on a distance function $d(\hat{s}_i, \hat{s}_j)$ for aligned sequences.

D. Sankoff, J.B. Kruskal (1983). *Time Warps, String Edits and Macromolecules: the Theory and Practice of Sequence Comparison*, Addison Wesley, Reading, MA.

This is a book entirely dedicated to sequence comparison.

M.S. Waterman (1984). General methods of sequence comparison. *Bull. Math. Biol.* 46, 473–500.

M.S. Waterman (1989). Sequence alignments. *Mathematical Methods for DNA Sequences.*, CRC Press, Boca Raton, Fl.

E. Myers (1991). *An overview of sequence comparison algorithms in molecular biology*, Technical Report 29, Department of Computer Science of the University of Arizona at Tucson Arizona, (also Technical Report TR91-29, http://www.cs.arizona.edu/people/gene/PAPERS/compbio.survey.ps.)

These are reviews on sequence alignment.

2.1 Pairwise sequence alignment

Assume that $k = 2$, i.e., only two strings s_1 and s_2 are given. The *Pairwise Sequence Alignment Problem* can be formulated as follows: Compute the alignment $\hat{S} = \{\hat{s}_1, \hat{s}_2\}$ of s_1 and s_2 that minimizes the distance $d(\hat{s}_1, \hat{s}_2)$. An overview of pairwise sequence alignment need not be given here since excellent reviews are available: The surveys listed above all cover pairwise sequence alignment thouroughly. Only certain special topics are not treated in depth in these general reviews.

K.-M. Chao, R.C. Hardison, W. Miller (1994). Recent developments in linear-space alignment methods: a mini survey. *J. Comput. Biol.* 1, 271–291.

A naive implementation of a sequence alignment algorithm requires space quadratic in the sequence length. This has been a major obstacle to the routine use of sequence alignment programs. Linear space alignment techniques have contributed greatly to the acceptance of alignment software by biologists.

M. Vingron (1996). Near-optimal sequence alignment. *Curr. Opin. Struct. Biol.* 6, 346–352.

Here, an overview of techniques for computing suboptimal sequence alignments is given.

M.S. Waterman, M. Eggert, E.S. Lander (1992). Parametric sequence comparison. *Proc. Natl. Acad. Sci. USA* 89, 6090–6093.
D. Gusfield, K. Balasubramanian, D. Naor (1994). Parametric optimization of sequence alignment. *Algorithmica* 12, 312–326.
M. Waterman (1994). Parametric and ensemble sequence alignment algorithms. *Bull. Math. Biol.* 56, 743–767.

Another special topic that has recently developed is parametric sequence comparison dealing with the computation of all solutions to the sequence alignment problem as scoring parameters are varied. The above articles describe algorithms for parametric sequence comparison.

M. Vingron, M.S. Waterman (1994). Sequence alignment and penalty choice: review of concepts, case studies and implications. *J. Mol. Biol.* 235, 1–12.
D. Gusfield, P. Stelling (1996). Parametric and inverse-parametric sequence alignment with XPARAL. *Meth. Enzymol.* 266, 481–494.

These papers study patterns in the tesselation generated by parametric alignment and their biological interpretation.

D. Eppstein, Z. Galil, R. Giancarlo, G. Italiano (1992). Sparse dynamic programming I: linear cost functions. *J. ACM* 39, 519–545.
D. Eppstein, Z. Galil, R. Giancarlo, G. Italiano (1992). Sparse dynamic programming II: concave, convex cost functions. *J. ACM* 39, 546–567.

In these papers the authors study sparse dynamic programming for sequence alignment.

G. Stephen (1994). *String Searching Algorithms*, World Scientific, Singapore.
M. Crochemore, W. Rytter (1994). *Text Algorithms*, Oxford University Press, New York.
D. Gusfield (1997). *Algorithms on Strings, Trees, and Sequences: Computer Science and Computational Biology*, Cambridge University Press, Cambridge.

These books contain extensive material on approximate string matching and are thus in a wider sense related to our topic.

W. Chang, E. Lawler (1990). Approximate matching in sublinear expected time. *Proc. 31st IEEE Symp. on Foundations of Computer Science*, 116–124.
E. Myers (1994). A sublinear algorithm for approximate keyword searching. *Algorithmica* 12, 345–374.

A problem of great importance in molecular biology is the search in a DNA or protein sequence database for sequences similar to a given query sequence. The running time of the algorithms given in the above papers is sublinear in the length of the database.

S.F. Altschul, W. Gish, W. Miller, E.W. Myers, D.J. Lipman (1990). Basic local alignment search tool. *J. Mol. Biol.* 215, 403–410.
W.R. Pearson, D.J. Lipman (1988). Improved tools for biological sequence comparison. *Proc. Natl. Acad. Sci. USA* 85, 24444–24448.
The programs BLAST and FASTA are widely used implementations of heuristic methods to quickly identify sequences similar to a query sequence in a database.

Much like sequence alignment, deriving the optimal fold of an RNA molecule is usually done by dynamic programming. This topic, too, is summarized in Waterman's "Introduction to computational biology".

2.2 Multiple sequence alignment

S. Chan, A. Wong, D. Chiu (1992). A survey of multiple sequence comparison methods. *Bull. Math. Biol.* 54, 563–598.
The biological task of comparing several sequences simultaneously has been formalized in different ways. An overview is given in [Doolittle 1996] (see §1).
Many practical implementations of multiple sequence alignment circumvent the difficulties of the optimization problems discussed below by using iterative pairwise alignments. This book contains several articles on these so-called progressive alignment methods.

S.R. Eddy (1996). Hidden markov models. *Curr. Opin. Struct. Biol.* 6, 361–365.
This article reviews a stochastic modeling approach to multiple sequence alignment that has recently become very influential.

2.2.1 Sum-of-pairs multiple alignment

The *Sum of Pairs Multiple Alignment Problem (SPMA)* is defined as follows: Compute the alignment $\hat{S}$ of S that minimizes the sum of the distances of all pairs $\hat{s}_i, \hat{s}_j$:

$$\text{SPMA}(S) := \min_{\hat{S}} \left[\sum_{i,j} d(\hat{s}_i, \hat{s}_j) \right] .$$

L. Wang, T. Jiang (1994). On the complexity of multiple sequence alignment. *J. Comput. Biol.* 1, 337–348.
The authors prove that the SPMA-problem is NP-complete by reduction from the *Shortest Common Supersequence Problem* (see Garey, M. and Johnson, D. (1979). *Computers and Intractability: A Guide to the Theory of NP-Completeness*, W.H. Freeman, New York).

The SPMA-problem can be formulated as a shortest-path problem: assume for simplicity that all strings in S have length n. The set of possible alignments of S can be represented by a k dimensional mesh-shaped graph with n^k vertices and $O(n^k 2^k)$ directed edges. Each path in the graph from the source to the sink represents a possible multiple alignment of S. The sequence of edges of a path represents the sequence of columns of the alignment, i.e., each edge codes for one column of the alignment. A vertex in the graph can be interpreted as a "frontier" in the string set S and thus also represents a set of prefixes of S ending at this frontier. Hence the set of all paths from the source to a vertex encode the set of all possible alignments of the prefixes represented by the vertex. Solving the SPMA-problem for S means computing the (shortest) path that minimizes the cost function.

M. Waterman, T. Smith, W. Beyer (1976). Some biological sequence metrics. *Adv. Math.* 20, 367–387.

In this article it is shown that the SPMA-problem can be solved using dynamic programming.

R. Jue, N. Woodbury, R. Doolittle (1980). Sequence homologies among E. Coli ribosomal proteins: evidence for evolutionary related groupings and internal duplications. *J. Mol. Evol.* 15, 129–148.
M. Fredman (1984). Algorithms for computing evolutionary similarity measures with length independent gap penalties. *Bull. Math. Biol.* 46, 553–566.
M. Murata, J. Richardson, J. Sussman (1985). Simultaneous comparisons of three protein sequences. *Proc. Natl. Acad. Sci. U.S.A.* 82, 3037–3077.
O. Gotoh (1986). Alignment of three biological sequences with an efficient traceback procedure. *J. Theor. Biol.* 121, 327–337.

Since the size of the graph grows exponentially with the number of strings k ($O(n^k)$ vertices and $O(n^k 2^k)$ edges) the dynamic programming approach works only for very small k. The above articles deal with dynamic programming alignment for three sequences.

H. Carrillo, D.J. Lipman (1988). The multiple sequence alignment problem in biology. *SIAM J. Appl. Math.* 48, 1073–1082.

Here, an elegant branch-and-bound approach is given based on a new technique for reducing the part of the graph that has to be examined. Only the paths that are contained in a certain "polytope" around the shortest path will be explored by their algorithm.

D. Lipman, S. Altschul, J. Kececioglu (1989). A tool for multiple sequence alignment. *Proc. Natl. Acad. Sci. U.S.A.* 86, 4412–4415.

The authors implemented a slightly modified version of the algorithm of Carrillo and Lipman. Since the bounds of Carrillo and Lipman are not sufficiently tight for solving "real world" multiple sequence alignment instances, Lipman et al. propose heuristics to improve bounds.

S. Gupta, J. Kececioglu, A. Schaeffer (1995). Improving the practical space and time efficiency of the shortest-paths approach to sum-of-pairs multiple sequence alignment. *J. Comput. Biol.* 2, 459–472.

Further significant reduction in space requirements can be achieved by introducing new algorithmic invariants determining when edges and vertices of the graph first need to be created and when they can be safely destroyed. Speedup results from the usage of more efficient data structures for the shortest-path problem.

The above algorithms for computing optimal multiple sequence alignments can handle at most a dozen sequences of length 200–400. Practical problem instances, however, may contain hundreds of sequences and sequence length may be above 1000. Such instances cannot be expected to be solved to optimality. Approximation algorithms and heuristics are required.

D. Gusfield (1993). Efficient methods for multiple sequence alignment with guaranteed error bounds. *Bull. Math. Biol.* 55, 141–154.

This article presents the first approximation algorithm for the SPMA-problem. The approximation factor is $2 - 2/k$, where k is the number of sequences. The main idea of Gusfield's approach is to consider alignments that are derived from the *star trees* of S.

P. Pevzner (1992). Multiple alignment, communication cost, and graph matching. *SIAM J. Appl. Math.* 52, 1763–1779.

The approximation bound can be improved to $2 - 3/k$ by deriving multiple alignments from so-called *3-stars*. Using the elegant concept of communication cost, Pevzner proved that there are 3-stars whose derived multiple alignments are at most a factor $2 - 3/k$ from optimal. Pevzner's algorithm computes an alignment in time $O(n^3k^3 + k^4)$, where n is the (maximal) length of the sequences.

V. Bafna, E. Lawler, P. Pevzner (1994). Approximation algorithms for multiple sequence alignment. *Proc. 5th Annual Symp. Combinatorial Pattern Matching, Lecture Notes in Comp. Sci.* 807, Springer-Verlag, Berlin, 43–53.

By exploring cliques of l sequences with one common center (l-stars instead of 3-stars) Bafna, Lawler and Pevzner achieve an approximation bound of $2 - l/k$ for any fixed $l < k$ and with the computation time increasing exponentially in l.

2.2.2 Maximum weight trace

The input to the *Maximum Weight Trace (MWT)* alignment problem is a set S of k strings and a graph $G = (V, E)$. The letters of the strings s_i of S are the vertices V of the graph. Every edge $e \in E$ of the graph has a positive weight w_e and connects two vertices (letters) that belong to different strings (i.e., there are no edges that connect two letters of the same string). An alignment $\hat{S}$ realizes an edge e if the two letters connected by the edge are placed in the same column of the alignment. The set of edges realized by an alignment $\hat{S}$ is called the *trace* (trace($\hat{S}$)) of $\hat{S}$. The MWT problem is defined as follows: Compute an alignment $\hat{S}$ that realizes a trace with maximal weight:

$$\mathrm{MWT}(S) := \max_{\hat{S}} \left[\sum_{e \in \mathrm{trace}(\hat{S})} w_e \right].$$

J. Kececioglu (1993). The maximum weight trace problem in multiple sequence alignment. *Proc. 4th Symp. Combinatorial Pattern Matching, Lecture Notes in Comp. Sci.* 684, Springer-Verlag, Berlin, 106–119.

The MWT-problem has been introduced by Kececioglu. He proves that the MWT-problem contains the SPMA-problem under certain conditions, as a special case, and that the MWT-problem is NP-complete by reduction from the *Feedback Arc Set Problem* [Garey, M. and Johnson, D. (1979). *Computers and Intractability: A Guide to the Theory of NP-Completeness.* W.H. Freeman, New York.] Kececioglu presents a branch-and-bound algorithm for the MWT-problem whose implementation could optimally align six sequences of length 250 in a few minutes. Here, a heuristic alignment of the k sequences yields a lower bound. Upper bounds for the branch-and-bound approach are calculated by adding up the weights of all MWT-optimal pairwise alignments of suffix sets of S.

2.2.3 Chaining multiple-alignment fragments

A fragment of a set S of k strings is a set $S' = \{s'_1, s'_2, \ldots, s'_k\}$, where s'_i is a non-empty substring of s_i (for all i). A fragment f_1 is smaller than a fragment f_2 ($f_1 < f_2$), if the substrings of f_1 and f_2 do not overlap and the substrings of f_1 are to the left of the substrings of f_2. The set of letters of S that lie between the substrings of two fragments f_1 and f_2 with $f_1 < f_2$ are called the gap of f_1 and f_2. We call a set $\{f_1 < f_2 < \cdots < f_l\}$ of ordered fragments a chain of fragments. The *Chaining Multiple-Alignment Fragments Problem* is defined as follows: Let F be a set of fragments of S, where each fragment $f \in F$ has a positive score (score(f)). Let gap_cost($*, *$) be a "gap" penalty function that assigns a cost to a gap between two ordered fragments. Compute a chain of fragments $\hat{F}$ that maximizes the following function:

$$\mathrm{CMAF}(S) := \max_{\hat{F}} \left[\sum_{f \in \hat{F}} \mathrm{score}(f) - \mathrm{gap\_cost}(f, \mathrm{successor}(f)) \right].$$

The CMAF-problem can be interpreted as an optimal-path problem: The fragments of S are the vertices of the graph. Each vertex is labeled with the score of its fragment. For each ordered pair $f_1 < f_2$ of fragments a directed edge from f_1 to f_2 will be added to the graph. The edge will be labeled with the penalty "$-$gap_cost(f_1, f_2)" for the gap between f_1 and f_2. The CMAF-problem can now be formulated as follows: compute the path in the directed graph that maximizes the sum of the vertex and edge labels.

W.J. Wilbur, D.J. Lipman (1984). The context dependent comparison of biological sequences. *SIAM J. Appl. Math.* 44, 557–567.

The above viewpoint in the context of pairwise alignment is introduced in this article.

E. Sobel, R. Martinez (1986). A multiple sequence alignment program. *Nucl. Acids Res.* 14, 363–374.

Here, it is applied to multiple sequence alignment. Assume that computing the gap cost takes time $O(g)$, then the CMAF-problem can be solved in time $O(|F|^2 g)$ by dynamic programming. Here, $|F|$ is the number of fragments in F. The algorithm requires $O(k|F|)$ space and works for any arbitrary gap cost function.

Z. Zhang, B. Raghavachari, R. Hardison, W. Miller (1994). Chaining multiple-alignment blocks. *J. Comput. Biol.* 1, 217–226.

The authors implemented a practical algorithm for the CMAF-problem that uses kD-trees to compute the optimal chain.

E. Myers, W. Miller (1995). Chaining multiple-alignment fragments in sub-quadratic time. *Proc. 6th Annual ACM-SIAM Symp. Discr. Algorithms*, 38–47.

Here, the first sub-quadratic time algorithm for special gap cost functions is presented. Myers and Miller prove that the maximal fragment chain can be computed in time $O(|F|(\log|F|)^k)$ with $O(k|F|(\log|F|)^{k-1})$ space.

2.3 Evolutionary trees

Molecular sequences are used to reconstruct the course of evolution. Since evolution is assumed to have proceeded from a common ancestral species in a tree-like branching of species (molecules), this process is generally modeled by a tree. When the most ancestral species is known, the model will be a rooted tree. The leaves of the tree are labeled with contemporary species while the inner nodes correspond to hypothetical ancestors. The key question is the reconstruction of this tree based on contemporary data. These data may come in one of two forms: As a multiple alignment with the sequences corresponding to leaves or as a matrix of distances between leaf-labels. Methods are thus divided into character-based methods and distance-based methods.

2.3.1 Character-based methods

F.K. Hwang, D.S. Richards, P. Winter (1992). *The Steiner Tree Problem*, North-Holland, Amsterdam.

A good introduction to character-based methods is contained in this book.

An idealized but interesting model of evolution is embodied in the *Perfect Phylogeny Problem*. Let the number of species be k. Let a set of m characters (e.g., the columns of the multiple alignment), each character having r possible states (e.g., the four nucleotides A, C, G, and T), be given. We say that a character is compatible with a tree when the inner nodes of the tree can be labeled such that each character state induces a subtree. A tree is said to be a perfect phylogeny when all characters are compatible with it. The perfect phylogeny problem is to decide whether a given set of characters has a perfect phylogeny and if so construct it.

H. Bodlaender, M. Fellows, T.J. Warnow (1992). Two strikes against perfect phylogeny. *Proc. 19th Int. Coll. Automata, Lang. and Program., Lecture Notes Comp. Sci.* 623, Springer-Verlag, Berlin, 273–283.
M.A. Steel (1992). The complexity of reconstructing trees from qualitative characters and subtrees. *J. Classification* 9, 91–116.

In both articles NP-completeness for arbitrary number of character states is shown.

D. Gusfield (1991). Efficient algorithms for inferring evolutionary trees. *Networks* 21, 19–28.

The perfect phylogeny problem for binary characters ($r = 2$) can be solved in linear time.

A. Dress, M. Steel (1992). Convex tree realizations. *Appl. Math. Lett.* 5, 3–6.

A solution for $r = 3$ can be found in time $O(km^2)$.

S. Kannan, T.J. Warnow (1994). Inferring evolutionary history from DNA sequences. *SIAM J. Comput.* 23, 713–737.

This paper presents an $O(k^2m)$ algorithm for $r \leq 4$ which is especially important because it allows the modeling of evolution of DNA sequences.

F.R. McMorris, T.J. Warnow, T. Wimer (1993). Triangulating vertex-colored graphs. *SIAM J. Discr. Math.* 7, 296–306.

The perfect phylogeny problem is polynomially equivalent to coloring triangulated graphs. This observation is used to get a perfect phylogeny algorithm for arbitrary r running in $O((rm)^{m+1} + km^2)$.

R. Agarwala, D. Fernández-Baca (1993). A polynomial time algorithm for the perfect phylogeny problem when the number of character states is fixed. *Proc. 34th Annual IEEE Symp. Found. Comput. Sci.*, 140–147. Also to appear in *SIAM J. Comp.*

S. Kannan, T.J. Warnow (1995). A fast algorithm for the computation and enumeration of perfect phylogenies when the number of character states is fixed. *Proc. 6th Annual ACM-SIAM Symp. Discr. Algorithms*, 595–603.

These papers improve on the earlier work on perfect phylogeny using dynamic programming. Kannan and Warnow give an $O(2^{2r}km^2)$ algorithm.

T.J. Warnow (1994). Tree compatibility and inferring evolutionary history. *J. Algorithms* 16, 388–407.

For a given alignment, the question of identifying the maximal number of alignment columns that allow a perfect phylogeny is addressed. It is mapped to a maximum clique problem.

The *Parsimony Problem*, famous in molecular biology, can be thought of as a relaxation of the perfect phylogeny problem. When a character state cannot be mapped to a subtree, it will induce a forest. For a given tree, finding the inner node assignment such that the number of trees in all the forests induced by the states of the characters is minimized is the parsimony problem. This number equals the minimal number of mutations required to explain the leaf-labeling of a given tree.

W. Fitch (1971). Toward defining the course of evolution: minimum change for specific tree topology. *Systematic Zoology* 20, 406–416.

A dynamic programming algorithm is suggested for solving the parsimonious inner node assignment problem in linear time.

J.A. Hartigan (1973). Minimum mutation fits to a given tree. *Biometrics* 29, 53–65.

Here, the correctness of Fitch's algorithm is proven. After minimizing the number of necessary mutations over all trees one finds the "most parsimonious tree".

L.R. Foulds, R. Graham (1982). The Steiner problem in phylogeny is NP-complete. *Adv. Appl. Math.* 3, 43–49.

The authors show NP-completeness of the problem of finding the most parsimonious tree. This is based on an analogy to Steiner trees: the most parsimonious tree is a minimal Steiner tree linking the given sequences.

L.R. Foulds, M.D. Hendy, D. Penny (1979). A graph theoretic approach to the development of minimal phylogenetic trees. *J. Mol. Evol.* 13, 127–149.

This article gives an approximation algorithm for the most parsimonious tree based on the minimum spanning tree heuristic for Steiner trees.

M.D. Hendy, D. Penny (1982). Branch and bound algorithms to determine minimal evolutionary trees. *Math. Biosci.* 59, 277–290.

Branch-and-bound algorithms for calculating the most parsimonious tree can be applied.

2.3.2 Distance-based methods

Distance-based methods attempt to approximate a given set of distances on the leaf-labels of the tree by the path-metric of an edge-weighted tree. A distance matrix that coincide with the path-metric of a tree is called an additive matrix.

P. Bunemann (1971). The recovery of trees from measures of dissimilarity. F. Hodson, D. Kendall, P. Tautu (eds.). *Mathematics in the archaeological and historical sciences*, Edinburgh University Press, 387–395.

Bunemann gives a characterization of such matrices.

M.S. Waterman, T.F. Smith, M. Singh, W.A. Beyer (1977). Additive evolutionary trees. *J. Theoret. Biol.* 64, 199–213.
H.-J. Bandelt, A. Dress (1986). Reconstructing the shape of a tree from observed dissimilarity data. *Adv. Appl Math.* 7, 309–343.
N. Saitou, M. Nei (1987). The neighbor-joining method: a new method for reconstructing phylogenetic trees. *Mol. Biol. Evol.* 4, 406–425.
J. Culberson, P. Rudnicki (1989). A fast algorithm for constructing trees from distance matrices. *Inform. Process. Lett.* 30, 215–220.
H.-J. Bandelt (1990). Recognition of tree metrics. *SIAM J. Discr. Math.* 3, 1–6.

These papers present algorithms to compute for a given additive matrix the tree whose path-metric coincides with the given matrix. For general matrices most of these algorithms produce trees which approximate the given matrix.

The idea that for all species the same amount of time has passed since the existence of some common ancestor has led to the study of rooted, edge-weighted trees with all leaves being the same distance away from the root. Such a tree is called an ultrametric tree and its corresponding path-metric is called an ultrametric. A simple clustering method like single linkage clustering [Duda, R.O. and Hart, P.E. (1973). *Pattern Classification and Scene Analysis*, John Wiley & Sons, New York.] suffices to recognize such metrics and compute the tree.

Let $(A)_{i,j}$ be the (additive) path-metric of a tree and let $(B)_{i,j}$ be an arbitrary distance matrix. To judge how well the additive matrix (and thus the tree that goes with it) approximates the distance matrix, a distance between matrices is used. Usually it is defined as $\sum_{i,j} |A_{ij} - B_{ij}|^\alpha$, with $\alpha = 1$ corresponding to the L_1-norm, $\alpha = 2$ corresponding to the L_2-norm. Some authors use the L_∞-norm ($\max_{i,j} |A_{ij} - B_{ij}|$).

M. Krivánek, J. Morávek (1986). NP-hard problems in hierarchical-tree clustering. *Acta Inform.* 23, 311–323.
W.H.E. Day (1987). Computational complexity of inferring phylogenies from dissimilarity matrices. *Bull. Math. Biol.* 49, 461–467.

The problem of finding the closest tree under either an L_1 or and L_2-norm has been proven NP-complete by Day using results of Krivánek and Morávek.

M. Farach, S. Kannan, T. Warnow (1995). A robust model for finding optimal evolutionary trees. *Algorithmica* 13, 155–179.

The authors develop guaranteed error bound algorithms for approximation of a distance matrix by ultrametric or additive trees.

R. Agarwala, V. Bafna, M. Farach, B. Naryanan, M. Paterson, M. Thorup (1996). On the approximability of numerical taxonomy (fitting distances by tree metrics). *Proc. 7th Annual ACM-SIAM Symp. Discr. Algorithms*, 365–372.

In this paper the approximation is measured in L_∞-norm.

The reader may have noted a certain abundance of algorithms to construct trees from additive matrices. The importance of having several algorithms on hand for this purpose lies in the fact that they also constitute a source of ideas for heuristics.

D.L. Swofford, G.J. Olsen (1990). Phylogeny reconstruction. D.M. Hillis, C. Moritz (eds.), *Molecular Systematics*, Sinauer Associates, Sunderland, MA, 411–501.

This review gives an overview of heuristics for tree approximation as well as many other approaches to phylogeny reconstruction that are in practical use.

J. Felsenstein (1981). Evolutionary trees from DNA sequences: a maximum likelihood approach. *J. Mol. Evol.* 17, 368–376.

This widely used method applies maximum likelihood estimation to judge the quality of a tree.

L. Székely, M. Steel, P. Erdös (1993). Fourier calculus on evolutionary trees. *Adv. Appl. Math.* 14, 200–216.
S.N. Evans, T.P. Speed (1993). Invariants of some probability models used in phylogenetic inference. *Ann. Statist.* 21, 355–377.
H. Bandelt, A. Dress (1992). A canonical decomposition theory for metrics on a finite set. *Adv. Appl. Math.* 92, 47–105.

In the last few years, several other interesting approaches to phylogeny reconstruction have emerged which are not based on discrete optimization. These are some of the key papers.

2.4 Trees and alignment

Let T be a tree whose k leaves are labeled with the k sequences of S. The m internal nodes represent sequences of hypothetical ancestral species and are not labeled. A tree alignment $\hat{S}(T) = \{\hat{s}_1, \hat{s}_2, \ldots, \hat{s}_k, s_1^a, \ldots, s_m^a\}$ is a set of $k + m$ strings over the alphabet $\hat{A}$ with the following properties: (1) All $k + m$ strings have the same length. (2) Ignoring dashes, the first k strings are identical with the k strings in S. The m last strings represent possible ancestral sequences, the labels for the internal nodes. The cost "$\mathrm{cost}_{\hat{S}}(e)$" of an edge e in a tree alignment $\hat{S}$ is defined as the cost of the "projected" pairwise alignment of the two sequences that are stored in the nodes connected by the edge. The cost of a tree alignment is the sum of the costs of all edges in the tree. The *Multiple Sequence Tree Alignment Problem* is defined as follows: Compute the tree alignment $\hat{S}(T)$ that minimizes the total sum of the edge costs:

$$\mathrm{MSTA}(S) := \min_{\hat{S}(T)} \left[\sum_{e \in T} \mathrm{cost}_{\hat{S}(T)}(e) \right] .$$

There is a more general variant of the MSTA-problem that is called *Generalized Multiple Sequence Tree Alignment GMSTA*. In this more difficult variant of the tree alignment problem, only the k sequences are given and the tree as well as the hypothetical ancestral sequences have to be constructed. Note that for a given alignment, finding the tree that minimizes Hamming distance along the tree edges is the parsimony problem of §§2.3.

T. Jiang, E.L. Lawler, L. Wang (1994). Aligning sequences via an evolutionary tree: complexity and approximation. *Proc. 26th Annual ACM Symp. Theory of Comput.*, 760–769.

The authors show that the MSTA-problem is NP-hard and the GMSTA-problem is MAX SNP-hard. Furthermore they present an approximation algorithm with performance ratio 2 for the MSTA-problem.

L. Wang, T. Jiang (1994). On the complexity of multiple sequence alignment. *J. Comput. Biol.* 1, 337–348.

The MSTA-problem is MAX SNP-hard if the given phylogeny (tree) is a star tree.

H.T. Wareham (1995). A simplified proof of the NP- and MAX SNP-hardness of multiple sequence tree alignment. *J. Comput. Biol.* 2, 509–514.

Wareham offers an alternative proof that the GMSTA-problem is both NP-complete and MAX SNP-hard.

D. Sankoff (1975). Minimal mutation trees of sequences. *SIAM J. Appl. Math.* 28, 35–42.

D. Sankoff, R.J. Cedergreen (1983). Simultaneous comparison of three or more sequences related by a tree. D. Sankoff, J.B. Kruskal (eds.). *Time Warps, String Edits and Macromolecules: The Theory and Practice of Sequence Comparison*, Addison-Wesley, Reading, MA, 93–120.

Sankoff raised the question of multiple sequence alignment and introduced the MSTA-problem. It can be formulated as a shortest-path problem in a k-dimensional mesh-shaped graph and can be optimally solved using dynamic programming. The edges of the graph represent possible columns of alignments. We have to assign letters to the lower m rows of each column (letters for the inner vertices), such that the cost of each column is minimized. The m lower letters of a column can be computed using a dynamic programming approach (the Fitch-Hartigan algorithm to construct a parsimonious tree, see §§2.3). This "inner minimization" has to be carried out for each edge (possible column) of the graph. The shortest-path from the source to the sink codes the optimal tree alignment.

S. Altschul, D.J. Lipman (1989). Tree, stars, and multiple biological sequence alignment. *SIAM J. Appl. Math.* 49, 179–209.

Here, a branch-and-bound algorithm for the MSTA-problem is presented that is an extension of Carrillo and Lipman's algorithm ([Carrillo and Lipman 1988], see Subsection 2.2). Altschul and Lipman describe a new approach to compute suitable bounds for the branch-and-bound algorithm.

D. Sankoff, R.J. Cedergren, G. Lapalme (1976). Frequency of insertion-deletion, transversion, and transition in evolution of 5S ribosomal RNA. *J. Mol. Evol.* 7, 133–149.

The authors designed an iterative procedure for local optimization in the tree. The tree is decomposed into small overlapping subtrees (star trees). In each iteration step all subtrees will be locally optimized starting with the more "peripheral" subtrees. The algorithm stops when an iteration step has been performed without change in the subtree cost.

J. Hein (1989). A new method that simultaneously aligns and reconstructs ancestral sequences for any number of homologous sequences, when the phylogeny is given. *Mol. Biol. Evol.* 6, 649–668.
J. Hein (1989). A tree reconstruction method that is economical in the number of pairwise comparisons used. *Mol. Biol. Evol.* 6, 669–684.

Hein designed and implemented a heuristic method that yields good approximations for tree alignment and suggests an algorithm for the GMSTA-problem. Hein introduced the concept of *sequence graphs* for storing large sets of sequences. A sequence graph is a directed, acyclic and connected graph with a source and a sink. Each edge represents a letter or even a subsequence of a sequence. Each path from the source to the sink codes a sequence. A sequence graph represents the set of sequences that is coded by all source-to-sink paths.

D. Gusfield (1993). Efficient methods for multiple sequence alignment with guaranteed error bounds. *Bull. Math. Biol.* 55, 141–154.

Since the MSTA-problem can be formulated as a Steiner tree problem on graphs, approximation approaches for Steiner minimal trees have been successfully applied to the MSTA-problem. Gusfield has shown that the minimum spanning tree approach yields an approximation ratio of 2. By using better approximation techniques for the Steiner tree problem, for instance the approach of Zelikovsky [Zelikovsky, A. (1993). The 11/6 approximation algorithm for the steiner problem on networks. *Algorithmica* 9, 463–470] or of Berman and Ramaiyer [Berman and Ramaiyer (1994). Improved approximations for the steiner tree problem. *J. Algorithms* 17, 381–408], better approximation ratios can be obtained.

R. Ravi, J. Kececioglu (1995). Approximation algorithms for multiple sequence alignment. Z. Galil, E. Ukkonen (eds.). *Proc. 6th Symp. Combinatorial Pattern Matching, Lecture Notes in Comp. Sci.* 937, Springer-Verlag, Berlin, 330–339.

Here, an approximation algorithm for regular deg-ary trees (each internal node has exactly deg children) is given that finds solutions with an approximation ratio of $\frac{\text{deg}+1}{\text{deg}-1}$.

T. Jiang, E.L. Lawler, L. Wang (1994). Aligning sequences via an evolutionary tree: complexity and approximation. *Proc. 26th Annual ACM Symp. Theory of Comput.*, 760–769.

This is the first polynomial time approximation scheme (PTAS) for the MSTA-problem. It computes for any given $t > 1$ an approximation with ratio smaller than $1 + 3/t$. The running time of the algorithm is exponential in deg^{t-1}.

L. Wang, D. Gusfield (1996). Improved approximation algorithms for tree alignment. *Proc. 7th Annual Symp. Combinatorial Pattern Matching, Lecture Notes in Comp. Sci.* 1075, Springer-Verlag, Berlin, 168–185.

The authors designed a polynomial time approximation scheme (PTAS) for regular deg-ary trees (each internal node has exactly deg children). For a fixed $t > 1$, the approximation ratio of the PTAS is $1 + \frac{2}{t} - \frac{2}{t2^t}$.

3 Tree Comparison

Given two or more evolutionary trees computed from different gene families, the problem of comparing these phylogenies arises. Several notions of similarity between trees have been suggested, some of which are merely a similarity measure while others compute a consensus tree. We give only a few references to some prominent approaches.

M.S. Waterman, T.F. Smith (1978). On the similarity of dendrograms. *J. Theor. Biol.* 73, 784–900.
K. Culik II, D. Wood (1982). A note on some tree similarity measures. *Inform. Process. Lett.* 15, 39–42.

The above articles introduce the Nearest Neighbor Interchange (NNI) metric on trees. The metric counts the number of elementary operations, the NNIs, required to transform one tree into the other. For an edge with two subtrees at either node, two NNIs are possible, representing the two alternative topologies.

E.K. Brown, W.H.E. Day (1984). A computationally efficient approximation to the nearest neighbor interchange metric. *J. Classification* 1, 93–124.

The authors give a heuristic approximation algorithm for calculation of NNI distance.

M. Krivánek (1986). Computing the nearest neighbor interchange metric for unlabeled binary trees is NP-complete. *J. Classification* 3, 55–60.

The existence of a polynomial time algorithm for NNI is still open. NP-completeness has been shown so far only for unlabeled trees.

C.R. Finden, A.D. Gordon (1985). Obtaining common pruned trees. *J. Classification* 2, 255–276.

The *Maximum Agreement Subtree* is introduced as the homeomorphous subtree common to two trees spanning the maximum number of leaves.

M. Steel, T. Warnow (1993). Kaikoura tree theorems: computing the maximum agreement subtree. *Inform. Process. Lett.* 48, 77–82.
M. Farach, M. Thorup (1994). Fast comparison of evolutionary trees. *Proc. 5th Annual ACM-SIAM Symp. Discr. Algorithms*, 481–488.

The above articles have successively improved the corresponding algorithms.

A. Amir, D. Keselman (1994). Maximum agreement subtree in a set of evolutionary trees – metrics and efficient algorithms. *Proc. 35th Annual IEEE Symp. Found. Comput. Sci.*, 758–769.
J. Hein, T. Jiang, L. Wang, K. Zhang (1995). On the complexity of comparing evolutionary trees. *Proc. 6th Annual Symp. Combinatorial Pattern Matching, Lecture Notes in Comp. Sci.* 937, Springer-Verlag, Berlin, 177–190.

These articles consider maximum agreement subtrees for more than two trees.

4 Genome Rearrangements

The order in which genes are arranged on the DNA molecule is the result of an evolutionary process. Over time, a gene order formerly present in an ancient species may, due to certain rearrangements in the genome, have evolved into a gene order we can observe today. The computational task lies in reconstructing the changes that may have occured to transform one gene order into another. A much studied elementary operation in these transformations is, e.g., the reversal of subsets of genes. The choice of elementary operation depends on the organism or cell organelle under study. More sophisticated scenarios model the evolution of sets of chromosomes which can exchange genes among each other.

S. Hannenhalli, P.A. Pevzner (1995). Towards a computational theory of genome rearrangements. J. van Leeuwen (eds.). *Computer Science Today*, Springer Verlag, Berlin, Heidelberg, 183–202.

This is an excellent overview of the field of genome rearrangements.

Let an ordered set of n genes be given. A *reversal* is the operation which cuts out a contiguous subset, inverts the order of genes in this subset, and reinserts it again in its original position. *Reversal distance* between two permutations is the minimal number of reversals required to transform one permutation into another. Since genes also have a direction, a more accurate model introduces signs for the genes. A given set of genes is represented by a permutation with signs on each entry. Reversing a subset of genes then has the additional effect of changing all signs of the affected genes. The problem is to calculate the minimal number of *signed reversals* necessary to transform one signed permutation into another.

G.A. Watterson, W.J. Ewens, T.E. Hall, A. Morgan (1982). The chromosome inversion problem. *J. Theor. Biol.* 99, 1–7.

This is the first paper to raise the formal problems of comparing gene orders.

D. Sankoff, G. Leduc, N. Antoine, B. Paquin, B.F. Lang, R. Cedergren (1992). Gene order comparisons for phylogenetic inference: evolution of the mitochondrial genome. *Proc. Natl. Acad. Sci. USA* 89, 6575–6579.

Heuristics can be used to estimate the number of rearrangements that occured between two contemporary species. The resulting distance measure is used to reconstruct an evolutionary tree.

J.D. Kececioglu, D. Sankoff (1995). Exact and approximation algorithms for the inversion distance between two permutations. *Algorithmica* 13, 180–210.

The authors give an approximation algorithm for the number of (unsigned) reversals with a guaranteed error bound of 2. Moreover, they devise a branch-and-bound algorithm. They speculate that the problem is NP-complete.

V. Bafna, P.A. Pevzner (1996). Genome rearrangements and sorting by reversals. *SIAM J. Comp.* 25, 272–289.

The error bound is improved to 1.75. It is shown that it may take up to $n - 1$ reversals to transform one permutation into another. Furthermore, a factor 1.5 approximation algorithm for signed reversals is given.

A. Caprara, G. Lancia, S.K. Ng (1995). *A column-generation based branch-and-bound algorithm for sorting by reversals*, Technical report, OR/95/7 DEIS, University of Bologna.

This paper introduces an integer programming formulation of sorting by reversals.

J.D. Kececioglu, D. Sankoff (1994). Efficient bounds for oriented chromosome inversion distance. *Proc. 5th Annual Symp. Combinatorial Pattern Matching, Lecture Notes in Comp. Sci.* 807, Springer-Verlag, Berlin, 307–325.

Here, a branch-and-bound algorithm for signed reversals is given.

S. Hannenhalli, P.A. Pevzner (1995). Transforming cabbage into turnip (polynomial algorithm for sorting signed permutations by reversals). *Proc. 27th Annual ACM Symp. Theory of Comput.*, 178–189.

The authors found a duality theorem for the number of signed reversals. Based on this theorem they devised an $O(n^4)$ algorithm.

S. Hannenhalli, P.A. Pevzner (1996). To cut ... or not to cut. *Proc. 7th Annual ACM-SIAM Symp. Discr. Algorithms*, 304–313.
Using connections between the two problems and the improved algorithm for signed permutations, the authors devise a practically efficient algorithm for unsigned reversals.

V. Bafna, P.A. Pevzner (1995). Sorting permutations by transpositions. *Proc. 6th Annual ACM-SIAM Symp. Discr. Algorithms*, 614–623.
In this article another operation on permutations, so-called transpositions, are studied. In this context a transposition does not exchange two single elements but two adjacent stretches from a permutation.

Genome rearrangements are more complicated when genes are distributed over different chromosomes. Chromosomes can exchange genetic material. A *translocation* is the process where a contiguous set of genes from an end of a chromosome is exchanged with a contiguous set of genes from an end of another chromosome. Fusion, the combination of two chromosomes, and fission, the breakage of one chromosome into two new ones, are other relevant processes.

J.D. Kececioglu, R. Ravi (1995). Of mice and men: evolutionary distances between genomes under translocation. *Proc. 6-th Annual ACM-SIAM Symp. Discr. Algorithms*, 604–613.
S. Hannenhalli (1995). Polynomial-time algorithm for computing translocation distance between genomes. *Proc. 6th Annual Symp. Combinatorial Pattern Matching, Lecture Notes in Comp. Sci.* 937, Springer-Verlag, Berlin, 162–176.
These articles study distances between genomes based on the above operations.

J.D. Kececioglu, D. Gusfield (1994). Reconstructing a history of recombinations from a set of sequences. *Proc. 5th Annual ACM-SIAM Symp. Discr. Algorithms*, 471–480.
Here, related operations in the context of reconstructing evolutionary history are studied.

5 Sequencing and Mapping

The goal of the Human Genome Project is the determination of the location of the human genes on the DNA and the sequencing (the determination of the sequence of nucleotides A,C,G, and T) of the entire human genome. In total, the human genome contains about 3 billion letters distributed over 23 chromosomes. Biochemical techniques, however, only allow the researcher to handle comparatively small segments of DNA at a time. Thus, a DNA molecule is broken into smaller pieces. The information one obtains on small pieces needs to be put into a larger context again. To this end one has to infer the order in which the segments occur on the underlying DNA. This task is called mapping. Depending on the scale at which it is done the techniques and the corresponding computational problems vary.

R.M. Karp (1993). Mapping of the genome: some combinatorial problems arising in molecular biology. *Proc. 25th Annual ACM Symp. Theory of Comput.*, 278–285.
This article gives an overview of the combinatorial problems arising in DNA mapping and sequencing.

P.A. Pevzner (ed.) (1995). *Combinatorial Methods for DNA Mapping and Sequencing, J. Comput. Biol.* 2.
This special issue contains eleven presentations of the DIMACS workshop "Combinatorial Methods in DNA Mapping and Sequencing", that opened the DIMACS computational molecular biology year 1994/1995. The articles give a strong emphasis on applications of combinatorial methods in molecular biology.

See also [Myers 1995] and [Pevzner and Waterman 1995] (see §1).

5.1 Sequence assembly

Current technology permits experimentalists to directly determine the sequence of a DNA strand of approximately 500 nucleotides in length. To sequence a long piece of DNA many such reads are taken and subsequently re-assembled to produce the original sequence. Computationally, this gives rise to the problem of assembling the fragments using the overlap information among them. Overlaps are deduced from sequence similarity. Since 1–10% of the nucleotides in the fragment data are missing or incorrect, and since a fragment's sequence can be reversed with respect to the others these overlaps cannot be perfectly determined. The formal *Sequence Reconstruction Problem (SR)* can be stated as follows: Given a collection of fragment sequences $\mathcal{F}$ and an error rate $0 \leq \epsilon < 1$, find a shortest sequence S such that every fragment $F \in \mathcal{F}$, or its reverse complement, matches a substring of S with at most $\epsilon|F|$ errors.

H. Peltola, H. Söderland, J. Tarhio, E. Ukkonen (1983). Algorithms for some string matching problems arising in molecular genetics. *Proc. 9th IFIP World Computer Congress*, 59–64.
H. Peltola, H. Söderland, E. Ukkonen (1984). SEQUAID: A DNA sequence assembly program based on a mathematical model. *Nucl. Acids Res.* 12, 307–321.
The SR-problem is formalized in the first paper. The second paper describes a decomposition of this extremely hard problem into three combinatorial optimization problems: the overlap graph construction, the layout phase, and the alignment problem.

J.D. Kececioglu, E.W. Myers (1995). Exact and approximate algorithms for the sequence reconstruction problem. *Algorithmica* 13, 7–51.
Here, for each of the subproblems either exact algorithms or approximate algorithms are suggested.

J.D. Kececioglu (1991). *Exact and approximation algorithms for DNA sequence reconstruction.* PhD Thesis, University of Arizona.
This thesis gives a detailed overview of all aspects concerning the sequence reconstruction problem such as the biological aspects, the related combinatorial optimiza-

tion problems, the literature and the computer software available.

E.W. Myers (1995). Toward simplifying and accurately formulating fragment assembly. *J. Comput. Biol.* 2, 275–290.

In this paper the pressing practical problem of the existence of repeats in biological sequences is addressed.

M.J. Miller, J.I. Powell (1994). A quantitative comparison of DNA sequence assembly programs. *J. Comput. Biol.* 1, 257–269.

Eleven sequence assembly programs are compared for their accuracy and the reproducibility with which they assemble DNA fragments into a completed sequence.

A strongly idealized formalization of sequence assembly assumes that fragment orientation is known and no errors occur. The resulting problem is the *Shortest Common Superstring Problem.* Although this problem is NP-hard (see Gallant, J., Maier, D. and Storer, J. (1980). On finding minimal length superstrings. *J. Computer Syst. Sci.* 20, 50–58), simple greedy algorithms seem to do quite well. There is a large number of papers analyzing greedy procedures or developing approximation algorithms for the shortest common superstring problem. We list only a few of them.

J. Tarhio, E. Ukkonen (1988). A greedy approximation algorithm for constructing shortest common superstrings. *Theoretical Comput. Sci.* 57, 131–145.
J. Turner (1989). Approximation algorithms for the shortest common superstring problem. *Inform. and Computation* 83, 1–20.

These two papers independently show that a greedily constructed superstring is close to optimal in the sense that it achieves a compression of at least 1/2 that of a shortest superstring. The compression of a superstring is given by the number of symbols "saved" compared to simply concatenating all the strings. However, their results do not imply a performance guarantee with respect to optimal length.

M. Li (1990). Towards a DNA sequence theory. *Proc. 31st Annual IEEE Symp. Found. Comput. Sci.*, 125–134.

Here, the first nontrivial bound for the superstring problem is obtained. The suggested algorithm produces a superstring of at most $n \log n$, where n is the length of the shortest superstring containing the given fragments.

A. Blum, T. Jiang, M. Li, J. Tromp, M. Yannakakis (1994). Linear approximation of shortest superstrings. *J. ACM* 41, 630–647.

The authors give the first linear approximation factor with respect to length. It is shown that the greedy algorithm leads to a superstring with length at most $4n$. A variation of the algorithm is suggested that leads to an improved approximation factor $3n$. Moreover, the authors show that the shortest common superstring problem is MAX SNP-hard.

C. Armin, C. Stein (1995). Shortest superstrings and the structure of overlapping strings. *J. Comput. Biol.* 2, 307–332.

This is one of the most recent papers in a series of papers which subsequently improved the approximation factor for the shortest common superstring problem. Here, an approximation factor of 2.75 is obtained by using the structure of strings with large amounts of overlap. The algorithm runs in time $O(|S| + k^3)$, where k is the number of of given strings and S is the sum of the lengths of all the strings.

T. Jiang, Z. Jiang, D. Breslauer (1996). *Rotation of periodic strings and short superstrings*, Technical report, Max-Planck-Institut f. Informatik, Saarbrücken, Germany. To appear in *J. Algorithms.*
E.S. Sweedyk (1995). *A 2 1/2 appoximation Algorithm for shortest common superstring.* PhD Thesis, Department of Computer Science, University of California, Berkeley.
Recently, Jiang, Jiang and Breslauer suggested an approximation algorithm for the shortest common superstring with approximation factor 2.667 (2.596 in the full paper). Their algorithm is simpler than the previous approximation algorithms that lead to a factor less than three. The thesis by Sweedyk achieves a factor of 2.5.

Most of the above authors conjecture the existence of an approximation algorithm of factor two, and that the greedy algorithm itself is a candidate for it, because no example is known where the greedy algorithm does worse.

5.2 Sequencing by hybridization

A method for DNA sequencing that requires considerable computational support is *Sequencing by Hybridization* (SBH). Hybridization is the process whereby two single-stranded DNA molecules form a double helix because of complementary base sequences in them. In SBH the basic operation is hybridizing a short piece of DNA of length l, called a probe or l-tuple, to a single-stranded target DNA sequence. Thus, an SBH-probe will yield a hybridization signal if there is a substring of the target DNA that matches the probe. In SBH this is done for all probes of a given small length in parallel providing information regarding which l-tuples occur in the target sequence. The computational task is to reconstruct the target DNA sequence from this information.

P.A. Pevzner (1989). l-tuple DNA sequencing : a computer analysis. *J. Biomolecular Struct. Dynamics* 7, 63–73.
It is shown that the SBH reconstruction problem can be reduced to the Eulerian path problem in a subgraph of the de Bruijn graph. The author also considers the occurence of erroneous signals (false positives) and missing signals (false negatives), and the case when repeats of length l occur in the target DNA.

R.J. Lipshutz (1993). Maximum likelihood DNA sequencing by hybridization. *J. Biomolecular Struct. Dynamics* 11, 637–653.
The author suggests a maximum likelihood method for the SBH reconstruction problem in the presence of false positives and false negatives and reduces SBH reconstruction to the graph matching problem.

P.A. Pevzner, R.J. Lipshutz (1994). Towards DNA sequencing chips. *Proc. 19th Int. Conf. Math. Found. Comp. Sci., Lecture Notes in Comp. Sci.* 841, Springer-Verlag, Berlin, 143–158.

This article gives a survey of the state of the art in sequencing by hybridization through 1994.

Recently, modifications of sequencing by hybridization have been proposed to reduce ambiguities in sequence reconstruction:

S. Hannenhalli, W. Feldman, H.F. Lewis, S.S. Skiena, P.A. Pevzner (1996). Positional sequencing by hybridization. *CABIOS* 12, 19–24.

The authors consider the case where additional information about the approximate position of each l-tuple in the unknown DNA fragment is given. No polynomial algorithms for the *Positional Sequence Reconstruction Problem* are known. A special case is the *Bounded Positional SBH Reconstruction Problem*, where the range of positions for each l-tuple is bounded by a small fixed number. For this case the authors present two polynomial time algorithms.

R.M. Idury, M.S. Waterman (1995). A new algorithm for DNA sequence assembly. *J. Comput. Biol.* 2, 291–306.

This article suggests the application of the SBH-related computational techniques to fragment assembly.

5.3 Digest mapping

Restriction enzymes cleave a DNA molecule at an instance of a short specific pattern. Although not very precisely, the lengths of the fragments resulting from digestion by a restriction enzyme can be measured. Using two enzymes, say A and B, first individually and then using A and B together, the experimentalist obtains three sets of fragments. These constitute a source of information for determination of the layout of the fragments on the underlying DNA strand. Let the cut sites for enzyme A and B be $a_1 < a_2 < \cdots < a_k$ and $b_1 < b_2 < \cdots < b_l$, respectively. Applying enzymes A and B simultaneously will cut at all these sites, say $c_1 < c_2 < \cdots < c_{k+l}$. This is the so-called double digest. The biological experiment yields the sets of fragment lengths $\bar{A} = \{a_1, a_2 - a_1, \ldots, N - a_k\}$, $\bar{B} = \{b_1, b_2 - b_1, \ldots, N - b_l\}$, and $\bar{C} = \{c_1, c_2 - c_1, \ldots, N - c_{k+l}\}$, where N is the length of the target DNA. The task is now to reconstruct the restriction sites from the given sets $\bar{A}$, $\bar{B}$ and $\bar{C}$. More precisely, the *Double Digest Problem* is the following: Find an ordering of the elements in $\bar{A}$ and an ordering of the elements in $\bar{B}$ such that the double-digest implied by these orderings is $\bar{C}$. The papers quoted below not only show that it is difficult to produce a solution to this problem but that, additionally, the number of solutions generally is huge.

L. Goldstein, M.S. Waterman (1987). Mapping DNA by stochastic relaxation. *Adv. Appl. Math.* 8, 194–207.

It is shown that the problem to decide if any solution of a given double digest instance exists is NP-complete. Furthermore, the authors show that the number of solutions that produce the same single and double digests increases exponentially with the length of the DNA segment when the enzyme sites are modeled by a random process.

W. Schmitt, M.S. Waterman (1991). Multiple solutions of DNA restriction mapping problem. *Adv. Appl. Math.* 12, 412–427.

The authors suggest partitioning the entire set of maps into equivalence classes. But still the number of equivalence classes grows very fast, as observed by the following author.

P.A. Pevzner (1995). DNA physical mapping and alternating Eulerian cycles in colored graphs. *Algorithmica* 13, 77–105.

The author studies the combinatorics of multiple solutions to the double digest problem and shows that the solutions are closely associated with alternating Eulerian cycles in colored graphs. Furthermore he gives a complete characterization of equivalent maps as introduced by Schmitt and Waterman.

S. Ho, L. Allison, C.N. Yee (1990). Restriction site mapping for three or more enzymes. *Comp. Appl. Biosci.* 6, 195–204.

The double digest problem is generalized to more than two enzymes.

G. Zehetner, A. Frischauf, H. Lehrach (1987). Approaches to restriction map determination. M.J. Bishop and C.J. Rawlings (eds.), *Nucleic Acid and Protein Sequence Analysis: A Practical Approach*, IRL Press, Oxford, 147–164.

This article gives a survey on commonly used methods for attacking this problem.

In the *Partial Digest Problem* only a single enzyme, say A, is used to cleave several clones of a segment. Let $a_1 < a_2 < \cdots < a_k$ be the set of the restriction sites. In contrast to a double digest experiment, the digestion process is stopped earlier, such that not all of the restriction sites are cut. Again, the task is to reconstruct the restriction sites from the given fragment length data. The *Probed Partial Digest Problem* is similar, except that a site p on the DNA segment is labeled, and the sizes of only those restriction fragments are measured that contain this site. Both problems can be reduced to polynomial factorization (Rosenblatt, J. and Seymour, P.D. (1982). The structure of homometric sets. *SIAM J. Algebraic Discr. Methods* 3, 343–350), which is "unlikely to be NP-complete" ([Karp 1993], see above).

S.S. Skiena, W.D. Smith, P. Lemke (1990). Reconstructing sets from interpoint distances. *Proc. 6th Ann. Symp. Computational Geometry*, 332–339.

L.A. Newberg, D. Naor (1993). A lower bound on the number of solutions to the exact probed partial digest problem. *Adv. Appl. Math.* 14, 172–185.

Like in the double digest problem the number of solutions can grow fast for these problems. The above two papers deal with this question. The authors show how the degree of ambiguity can grow as a function of the number of restriction sites.

5.4 Mapping using hybridization data

For large pieces of DNA digest mapping has mostly been superseded by mapping techniques that are based on experimental determination of overlap between fragments (also called *clones*). This overlap data can be represented as a graph in which vertices are clones and there is an edge between two vertices if the corresponding clones overlap. In the absence of error, this graph is an interval graph.

Several experimental methods to obtain the required overlap information are in use. One approach detects certain, very specific sequence segments in the fragments. This is done using probes hybridizing to specific bits of the sequence. The hope is that two clones share a probe if and only if they overlap. If a probe is absolutely unique it is called a *Sequence Tagged Site* (*STS*). Ideally, the matrix describing which clones contain such a probe will have the consecutive ones property [Golumbic, M.C. (1980). *Algorithmic Graph Theory and Perfect Graphs*, Academic Press, New York.]. Hence, the correct orderings can be found very easily using the PQ-tree data structure [Booth, K.S. and Lueker, G.S. (1976). Testing for consecutive ones property, interval graphs and planarity using PQ-tree algorithms. *J. Computer Syst. Sci.* 13, 335–379] to generate the set of all arrangements of probes consistent with the data. Alternatively, there is another experiment that allows determination of overlap between two clones directly.

However, both techniques are error prone. In practice, false positive overlaps will be reported and overlaps may not be recognized, i.e., there are false negatives. A further complication is that sometimes two clones are merged into one. This phenomenon is called chimerism.

M.S. Waterman, J.R. Griggs (1986). Interval graphs and maps of DNA. *Bull. Math. Biol.* 48, 189–195.

The authors study a model that combines digest mapping with information on overlap between the fragments. They introduced the representation of maps as interval graphs and characterize the interval graphs arising in the specific experimental setup modeled in this paper. A simple linear time algorithm for recognizing and representing the data in interval representation is given.

Since in practice the overlap data is error prone the question arises if there exists an interval graph that is in some sense "close" to the given overlap data.

M.C. Golumbic, H. Kaplan, R. Shamir (1994). On the complexity of DNA physical mapping. *Adv. Appl. Math.* 15, 251–261.

Given some "noisy" and some correct part of the overlap data, the authors show that the following problem is NP-complete. Is there an interval graph induced by E_1 satisfying $E_0 \subseteq E_1 \subseteq E_2$ for given edge sets E_0 and E_2 ($E_0 \subseteq E_2$) (*Interval Sandwich Problem*)?

H. Kaplan, R. Shamir, R.E. Tarjan (1994). Tractability of parameterized completion problems on chordal and interval graphs: minimum fill-in and physical mapping. *Proc. 35-th IEEE Symp. Found. Comp. Sci.*, 780–791.

First, the authors consider the case of unidentified overlaps. Although the problem of building a map with fewest errors is NP-hard (*Proper Interval Graph Completion Problem*), the authors present a linear time algorithm which gives an augmenting set with no more than k edges (k fixed) if one exists. Observing that the arising interval graphs have small clique size, the authors use this fact to present a polynomial time algorithm for the proper interval graph completion problem with bounded clique size.

M.R. Fellows, M.T. Hallett, W.T. Wareham (1993). DNA physical mapping: three ways difficult. *Proc. European Symp. Algorithms, Lecture Notes Comp. Sci.* 726, Springer-Verlag, Berlin, 157–168.

The authors consider the case of digest mapping when k enzymes are used and noisy overlap data between the fragments are given. This is a generalization of the model used by Waterman and Griggs (1986). They study the following problem: given a graph G and a coloring c of the vertices to k colors, is there a supergraph of G which is properly colored by c and which is an interval graph? It is shown that this problem is NP-complete, and is not fixed-parameter tractable, i.e., it cannot be solved in time $f(k)n^{\alpha}$, where α is independent of k (unless an apparently resistant problem can be solved).

The following articles give algorithms for constructing maps using STS probes.

E.D. Green, P. Green (1991). Sequence-tagged site (STS) content mapping of human chromosomes: Theoretical considerations and early experiences. *PCR Methods and Appl.*, 1, 77–90.

In this article some background information on STS mapping is given. Furthermore, the authors discuss a strategy for developing clone-based STS maps of chromosomes.

D.S. Greenberg, S. Istrail (1995). Physical mapping by STS hybridization: algorithmic strategy and the challenge of software evaluation. *J. Comput. Biol.* 2, 219–273.

The authors develop some algorithmic theory of the mapping process, and propose a performance evaluation procedure. Furthermore, they suggest various combinatorial optimization problems such as the hamming distance travelling salesman problem that could be useful for solving the practical mapping problem.

F. Alizadeh, R.M. Karp, D.K. Weisser, G. Zweig (1995). Physical mapping of chromosomes using unique probes. *J. Comput. Biol.* 2, 159–184.

Several combinatorial methods for reconstructing a DNA fragment in the presence of errors are presented. The methods include techniques for the hamming distance travelling salesman problem, and simulated annealing as well as screening methods for detecting errors in the given data.

Several experimenters have approached mapping using probes that do not satisfy the uniqueness condition.

F. Alizadeh, R.M. Karp, L.A. Newberg, D.K. Weisser (1995). Physical mapping of chromosomes: a combinatorial problem in molecular biology. *Algorithmica* 13, 52–76.

This is the first result that effectively addresses the ordering problem for mapping with hybridization fingerprints and non-unique probes. The authors introduce approximations to a likelihood function quantifying overlap information. This leads to optimization problems that are reasonably tractable in practice, although they are NP-hard.

R. Mott, A. Grigoriev, J.H.E. Maier, H. Lehrach (1993). Algorithms and software tools for ordering clone libraries: application to the mapping of the genome of schizosaccharomyces pombe. *Nucl. Acids Res.* 21, 1965–1974.

The authors describe a complete set of software tools for mapping of a genome that has been successfully applied to the genome of fission yeast.

6 Protein Threading and Lattice Models

The protein folding problem is the problem of predicting the three-dimensional fold that a given one-dimensional amino acid chain assumes. There is a vast amount of literature on the protein folding problem.

G.D. Fasman (1989). *Prediction of protein structures and the principles of protein conformation*, Plenum Press, New York.
K.M. Merz, S.M.L. Grand (1994). *The protein folding problem and tertiary structure prediction.* Birkhäuser, Boston.

The above books give an overview.

6.1 Threading

The *Threading Problem* is a generalization of the pairwise sequence alignment problem where for one of the two proteins a three-dimensional structure is given. An alignment thus implies spatial proximity between certain amino acids which is in turn weighted by a so-called pair-potential. The resulting optimization problem attempts to find an alignment that minimizes the sum over all implied pair-potentials.

R.H. Lathrop (1994). The protein threading problem with sequence amino acid interaction preferences is NP-complete. *Protein Engineering* 7, 1059–1068.

Here, it is shown that the Threading Problem is NP-complete (by reduction to One-in-Three 3SAT).

R.Z.R. Thiele, T. Lengauer (1995). Recursive dynamic programming for adaptive sequence and structure alignment. C. Rawlings, D. Clark, R. Altmann, L. Hunter, T. Lengauer, S. Wodak (eds.). *3rd International Symposium on Intelligent Systems for Molecular Biology*, 384–392.

The algorithm introduced in this paper recursively breaks down the task of aligning a sequence into a structure. It thus accounts for alterations in structural environments due to early alignment decisions.

R.H. Lathrop, T.F. Smith (1996). Global optimum protein threading with gapped alignment and empirical pair score functions. *J. Mol. Biol.* 255, 641–665.

The authors present a branch-and-bound algorithm for a slightly modified version of threading.

M. Sippl (1995). Knowledge-based potentials for proteins. *Current Opinion in Structural Biology* 5, 229–235.
This article summarizes other heuristic approaches to threading.

6.2 Lattice models

Lattice Models are discrete, in most cases strongly simplified versions of the protein folding problem.

R. Unger, J. Moult (1993). Finding the lowest free energy conformation of a protein is an NP-hard problem: proof and implications. *Bull. Math. Biol.* 55, 1183–1198.
A.S. Fraenkel (1993). Complexity of protein folding. *Bull. Math. Biol.* 55, 1199–1210.
The above articles give NP-completeness proofs for discretized versions of protein folding.

K. Yue, M. Fiebig, P.D. Thomas, H.S. Chan, E.I. Shakhnovich, K.A. Dill (1995). A test of lattice protein folding algorithms. *Proc. Natl. Acad. Sci. USA* 92, 325–329.
Here, algorithms for finding minimal energy structures for lattice models in proteins are summarized and tested.

W.E. Hart, S. Istrail (1995). Fast protein folding in the hydrphobic-hydrophilic model within three-eights of optimal. *Proc. 27-th Annual ACM Symp. Theory of Comput.* 157–168.
The authors give a factor $\frac{3}{8}$ approximation algorithm for lattice models.

Acknowledgment

The authors thank Dan Gusfield, Tao Jiang, Gene Myers, Dalit Naor and Pavel Pevzner for helpful comments on the manuscript.

Index

Index compiled by Geoffrey C. Jones

Continued from the front of this volume

GLOVER, KLINGHAM, AND PHILLIPS
Network Models in Optimization and Their Practical Problems

GOLSHTEIN AND TRETYAKOV
Modified Lagrangians and Monotone Maps in Optimization

GONDRAN AND MINOUX
Graphs and Algorithms
(Translated by S. Vajdã)

GRAHAM, ROTHSCHILD, AND SPENCER
Ramsey Theory
Second Edition

GROSS AND TUCKER
Topological Graph Theory

HALL
Combinatorial Theory
Second Edition

JENSEN AND TOFT
Graph Coloring Problems

LAWLER, LENSTRA, RINNOOY KAN, AND SHMOYS, EDITORS
The Travelling Salesman Problem: A Guided Tour of Combinatorial Optimization

LEVITIN
Perturbation Theory in Mathematical Programming Applications

MAHMOUD
Evolution of Random Search Trees

MARTELLO AND TOTH
Knapsack Problems: Algorithms and Computer Implementations

MINC
Nonnegative Matrices

MINOUX
Mathematical Programming: Theory and Algorithms
(Translated by S. Vajdã)

MIRCHANDANI AND FRANCIS, EDITORS
Discrete Location Theory

NEMHAUSER AND WOLSEY
Integer and Combinatorial Optimization

NEMIROVSKY AND YUDIN
Problem Complexity and Method Efficiency in Optimization
(Translated by E.R. Dawson)

PACH AND AGARWAL
Combinatorial Geometry

PLESS
Introduction to the Theory of Error-Correcting Codes
Second Edition

ROOS, TERLAKY, AND VIAL
Theory and Agorithms for Linear Optimization: An Interior Point Approach

SCHRIJVER
Theory of Linear and Integer Programming

TOMESCU
Problems in Combinatories and Graph Theory
(Translated by R.A. Melter)

TUCKER
Applied Combinatorics
Second Edition

ZAK AND CHONG
An Introduction to Optimization